Water Industry Systems:
modelling and optimization applications

VOLUME 1

WATER ENGINEERING AND MANAGEMENT SERIES

Series Editors: **Professor Bryan Coulbeck**
Professor of Water Control Systems
De Montfort University, Leicester, UK

and

Dr Bogumil Ulanicki
Research Director of Water Software Systems
De Montfort University, Leicester, UK

1. Performance in Water Distribution: A SYSTEMS APPROACH
   Sérgio Teixeira Coelho

2. Computing and Control for the Water Industry
   Edited by **Roger Powell** *and* **K S Hindi**

3. Water Industry Systems: modelling and optimization applications
   VOLUME 1
   Edited by **Dragan Savic** *and* **Godfrey Walters**

4. Water Industry Systems: modelling and optimization applications
   VOLUME 2
   Edited by **Dragan Savic** *and* **Godfrey Walters**

Water Industry Systems:
modelling and optimization applications

VOLUME 1

Edited by

Dragan Savic

Director of Centre for Water Systems, University of Exeter, UK

and

Godfrey Walters

Director of Centre for Water Systems, University of Exeter, UK

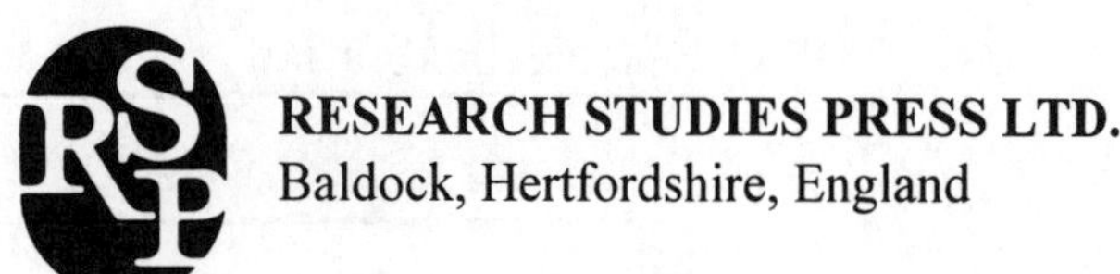

RESEARCH STUDIES PRESS LTD.
Baldock, Hertfordshire, England

RESEARCH STUDIES PRESS LTD.
15/16 Coach House Cloisters, 10 Hitchin Street, Baldock, Hertfordshire, England, SG7 6AE

and

325 Chestnut Street, Philadelphia, PA 19106

Marketing:

Research Studies Press Ltd.
15/16 Coach House Cloisters, 10 Hitchin Street, Baldock, Hertfordshire, England, SG7 6AE

Distribution:

NORTH AMERICA
Taylor & Francis Inc.
47 Runway Road, Suite G, Levittown, PA 19057 - 4700, USA

ASIA-PACIFIC
Hemisphere Publication Services
Golden Wheel Building, 41 Kallang Pudding Road #04-03, Singapore

EUROPE & REST OF THE WORLD
John Wiley & Sons Ltd.
Shripney Road, Bognor Regis, West Sussex, England, PO22 9SA

Library of Congress Cataloging-in-Publication Data

Available

British Library Cataloguing in Publication Data
A catalogue record for this book is available from the British Library.

ISBN 0 86380 248 6

Printed in Great Britain by SRP Ltd., Exeter

Editorial Preface

This series of books is intended to present reviews of recent developments and associated research, and the application of that work, to engineering, operation and management of water systems.

The basic methods of engineering, operating and managing water systems are well established and well documented. However, the increasing availability and application of computer based methods has had a revolutionary effect on these traditional methods. Significant advances have been made, and are continuing, in such broad areas as: Information technologies, Operational modelling, and Optimal decision strategies.

Very efficient and effective SCADA and GIS systems are now available; these can collect, store and organize large quantities of data. Water companies have taken advantage of these systems to achieve comprehensive information coverage and integration. In turn, this has enabled the development and application of more sophisticated engineering software. The most recent computational modules are based on some new and interesting mathematical and heuristic techniques.

These two books are representative of such advances, and contain contributions from leading experts. The material is expected to be of significant interest to practitioners and researchers in the subject area.

Volume 1 is the third book in the series. This is mainly concerned with analysis and modelling of water systems and includes such topics as:

- Analysis, simulation and design
- Integration and emerging technologies
- Model calibration
- Monitoring and control
- Reservoir modelling and control
- Simulation and modelling
- Transient analysis and design

Volume 2 is the fourth book in the series. This is mainly concerned with operation and management of water systems and includes such topics as:

- Data acquisition and information
- Decision support systems
- Genetic algorithm development and applications
- Geographic information systems

- Operational optimisation
- Optimisation applications
- Quality management and control
- Reliability and failures

The two companion books are based on the proceedings of the Exeter University International Conference on Computing and Control for the Water Industry, 1999. This event is a continuation of the series of conferences which started at De Montfort University in 1981.

Bryan Coulbeck
Bogumil Ulanicki
August 1999

Contributing Authors

Agudelo R. J.J., *Empresas Publicas de Medellin, Colombia*
Ahmed I., *Dept. of Civil Engineering and Engineering Mechanics, University of Arizona, U.S.A.*
Alonso J.M., *Universidad Politecnica de Valencia, Dept. de Sistemas Informaticos y Computacion, Spain*
Alvarruiz F., *Universidad Politecnica de Valencia, Dept. de Sistemas Informaticos y Computacion, Spain*
Anderson J.H, *Water Operational Research Centre, Dept of Manufacturing and Engineering Systems, Brunel University, U.K.*
Andrés M., *Aguas de Valencia, S.A., Spain*
Angel R. P.L., *Universidad Nacional de Colombia, Colombia*
Araujo J., *Dept. of Civil Engineering, Oklahoma State University, U.S.A.*
Babovic V., *Danish Hydraulic Institute, Denmark*
Balmaseda C., *Fluid Mechanics Group, Universidad Politécnica de Valencia, Spain*
Barrera Buteler G., *University of Cordoba, Argentina*
Bascia A., *Dipartimento di Meccanica e Materiali, Facoltà di Ingegneria, Università di Reggio Calabria, Italy*
Biggs M.J., *Department of Chemical and Process Engineering, University of Surrey, U.K.*
Bloom L.G., *Department of Civil and Environmental Engineering, University of Cincinnati, U.S.A.*
Boldy A.P., *HYDROSIM Consultants Limited, U.K.*
Borga A., *Civil Engineering Department, Technical University of Lisbon, Portugal*
Bouchart F.J.-C., *Dept. of Civil and Offshore Engineering, Heriot-Watt University, U.K.*
Boulos P.F., *MW Soft, Inc, U.S.A*
Bounds P.L.M, *Water Software Systems, De Montfort University, U.K*
Brandt M.J., *Binnie Black & Veatch, U.K*
Bremond B., *Hydraulics and civil engineering division Research Unit, Cemagref, France*
Brunone B., *University of Perugia, Italy*

Buchberger S.G., *Department of Civil and Environmental Engineering, University of Cincinnati, U.S.A.*
Burrows R., *Department of Civil Engineering, University of Liverpool, U.K.*
Butler D., *Department of Civil and Environmental Engineering, Imperial College of Science and Technology, U.K.*
Cabrera E., *Fluid Mechanics Group, Universidad Politécnica de Valencia, Spain*
Cabrera E.Jr., *Fluid Mechanics Group, Universidad Politécnica de Valencia, Spain*
Chaudhry F.H., *São Carlos School of Engineering, University of São Paulo, Brazil*
Chopard P., *Syndicat des Eaux d'Ile de France, Paris, France*
Conejos P., *Aguas de Valencia, S.A., Spain*
Covas D., *Civil Engineering Department, Technical University of Lisbon, Portugal*
Creasey J.D., *WRc plc, U.K*
Crowder G.S., *Crowder and Company Limited, U.K.*
de Schaetzen W., *Optimal Solutions, University of Exeter, U.K.*
Delgadino F.A., *University of Cordoba, Argentina*
Depauli C.P., *University of Cordoba, Argentina*
Dunlop E.J., *Local Government Computer Services Board, Dublin, Ireland*
Edwards K., *Anglian Water UK*
Elain C., *Stoner Associates Europe, Grenoble, France*
George D.G., *Institute of Freshwater Ecology, NERC, U.K.*
Graham N., *Department of Civil and Environmental Engineering, Imperial College of Science and Technology, U.K.*
Green A., *Stoner Associates Europe, Birmingham, U.K.*
Green A.P.E, *Halcrow Water, U.K.*
Guerrero D., *Universidad Politecnica de Valencia, Dept. de Sistemas Informaticos y Computacion, Spain*
Haarhoff J., *Rand Afrikaans University, South Africa*
Hague J., *Department of Chemical and Process Engineering, University of Surrey, U.K.*
Hamon J., *Général des Eaux, Paris, France*
Hampartzoumian E., *Dept. of Civil and Offshore Engineering, Heriot-Watt University, U.K.*
Hedger R.D., *Department of Geography, University of Edinburgh, U.K.*
Helmbrecht E., *University of Cordoba, Argentina*
Hernandez R. J.A., *Universidad Nacional de Colombia, Colombia*
Hernández V., *Universidad Politecnica de Valencia, Dept. de Sistemas Informaticos y Computacion, Spain*
Herz M., *University of Cordoba, Argentina*
Heslop S.E., *Department of Civil Engineering, University of Bristol, U.K.*
Inglesias P., *Fluid Mechanics Group, Universidad Politécnica de Valencia, Spain*
Juarez M.L., *University of Cordoba, Argentina*
Karney B.W., *Dept. of Civil Engineering, University of Toronto, Canada*

Keijzer M., *Danish Hydraulic Institute, Denmark*
Kirby R., *Anglian Water UK*
Lansey K., *Dept. of Civil Engineering and Engineering Mechanics, University of Arizona, U.S.A.*
Lee Y., *Department of Civil and Environmental Engineering, University of Cincinnati, U.S.A.*
Liu P.E.P., *Los Angeles Department of Water and Power, U.S.A.*
Lobbrecht A.H., *International Institute for Infrastructural, Hydraulic and Environmental Engineering (IHE), Netherlands*
Maksimovic C., *Department of Civil and Environmental Engineering, Imperial College of Science and Technology, U.K.*
Martínez F., *Universidad Politecnica de Valencia, Dept. de Ingenieria Hidraulica y Medio Ambiente, Spain*
Morel P., *Mathematics and Computer Science Research and Formation Unit, Bordeaux University, France*
Olsen N.R.B., *Department of Hydraulic and Environmental Engineering, Norwegian University of Science and Technology, Norway*
Oroná C., *University of Cordoba, Argentina*
Orr C., *MW Soft, Inc, U.S.A*
Pendlebury M., *Dept. of Civil Engineering, University of Toronto, Canada*
Perkins R., *Distribution Development, Severn Trent Water Ltd, U.K.*
Piller O., *Hydraulics and civil engineering division Research Unit, Cemagref, France*
Politaki S., *Athens Water Company EYDAP, Greece*
Ponz R., *Aguas de Valencia, S.A., Spain*
Pothof I., *WL | DELFT HYDRAULICS, Netherlands*
Powell R.S., *Water Operational Research Centre, Dept of Manufacturing and Engineering Systems, Brunel University, U.K.*
Ramos H., *Civil Engineering Department, Technical University of Lisbon, Portugal*
Rance J.P., *Water Software Systems, De Montfort University, U.K.*
Randall-Smith M., *Optimal Solutions, University of Exeter, U.K.*
Reis L.F.R., *São Carlos School of Engineering,University of São Paulo, Brazil*
Reyna S.M., *University of Cordoba, Argentina*
Reynolds L., *Thames Water Utilities Ltd, U.K.*
Ribelles J.V., *Fluid Mechanics Group, Universidad Politécnica de Valencia, Spain*
Rodriguez A., *University of Cordoba, Argentina*
Rolf B.W., *Department of Civil and Environmental Engineering, University of Cincinnati, U.S.A.*
Ruiz P.A., *Universidad Politecnica de Valencia, Dept. de Sistemas Informaticos y Computacion, Spain*
Savall R., *Department of Hydraulic Engineering, Polytechnic University of Valencia, Spain*

Savic D.A., *Optimal Solutions, University of Exeter, U.K.*
Sheronosova T.Y., *Novosibirsk State University of Architecture and Civil Engineering, Russia*
Signes M., *Department of Hydraulic Engineering, Polytechnic University of Valencia, Spain*
Solomatine D.P., *International Institute for Infrastructural, Hydraulic and Environmental Engineering (IHE), Netherlands*
Souza R.S., *Department of Hydraulics and Transportation, Federal University of Mato Grosso do Sul, Brazil*
Stern, C.T., *Los Angeles Department of Water and Power, U.S.A.*
Ta C.T., *Research and Development, Thames Water Utilities Ltd, U.K.*
Tang K., *Dept. of Civil Engineering, University of Toronto, Canada*
Tarasevich V.V., *Novosibirsk State University of Architecture and Civil Engineering, Russia*
Termini D., *Dipartimento di Ingegneria Idraulica ed Applicazioni ambientali, Facoltà di Ingegneria, Università di Palermo, Italy*
Thompson G., *Binnie Black & Veatch, Redhill, U.K.*
Tiburce V., *Général des Eaux, Paris, France*
Todini E., *Department of Earth and Geo-Environmental Sciences, University of Bologna, Italy*
Tomic S., *Haestad Methods, Waterbury, U.S.A.*
Torres J., *University of Cordoba, Argentina*
Tucciarelli T., *Dipartimento di Meccanica e Materiali, Facoltà di Ingegneria, Università di Reggio Calabria, Italy*
Ulanicki B., *Water Software Systems, De Montfort University, U.K.*
Vamvakeridou-Lyroudia L.S., *Department of Water Resources, Hydraulic and Maritime Engineering, National Technical University of Athens, Greece*
van der Walt J.J., *Magalies Water, South Africa*
Vercher J., *Universidad Politecnica de Valencia, Dept. de Ingenieria Hidraulica y Medio Ambiente, Spain*
Vidal A.M., *Universidad Politecnica de Valencia, Dept. de Sistemas Informaticos y Computacion, Spain*
Walters G.A., *Optimal Solutions, University of Exeter, U.K.*
Zhang F., *Dept. of Civil Engineering, University of Toronto, Canada*
Zhang J., *Crowder and Company Limited, U.K.*

Contents

Introduction

Water Industry Systems: Modelling and Optimization Applications is based on the proceedings of the Fourth International Conference on Computing and Control for the Water Industry (CCWI'99) hosted by the Centre for Water Systems at Exeter University in September 1999. Eighty papers were presented by industry practitioners and leading academics from over twenty countries, providing an informed international view of leading edge computer technology applied to the water supply and waste water disposal industry.

The aims of the conference were:

- To facilitate continued co-operation between academic institutions and industry.
- To examine the current state-of-the-art in computing and control techniques as applied to the water industry.
- To provide a forum for discussion and the dissemination of ideas on applied computing and control for the water industry, with particular emphasis on:
  - Provider's perspective: recent developments in research.
  - User's perspective: industry's experience of the latest techniques.
  - Future needs: current and future planning and operational requirements.

The first volume of the book starts with an industry overview, in the form of five invited papers from internationally recognised authorities. The volume then concentrates on the general areas of hydraulic analysis and control. The second volume covers hydroinformatics, optimisation and quality management. The balance of papers between the two volumes reflects the steady growth in the use of computers, not just for analysis and modelling, important though these applications are, but also for information handling and informed decision making.

In the UK there have been immense changes to the water industry over the last 25 years, with the formation and subsequent privatisation of regional water utilities. In parallel, there has been a workplace revolution led by the availability of cheap and powerful desk-top computers, providing both analytical power and technical communication at the push of a button. Yet the basic hydraulic behaviour of water and most of the engineering design principles remain the same, technical practice now being based on many decades of research, development and experience. We have successfully transformed many of the simulation and analysis processes into computer software applications, improving efficiency, accuracy, and, hopefully, understanding of water system analysis and design.

However, computers now offer a far wider range of opportunities for industry – not just as tools for performing established processes more efficiently, but as ways of exploring and exploiting previously intractable areas such as:

- On-line optimal control of complex systems
- On-line state estimation
- Optimal design of large systems
- Data-mining for information retrieval
- Intelligent decision support systems

As we know, information gained and decisions made are, at best, only as good as the models used and the data available. Hence there is still a need for fundamental research to improve our understanding of the processes involved, particularly in areas such as water quality, otherwise flawed models will lead to unsatisfactory and ultimately expensive solutions, with rejection by industry of the software that delivered them. It is equally important that software developers and users have a clear understanding of the models incorporated in design and analysis packages, particularly concerning limitations and validity, so that models are not called on to simulate situations for which they were not intended.

We hope that these two volumes will go some way not only in keeping industry practitioners in touch with the latest research developments, but also in keeping academics in touch with the practical requirements of a rapidly changing industry.

Finally the editors would like to thank the authors for their excellent papers, the technical committee for their efforts in reviewing the work, and the invited speakers for their inspiring addresses. We look forward to a future water industry in which all engineering and management decisions are supported by well-developed and properly applied computer technology, to the benefit of consumers and industry alike.

Dr Godfrey Walters
Dr Dragan Savic

Centre for Water Systems
University of Exeter
United Kingdom

September 1999

PART I

KEYNOTE PAPERS

Data to Knowledge — The New Scientific Paradigm

V. Babovic *and* M. Keijzer

ABSTRACT

Present day instrumentation networks already provide immense quantities of data, very little of which provides any insights into the basic physical processes that are occurring in the measured medium. This is to say that the data by themselves contribute little to knowledge of such processes. The presently ongoing data mining project at Danish Hydraulic Institute aims at changing this situation by providing tools that will greatly facilitate the mining of data for knowledge.

Knowledge Discovery in Databases (KDD) is concerned with extracting useful information from data stores. *Data mining* is the step (be it automated or human-assisted) in this larger process called the *KDD process*. The data-mining step fits models to, or extracts patterns from, the pre-processed data.

The role of human expert is to provide domain knowledge, interpret models suggested by computer and devise further experiment that will provide even better data coverage. Clearly, there is an enormous amount of knowledge and understanding of physical processes that should not be just thrown away. Consequently, we strongly believe that the most appropriate way forward is to combine the best of the two approaches: the theory-driven, understanding-rich process and the data-driven discovery process.

This paper describes a practical experiment that utilized real-world sediment transport data set. The result is a computational environment which supports interaction of domain specialists with scientific discovery computer systems effectively creating a *knowledge discovery environment.*

1 INTRODUCTION

The formation of modern science is grounded in the period between the late XV century and the late XVIII century. The new foundations were based on the utilisation of the concept of a physical experiment and the applications of a mathematical apparatus in order to describe these experiments. The works of Brahe, Kepler, Newton, Leibniz, Euler and Lagrange clearly exemplify such an approach. Prior to these developments, scientific work primarily consisted only of collecting the observables, or recording the 'readings of the book of nature itself'.

This novel scientific approach was principally characterised by two stages: a first one in which a set of observations of the physical system are collected, and a second one in which inductive assertion about the behaviour of the system - hypothesis - is generated. Observations represent specific knowledge, whereas hypothesis represents a generalisation of these data which implies and describes observations. One may argue that through this process of hypothesis generation, one fundamentally economises thought, as more compact ways of describing observations are proposed.

Today, in the late XX century, we are experiencing yet another change in a scientific process as just outlined. This latest scientific approach is one in which information technology is employed to assist the human analyst in the process of hypothesis generation. This computer-assisted analysis of large, multi-dimensional data sets is sometimes referred to as a process of Data Mining and Knowledge Discovery. Data Mining and Knowledge Discovery aims at providing tools to facilitate the conversion of data into a number of forms that convey a better understanding of process that generated or produced these data. These new models combined with the already available understanding of the physical processes - the theory - can result in an improved understanding and novel formulations of physical laws and an improved predictive capability.

2 MODEL INDUCTION

One particular mode of data mining is that of model induction. Inferring models from data is an activity of deducing a closed-form explanation based solely on observations. These observations, however, always represent (in principle only) a limited source of information. The question emerges how this, a limited flow of information from a physical system to the observer, can result in the formation of a model that is complete in the sense that it can account for the entire range of phenomena encountered within the physical system in question - and to even describe the data that are outside the range of previously encountered observations (Babovic and Abbott, 1997). The confidence in model performance can not be based on data alone, but might be achieved by grounding models in the domain so that appropriate semantic content is obtainable. This should be the ultimate goal of knowledge discovery.

Thus, we need model induction algorithms that produce models amenable to interpretation next to the ability to fit data. Clearly, every model has its own syntax. The question is whether such syntax can capture the semantics of the system it attempts to model. Certain classes of model syntax may be inappropriate as a representation of a physical system. One may choose the model whose representation is complete, in the sense that a sufficiently large model can capture the data's properties to a degree of error that decreases with an increase in the model size. Thus, one may decide to expand Taylor or Fourier series to a degree that will decrease the error to a certain, arbitrarily given degree. However, completeness of representation is not the issue. The issue is in providing an adequate representation amenable to interpretation.

3 GENETIC PROGRAMMING

In genetic programming (GP) the evolutionary force is directed towards the creation of models that take symbolic form. In this evolutionary paradigm, evolving entities are presented with a collection of data and the evolutionary process is driven towards finding closed-form symbolic expressions describing the data. In principle, GP evolves parse trees. The types of nodes in this tree structure are user-defined. This means that they can be algebraic operators, such as sin, log, +, -, etc., but they can also take the form of logical rules, making use of operators such as OR, AND, etc.

GP lends itself quite naturally to the process of induction of mathematical models based on observations: GP is an efficient search algorithm that need not assume the functional form of the underlying relationship. Given an appropriate set of basic functions, GP discovers a (sometimes very surprising) mathematical model that approximates the data well. At the same time, GP-induced models come in a symbolic form that can be interpreted (see for example, Babovic, 1996).

However, the application of standard GP in a process of scientific discovery does not always guarantee satisfactory results. In certain cases, GP-induced relationships are too complicated and provide little new insight on the process that generated the data. One may argue that GP, in such situations, blindly fits parse trees to the data (in almost the same way as in Taylor or Fourier series expansion). It can be argued that GP then results in a model with accurate syntax, but with meaningless semantics. In these cases, the dimensions of the induced formulae often do not fit, pointing at the physical uselessness of induced relationships.

4 UNIT TYPING IN GP

The present contribution utilises an augmented version of GP - dimensionally aware GP - which is arguably more useful in the process of scientific discovery (Keijzer and Babovic, 1999).

Throughout science, the units of measurement of observed phenomena are used to classify, combine and manipulate experimental data. Exploitation of this information within GP makes this algorithm somewhat stronger as a search method. At the same time its applicability is broadened to encompass problems containing information on units of measurement. Standard GP is ignorant of the dimensionality of its terminals and can safely be applied to problems composed of dimensionless numbers only. Given the symbolic nature of GP and its ability to manipulate the structure of functional relationships it seems strange that information contained in units of measurement is not used as an aid in search process. After all, the dimensional correctness as used in science acts as a syntactic constraint on any formula it induces. It is therefore expected that the introduction of dimensions in GP paradigm might result in improved search efficiency.

4.1 Extending GP

GP, as an instance of the evolutionary algorithm family, iteratively applies variation and selection on a population of evolving entities. Standard variation operators in genetic programming are subtree mutation (replace a randomly-chosen subtree with a randomly generated subtree) and subtree crossover (replace a randomly chosen subtree from a formula with a randomly chosen subtree from another formula). In order to accommodate the additional information available through units of measurement, extensions of standard GP were proposed (Keijzer and Babovic, 1999). These extensions are briefly described in the following sections.

4.2 Introduction of Units in GP

In the dimension-augmented setup, every node in the tree maintains a description of the units of the measurement it uses. These units are stored as arrays of real-valued exponents of the corresponding dimensions. In the present set of experiments only the dimensions of length, time and mass (LTM) are used, but the setup may be trivially extended to include all other SI-dimensions (amount of substance, electric current, thermodynamic temperature and luminous intensity). Square brackets are used to designate units, for example [1, -2, 0] corresponds to a dimension of acceleration ($L^1T^{-2}M^0$). Similarly, [0,0,0] defines a dimensionless quantity.

4.3 Definition of the Terminal Set

The definition of the terminal set is straightforward, in that variables and constants are accompanied with the exponents of their respective units of measurement:

$$T = \{ v_1[x_{v1},y_{v1},z_{v1}], \ldots, v_n[x_{vn},y_{vn},z_{vn}], c_1[x_{c1},y_{c1},z_{c1}], \ldots, c_m[x_{cm},y_{cm},z_{cm}] \} \quad (1)$$

where v designates a variable, c a constant and [x,y,z] the corresponding array containing the units of measurement. For example, $v_1[0,0,1]$ designates a variable – v_1 – with a dimension of mass. User-defined constants can be defined along with their dimensions, such as 9.81[1, -2,0] defining the earth's gravitational acceleration. Randomly generated constants are allowed only as dimensionless quantities ([0,0,0]). There is a definitive reason for allowing random numbers to be dimensionless only. Should random constants with random dimensions be allowed, GP would have an easy way to correct the dimensions by introducing transformation from one arbitrary unit of measurement to another. Some form of pressure should be applied on the application of unit transformation. This issue is addressed during the experimentation.

4.4 Definition of the Function Set

Application of arithmetic functions on dimension-augmented terminals violates the closure property for these functions (Koza, 1992). For example, adding meters to seconds renders a dimensionally incorrect result of the operation. Therefore, the definition of arithmetic operators is augmented to: (a) specify the transformation of units of measurement, (b) accommodate units of measurement-related constraints on the application of functions and (c) introduce additional functions that repair trees and provide the benefits of closure.

Table 1 summarises the effects of the application of functions on units of measurement and specifies constraints on the applications of functions. For example, exponentiation of a value can only take place when the operand is dimensionless, in which case the result of the operation is also a dimensionless value. Similarly, addition and subtraction are constrained so that their operands must have the same dimensions. Multiplication and division combine the exponents by adding and subtracting the dimension exponents respectively. The standard `Power` function can be applied to dimensionless values only, whereas `PowScalar` can be applied to dimensional operands, affecting their dimensions correspondingly. Other functions can be defined in similar ways.

Table 1 Effects and constraints that units of measurement impose on function set

Function	Operand Dimensionality	Result
Exponentiation:	[0,0,0]	[0,0,0]
Logarithm:	[0,0,0]	[0,0,0]
Square Root:	[x,y,z]	[x/2, y/2, z/2]
Addition:	[x,y,z], [x,y,z]	[x,y,z]
Subtraction:	[x,y,z], [x,y,z]	[x,y,z]
Multiplication:	[x,y,z], [u,v,w]	[x + u ,y + v, z + w]
Division:	[x,y,z], [u,v,w]	[x - u ,y – v, z - w]
Power:	[0,0,0], [0,0,0]	[0,0,0]
PowScalar (c):	[x,y,z]	[x*c, y*c, z*c]
If less than zero:	[0,0,0], [x,y,z], [x,y,z]	[x,y,z]

4.5 DimTransform

As mentioned earlier, an additional function should be introduced in order to guarantee closure.

Table 2 Effects of `DimTransform` on units of measurement

Function	Operand Dimensionality	Result
`DimTransform` c[x,y,z]:	[u,v,w]	[x + u ,y + v, z + w]

This transformation operator multiplies its operands with the constant value *c* and also affects dimensions through applying values for *x*, *y* and *z* as indicated in **Table 2**. `DimTransform` can be used to resolve dimensional violations that will inevitably arise when using standard-GP-style randomized crossover on formulae with dimensional variables. For example, when meters are added to seconds, the second operand can be transformed into meters by wrapping it with a transformation of magnitude 1.0 and unit description [1,-1,0]. `DimTransform` can therefore transform a quantity expressed in one unit into a quantity expressed in a completely different unit.

At first sight, it appears that the application of `DimTransform` eliminates all the problems related to dimensions. It is evident, however, that the application of `DimTransform` can 'fix' any dimensionality-related problem by introducing physically meaningless transformations into evolving formulae.

Initialisation, crossover and mutation impose heavy demands on the use of this transformation. Whenever a constraint violation occurs (for instance after an insertion of a subtree), a `DimTransform` function which solves this violation is inserted. At present, this is done in an arbitrary manner *i.e.* no attempt is made to find the optimal sequence of transformations for a given formula.

Therefore, the application of `DimTransform` does not always contribute to an interpretable solution to the problem. To control its application an additional objective is introduced (see section 4.6).

The `DimTransform`-based mechanism is not a form of strongly typed GP. A strongly typed GP would initialise and keep all expressions dimensionally correct throughout evolutionary process. For ill-posed and incomplete problems, a strongly typed GP would fail even at initialisation. The current approach allows dimensionally incorrect solutions, but introduces evolutionary pressure on dimensional incorrectness. Therefore, the present version of GP is based on promotion of type.

4.6 Definition of Goodness-of-Dimension Calculation

Apart from the usual *goodness-of-fit* statistic, a second objective — *goodness-of-dimension* — is introduced. As stated earlier, the application of the `DimTransform` operator can result in an arbitrary transformation of units. This is not necessarily a desirable behaviour, as multiple occurrences of this operator do not enhance the interpretability of the resulting formulae. When no constraints are placed on the frequency of `DimTransform`, the approach reduces to a standard GP with the added 'feature' that the resulting formulae grow proportionately to the number of unit violations.

One of the approaches used to reduce the number of applications of `DimTransform` takes full advantage of the explicit representation of the dimensions as a vector of real valued exponents. The distance of an expression from a dimensionally correct formulation can be calculated as the number of required transformations, *i.e.*

$$\textit{Goodness-of-Dimension} = \sum |x_i| + |y_i| + |z_i| \qquad (2)$$

where the subscript i ranges over all applications of the `DimTransform` operator in the formula, whereas x, y and z are the components of the dimension vector. The goodness-of-dimension acts as an effective metric of distance from desired dimensions and can be treated as an additional measure of fitness. Goodness-of-dimension can be combined with the goodness-of-fit statistic in a multi-objective optimization fashion .

4.7 Dimension-Based Brood Selection

Brood selection (Tackett, 1994) is a technique where the parents produce a large number of offspring. These offspring are evaluated against a 'cheaper' fitness function (often referred to as the culling function). The best of these offspring are moved to the next generation. The evaluation of a large collection of the offspring by a culling function improves the overall performance because little effort is wasted on the evaluation of bad offspring on the expensive complete fitness function.

The culling function used here is the goodness-of-dimension of the formula. This evaluation is very cheap as it can be calculated independently of the training set and it requires a single pass through the parse tree.

The present implementation reads as follows: two parents are chosen for crossover; they produce *m* offspring by repeated application of the random sub-tree crossover operation; constraint violations are corrected for dimensions in the manner outlined in 4.5 DimTransform; and, finally, the best among the *m* offspring with respect to goodness-of-dimension are added to the intermediate population.

Therefore, this operation can be best understood as a unit-informed crossover. Although the use of a culling function equates to selection, it is applied immediately after individual crossovers, thus modifying the results of crossover to produce formulae that are dimensionally more correct.

In addition to the usual goodness-of-fit statistic, a supplementary measure of fitness - goodness-of-dimension - is introduced. Goodness-of-dimension is effectively a metric of distance of induced formula from its desired dimensionality. Goodness-of-fit and goodness-of dimension are then utilised in a multi-objective setting to drive the evolutionary process towards expressions of both high accuracy and correct dimensionality.

Fundamentally, augmentation of GP with dimensional information adds a descriptive, semantic component to the algorithm. This is in addition to the functional semantics that defines the manipulation on numbers. While functional semantics grounds formulae in mathematics, the dimensional semantics grounds them in the physical domain.

5 DATA

To test the performance of unit typing in GP, experimental flume data utilised by Zyserman and Fredsøe (1994) were analysed. The experimental data consisted of total, steady state sediment load for a range of discharges, bed slopes and water depths. Zyserman and Fredsøe used the Engelund-Fredsøe and Einstein formulation to calculate the bed concentration of suspended sediment c_b and used these values in conjunction with hydraulic parameters to perform system identification and formulate the expression for bed concentration of suspended sediment c_b. The hydraulic conditions were represented by Shields parameter θ, defined as:

$$\theta = \frac{u_f}{(s-1)gd} \quad \textit{and} \quad \theta' = \frac{u_{f'}}{(s-1)gd} \qquad \text{(2) and (3)}$$

where:

u_f -shear velocity $=(gDI)^{0.5}$
s -relative density of sediment

d_{50} -median grain diameter
D -average water depth
I -water surface slope
u_f' -shear velocity related to skin friction $=(gD'I)^{0.5}$
D' -boundary layer thickness defined through:

$$\frac{v}{u_{f'}} = 6 + 2.5\ln\left(\frac{D'}{k_N}\right) \tag{4}$$

v -mean flow velocity
k_N -bed roughness =2.5d

The study makes use of the so-called Rouse number z, defined as:

$$z = \frac{w_s}{\kappa u_f} \tag{5}$$

where:
w_s -settling velocity of suspended sediment
κ -von Karman's constant (≈ 0.4)

An interesting observation is that all 'directly measurable' quantities do not correlate as well with the concentration of sediment cb as derived dimensionless quantities $\theta$and θ (**Table 3**).

This is a perfect example of standard scientific practise: units of measurements accompanying direct observations are effectively eliminated through introduction of dimensionless ratios (other well known examples are the Froude number and the Reynolds number). Once the dimensionless numbers are used instead of the original dimensional values the problem of dimensional correctness is conveniently avoided, as all analysed quantities are dimension-free. It is also argued that dimensionless ratios collapse the original dimensional search space, making it more compact, thus resulting in a more effective behaviour of algorithms that fit models to the data. At the same time, the information contained in units of measurement is ignored entirely.

Table 3 Correlation coefficients for directly observable and derived, dimensionless quantities

Directly Observable		Derived quantities	
Correlation pairs	Correlation coefficient	Correlation pairs	Correlation coefficient
cb – Uf'	0.784	cb – θ'	0.894

cb – Uf	0.628	cb - θ	0.711
cb – ws	0.430		
cb – ν	0.152		
cb – d50	-0.232		

Bearing the strong correlations in mind, it is of little surprise that, after dimensional analysis, Zyserman and Fredsøe (1994) formulated the following expression:

$$c_b = \frac{0.331(\theta' - 0.045)^{1.75}}{1 + \dfrac{0.331}{0.46}(\theta' - 0.045)^{1.75}} \qquad (6)$$

so that c_b is a function only of θ'. Comparative analysis with some other and more complex expressions involving more variables presented in their 1994 paper, has shown that formula (7) is of comparable, if not higher accuracy.

5.1 Formulations induced by the means of Genetic Programming

By way of comparison, a genetic programming environment was set-up in such a way as to comprehend all measured data and not the corresponding dimensionless parameters based on the measurements. The purpose for conducting such experiment was to test whether such GP setup is capable of creating a dimensionally correct and still accurate formulation. Since the pre-processing of raw observations (formation of dimensionless θ and θ' was not employed here, it can be argued that GP was confronted with a problem of trying to formulate a solution from first principles. The evolutionary processes resulted in a number of expressions, of which only the most interesting one is presented here:

$$c_b = 1.12\,10^{-5}\,\frac{\left(u_f' - w_s\right)\left(1 + 100\dfrac{u_f' w_s}{g d_{50}}\right)}{u_f + u_f'} \qquad (7)$$

The degree of accuracy of the induced expression is quite satisfactory. A statistical measure of conformity, such as the coefficient of determination, gives a value of 0.82. This provides an improvement over the value of 0.80 based on the Zyserman-Fredsøe relationship (Eq.7). The total error over the data set is reduced and all other statistical measures of accuracy such as average deviation, coefficient of efficiency, robustness and 95% confidence disclose improvements.

At the same time, the formula is dimensionally correct, it uses the most relevant physical properties in the relevant context. For example, the dimensionless term

$\frac{u_f' w_s}{g d_{50}}$ is effectively a ratio of shear and gravitational forces. Shear forces are represented by u_f', 'responsible' for elevating sediment particles into the stream, while the gravitational term $\frac{g d_{50}}{w_s}$ is 'responsible' for settling the particles. The remaining group $\frac{(u_f' - w_s)}{u_f + u_f'}$ is a ratio of resultant energy near the bed and of the total available energy in the flow transporting the particles.

In the present case, evolution produced a dimensionally correct, meaning-rich formulation which is very different in form from the benchmark formulation and still more accurate. It did so without referring to any of the background knowledge, the process operated on raw observations only. Such results may open new avenues of research in this (and for that matter any other) domain so that we can talk about knowledge-discovery-driven domain research. Domain experts can be exposed to a completely new set of formulations, off of the beaten track, yet within the domain of physical validity.

One question still remains: which formula to choose? Fortunately, the primary purpose of this text is not to propose an improved formulation for the concentration of suspended sediment near bed c_b , but to demonstrate an alternative discovery process. The physical interpretation of formulations (7) and (8) requires a wealth of knowledge in sediment transport and the authors feel ill-equipped to perform this task in a qualified way. At the same time, such discussion would fall far from the scope of the present contribution.

Thus, the choice of the best formula is not made here. The choice should be made by the experts in the field, by the people who can competently judge the quality of the data sets used, interpret the induced model and in the end choose the one which makes most sense. The human experts should make the ultimate step in this process.

The presently described work is part of a research effort aiming at providing new (and sometimes provocative) hypotheses built from data alone. The ultimate objective is to build models which can be interpreted by the domain experts. Once a model is interpreted, it can be used with confidence. It is only in this way that one can take full advantage of knowledge discovery and advance our understanding of physical processes.

ACKNOWLEDGEMENTS

This work was in part funded by the Danish Technical Research Council (STVF) under the Talent Project N° 9800463 entitled "Data to Knowledge - D2K". Their support is greatly appreciated. For more information on the project, visit http://projects.dhi.dk/d2k

REFERENCES

Babovic, V., 1996, *Emergence, Evolution, Intelligence; Hydroinformatics*, Balkema, Rotterdam

Babovic, V., and Abbott, M.B., 1997, The evolution of equations from hydraulic data, Part I: Theory, *Journal of Hydraulic Research*, Vol 35, No 3, pp. 1-14

Keijzer, M., and Babovic, V., (1999), Dimensionally aware genetic programming, in Banzhaf, W., Daida, J., Eiben, A. E., Garzon, M. H., Honavar, V., Jakiela, M., & Smith, R. E. (eds.)., *GECCO-99: Proceedings of the Genetic and Evolutionary Computation Conference*, July 13-17, 1999, Orlando, Florida USA. San Francisco, CA: Morgan Kaufmann.

Koza J R., 1992., *Genetic Programming: On the Programming of Computers by Natural Selection*, MIT Press, Cambridge, MA

Tackett, W. A.,1994., *Recombination, Selection, and the Genetic Construction of Computer Programs,*. Ph. D. thesis, University of Southern California

Zyserman, J.A., and Fredsøe, J., (1994), Data analysis of bed concentration of suspended sediment, *Journal of Hydraulic Engineering*, ASCE, 120, No.9, pp.1021-1042

Making Decisions that Matter with Water Quality Models

M. Brandt, J. Creasey *and* R. Perkins

ABSTRACT

As part of a drive to reduce taste complaints, Severn Trent Water decided to reduce chlorine residual concentrations and their variability in the distribution system. They commissioned Binnie Black and Veatch, and WRc to investigate the effects of this decision and to propose solutions to any problems found.

The approach included the following steps:

1) data collection
2) network model update
3) development of chlorine decay models
4) simulation of chlorine residuals
5) investigation of residuals and the effect on microbiological compliance
6) recommendation of solutions where results are unacceptable
7) simulation of recommendations
8) appraisal with local STW staff

The work involved the appraisal of some 180 chlorine control zones. In most cases, it was decided to calibrate the large number of models against recent historic chlorine sample data.

The network models were then used to simulate conditions in the systems under current dosing regimes and when the set point at the chlorinator was reduced. The results of the simulations plus sample data were used to investigate the acceptability of the new regimes. With such a large number of zones to process, it was decided to draw up a decision tree to guide the decisions. Other parts of the decision tree guided the user towards an explanation for any problems and hence to possible solutions. Where possible, solutions were tested using the network models.

A series of meetings was held with local staff. The combination of local knowledge with network modelling and expert appraisal proved to be very powerful.

1 INTRODUCTION

Software to simulate the behaviour of water quality parameters in distribution systems has been available for more than ten years. Standard capability includes calculating the age of water and the chlorine concentration at all points in the network. Unfortunately, examples of the use of water quality models to answer design or operations questions are all too rare. This paper reports on a project in which models played a vital role. It was carried out for Severn Trent Water (STW) by Binnie Black and Veatch (BBV) and Water Research Centre (WRc). It is also remarkable that a large number of models were used to arrive at a large number of implementable decisions in a short time.

2 SEVERN TRENT'S TASTE ACTION PLAN

2.1 Background

Severn Trent Water supplies approximately 2,000 Ml/day of water to a population in excess of eight million people. The water is disinfected with chlorine at the treatment works and receives further chlorination at chlorine booster stations to protect the quality of the water throughout the complex distribution network. The chlorination strategy has contributed to an excellent record for delivering high quality drinking water. However, chlorine residuals within parts of the network can be relatively large and variable leading to customer dissatisfaction. The impact of high and/or variable chlorine residual has been confirmed by an analysis of customer complaints and customer surveys.

2.2 Objectives

The introduction of a methodology for setting chlorine residual targets led to a reduction in chlorine residuals within the distribution system. However, it was apparent that capital works would be required to enable chlorine residuals to be lowered and stabilised in some areas. A Taste Action Plan was developed by STW with the principal objective of reducing the number of taste related customer complaints. Chlorine residual was identified as a surrogate for taste and the following chlorine related objectives were developed:

- Ex-works free chlorine residuals to be reduced to a maximum of 0.4 mg/l. (For the purpose of the study described in this paper this was interpreted as 0.3 mg/l $^{+}/_{-}$ 0.1 mg/l)
- Variation in chlorine residual at any point to be reduced to less than or equal to 0.2 mg/l.
- Chlorine residual at the customers tap to be at or below 0.3 mg/l.
- Minimum chlorine residual to be 0.05 mg/l
- Bacteriological compliance not to be compromised.

2.3 Detailed Plan

The detailed plan comprised the following steps:

1) A study was made of the Company's quality data to identify chlorine control areas where the residual at the control point was greater than 0.4 mg/l and also those areas which had experienced any bacteriological problem during the previous 12 months. "Areas which had experienced bacteriological problems" were defined, for the purpose of the study, as those areas where coliforms had been identified in any samples or where

there was evidence of high or rising plate counts which might lead to a reluctance to reduce chlorine residuals.

2) These areas were matched to identify:
    - 'Red' areas which experienced 'high' chlorine residual at the control point together with bacteriological problems.
    - 'Amber 1' areas which experienced 'high' chlorine residuals at the control points but no bacteriological problems.
    - 'Amber 2' areas which experienced 'low' chlorine residuals at the control points together with bacteriological problems.
    - 'Green' areas which experienced 'low' chlorine residuals at the control points and no bacteriological problems.
3) Undertake a study of the Red and Amber 1 areas (high chlorine / potential taste problems) to identify the need for, type and location of capital works, or the operational improvements required in order to achieve the objectives
4) In parallel with 3 above, identify areas with chlorine residuals varying by more than 0.2 mg/l and identify solutions to these problems.
5) Implement the operational changes and prioritise the capital schemes identified.

Steps 1 and 2 above were undertaken by STW prior to BBV and WRc being appointed to undertake the studies, defined in steps 3 and 4, which are the subject of this paper.

3 DATA COLLECTION

The project programme was tight and the volume of data needed to analyse 184 control areas was large. Although it was realised that there would be the inevitable repeat visits to some STW offices, the plan was to obtain the majority of the required data in a single extended visit during the first phase of the work. Therefore the exercise needed to be systematic, comprehensive and effective.

For the Red and Amber 1 areas, the data requirements included:

- Drawing of the distribution system network:
  - configuration of the network
  - system hydraulics, operation and control
  - mains materials and condition
  - areas and details of network rehabilitation in previous three years
  - planned rehabilitation, routine mains flushing and plans to improve water quality
  - known system problems particularly with reference to water quality
  - reservoir details; internal layout and inlet and outlet arrangements
- Stoner model, where available, including identifying any network changes since the model was built
- Fixed and random location sample data for:
  - source water
  - locations of chlorine and bacteriological sample points
  - sample results for at least the last 12 months minimum (QUIS data)

- Sundry data;
  - leakage results (DILIS report)
  - water supplied (SUMIS report) for last 12 months
  - Blue Book schematics for each water quality zone

The data collection teams were provided with a six-page general data collection pro-forma and an eight-page bacteriological data form for the compliance modelling. The forms included checklists of data needs and guidelines on where to obtain the data, decision criteria and outcome for that item.

A small section of the data proforma is shown in Figure 1.

DATA COLLECTION PROFORMA - Sheet 2

<table>
<tr><td>County</td><td colspan="3"></td></tr>
<tr><td>Area classification</td><td></td><td>Area ref</td><td></td></tr>
<tr><td>Site</td><td colspan="3"></td></tr>
<tr><td colspan="4">Drawing of distribution network</td></tr>
<tr><td colspan="3">Tick to confirm a drawing/schematic of the distribution network has been obtained</td><td></td></tr>
<tr><td colspan="3">Obtain drawings showing the internal details of all reservoirs directly feeding the network or actually within it with any chlorine sample points marked - tick if obtained</td><td></td></tr>
<tr><td colspan="3">Tick to show locations with refs of vulnerable sample points have been shown</td><td></td></tr>
<tr><td colspan="3">Tick to show locations with refs of sensitive sample points have been shown</td><td></td></tr>
<tr><td colspan="3">Tick to show location with ref of chlorine control point has been shown</td><td></td></tr>
<tr><td colspan="3">Tick to show locations with refs and descriptions of any reservoir sample points</td><td></td></tr>
<tr><td colspan="3">Tick to show locations with refs of bacteriological sample points have been shown</td><td></td></tr>
<tr><td>Any comments?</td><td colspan="3"></td></tr>
<tr><td colspan="4"></td></tr>
</table>

Figure 1 Part of data collection proforma

Data collection started in December 1997 and lasted for about six weeks. Free chlorine, temperature, location address and date of sampling data was obtained centrally from the Quality Information System (QUIS) in conjunction with the data collection for compliance modelling. In practice, as the detailed analyses proceeded, additional information was needed and anomalies in the data had to be resolved, which involved further visits and numerous telephone calls. Although not required for this study phase, during the initial data collection visits, similar information was obtained for the Amber 2 areas at the same time.

For the Green areas, where records indicated excessive chlorine variation, (that is concentration fluctuations greater than 0.2 mg/l), it was intended that representative company wide sample data would be used to select 15 sites for detailed examination. In practice, all Plant Monitoring and Control System (PMCS) records and available chart recorder records were reviewed for chlorine variation outside the acceptable criteria. Because processing and plotting individual PMCS records was time consuming, the data was reviewed on the screen and only typical plots of 'excessive fluctuation' were abstracted for detailed analysis.

Although it was not used, there was also an option to undertake site sampling where it was felt that there was insufficient available chlorine data to achieve a coarse calibration of the Stoner model.

4 WATER QUALITY MODEL DEVELOPMENT

The objective of the water quality modelling was to use coarsely calibrated chlorine models to predict concentrations within distribution resulting from changes in dosing rates at the control points. The means of achieving the objective were less straightforward.

4.1 Spreadsheet analysis

Spreadsheet analysis techniques were used to predict the time of travel and chlorine decay in simple linear networks. Example results are shown in Figure 2.

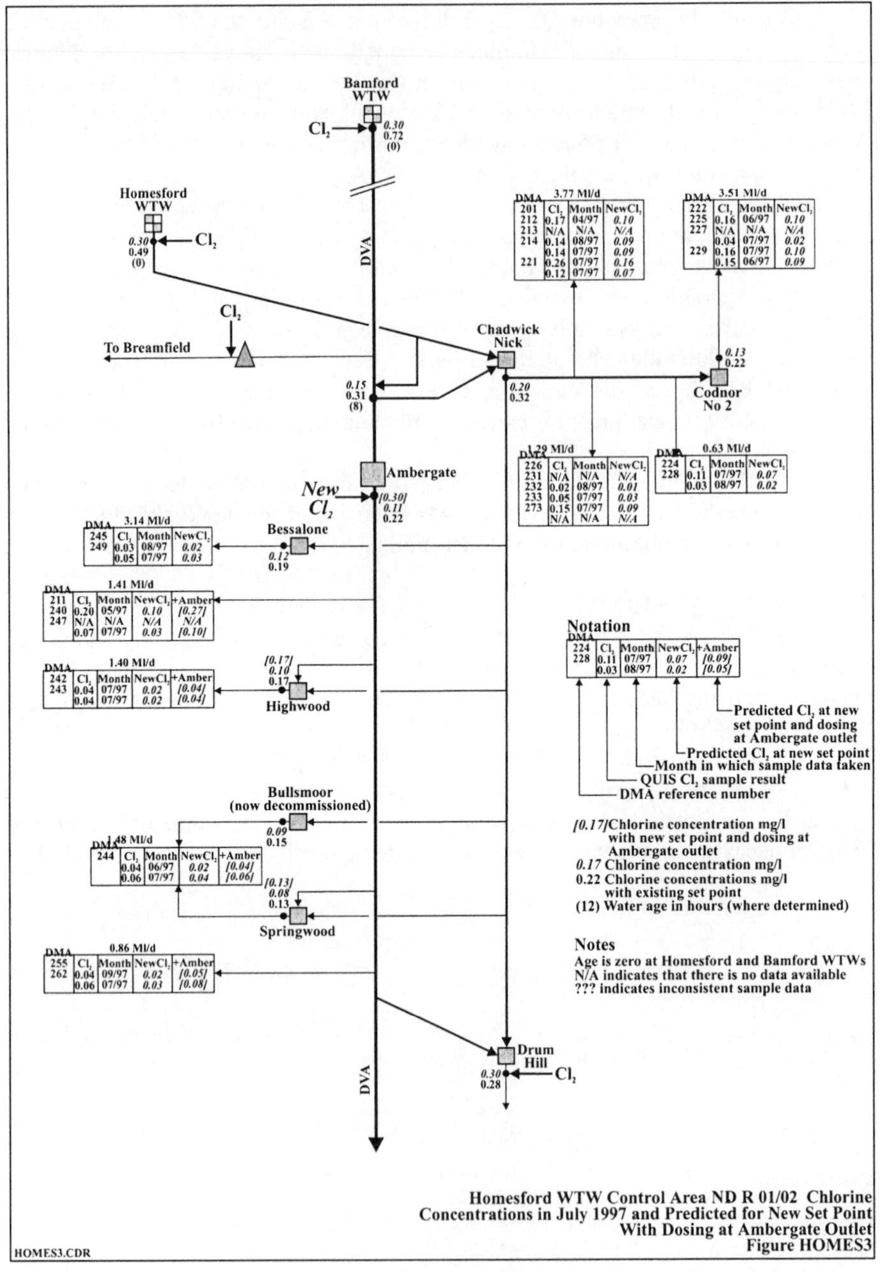

Figure 2 Example of spreadsheet analysis output

4.2 Stoner hydraulic models

For more complex networks and hydraulic conditions, STW Stoner hydraulic models were used. Ideally, it would have been preferable to work with all-mains models which were set up for water quality analysis. In practice the available models were of varying vintage and to different construction standards.

The stages of the analysis were:

1 Update existing models to reflect current supply arrangements and network changes.
2 For areas not covered by existing models, build new model or extend existing model.
3 Set the Stoner model to run in Age Mode
4 Coarse calibrate the chlorine model for winter and summer conditions
5 Carry out predictive analysis, test options and report results

The library models generally represented the average day demand condition for the year of hydraulic calibration. For the majority of areas the hydraulic models still provided a good or reasonable indication of performance without needing to raise demands to current levels. However, it was necessary to update the models to reflect physical changes in the network since their construction. Changes included mains rehabilitation, replacements or extensions and system operational changes.

Although STW has extensive model coverage of their networks, there were parts of their areas not covered by models. For these areas, simple Stoner models were constructed based on the DMA schematics or mains record drawings. Demands were derived from DILIS reports and standard diurnal profiles. The DILIS report gave the daily flow in m^3/day for each DMA for a particular week. The demand was adjusted to give an annual average value by applying an appropriate factor derived from SUMIS data. Demand was allocated to model demand nodes equally. The models were constructed to a sufficient level of detail to ensure that the model represented the hydraulics of the system as far as was required to complete the quality analyses. Areas not required for this study were represented only as far as they impinged on the hydraulics of the study area.

The hydraulic models were initially tested to ensure they ran uninterrupted through a 24-hour cycle and beyond and that the demands and reservoir performances hydraulically balanced over the analysis period. Problems at this stage, which had to be resolved before progress to the quality analyses, generally centred around replicating system performance beyond 24 hours and resolving infeasible solutions due to excessive hydraulic losses.

4.3 Age analysis

Each model was then set to operate in water quality mode. Once the model would run successfully to 48 hours and produced repeatable performance, the age of water was calculated. This involved a number of intermediate steps to reset ages in storage facilities until they exhibited repeating age profiles. The age analysis results give an indication of possible water quality problems in distribution.

4.4 Chlorine models

The Stoner software represents chlorine decay by a first order equation and does not distinguish between bulk and pipe wall decay coefficients. There has been considerable debate in the industry about the chlorine decay mechanism in distribution systems and how it should be represented mathematically. Recent work at Birmingham University (Powell et al, 1999) on the performance of a number of algorithms concluded that the additional benefits of more complex algorithms were small and that they all gave sufficient accuracy for network modelling purposes.

Winter and summer chlorine models were produced to reflect the effect of temperature on decay rates. The chlorine data was assembled to cover the period December to March for winter and June to September for summer by sample date, location, temperature and free chlorine concentration. Although ideally an average seasonal temperature was used to select the data set for calibration, in practice preference was given to sets of data with the largest amount of information and no obvious anomalies. The locations of random daytime samples between winter and summer models posed interesting problems when trying to resolve anomalies between calibration result.

South Warwickshire Rowington Model
Free Chlorine and Temperature

SITE NAME / Free CL^2	JAN	FEB	MAR	APR
HAMPTON TOWER Outlet	0.18	0.19	0.19	0.19
LAPWORTH TOWER Outlet	0.10	0.05	0.25	0.42
STRENSHAM TREATMENT WORKS				
FINAL NO.1 MAIN	0.76	0.69	0.70	0.68
FINAL NO.2 MAIN	0.75	0.70	0.68	0.68
BUDBROOKE BOREHOLE final	0.25	0.31	0.23	-
HEATH END BOREHOLES Blended final	0.39	0.45	0.43	0.44
HEATH END CONTACT TANK	0.37			
HENLEY IMPORT WARWICK ROAD BOREHOLE Blended final	0.37	0.37	0.32	0.27
SHREWLEY BOREHOLE Final	0.28	0.33	0.39	0.21
LYE GREEN TOWER Outlet+J17	0.37	0.26	0.27	0.29
ROWINGTON BOREHOLE Final	0.42	-	0.44	0.56
THELSFORD BH	0.37	0.29	0.39	0.39

Figure 3 Example of chlorine sample data

4.5 Chlorine model calibration

From the weekly SUMIS data, the average daily supply for the year was then calculated. By comparing the SUMIS supply for the month in which the chlorine data was selected with the SUMIS yearly average, factors were generated to relate the calibration months to the annual average demand. The two chlorine models were then created for their respective demands. Once each model had been adjusted to ensure that it repeated hydraulically, measured chlorine injection points were set up together with the target sampling points and defined target profiles.

Calibration was achieved by applying global adjustments to pipe and tank reactive coefficients until a best match was achieved between the modelled and measured data. Where there was sufficient information for a limited number of cases, single feed pipes could be calibrated separately. Ranges of pipe coefficients were typically between 0.001 and 0.3 (1/hr) depending on the pipe wall material and condition. Tank reactive coefficients started at 0.01 (1/hr). As one would expect, generally the reactive coefficients were higher for the summer models corresponding to the higher water temperatures and more rapid decay. A similar procedure to that used for the age calculation was necessary during the calibration process to ensure repeatable chlorine concentrations.

5 INVESTIGATING PROBLEMS AND IDENTIFYING SOLUTIONS

5.1 Model results

The work was carried out by two contractors on four sites. The transfer of large amounts of data and results had to be done in a disciplined way. A procedure was drawn up which specified the content and form of the results for each model which were to be passed from BBV to WRc.

These included:

- a schematic showing the age of water at nodes throughout the control zone
- a similar schematic showing chlorine concentrations
- the chlorine concentration profile over 24 hours at the node with the lowest concentration
- the chlorine concentration profile over 24 hours at the node with the most variable concentration

The results were produced for existing chlorinator set points and also with set points reduced to comply with the new maxima. Each situation was simulated for Winter and Summer conditions. Figure 4 gives an example schematic showing age of water and chlorine concentrations.

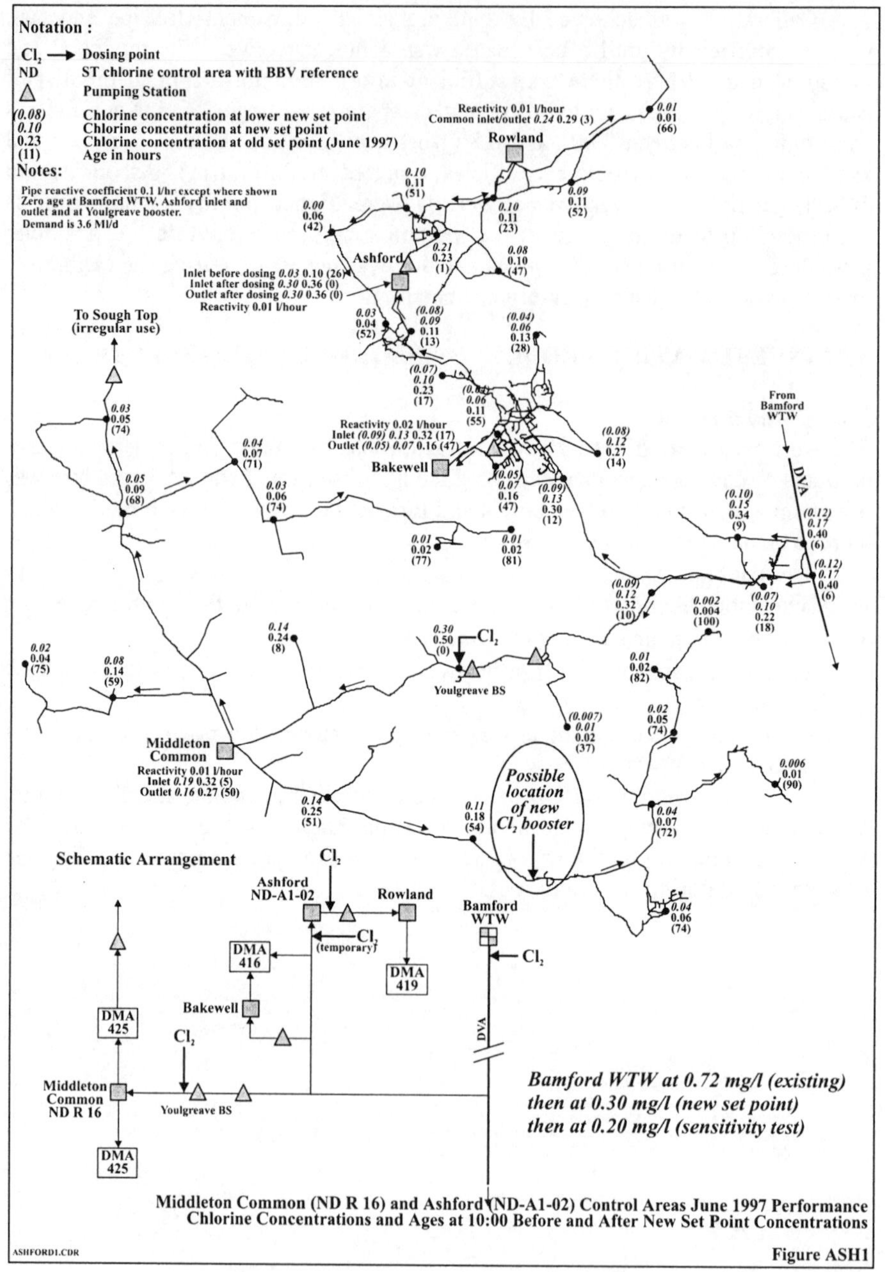

Figure 4 Example of results on schematic

The model results with the new set points for all zones had to be assessed against a number of criteria:

- Do chlorine concentrations at all nodes lie within the allowable band?
- Does concentration vary by more than the acceptable range at any property?
- Will the probability of microbiological compliance failure be acceptable?

If the answers to these questions were all "Yes", the new set points were accepted. Otherwise, reasons for the problem were needed.

5.2 Microbiological considerations

As was indicated in the previous section, it is not acceptable merely to maintain chlorine concentrations within prescribed limits. It was necessary to decide whether the new regime would produce an unacceptable risk of microbiological failure. This was particularly important because, in the zones studied, chlorine concentration reduction was being considered.

The microbiologists on the team considered past failure data and modelled the relationship between failure probability and chlorine levels when deriving their assessment of future risk. This was fed into the decision process.

5.3 The decision chart

There are a number of reasons why the results may fail one of the criteria. In the case of low chlorine at a point in the zone, they include:

- the travel time in the pipes is long
- the residence time in a reservoir is high
- the chlorine decay rate constant (reactive coefficient) is high for some of the pipework
- the drop in concentration across a reservoir is high although the residence time is low.

The information needed to make these decisions is all available in the model results. However, since more than one person would be involved in the investigation and since there were many zones to consider, a decision chart was developed to ensure consistency. A section of the chart is shown as Figure 5.

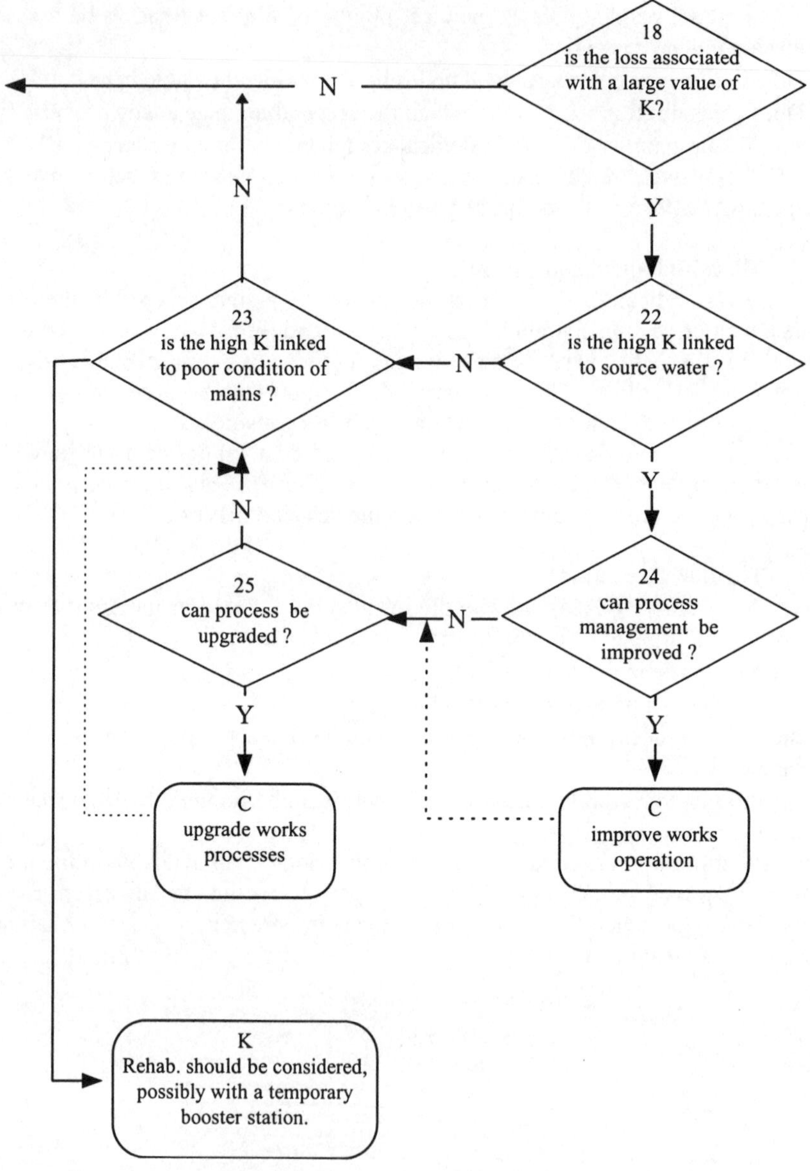

Figure 5 Section of decision chart

The chart was an aid to understanding the root cause of the problem and to indicate a solution. If a high rate constant has been calculated, then dirty or corroded pipes may be indicated. However, if the rate constant is low but the time of travel ("age") is high, pipe renovation is not an answer. It may be necessary to install booster chlorination at some point on the route to the properties suffering low chlorine.

Some recommendations could be made with confidence; others pointed to further investigation. For example, some reservoirs with reasonable volumes with respect to the local demand had, nevertheless, large concentration drops. Further questions were needed to determine if the feed and supply pipework could be improved or if the flow patterns within the reservoir were poor.

Reasons for unacceptable conditions are not as clear-cut as the decision chart idea suggests. What is a long travel time? What decision should be made if rate constant and travel times are both fairly large? Experience and expertise were needed throughout this investigation.

5.4 Testing the solutions

Recommendations were sent from WRc to BBV to be checked by modelling. Typical of this stage was a suggested site and set point for an additional booster chlorinator. The model was rerun with the chlorinator added and the results checked against the criteria. On some occasions this resulted in at least one modification to the recommendation and another modelling exercise.

Figure 6 shows an unusual case where chlorine is low on one side of a boundary valve (shown as "Node with lowest concentration" in the bottom left hand corner). The solution is to move that boundary valve a short distance. This produces the result in Figure 7.

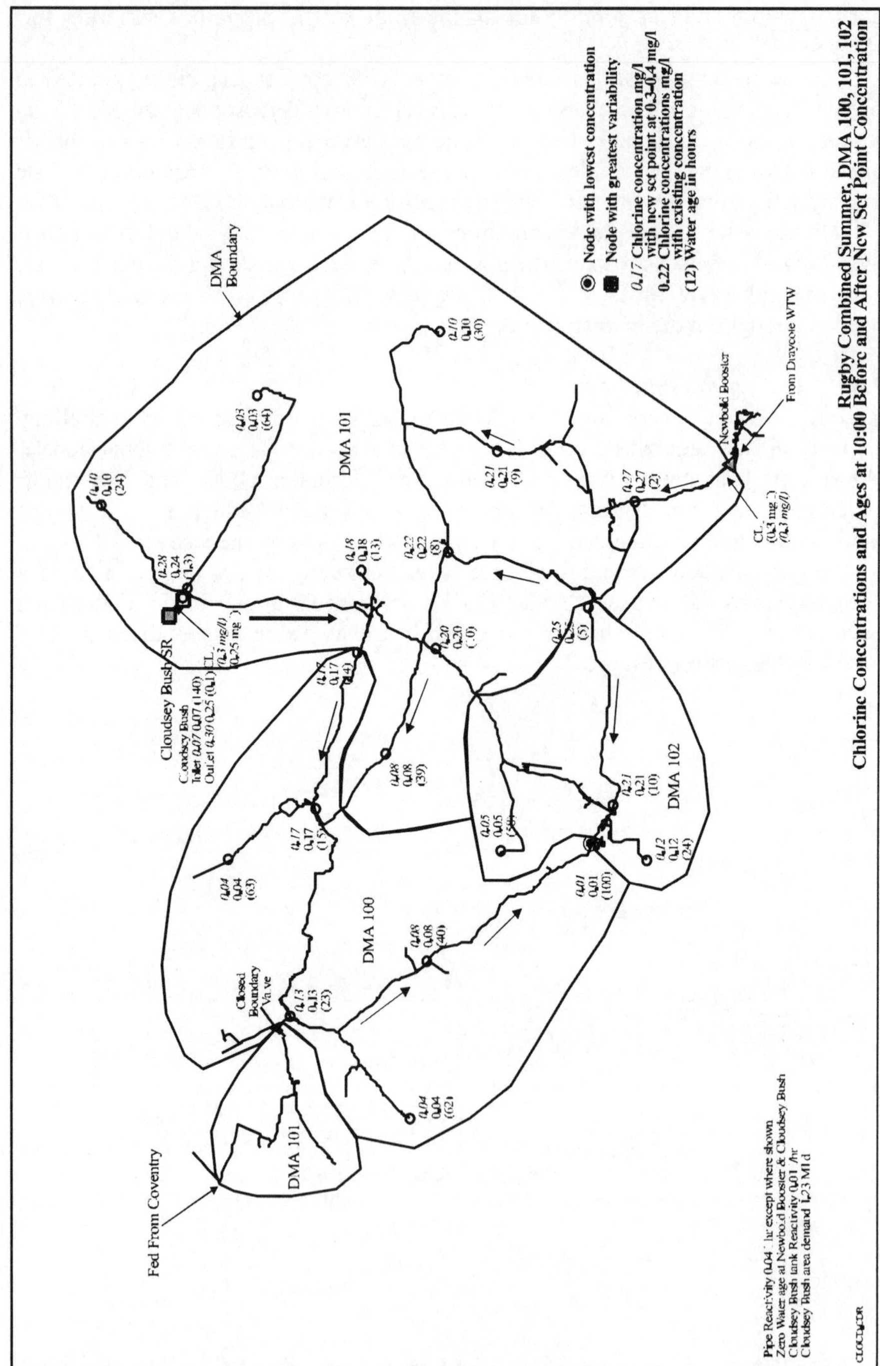

Figure 6 Low chlorine at a boundary valve

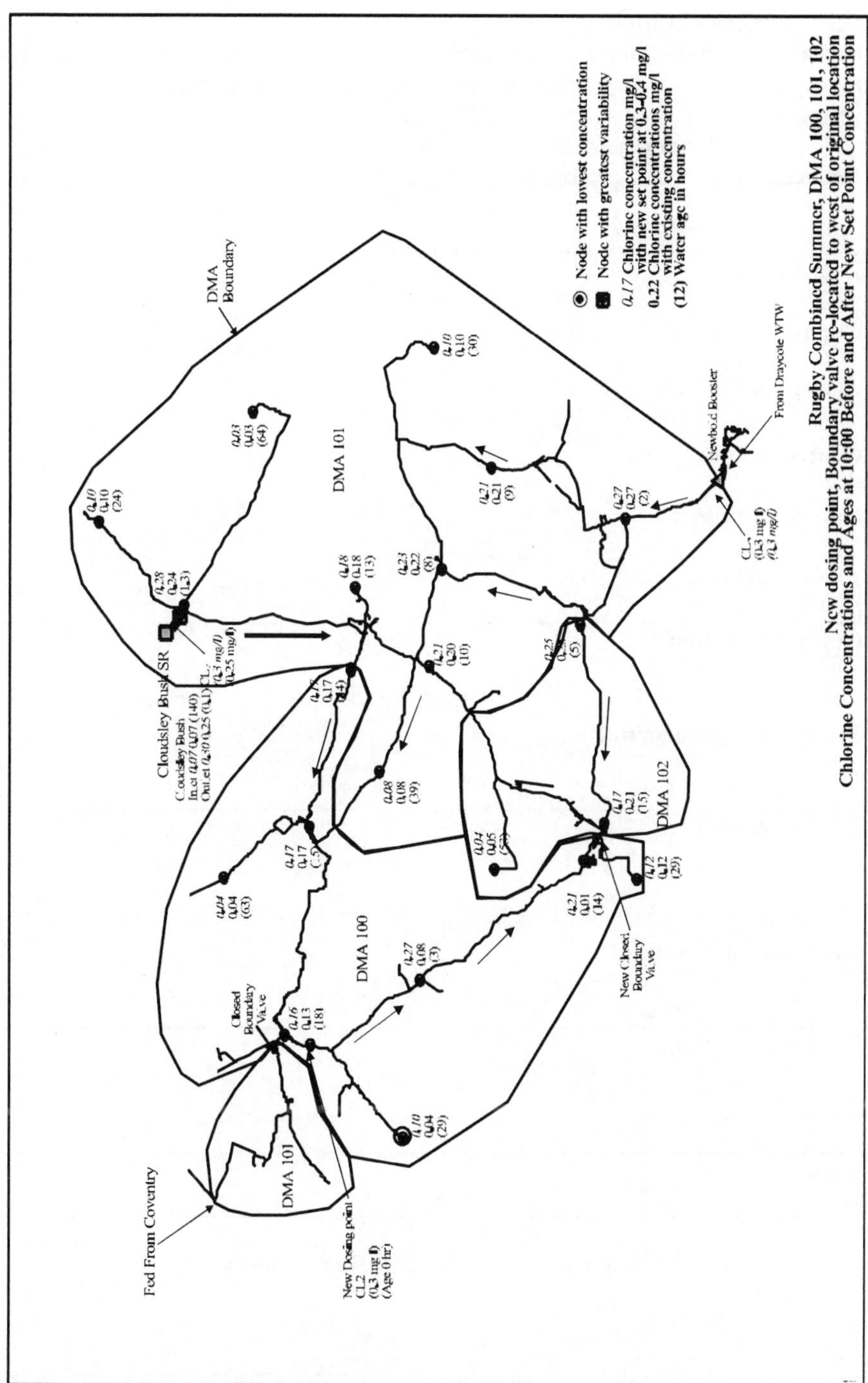

Figure 7 Result of moving the boundary valve

5.5 Analysis output

For some areas deriving an acceptable solution was an iterative process between the hydraulic modelling and microbiological teams, with options being tested before a recommended solution could be finalised.

When all recommendations had been made and, where possible, checked, they were entered on a proforma as in Figure 8.

Zone reference **County** **Area type** **Control point** **Current set point** **New set point**
Current situation   **Variation at chlorinator**
New set point   **Chlorine variability**
Recommendations
Signed Date

Figure 8 Proforma for recommendations

The additional documentation for the recommended solution included:

- Spreadsheet of the control and sample data
- Schematic of the control area
- Plots (winter and summer models) of the control area showing reactive coefficients
- Individual plots (winter and summer) of DMAs or hydraulic areas showing chlorine concentrations with the existing and proposed dosing set points.
- Time based graphs (winter and summer) of age and concentration with the new dosing rate/injection point for the node in each DMA with the lowest concentration.
- Time based graphs (winter and summer) of age and concentration with the new dosing rate/injection point for the node in each DMA with the highest concentration.
- Time based graphs (winter and summer) of age and concentration with the new dosing rate/injection point for the node in each DMA with the most variable concentration.
- Record sheet of key aspects of the calibration and analysis (Audit trail).

Recommendations included "operate at new set point with no further work", "install new booster chlorination", "rehabilitate mains", "improve water quality or hydraulics", "change the method of operation of a service reservoir or borehole" and "further investigation required".

6 DISSEMINATION MEETINGS

A series of meetings were held in different parts of the Company in which modellers and microbiologists explained their results to operations, planning and water quality staff. These long and arduous meetings were a vital component of the process towards solutions that could be implemented.

Local STW staff had been involved in the early information gathering. Consultants used their experience and expertise to build models, investigate issues and make recommendations. However, it was only at this stage that some practical difficulties or misunderstandings come to light. These were typically of the form "We cannot do that because...." or "That reservoir is being decommissioned in the near future".

The meetings also proved valuable because of some alternative solutions produced by local staff but stimulated by the results produced by the project team. The recommendations would have been far less acceptable if this final stage had not taken place.

7 SUMMARY

- Severn Trent Water Ltd. sought to optimise chlorine concentrations to improve customer satisfaction.
- This was to be done without reducing microbiological water quality.
- Network water quality models were used to examine chlorine concentrations before and after reducing the chlorine dose.
- These were calibrated using existing chlorine measurements.

- Model results and data were examined to determine if remedial measures were needed.
- Recommendations were tested using the network models.
- A large number of models were used to produce implementable results in a short time.
- Dissemination meetings to discuss the results with Company staff were very valuable and resulted in their endorsement and ownership of the final recommendations.

ACKNOWLEDGEMENTS

The authors wish to thank Mr J K Banyard, Director of Asset Management, Severn Trent Water Ltd., for his permission to publish this paper. The results described are based on the combined activities of Binnie Black and Veatch, WRc plc and Severn Trent Water Ltd. Acknowledgements are made to members of each organisation. The opinions expressed are those of the authors and not necessarily of Severn Trent Water Ltd.

REFERENCES

Powell, J. West, J. Hallam, N and Simms, J (1999) *Durability and maintenance of chlorine network models* presented at the American Water Works Association conference, April 1999.

Water Hammer in Distribution Networks

B. W. Karney

ABSTRACT
Although the basic phenomena of water hammer has been investigated for many years, in many ways the integration of transient considerations into the routine analysis and design of water distribution systems has been slow in coming. This paper explores the context of the modelling question and speculates on the challenges and opportunities facing the transient modeller. In particular, the role of transients in system optimization, control and water quality transformations are emphasized.

1 INTRODUCTION

Traditionally, the key issue in the design and analysis of water distribution systems has been to properly size the various pipes, pumps and reservoirs so as to achieve a specified minimum pressure while meeting a set of steady or constant demands. Since the system is to be in existence for many years, the system must function well not simply for present flows but also for future demands. As a direct result, these design values of the steady demands are seldom actual flows that could be measured in the real system. Rather, "demands" are estimated by projecting into the future expected changes in the population served and associated water use.

This context for the conventional hydraulic design problem has justified many simplifications used in the analysis and design of water distribution systems. For example, one key assumption is that the demands are steady in time (or, at most, changing slowly) and independent of the fluid pressure at the demand location. These simplifications can be justified in three ways. (i) Since demand projections are actually uncertain, there is no point in pretending that the hydraulic predictions are accurate in detail, so a simple steady state hydraulic model is justified. (ii) The system is designed with "extra" or redundant pipes, and is quite forgiving of small data errors since the looped arrangement is an excellent mechanism for coupling the pipe and demand characteristics into an overall (average) solution. And finally, (iii) there have been few practical alternatives to this modelling problem, since even the steady state solution was not computationally trivial and more complex analyses were prohibitively expensive.

However, recent developments should be slowly changing our perceptions of the hydraulic analysis problem. For example, it is now more common to monitor, track and optimize the current performance of a water distribution system. In fact, from this perspective the set of three assumptions above can be countered point-by-point. (i) For the analysis of current flows and pressures (i.e., today's problem as opposed some future problem), better estimates (e.g., measurements) of demand should be available and thus one should expect more accurate solutions. (ii) Although average behaviour is insensitive to the specifics of the system, a number of decisions and predictions, particularly those related to water quality and economic optimization, require a more accurate accounting of the system details. (iii) Recent developments in high speed computers, sensors, data acquisition systems, remote monitoring devices and computer programs all make more accurate and detailed analysis both more possible and more practical than it has ever been before.

The issue is, so what? What is involved in the more realistic assessment of water distribution system performance? What are the key areas of activity and what key gains can be expected? It is the contention of this paper that a missing component (and a missing opportunity!) in a great deal of modern practice is a careful and realistic consideration of transient conditions in the water distribution system.

2 TRANSIENT CONDITIONS

Although water distribution systems have been in use for centuries, interest in transient or water hammer conditions that occur within these systems is a relatively recent development. However, if one considers any medium-to-large sized water distribution, having thousands of kilometres of pipe and tens of thousands of users, it is steady state conditions that must be viewed as the exception. Users are constantly changing their water requirements, pumps are being turned on or shut down, valves are being adjusted and water is accumulating in (or being withdrawn from) reservoirs. The normal state of a water distribution system is one of continual adjustment.

The mechanism of transient adjustment in a water distribution system involves little more than an information exchange between various parts of the system. This "information" is primarily carried through the fluid contained in the pipe system, particularly its pressure and velocity. More specifically, when demand is changed, say by opening a hydrant to fight a fire, the hydrant immediately accesses (taps in to) the high pressure that is present before the valve is opened, causing the flow to take place. But the pipe under the hydrant is not a reservoir with a free surface and a significant stored volume. Rather, the flow that is created causes the pressure to decrease owing to the depressurization of the pipe in the immediate vicinity of the valve. This depressurization, in turn, influences adjacent areas, thus causing a low pressure wave to propagate into the distribution system.

Eventually, through a process of pressure and flow adjustments, the low pressure wave will reach the supply locations, thus bringing news of the now open valve to the pumps and reservoirs that feed the system. The pumps and reservoirs respond by increasing the inflow as a direct and natural response to the decrease in pressure they experience. In this way, the system hunts for a new equilibrium, often in a convoluted and jerking manner as a result of inertia and system interaction effects.

Of course, all this is routine and well known. So too is the fact that these transient events can be significant, flow sometimes creating pressure changes that are large enough to fracture or weaken the pipe or its supports. Yet these events have consequences that are often forgotten or too little appreciated. I focus on a few of these issues next.

3 NEED FOR INTEGRATION

One of the greatest challenges, and perhaps one of the greatest opportunities, for the inclusion of water hammer considerations in hydraulic analysis is to achieve a smooth integration between the various levels of models. Currently, it is quite common to use a variety of different modelling tools for steady state analysis, for system optimization, for operation in real time, and for system calibration. As computers have become faster, there is less need for this artificial separation. Several examples are presented to illustrate this point.

First, it is well known that steady state conditions are special cases of the transient equations. Thus, aside from computer time (which is becoming less and less of a constraint), there is little need in many systems to construct a steady state model at all, as the transient model serves well for this purpose. This approach has the added advantage of giving the analyst a natural feel for how the network responds to changing demands. In rare cases, it has the added bonus of identifying "impossible" steady state solutions. In these cases, a steady state solver may find a "solution" that is physically unstable due to the dynamic interaction of two or more automatic control valves. The issue here is that if both models are not needed, why not drop the model that can easily be replaced by the other. Then only one model needs to be understood and maintained.

A related point is the issue of control system design. Since many steady state programs are beginning to incorporate issues of control and feedback, there appears to be a growing tendency for users to rely on the associated settings and strategies. However, a typical steady state program knows nothing about the way the real system interacts and responds to changes in valve and pump speed settings. As a result, it is possible to actually make the operation of a pipe network worse by the very actions that are designed to improve its performance. Except as a screening tool, steady state control actions are difficult to interpret.

Perhaps nowhere is the interaction between system characteristics more crucial than in system optimization. In particular, there is a widespread approach in optimization studies to assume that pipe cost is directly related to pipe diameter alone, completely forgetting the role of wall thickness in controlling costs. Now wall thickness directly influences the rated pressure of the system, while diameter directly influences the pressures experienced by a pipe under transient conditions. Clearly, there are interactions here that have a direct bearing on the issues of pipe optimization. A little thought will show that what is true for capital optimization is just as valid for considerations of optimal operation of pumps and reservoirs.

In fact, this raises an interesting point that is quite easy to resolve as well. Some will argue that the role of steady state models is most highlighted in "quasi-steady" analysis. In this approach, reservoir levels are tracked over an extended duration, typically 24 hours, to investigate issues of pump operation and system control. Here again there may be a case for the application of transient models, but not as they have traditionally been applied. It is actually quite straightforward to adapt a transient model to be applied in a kind of "quasi-steady" mode. All that is required is to accumulate fluid in reservoirs in a manner analogous to the steady models using a "jump" time step to keep the method efficient. With this accomplished a modern computer system can quite readily run through a 24 hour simulation using a transient model of small to medium sized distribution systems. Although currently impractical for large systems this method does have some promise.

However, using a transient model for a 24-hour simulation has an added bonus (aside from the advantages of joint capacity/strength consideration) to do with characterization or "worst-case" conditions in the network. Because of the non-linear interactions, it is often quite difficult in a transient model to determine what combination of demands, pumps and reservoir levels will produce the most severe transient response. This issue is quite vexing and not easily solved by simple considerations of maximum velocity or some similar criteria. But if a transient model is used to perform a 24-hour simulation, it is a simple matter to instruct the program to use each step profitably. In particular, the program can perform, say, a power failure run at each time step, and let the program keep track of which set of conditions is worst from the point of view of the system's transient response.

There is yet another facet to these considerations that in my view is probably the most exciting of all. A transient model opens up new possibilities not only for integrated analysis, but for integrated calibration as well. In essence, this approach is to create a transient condition and to use high frequency pressure transducers to listen to the system's response. Next an optimization model is used to bring the predicted and measured responses into reasonable agreement. This procedure, called inverse transient analysis, has the capacity to collect and compile many orders of magnitude more data than is typically available with a steady state approach. In addition, the method can more readily distinguish closed valves and reduced diameters from true friction factor effects. An introduction is provided in one of the other papers in this volume (Tang et al., 1999).

4 WATER QUALITY MODELLING

For many years the assumption was naively made that the quality of water that entered at one end of a distribution system was essentially equivalent to that leaving at the other. Those days are, of course, long past. However, there is still the widespread perception that water quality models can account for the key mechanisms of transformation by considering steady state considerations alone. It seems to me that this assumption is similarly optimistic.

To illustrate this point, consider that shear stresses under transient conditions can be an order of magnitude or more higher than those present under steady state conditions. This fact needs little proof for those who have conducted field hydrant tests in older parts of a distribution system and know the common complaints of red water. Rapid valve closures and power failure events also often create visible evidence of water quality disruptions.

A related issue arises due to inertial effects at devices such as storage reservoirs and air chambers. As a result of transients in the system, it is quite possible to set up surge conditions in the system where the water pressure and velocity oscillate for some time about a mean value before steady state is established. Dispersion and wall-exchange coefficients are likely to be poorly estimated by mean conditions in many of these flows.

However, perhaps even of greater concern is the possibility of cross-connections or other contamination that can enter pipelines under transient conditions. Since virtually all pipeline systems leak, and since leaks can pass fluid in both directions, low pressure transient waves offer considerable potential to draw untreated and possibly hazardous water into a pipeline system. Water quality issues are also discussed in the conference paper by Fernandes and Karney (1999).

5 UNRESOLVED ISSUES

Of course, it would be quite inappropriate to leave the impression that all the issues raised above are "done deals." Many issues remain to be solved or at least further investigated. There is still considerable work to be done in the following areas:

1. A more accurate long-term transient decay model is needed that accounts for unsteady friction effects. Such a model must account for the velocity profile as the system undergoes transitions from one state to another.
2. The coupling of quasi-steady models with transient models requires further development to make the interaction between components as efficient and logical as possible.
3. Optimization procedures that effectively integrate steady state and transient considerations must be shown to be practical and effective. This will require some flexibility both on the part of programmers and researchers and on the part of system owners and operators to be willing to implement new approaches.
4. While the inverse transient calibration procedure shows great promise, there are many challenges ahead. How many sensors are required? What accuracy can be expected? How robust is the procedure under a wide range of field conditions, operating approaches and system configurations? And the list goes on.
5. Integrated water quality models that account for multiple species interactions and transient considerations require careful development and extensive calibration and fieldwork. This is a huge and expensive undertaking that will not be solved quickly or easily. Yet its importance cannot be downplayed despite these challenges.
6. The detail of network topology creates some real challenges of analysis. Some very unusual superposition of waves can occur in a network that can produce some surprising results. This phenomenon creates a significant analysis problem associated with the level of detail required for an analysis. These issues are discussed in more detail in the conference paper by Karney and Brunone (1999).

Clearly there are other issues as well. Yet the purpose here is not to be exhaustive but to be suggestive. The point is that there is a great deal of interesting and important work yet to be completed that will require the integration of many skills and researchers to bring to completion. That is the challenge and the task before us, and represents the true value of a conference of this kind.

6 FINAL COMMENTS

Transient considerations play the key role in establishing flow and pressures in a water distribution system. However, despite this importance, water hammer analyses have too often been the preserve of a scant number of specialists. This is unfortunate for the basic mechanism can be appreciated by anyone who takes the time to think physically about one key question: how does one end of a pipe know what the other end is doing? The answer is of course, it doesn't, unless the pipe conducts this information. The sequence of waves that communicates between devices has great implications for issues of system control, optimization, water quality modelling and system calibration.

REFERENCES

K. Tang, B. Karney, M. Pendlebury and F. Zhang (1999) Inverse transient calibration of water distribution systems using genetic algorithms, *Water Industry Systems Volume 1*, RSP, UK.

C. Fernandes and B. Karney (1999) Assessing water quality issues in water distribution systems from source to demand, *Water Industry Systems Volume 2*, RSP, UK.

B. Karney and B. Brunone (1999) Water hammer in pipe networks: two case studies, *Water Industry Systems Volume 1*, RSP, UK.

Emerging Technologies in the Water Industry

C. Maksimovic, D. Butler *and* N. Graham

ABSTRACT

The rapid rise in urban population increases pressure on the water systems serving city dwellers, and the challenges facing both water industry and society have to be addressed. This paper is based on the concept that the real life problems determine the technology to be applied in dealing with them and providing the solutions, rather than other way round. Since this conference deals with the application of computers and control in the water industry, this paper emphasises the range of problems that pose challenges to the water industry, indicates some possible means of resolving the emerging conflict, and points out the innovative technologies which still need additional development before being fully applied on a routine basis. The paper does not cover all possible problems but rather concentrates on those in which the authors are active in research. These include sustainable water management, demand management, domestic water recycling, data needs for better assessment of water balance in distribution systems, emerging technologies in drinking water treatment, automatic meter reading, GIS applications in storm drainage studies, and integrated system modelling. The paper concludes with proposals for generic priority areas, indicating some research and development needs, and introduces several concrete initiatives in this direction including Y2K2C - the **Y**ear **2**00**2** **C**hallenge.

1 INTRODUCTION

Although it may be considered a conventional engineering discipline, urban water re-emerges as an important topic for the years to come. By the middle of the next century, it is confidently predicted that 70% of the global population will live in urban areas occupying only 2% of the land surface. The number of mega cities (> 10 million inhabitants) will increase to over 20, 80% of which are in developing countries (Niemcynowicz, 1996). While this trend will be more pronounced in developing countries, developing and developed states alike are faced with enormous urban challenges – and one of the greatest challenges of all concerns water, the most precious of all natural commodities. Urban water systems are essential to support and maintain healthy city life. Although, over the last century, we have engineered clean water on demand from our taps, sanitary waste removal from our toilets and largely flood-free roads in rich western cities, many problems remain to be addressed and sustainable solutions to be found in the years to come.

In this paper, we present in outline the innovative trends in several important urban water areas, drawing on experience and results from past and present studies at Imperial College's Urban Water Research Group. These include sustainable water management, demand management, domestic water recycling, data needs for better assessment of water balance in distribution systems, emerging technologies in drinking water treatment, automatic meter reading, GIS applications in storm drainage studies, and integrated system modelling. The paper concludes with the proposal that new research and informatic tools and a new generation of software products tested against reliable sets of data are greatly needed. To do this effectively, five broadly-based technical priority areas are suggested, under the headings: interaction, interfacing, instrumentation, integration, and information. Finally, the Y2K2C project is briefly introduced.

2 SUSTAINABLE URBAN WATER MANAGEMENT

2.1 Sustainable Water Systems

The concept of sustainable development is provoking a profound rethinking in our approach to urban water management (ASCE/UNESCO-IHP, 1998). Sustainable development is that which "meets the needs and aspirations of the present generation without compromising the ability of future generations to meet their own needs" (WCED, 1987). An alternative definition (IUCN-UNEP-WWF, 1991) asserts that sustainable development is that which "improves the quality of human life while living within the carrying capacity of supporting ecosystems". Here, the emphasis is placed on mankind's demand for and impact upon earth resources and the environment. Sustainable services must be environmentally friendly, socially acceptable and financially viable into the next millennium.

We suggest that conventional urban water systems are inherently unsustainable in that they have been designed to ameliorate the unsustainable practices associated with urban development (e.g. excessive use and contamination of water, sealing of catchment surfaces). These engineering methods, in turn, create their own problems. For example, large quantities of water are required for waste transportation; sewage that could be better reused as a crop fertiliser is contaminated by heavy metals and other toxic compounds. Pipe networks concentrate essentially solid wastes from large areas into point sources of liquid pollution in receiving waters. In developing countries sewerage and drainage have a very low priority; these systems are either underdeveloped, inadequate or do not exist at all. Waste water is spread and it easily finds its way to intermittently operated potable water systems, thus creating a significant threat to public health.

Larsen & Guyer (1997) consider management of resources to be a vital component of sustainable urban water management. They see four types of resources. Primary resources are water and nutrients such as nitrogen and phosphorus. Secondary resources are energy and construction materials.

Natural resources are the earth media of receiving water, land and air. Finally, and perhaps most controversially, are the anthropogenic resources, identified as knowledge, creativity and wealth to which we have added information (that is, processed data). Currently, little is known about how efficient present systems actually are, in terms of these resources. Sustainable management will require efficient use of all such precious resources, for example, recycling of nutrients and 'slim' construction.

Butler & Parkinson (1997) define a sustainable urban drainage system as one which:

- maintains an effective public health barrier and provides adequate protection from flooding,
- avoids (local or distant) pollution of the environment,
- minimises the utilisation of non-renewable natural resources,
- limits the utilisation of other resources to their rate of regeneration/recycling,
- is operable in the long-term and adaptable to future requirements.

They advocate an incremental strategy to achieve such a system by:

- reducing the use of potable water as a waste carriage medium,
- avoiding mixing certain industrial wastes with domestic sewage,
- separating storm runoff from flows of polluted wastewater.

The types of problem associated with each of these three urban flows, together with the benefits of the proposed strategy, are summarised in Table 1. Proposed practical solutions are also included. The problem of storm drainage source control is further analysed in detail.

Regardless of which approach to sustainable urban water management is adopted, an integrated and multi-disciplinary framework will be required. In particular, it is most important to assess the applicability of technical solutions in their local (environmental, social and economic) context. Individual cases must be decided on their merits and no one technique or technology is suitable in all situations.

An incremental approach containing both high-tech and low-tech solutions is the most likely approach to be implemented. Many of the solutions advocated in Table 1 demand small-scale or local provision. It is clear that if such solutions are adopted, the active participation of users is essential requiring public education and co-operation.

Table 1 Strategy towards sustainable urban drainage (Butler & Parkinson, 1997)

COMPONENT	PROBLEMS	SOLUTIONS	POTENTIAL BENEFITS
POTABLE WATER	Unnecessary use of potable water Dilution of wastes Expensive end of pipe treatment	Reduce consumption Use water conservation/ reuse techniques Seek alternative means of waste conveyance	Conservation of water resources Improved efficiency of treatment processes
INDUSTRIAL WASTEWATER	Disrupts conventional biological treatment Increases cost of Wastewater treatment Causes accumulation of toxic chemicals in the environment Renders organic wastes unsuitable for agricultural reuse	Remove from domestic waste streams Pre-treat to remove /reduce concentration of problematic chemicals Promote alternative industrial processes using biodegradable substances	Improved treatability of wastewaters Improved quality of effluents and sludges Reduced environmental damage Cost savings associated with re-use of recovered chemicals
STORMWATER	Requires expensive sewerage systems Transient flows cause disruption at treatment works Discharge at overflows and outfalls causes pollution Causes flooding	Separate storm runoff from sewer flows Provide attenuation / storage facilities Provide local infiltration or reuse Promote ecologically sensitive engineering	Reduced pollution from outfalls Improved efficiency of treatment Groundwater recharge Reduced demand for potable water

2.2 Demand management

Modern water supply systems are facing a change: demand driven systems will be gradually replaced by demand managed systems although estimates suggest unabated increase in demand for water services into the future. According to Zehnder (1997) by the middle of the next century humankind could need to consume more fresh water than is being replenished by rainfall. For this and other reasons, attention is focusing away from supply-side measures to demand-side management measures concerned with restraining rather than satisfying runaway consumption.

In order to assess the economic and environmental merits of demand management a new methodology has been developed known as the Reference Sustainability System (RSS) (Pearson et al, 1998). This can represent the resource and material flows that govern the sustainability of cities, thus enabling systematic assessment of the potential of technologies and resource management strategies to enhance urban sustainability. The model considers the flow, over the life-cycle of supply, demand and waste management, needed to meet the end-use service demands for any resource or material by urban households. For a more detailed presentation of the model see Butler et al, 1999. Foxon et al (1999a) and Pearson et al (1998) give background information on the modelling approach and details of the other services and commodities studied.

The model was applied to London as a case study. It is driven by the supply of water needed to meet end-use service demands for water by households, with non-household demand assumed to remain at a constant level. The water flow to meet different end-use services - bath, shower, dish washing, drinking and cooking, laundry, hand washing, WC flushing and external (garden and car washing) use - is disaggregated at this stage in the model to allow variations in each individual type of use to be studied. The model allows for different 'directions' of flow, such as the recycling of greywater from bath, shower and laundry use for re-use in WC flushing (see Figure 1). Scenarios for the gradual introduction of each demand management measure over a 20 year period were compared to a 'baseline' scenario that incorporated two main elements - increasing per capita demand from shower use and garden watering and a projected 23% increase in the number of households over the 20-year period. Sensitivities to different rates of penetration or implementation of the measure were explored for each alternative scenario.

A number of sustainability indicators were used to compare the measures considered in the different scenarios. The greatest water saving was found in the leakage reduction scenario, in which the central penetration rate gave a saving of 915 Ml/day (27% of the projected baseline distribution input in 2016, at the end of the 20-year period). The greatest wastewater discharge avoided to meet the same end-use level of service demand was found in the grey water (all the wastewater from a house except that produced by the toilet) recycling scenario, in which 373 Ml/day were saved (28% of baseline wastewater discharge in 2016).

The most cost-effective options were found to be leakage reduction, with a cost of between 1 and 16 pence per cubic metre of water saved (p/m^3), WC conversion, with a cost of 7 to 16 p/m^3, and metering (with high penetration rates or high per capita reduction), at a cost of 5 to 47 p/m^3. These options all compare favourably with the current cost of supplying water in the London area of 49 p/m^3.

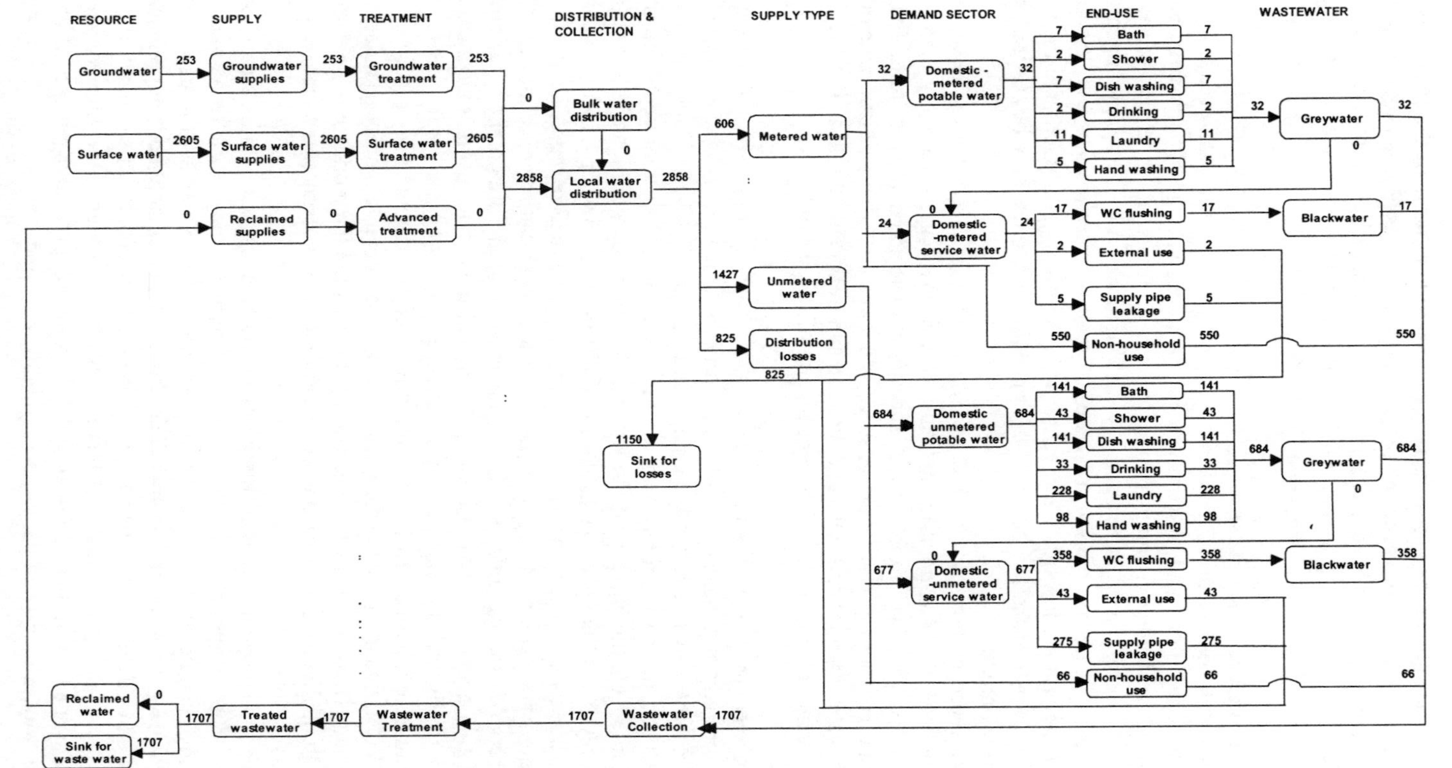

Figure 1 Possible Scenarios for Urban Water Directions

The RSS thus allows the assessment of system-wide impacts of the measures over the 'life-cycle' of the urban water system, the ability to assess measures relative to a baseline scenario, and the preliminary incorporation of environmental costs and benefits. In this way, rational, informed decisions can be made concerning the introduction of demand management techniques.

2.3 Domestic water reuse

One potential component of demand management is domestic water recycling. The concept is not new, and systems are in operation, particularly in Japan and in developed arid region countries. However, in most other areas, social and economic factors have prevented its development and widespread integration within the traditional urban water system. Progress in technology and discernible changes in attitude towards water reuse from industry, government and the public suggests to us that the full potential of domestic water reuse may be realised in the next millennium.

The reuse of grey water and rainwater potentially reduces the need to use potable water for non-potable applications with the water effectively being used twice before discharge to the sewer. The major reuse potential is for toilet flushing and garden watering that both conserve water resources and relieve demand on public water supplies and sewage collection and treatment facilities. The collection and storage of rainwater also reduces runoff flows into the storm drainage network.

An average annual rainfall of 750 mm, if collected from a 60 m^2 roof, would also approximately satisfy WC and garden demand. However, rainfall is significantly more erratically timed throughout the year than grey water production, and so relatively large storage tanks would be required - but again, how large? On the other hand, rainwater is known to be significantly less polluted than grey water (Grey water COD 200-6000 mg/l, Rainwater COD < 200 mg/l) and so may well require less treatment than grey water before reuse. Using this, the designer of a water reuse system (Butler et al, 1998) can explore both the trade-off's between the variables (such as storage volume and treatment type) and the interaction between system users (residents) and their water using appliances. The model is illustrated schematically in Figure 2 showing the major components including the inputs of grey water and rainwater and outputs of garden water use and WC demand. In assessing flow quality, acceptable output standards are important (Dixon et al, 1999). Each component within the model may be varied to simulate a diverse range of operating conditions over both short and long-term periods.

The model operates by considering the simultaneous supply and demand on the system due to appliance usage based on a typical house size (roof area) and dwelling occupancy. This calculation is undertaken with hourly time steps and can be continued over several years. The model storage component aggregates the inputs and outputs within that hour. If the storage capacity is exceeded then the excess is discharged to waste. If there is insufficient water in the store to meet the demands made of it then the mains potable water supply makes up the difference. In this way, a long-term picture can be built-up of the water-saving efficiency of that particular scenario. Also calculated are the actual potable water demand and the wastewater flow quantities.

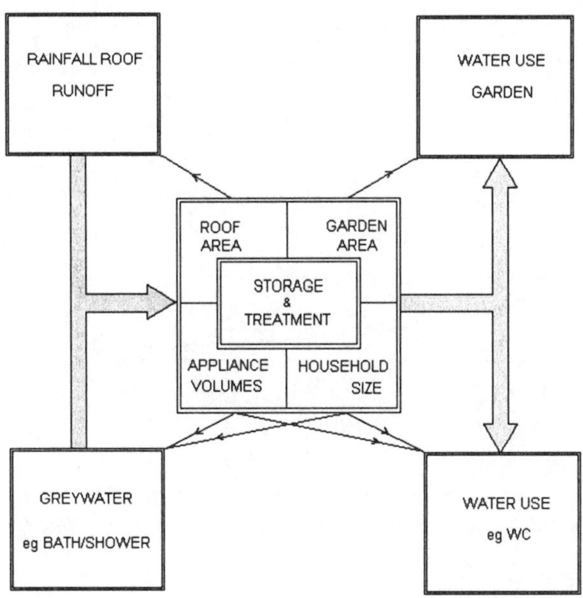

Figure 2 Conceptual diagram of water reuse model

Figure 3 shows results of a simulation run for a house with 3 residents, that reuses grey water from a range of domestic appliances and collects rainwater from a nominal roof area of 20 m^2. The model run is of 6 months duration. As can be seen the gain in water saving efficiency from increasing storage capacity becomes less significant from a storage capacity of 200 litres upwards. Even quite small storage volumes produce significant savings.

Small-scale, domestic water reuse has the potential to make a significant contribution to water conservation. Detailed analysis of grey water and rainwater flow quality and quantity in combination by computer simulation provides compelling evidence of the efficiency and effectiveness of grey and rainwater reuse systems. For such systems to be more widely used, wider considerations of social acceptance and financial viability have yet to be fully addressed. Findings suggest systems in individual dwellings will rarely be cost-effective unless wider economic benefits are included. Medium scale systems (such as described by Surendran & Wheatley, 1998) have shorter pay-pack periods.

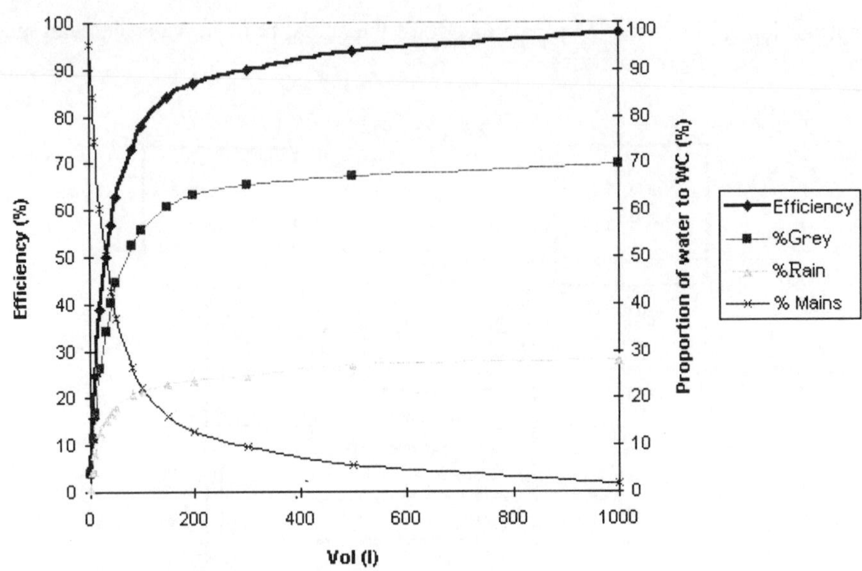

Figure 3 Water saving efficiency of a reuse system with storage capacity and individual contribution of mains, rainwater, and grey water

3 POTABLE WATER ISSUES

3.1 Water distribution

The management of water distribution systems is in a state of rapid change and development. Developments in sensors and data acquisition and processing technology coupled with improved representation of the hydrodynamic process in the networks is making it possible to introduce advanced asset management and operational management techniques, as shown in Figure 4. Rapid progress is being made in analytical techniques for steady, gradually varied and transient flow analysis and techniques for optimisation of network design and operational management (Savic and Walters, 1997). Application of transient flow analysis in identification and quantification of leakage developed by Liggett and Chen (1994) is subject to further research in a recently started EPSRC project (GR/M68213). The project aims at the development of a procedure for leakage location and quantification of leakage flow rate based on inverse transient flow analysis.

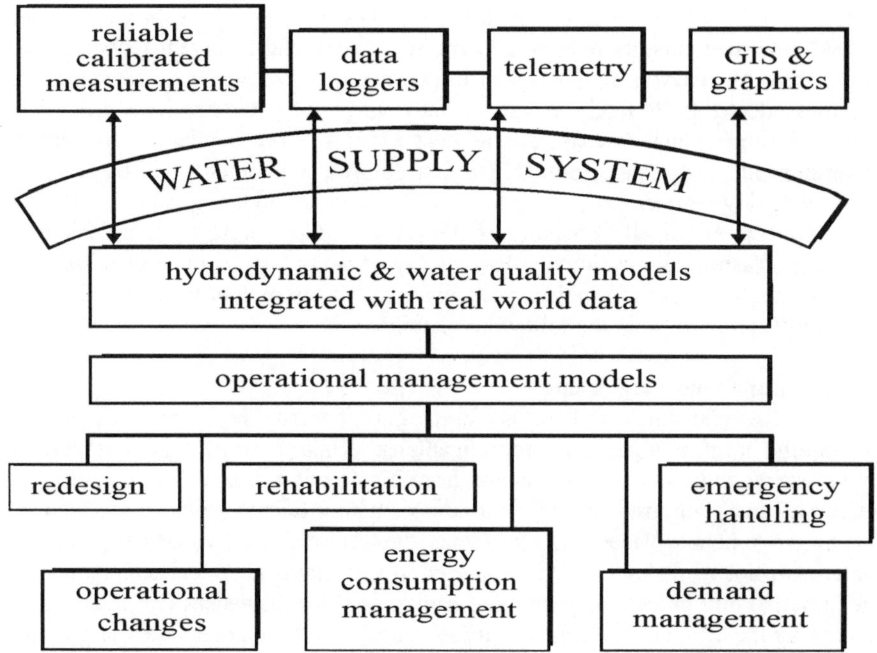

Figure 4 Data sources, modelling tools and management activities in water distribution (Maksimovic et al, 1996)

Distribution leakage control has already been mentioned as a fundamental element of demand management. To implement a control strategy, an unbiased insight is required into the water balance (with uncertainties quantified), both spatially (throughout the network) and temporally (daily and seasonal variations). Information from individual networks can then be used to form water company policy. Carmi and Maksimovic (1999) have shown how data derived from such hydrodynamic modelling can be combined with asset data to produce leakage vulnerability plots. Relevant arial data is stored in different 'layers' of a geographical information system (GIS). Information includes pipe material and diameter weighted against the statistics of past damage, "pressure load" obtained by integration of the network pressures (24 hours simulation), and the estimated combination of rheological properties of various pipe materials. Statistical analysis based on Boolean algebra is then used to estimate vulnerability to leakage and present the results in a spatially distributed manner.

3.2 Emerging developments in drinking water treatment

At present, there still remain a number of challenges to the water utilities in the developed world in terms of fully complying with prevailing water quality regulations. In the UK and in other countries, these include the prevention of cryptosporidium and pesticide breakthroughs, taste and odour incidents, and reducing the leaching of iron and polynuclear aromatic hydrocarbon concentrations within water distribution networks. Most of these problems can, and will, be overcome by new investment in conventional treatment technologies, and simply by renewal of the capital infrastructure. However, new technologies and methodologies are likely to offer the advantages of greater process efficiency and reliability, capital and operational cost savings, more compact plant, and a lower dependency on chemicals. In addition, future developments will be required to provide higher standards of treated water quality and a greater degree of *sustainability*, principally in terms of the reuse and recycling of chemicals and materials, waste minimisation and less energy consumption. A number of aspects of water treatment are the subject of innovation and research at present, and these are briefly summarised in the following sections.

3.2.1 Enhanced coagulation

In the USA and Europe there is continuing concern about the presence of potentially harmful organic micropollutants in drinking water, and particularly those arising as byproduct compounds from disinfection. Since disinfection must be carried out rigorously and without compromising microorganism inactivation, emphasis is being placed on improving the removal of byproduct precursor materials prior to disinfection. In many treatment plants chemical coagulation is the principal unit process for removing natural organic substances and methods of improving the efficiency of this (*enhanced coagulation*) are under study at present. New types of coagulant chemicals are of continuing interest and recent studies have concentrated on the development of low basicity, highly charged polymeric aluminium and ferric salts. Laboratory trials have demonstrated the superior treatment performance of particular forms of poly-ferric sulphate (Jiang et.al., 1996; Jiang and Graham., 1998) and poly-alumino-ferric sulphate sulphate (Jiang and Graham., 1998A) in terms of the removal of natural organic matter (including disinfection byproduct precursors), lower sensitivity to pH and temperature, and reduced sludge production. These improvements need to be replicated at pilot- and full-scale. The beneficial combination of new inorganic coagulants and organic polyelectrolytes is also being actively considered for particular types of water quality.

At present, coagulation and floc separation is widely achieved through the use of sludge blanket clarification or, more recently, dissolved air flotation. Both processes have their advantages and disadvantages in regard to cost, treatment performance and operational requirements. A new approach under study at the moment is that of using an electrocoagulation/flotation cell (ECF) to generate the coagulant and bring about floc separation by flotation (Cerisier et al.1996). This process system, currently under study at Imperial College in association with Thames Water, offers many potential advantages, particularly in providing a more compact process configuration, telescoping into one unit the three conventional stages of coagulant mixing, mechanical flocculation and flotation separation. In addition, it is possible that the ECF may provide other benefits in terms of lower energy demand, more efficient treatment and some microbiological inactivation.

3.2.2 Advanced oxidation processes

In recent years there has been a widespread interest in, and application of, oxidant chemicals for a range of water treatment purposes. Often this has been done to either replace, or reduce, the use of chlorine because of the concern over the formation of halogenated byproduct compounds. Whilst the predominant oxidant applied so far has been ozone, there have been successful applications of chlorine dioxide and potassium permanganate in the treatment of surface waters (Ma et al, 1997). Ozone has been of particular interest because of its ability to degrade pesticide compounds and other organic micropollutants. However, typical water treatment conditions limit the effectiveness of ozone treatment by minimising the generation from ozone of highly reactive radical species. Research interest is currently focused on methods of enhancing radical formation, including combinations of ozone with either hydrogen peroxide (Lambert et al. 1998), UV-irradiation, metal catalysts (Ma et al., 1999) or activated carbon (Jans et al., 1998). Recent studies of the ozonation of the herbicide, atrazine, have shown that the presence of a small concentration of manganese (~ 0.5 mg/l) can catalyse the degradation of atrazine (Ma et al., 1999). A possible application of this phenomenon is the ozone treatment of contaminated surface and ground waters that also contain low levels of manganese.

Other treatment chemicals that combine both oxidation and coagulation /precipitation capabilities are also under active study at present. Whilst permanganate, as mentioned above, can also assist coagulation through the precipitation of solid phase manganese dioxide (Ma et al.,1997), Fenton's reagent (ferrous ions and hydrogen peroxide) and Ferrate are also able to produce iron coagulating species as a result of powerful oxidation reactions (White et al., 1998).

3.2.3 Inorganic Adsorbents

The use of carbon in its activated state is widely established for removing trace organic substances in water (eg. taste and odour compounds, pesticides, disinfection byproduct precursors). Whilst this is a very effective and proven treatment process there are practical and economic disadvantages associated with the need to remove granular carbon and regenerate it regularly. As an alternative approach, inorganic materials have been studied recently in view of the ability of some materials to adsorb both cationic and anionic species. Examples of these materials are clays, metal oxides, bone char and zeolites. Studies of these have shown that some of them, in their activated form, are capable of removing both trace metals and organic compounds (Lambert et al., 1997), particularly after oxidation (Lambert et al., 1995). More work is needed to evaluate these materials but their application could provide the advantage of chemical regeneration *in situ*, without the need to remove and thermally regenerate.

3.2.4 Cryptosporidium

Managing the risks of cryptosporidium breakthrough is currently a pressing problem for water utilities and owing to the complexity of the problem this will continue to be an important topic for research and development in the future. At present, the problem is being managed by careful surveillance of the water source and the control of conventional treatment processes, but there is still a major interest in finding an effective method of cryptosporidium inactivation. Among the common alternative disinfectant chemicals being used, only ozone appears to be able to achieve an adequate degree of inactivation (say, 2 log reductions) at realistic doses, and formal design information is likely to become available very soon (Rice, 1999). However, very recent work is suggesting that UV-irradiation at the usual wavelengths of 197 and 254 nm, may be highly effective (4-5 log reduction) in dealing with cryptosporidium (Rice, 1999).

3.2.5 Process Simulation and Control

Currently, the level of process monitoring and direct computer control at water treatment works is fairly extensive. In contrast, real-time process simulation and optimal operation remain undeveloped due to the complexity and inadequacy of unit process models. Whilst considerable effort has been invested over the last 15 years in the particular aspect of in-line coagulation control (e.g. via streaming current detectors), the success of full-scale systems has been site-specific and partial. At present, a very limited number of water treatment process simulation models exist (e.g. 'OTTER' by WRc plc., UK), and these are being evaluated for their ability to simulate actual process performance, and as a predictive tool for future scenarios. These models are generally deterministic in nature and suffer from the lack of a sufficient range of on-line water quality data to employ them. Whilst developments and improvements in these models will continue in the future there is complementary and growing interest in the use of neural-network, 'black-box' type models of treatment processes (e.g. Adgar et al, 1995; Appleton et al, 1995).

3.3 Automatic Meter Reading for individual consumers

The water balance in distribution systems consists of three major components: input (inflow minus outflow), consumption, and losses (mainly leakage). It is becoming apparent that the benefits of model-based prediction and control can only be achieved if at least two of these three elements are measured with sufficient data reliability, accuracy and temporal resolution. Lonsdale and Obradovic (1998) illustrate a range of real world problems found in the water industry and provide their interpretation in both data rich and data poor systems.

Water companies are struggling with introduction of advanced metering and reading of individual consumers. Fig. 5 depicts a concept being developed by Logica for Anglian Water (Plunkett, 1999). Anglian Water's domestic Automatic Meter Reading (AMR) system will comprise up to 2,000,000 meters, linked by a variety of communication media to a central server. Acting in concert, their primary function is to collect one reading each week from each meter, as well as enabling more frequent readings to be made.

A new type of battery powered solid state meter is the basis for this AMR system. Each meter has a built-in 2-way 184MHz MPT1601 VHF radio for local communications (up to 0.4 kilometres) to MIUs (Management Interface Units). Most MIUs are solar-powered and typically fixed to the top of lamp-posts. These are termed 'daughter' MIUs as they have no communication interface other than VHF radio. The daughter MIUs use their VHF radio not only to communicate downwards with meters, but also upwards to 'master' MIUs (up to 6 kilometres away). These in turn communicate via UHF or PSTN to the central system. The UHF master MIUs share AW's system (ARTS2000) comprising 40 UHF radio-scanners and 8,000 outstations. All meters can also communicate with portable Hand Held Units (HHUs). Where communication via MIUs to meters is not possible, the server instructs an HHU operator to 'walk' to those meters when a reading for billing purposes is required.

Technology of this nature will not only enable better insight into consumption and billing reliability, but should also increase the reliability of overall network analysis. The problem of leakage detection and quantification will be reduced. Additional improvements in data acquisition and processing would complement the improved reliability of both water balance analysis and on-line operational management.

4 TOWARDS SUSTAINABILITY OF STORMWATER SYSTEMS - SOURCE CONTROL

In natural catchments, vegetation, natural depressions, flood plains, floodways and pervious land enhance temporary detention, infiltration and evapo-transpiration. Natural hydrological features attenuate the runoff before releasing it to receiving waters thus resulting in greater concentration times, lower discharges and smaller volumes of runoff as well as water quality enhancement. Andoh (1995) has called this "a natural distributed control system". Urbanisation has profound effects on the natural water balance of a catchment and the adverse effects are numerous. Construction of buildings, roads and pavements increases the impermeability of the catchment and thus the volume of water infiltrated into the subsoil and available for aquifer recharge is significantly reduced. The changes in land use also involve loss of vegetation, which consequently blocks the interception and evapo-transpiration mechanism of stormwater reduction. Improvements in the capacities of streams and drainage ditches will increase the velocity of flow. This reduction of the catchment response time will consequently increase the maximum rate of flow discharging to the drainage system or increase the frequency of significant floods.

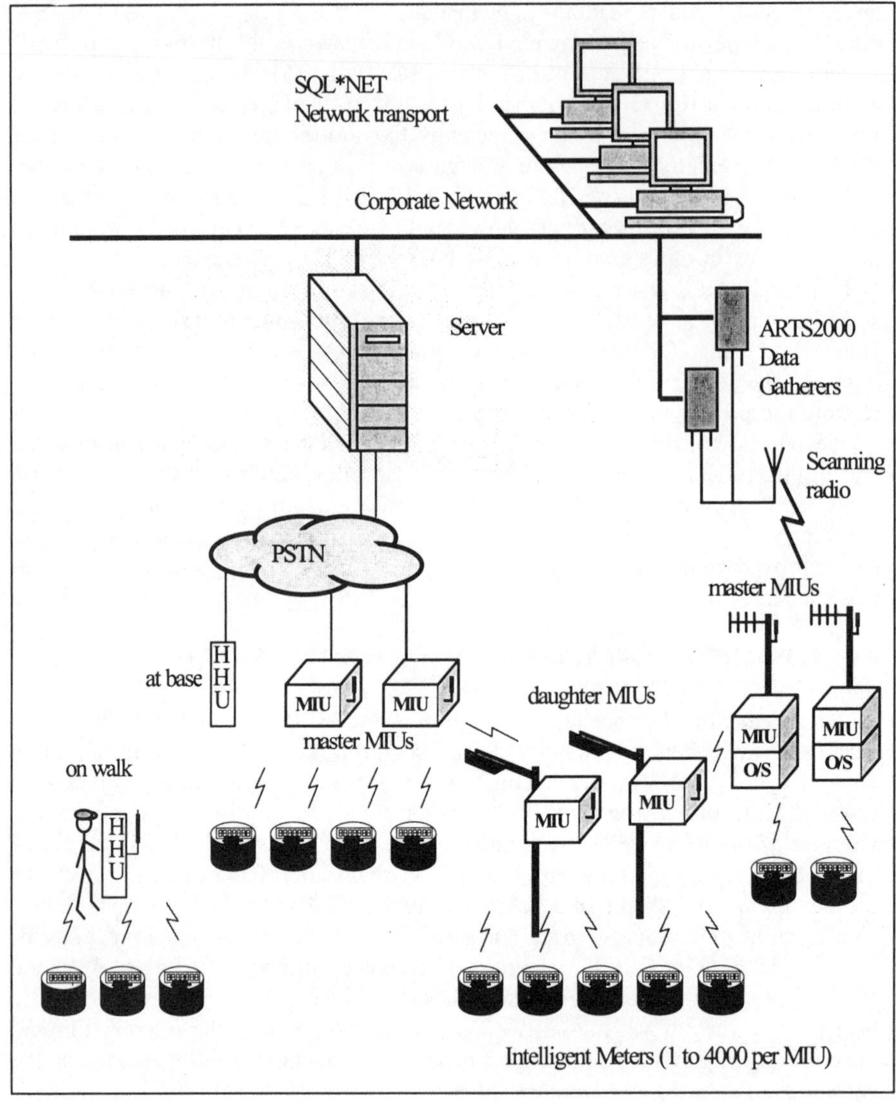

Figure 5 Automatic Meter reading System of Anglian Water (Plunkett, 1999)

Urban stormwater management has, until relatively recently, been focused on solving flooding problems by transferring large amounts of water downstream, as quickly as possible. This inevitably increases volumes and peak discharges causing damage not only to the downstream reaches but also to the natural receiving bodies as well, thus to the environment as a whole. Source control is an alternative strategy, which tries to emulate natural catchment conditions and so pre-empt or reverse the negative implications of urbanisation. The general philosophy is to reduce and attenuate the storm flows before they reach the drainage network and to improve water quality by allowing natural treatment to take place.

The applicability of a specific type of source control facility is dependent on a large number of factors: the level and sensitivity of the water table, the density of vegetation and the porosity of the soil. Successful and appropriate selection of source control measures is thus highly area dependent. Macropolous et al (1999) have developed a new method of selection again using areal information stored in GIS format. The suitability of individual techniques can be tested against pre-defined acceptance criteria at every point of the area in question using fuzzy sets. Fuzzy sets are used because they allow for situations where sharp boundaries between "membership" and "non-membership" do not exist, as in this case. Multi-criteria evaluation is used to combine information from several criteria to form a single index of evaluation and hence derive "suitability maps". Figure 6 shows the spatial distribution of peak flow reduction after a hypothetical application of source control measures.

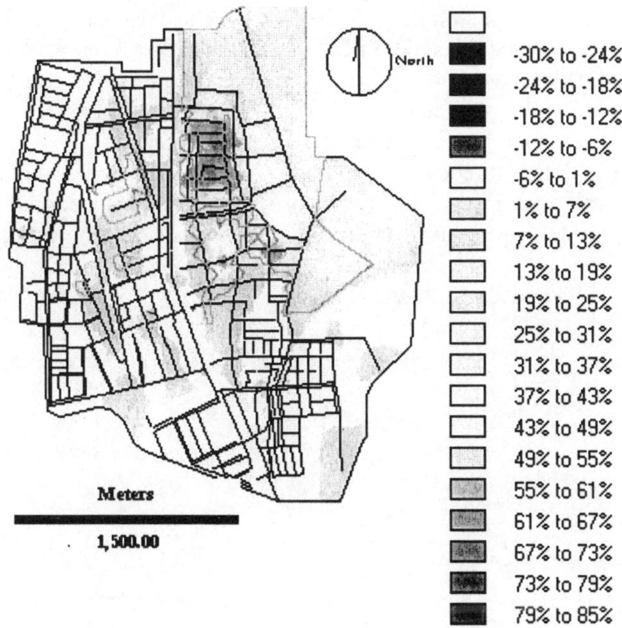

Figure 6 Reduction of peak flows for a 2 years return period input rainfall after implementation of source control measures (Macropoulos et al, 1999)

It is clear from both the previous examples that data management systems and techniques that can handle spatially variable information will become powerful tools for the future. Of course, in order to attempt this level of analysis, high quality site-specific data of sufficient resolution is also absolutely vital. For the method to be widely used, cost-effective means of data collection, storage, arrangement and analysis are needed. In this respect, use of an airborne LIDAR technique for obtaining high resolution digital terrain models with the level of details needed for urban storm drainage studies appears to be a step in the right direction.

5 INTEGRATED SYSTEM MODELLING

Conventionally, planning, design and operation of the urban wastewater system (sewerage, wastewater treatment plant and receiving water) is carried out separately for each of its components. Current practice considers these components as separate units neglecting interactions between them. The importance of analysis of the entire system has been realised for some time (Durchschlag et al., 1991; Tyson et al., 1993), however, detailed analysis of the system as a whole began only recently due to the lack of appropriate simulation tools.

A new simulation tool, capable of taking into account the main elements of the system, their interactions and their integrated control, has been assembled (Schutze et al, 1999a) and is shown in Figure 7. It consists of a sewer model (KOSIM), a treatment plant model based on the IAWPRC ASM No 1 (Henze et al, 1986) and a river model implemented in a DUFLOW shell program. A control module, an optimisation module and various auxiliary routines were added, thus creating a powerful simulation for the analysis of novel control scenarios as well as the comparative evaluation of conventional and integrated control.

The power and effectiveness of the model is illustrated by its application to a semi-hypothetical site based on a composite of several locations where the appropriate data was available. The case study consists of a combined sewer system with 5 storm detention tanks and overflows (including one at the treatment plant), a nitrifying activated sludge treatment plant and a simple, hypothetical, river.

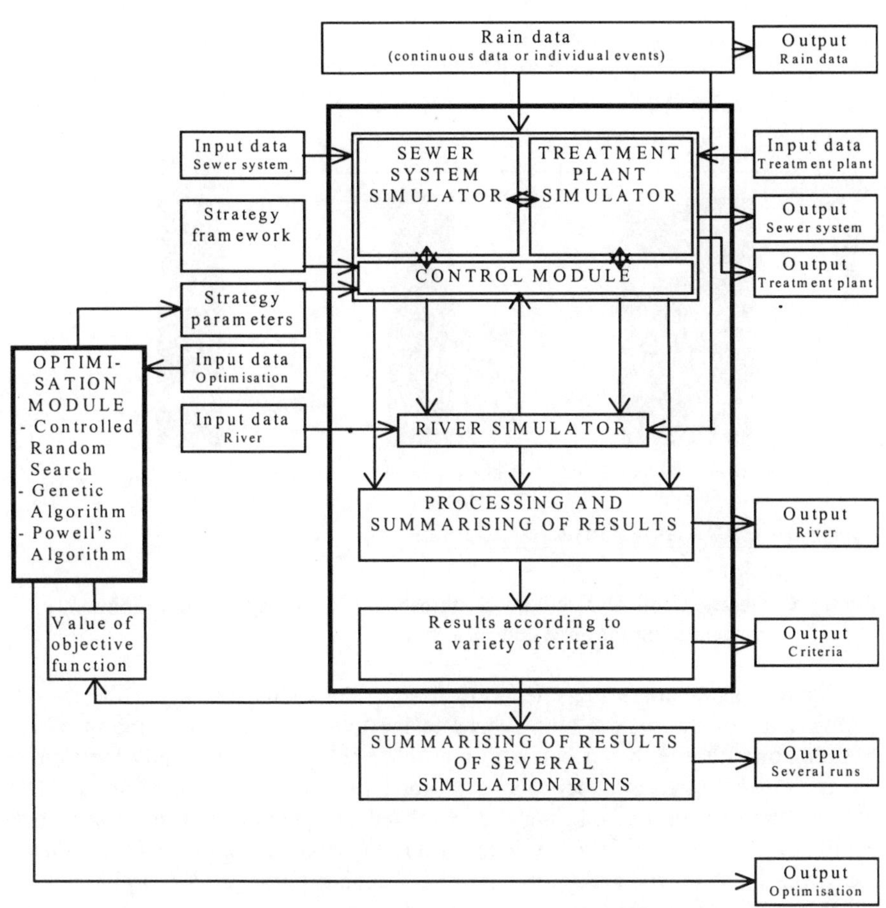

Figure 7 Overview of Integrated wastewater system model (Schutze et al, 1999b)

Three scenarios were investigated to assess the potential of various degrees of control (Schutze et al, 1999a). These were a base case (applying local control only, using default values for the parameter settings), an optimised variation of this setting (still using constant but optimised set-points) and an example of simple hierarchical, integrated control. Here, the local controllers are overridden in exceptional situations by a central control unit in order to balance loads between the sewer system, treatment plant and receiving river. For each of these scenarios, a rule-based strategy is defined and its numerical parameters (e.g. pump rates, threshold values) optimised against two river dissolved oxygen criteria.

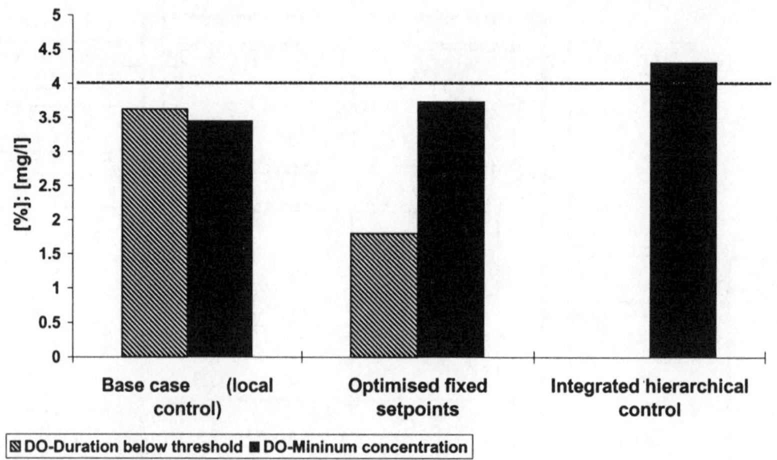

Figure 8 Duration of DO below a threshold of 4 mg/l and minimum DO concentration for three control scenarios

Figure 8 summarises the effects of these strategies by displaying the overall performance of the system with regard to the two criteria - the duration of the oxygen concentration being below a threshold value of 4 mg/l at any location in the river (expressed as a percentage of total run time) and the minimum DO concentration in the river during the simulated time period. It can be seen that control with optimised fixed set-points leads to improved performance: reduced time over threshold and greater minimum DO concentration. Further improvement is achieved for the simple form of integrated control described above, in this case providing complete protection against low DO.

It is clear that we are now in a position to improve and indeed optimise performance of the system as a whole. These tools will be increasingly developed and used in the future, coupled with more detailed and sophisticated data collection.

6 EMERGING TECHNOLOGIES AND THE FIVE Is

In the past, urban water systems have been analysed, designed, constructed and managed separately, often by different companies and under jurisdiction of different ministries. Sustainable development of these systems requires first of all that the mutual interactions of the systems are taken care of, thus fostering the holistic concept. Butler & Maksimovic (1999) propose generic (technical) priority areas that are encapsulated by five Is: Interactions, Integration, Interfacing, Instrumentation and Information. These five principles will guide future upgrade and development of urban water systems.

Mutual interactions of UW systems comprise mutual transfer of liquid and solid matter from one to the other, altering the mass balance and the physical, biological and chemical characteristics of the surrounding environment. For example, water leaking from mains can cause rising groundwater. Similarly, sewers or septic tanks may leak, polluting the aquifer. Infiltration from ground water into sewers (and into water mains in the case of intermittent water supply frequently met in developing countries), can create both change in groundwater table and water quality in networks. Illicit connections of wastewater to stormwater network and vice versa often result in both storm and wastewater systems behaving as if combined. Solid wastes, nutrients, sediments and pathogens easily find their paths to storm drainage especially in the case of open ditches frequently used in developing countries.

Integration will have to take place at both the level of individual UW systems and between various UW systems such as:

- intakes, treatment, distribution, consumers of potable water
- consumers, sewers, treatment, receiving surface and groundwater bodies
- solid waste and storm water collection, disposal, leachate control, ground water, urban streams.

This problem is addressed at various levels including research (Schutze et al, 1999) and real life projects (Milina et al 1999). The proposed EU water Framework directive (European Commission, 1997) seeks to integrate rather than separate. Water companies seem to have started to follow this route and software companies are developing integrated urban water models (e.g. DHI, 1998).

Interfacing of UW systems is observed through its public and environmental aspect. Environmental interfacing is strongly related to sustainable development. A balance needs to be struck between over-exploitation on the one hand, and pollution on the other. Technological development and careful management are key elements in this. Interfacing with the public should result in better mutual understanding between water industry and water users. For example while preserving or improving the level of services, the water industry should gain the co-operation of the users in implementing a demand-managed approach (replacing the current demand driven one). This requires a high level of public awareness and greater public participation in the decision making process. This will be a significant challenge in the future, requiring development and implementation of awareness-raising user-friendly training tools (e.g. Ostrowsky & Dumas 1998)

Instrumentation of tomorrow will have to be more “intelligent”, capable of responding to irregularities and signal perturbations or even anticipating them by using prediction and optimisation models run at speeds higher than real time. Reliability and quality of “point source” sensors has to be improved and, if possible, integrated into the system for gathering of spatially distributed data. Non invasive techniques (Reynolds 1999) have to be upgraded and broadly applied. Techniques of acquisition, target oriented data compaction and implementation of high-resolution spatial data have to be nurtured from their infancy to maturity. In this respect a high level of integration of point source data and remotely sensed spatial data with operational management models has to be achieved.

Information (informatic support), as we have already argued, will have to undergo a significant "quantum leap" in order to meet the needs of integrated urban water systems.

7 Y2K2C PROJECT INITIATIVE

Introduction and implementation of the 5 Is concept requires not only development of new software and data handling tools, but also calls for much stronger emphasis to be placed on physically-based models, data coverage and data quality. A new generation of water and environmental physically-based software products is needed for the new Millennium to boost the modelling and operational management capabilities of future generations by removing outdated concepts.

Present day software packages in urban water (as with other water and environmental disciplines) suffer from various "imperfections":

- Concepts dating back many decades. The development of information technology's computing power has not always been mirrored by improvements in the models,
- The outdated concepts are often hidden behind powerful graphics and presentation glamour
- Modelling of urban water interactions are almost non-existent and integrated modelling is in its infancy
- Many models lack modularity, transparency and transportability (automatic "scaling up and down"),
- Data quality and completeness is usually not properly addressed by software developers, its users often lack the knowledge of basic assumptions on which models are built,
- Data acquisition and processing are not compliant with model structure and complexity, or models are not capable of producing proper results from available data base (DRIPS Syndrome - Data Rich Information Poor Systems),
- Complete digital data on the urban infrastructure and on the spatial distribution of basic urban environment features (land use, DEM- digital elevation models etc.) is rarely found at an appropriate horizontal and vertical resolution,
- Thorough testing against high quality data sets is rarely exercised by developers or by users,
- High level of independent, international verification of new products is rarely performed. In-house verification tends not to reveal the weak points of the products,
- A proper educational component is often missing.

In an attempt to address the above shortcomings and to trigger a global initiative on developing a new generation of integrated multi-level, multi-user problem-solving tools, Maksimovic (1999) has recently introduced the "Year 2002 Challenge – Y2K2C project". The following time frame is foreseen:

- Year 1999 Bring together the key players of the Project
- Year 2000 (early): Learn by doing -Y2K-Seriousness of trivia

- Year 2000 (spring): Create the initial international framework and funding mechanism
- Year 2000 (fall): Workshop - Framework for the first series of pilot projects and verification mechanism
- Year 2001 (International Water Day) : Promote Project
- Year 2001 Formulate pilot projects
- Year 2002 Initiate project– Year 2002 Challenge

It is hoped that the project will serve as a pilot for narrowing the gap between the virtual cyber world and real world of (urban) water and environment in the next century.

8 CONCLUSIONS

In this paper we have identified some of the striking urban water problems caused mainly due to rapid urbanisation. A range of challenges to be faced in urban water management has been discussed and some of the possible paths for tackling the problems have been indicated. These have been expanded upon by reference to a number of examples drawn from studies carried out in the Urban Water Research Group at Imperial College. The main conclusion is the need for the urban water system to be considered and managed as an integrated whole. This will not be fully achieved until the technical priority areas of interaction, integration, interfacing, instrumentation and information (the five Is) have been tackled and the appropriate tools (new generation of models and reliable data sets) developed and implemented. The Y2K2C project has been announced.

As we have seen, the challenges in urban water management are many – it is no overstatement to say that it is a matter of global survival that we rise to them. The community of the CCWI Conference has a significant role to play in this endeavour

REFERENCES

Adgar, A. et al 1995: Improved Measurement and Control Philosophies for Water Treatment Plant Using Artificial Neural Networks. In Water and Wastewater Treatment (Ed. M. White), BHR Group Conference Series Publication No. 17, BHR Group Ltd, UK, pp 213-222.

Andoh R.Y.G. 1995: Alternative Urban Drainage Strategy - The Philosophy. Hydro Research and Development Seminar. Ilkley.

Appleton, T. et al 1995: The Development of a Knowledge Based System for the Advanced Control of a Water Treatment Works. In Water and Wastewater Treatment (Ed. M. White), BHR Group Conference Series Publication No. 17, BHR Group Ltd, UK, pp223-233.

ASCE/UNESCO-IHP 1998: Sustainability Criteria for Water Resource Systems, ASCE, Reston, Virginia, USA

Butler, D. and Parkinson, J. 1997: Towards sustainable urban drainage. Water Science and Technology 35, 9, 53-63.

Butler, D. et al 1998: Local Water Conservation, Re-Use and Renovation: Combined Greywater and Rainwater Recycling. EPSRC Final Report, Grant Nos. GR/K63450 & GR/K61616.

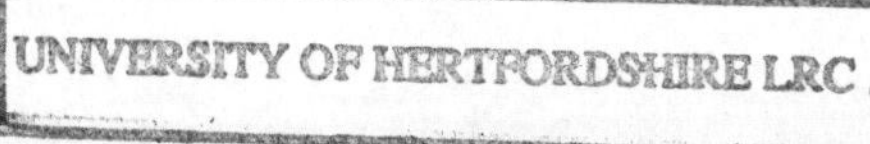

Butler D. & C. Maksimovic 1999: Urban Water Management- Challenges for the Third Millennium. Accepted for: Progress in Environmental Sciences

Carmi N. & Maksimovic, C. 1999: GIS Supported analysis of pressure dependent vulnerability of distribution networks to leakage, Water Industry Systems, Volume 2, RSP, UK.

Cerisier, S.D.M. & Smit, J.J. 1996: The electrochemical generation of ferric ions in cooling water as an alternative for ferric chloride dosing to effect flocculation, Water SA, 22, (4), 327-330.

DHI 1998; Technology Validation Project (TVP). CD-ROM by Danish Hydraulic Institute & WRc plc.

Dixon' A. et al 1999a: Guidelines for greywater reuse – health issues, Journal of Chartered Institution of Water and Environmental Management, accepted.

Dixon, A. et al 1999b: Water saving potential of domestic water re-use systems using greywater and rainwater in combination, Water Science & Technology, 39, 5, 25-32.

DoE 1992; Using Water Wisely. Department of Environment & Welsh Office.

Durchschlag, A. et al 1991: Total emissions from combined sewer overflow and wastewater treatment plants. European Water Pollution Control 48, 6, 13-23

European Commission 1997: Water Framework Directive Commission Proposal COM(97)49, COM(97)614 and COM(98)76.

EPSRC (GR/M68213 Grant): Inverse Transient Analysis in Pipe Network for Leakage Detection and Roughness Calibration, Imperial College and Exeter University

Foxon, T.J. et al 1999a: Useful indicators of urban sustainability: some methodological issues. Local Environment, 4, 2.

Foxon, T.J. et al 1999b: An assessment of water demand management options from a systems approach. Journal of Chartered Institution of Water and Environmental Management.

IUCN-UNEP-WWF. 1991: Caring for the Earth. 2nd Report on World Conservation & Development. Earthscan, London.

Grubb M. et al 1993: The Earth Summit Agreements: A Guide and Assessment. Earthscan, London.

Harremoes, P. 1997: Integrated water and waste management. Water Science and Technology 35, 9, 11-20.

Henze, M. et al 1986: Activated Sludge Model No. 1. IAWPRC Task Group on Mathematical Modelling for Design and Operation of Biological Wastewater Treatment, IAWPRC, London.

House M.A. et al 1994: Public perception of aesthetic pollution, FWR Report No. FR0439, Mar.

Jans, U. & Hoign, J. 1998. Activated Carbon and Carbon Black Catalyzed Transformation of Aqueous Ozone into OH-Radicals, Ozone Science and Engineering, 20, (1), 67-90.

Jiang, J. et al. 1996: Coagulation of Upland Coloured Water with Polyferric Sulphate Compared to Conventional Coagulants, JWSRT -Aqua, 45, (3), 143-154.

Jiang, J. & Graham, N.J.D. 1998: Preparation and Characterisation of an Optimal Polyferric Sulphate (PFS) as a Coagulant for Water Treatment, Journal of Chemical Technology and Biotechnology, 73, (4), 351-358.

Jiang, J. & Graham, N.J.D. 1998A: Evaluation of Poly-Alumino-Iron Sulphate (PAFS) as a Coagulant for Water Treatment, 8th International Gothenburg Symposium on Chemical Treatment, 7-9 September, Prague, Czech Republic.

Lambert, S.D. et al. 1996: Degradation of Selected Herbicides in a Lowland Surface Water by Ozone and Ozone-Hydrogen Peroxide, Ozone Science and Engineering, 18, (3), 251-269.

Lambert, S.D. et al 1997: Evaluation of Inorganic Adsorbents for the Remova. Problematic Textile Dyes and Pesticides, Water Science and Technology, 36,(2-3), 173-180.

Lambert, S.D. and Graham, N.J.D. 1995: Removal of Non-Specific Dissolved Organic Matter from Upland Potable Water Supplies. Part II. Ozonation and Adsorption, Water Research, 29, (10), 2427-2433.

Larsen, T.A. and Gujer, W. 1997: Sustainable urban water management – technological implications. EAWAG News 43E, Dec., 12-14.

Liggett, J.A. and Chen, L. C. 1994: Inverse Transient Analysis in Pipe Networks, Journal of Hydraulic Engineering, ASCE, 120(8), August, 934-955.

Ma, J. et al 1997: Effectiveness of Permanganate Preoxidation in Enhancing the Coagulation of Surface Waters - Laboratory Case Studies, J Water SRT- Aqua, 46, (1), 1-10.

Ma, J. & Graham, N.J.D. 1999: Degradation of Atrazine by Manganese-Catalysed Ozonation: Influence of Humic Substances, Water Research, 33, (3), 785-793.

Maksimovic, C., Calomino, F. & J. Snoxell, (Eds) 1996: Water Supply Systems - New Technologies, Springer,

Maksimovic, C. 1999: Urban Water Issues – Challenges for the Future, Scientific Lecture, Fifth WMO / UNESCO Intergovernmental Conference on Hydrology and Water Resources-Planning for the Future, Geneva, February 1999

Makropoulos, C., D. Butler, C. Maksimovic 1999: Accepted: GIS supported evaluation of source control applicability in urban areas. Water Science and Technology, 39, 9.

Marsalek, J. et al 1998: Hydroinformatic Tools for Planning, Design, Operation and Rehabilitation of Sewer Systems, Springer.

Milina, J., J. Lei, S. Sægrov, A. König, I. Selseth, L. Risholt, W. Schilling, A. Ellingson, J. Alex: 1999:Maximisation of Pollution Load Interception, Documentation 11th European Sewage and Refuse Symposium, Munich May 1999, 59-75

Niemczynowicz, J. 1996: Challenges and interactions in water future. Environmental Research Forum 3-4, 1-10, Transtec Publications, Switzerland.

Obradovic, D. And P. Lonsdale Public water supply: data, models and operational management, E & FN Spon, London: 1998, 0-419-23220-6

Ostrowski, M.W. and James, W. 1998: Requirements for group decision support systems for urban stormwater management. Fourth Int. Conf on Developments in Urban Drainage Modelling: UDM '98, London, September, 569-578.

Pearson, P.J.G. et al 1998: The Reference Sustainability System: A Systems Approach to Assessing The Sustainability Of Cities. EPSRC Final Report, Grant No. GR/K59637

Plunkett, J. (1999) Personal communication.

Reynolds, D. 1999: Prospect for 'star wars' monitoring of water and wastewater quality. Water Quality International Jan/Feb, 12-13.

Rice, R. (1999). Personal communication.

Savic D.A. & G.A. Walters 1997: Evolving Sustainable Networks, Hydrological Sciences Journal 42, 4, 549-564.

Schütze,M. et al 1999: Optimisation of control strategies for the urban wastewater system - an integrated approach. Water Science and Technology, 39, 9.

Schütze, M. et al 1999: Synopsis – a tool for the development and simulation of real-time control strategies for the urban wastewater system, Proc 8th Int. Conf. on Urban Storm Drainage, Sydney, Australia, August.

Surendran S. and Wheatley A. D. 1998: Grey-water reclamation for non-potable re-use. Journal of Chartered Institution of Water and Environmental Management 12, Dec., 406-413.

Tyson, J.M. et al 1993: Management and institutional aspects, Water Science and Technology 27, 12, 159-172

UNESCO-IHP 1997; Water, the City and Urban Planning, Paris, France, April.

Varcoe K. 1996; Consumers in a sustainable environment. In Nath, B., Hens L. and Devuyst, D., editors, Sustainable Development. VUB Press, Brussels.

White, D.A. and Franklin, G.S. (1998). A Preliminary Investigation into the Use of Sodium Ferrate in Water Treatment, Environmental Technology, 19, (11), 1157-1160.

World Commission on Environment and Development 1987; Our Common Future. Oxford University Press.

Zehnder, A.J.B. 1997: Water – a commodity in short supply, EAWAG News 43E,

Integrating the IT Environment

G. Thompson

ABSTRACT

IT plays an ever-increasing role in our lives. Success in many areas is dependant upon how we use IT, not just to mimic the tasks that we each carry out every day, but to reshape our working environments to capitalise on the true benefits of IT as it develops. We each see data / knowledge being re-handled, rediscovered and redeveloped continuously. This is during design, tendering, construction and continuously during operations and maintenance/ upgrades. This is inefficient and introduces errors. This paper discusses the requirements to overcome these inefficiencies. It considers the issues of communications, data standards and cultural issues, which are each as critical to success, if not more so than the actual technical advances in IT. However, this is not just an internal company issue, but an industry wide issue. The paper concludes that data / knowledge is the true commodity that we deal with, this needs to be proactively managed and preserved throughout the supply chain in an Integrated Environment to ensure more profitable business and a better served customer.

1 INTRODUCTION

Computers and IT are now an integral part of our business and home life. The recent issues related to Year 2000 compliance have made us all realise just how much of our everyday existence is dependant upon computer processors and embedded chips. Dependency on the micro-processor will increase continually throughout our lifetimes. The challenge that we all face is how to best utilise this rapidly developing power.

We recognise that the use of computing power is changing rapidly. No longer is it sufficient to simply automate existing tasks, but we need to redefine the way business is carried out in order to make best advantage of the available technology. Businesses are being built around technology rather than the technology being used to support business. Companies which not only fully embrace technology, but are shaped by technology are making major improvements in their performance. Drivers for such changes include: need for added value; rising costs of research; access to significant computing power and suppliers worldwide.

Our industry is itself fragmented. During the life of a project the same information and data are used by many diverse organisations and disciplines, with work carried out in many widely varying geographic locations. The IT systems and procedures generally being used are stand-alone applications dealing with specific parts of the design, construction, finance or operations process. The situation is further exacerbated by the large and diverse data sets involved and the amount of information that needs to be collected, stored, analysed and processed during the project life cycle (cradle to grave). Related to this, the lack of standardisation or commonality of data within existing environments limits the ability of our organisations to capture, communicate and share large amounts of information.

IT has the capacity to provide a common base for sharing information throughout our businesses. The information may relate to business aspirations or to detail characteristics of a transducer. It may relate to knowledge of a particular process or to customer information. All of which, if shared, allows focus on the use and application of that information, to which the human mind is so astute.

The technical problems of sharing data are being rapidly eroded with the major steps forward with web technology and access to it. Developments in high-speed information links and Wide Area Networks are further reducing the transport of information to being a transparent process.

However, we do have cultural problems to overcome in the use of IT. In general many groups of individuals and businesses are not comfortable with the concepts of sharing knowledge, either within their own organisations or within the market place. In many instances such concerns do have some substance, if it is felt that the information refers to a significant advance over other organisations in the same business. However, at present there appears to be a knee jerk response that such concerns apply to the sharing of any information and hence dialogue is inhibited. In order to generate the open sharing culture that is brought out in the Egan Report and a raft of other reports on the future of IT in the construction sector we do need to make a concerted effort towards the development of Integrated Environments.

2 THE INTEGRATED ENVIRONMENT

The design, construction, operation and regulation business is a fragmented, multi-disciplined, distributed process, heavily reliant on gathering and presenting information in a useful, timely and logical manner. Information management and communication are key areas where improvement would have a fundamental impact on performance and productivity.

IT is an enabling technology that could deliver major efficiencies and hence a competitive advantage. However, the successful utilisation of IT in this area would require the integration of IT systems in a virtual environment, accessing common data. By definition this involves:

- IT applications that are integrated or can communicate with each other,
- A distributed computing environment to support the communication and co-ordination between separate teams,
- An integrated project database.

If implemented successfully some of the benefits such a system could deliver are given below:

- Improved client integration in the whole project
- Improved quality through improved design communication and co-ordination
- Improved design efficiency by:
    - one time data entry by those responsible for a particular data type
    - immediate access to the latest data

  - minimised rework and elimination of duplication of effort
- Design improvements from evaluation of constructability
- Improved communication / understanding between design team and contractor.
- Improvements in the construction stage by:
  - Design problems detected early, before they reach the construction stage
  - Utilisation of Programme Management scheduling
  - Reduced material costs by standardising components and improved supply chain management
- Utilise all types of delivery systems
- Accurate and complete data store for use during operation and decommissioning
- Improved supply chain

Competitive business advantage is gained by the factors above contributing to the achievement of one or more of the following:

- Reduce costs and schedule
- Improve overall product
- On time completion
- Customer satisfaction
- Business efficiency
- Profitability

To achieve such redesign of the business several companies have adopted a data-centric approach to create an Integrated Environment. In this the accumulation of knowledge and information is key. Projects are regarded as data-sets, with the project being designed within the information held in the database. Reports and drawings are an output from the database once the project is complete. This architecture requires an extensive database and information handling system together with a wide range of analysis modules which are used to assist with design or evaluate impacts. Projects can be rapidly created. There is no double handling of data or information, information for analysis modules are abstracted from the database, and results from analysis modules go back into the database to be used in the design. With only a single system of project information, clash analysis and the impact of changes are readily identified across the project.

The nature and complexity planning, design, estimating and procurement, make the task of sharing and processing common information very difficult. Much of this information is traditionally provided from the CAD environment. However, geometry alone, derived from the CAD database, is limited in its usefulness for driving some applications where additional attribute information is required, such as in design procurement and maintenance. For this reason design elements should be exchanged as features and components (objects) rather than graphical primitives (points, lines and arcs).

An Integrated Environment will rely heavily on the implementation of object technology and its application to object modelling. In this environment, the real world (reality) is modelled as a collection of objects inter-reacting and influencing each other.

In a future Integrated Environment the project model will be more accessible via a virtual prototyping environment, utilising an immersive workbench and VR technology than a CAD system. Certainly during the feasibility stage a virtual library of standard design components will be manipulated in a constraint-based modelling environment to develop and visualise client requirements. The use of virtual reality models for feasibility design and client approval will improve client and engineer understanding of what is required and what can be delivered, thus reducing the need for costly re-work and editing. Despite ongoing research and development the 3D object orientated CAD tool will form the basis of the detail design "engine" for some time, until VR and CAD are fully integrated and indistinguishable as separate products.

3 CURRENT DEVELOPMENTS

At present our software supply industry is fragmented, their main concern is in providing solutions for a particular problem or client base. This is not surprising considering the complexity of providing IT solutions to a very disparate and diverse industry such as construction. For example just because a vendor can provide an IT product that assists with finite element analysis does not mean he can provide another application capable of providing all the features found within a steel reinforcement package. Whilst this is understandable, what is most disappointing is that in most instances two different yet discipline related applications have no effective means of communicating common or relevant data between them.

One approach has been to develop "all-singing-all-dancing" packages that address particular sectors of the construction industry. Instead of several different point solutions, organisations have developed software systems supporting limited degrees of integration via neutral file exchange or shared project databases. Examples include *Hummingbird* by Bovis, *JobMaster* by Lang and *Prospero* by UNISYS. It is interesting to note that of these examples, only one has been developed by a software house; the others have been developed by players in the construction industry. This is presumably because they believe that they needed to produce a bespoke application to fit their business needs, tailored to their specific method of operation, and that the development would provide them with a competitive advantage over others. All these applications tend to deal with, and are limited to, the co-ordination of the areas of estimating, procurement, project management and accounts during the construction stage of a project.

A separate approach, being developed by Binnie Black & Veatch, is Cygnet, a totally integrated IT based project support system applicable to all units throughout the world. Its objectives are to:

- Reduce costs and schedule
- Improve overall product
- Incorporate varying client quality expectations
- Improve client input processes
- Utilise all types of delivery systems

- Eliminate duplication of effort
- Integrate Program scheduling
- Minimise rework

This system moves on from *project co-ordination* to *project integration,* using technologies such as central data models, on-line data sharing and distributed systems. Working within this environment enables a move from the traditional sequential approach of a project life cycle (where each task is carried out in relative isolation by a specific engineering discipline), to a simultaneous life cycle (concurrent engineering). In this environment all project activities are carried out as early as possible and in parallel, with as much interaction between different professionals on a team as possible to ensure consideration of all elements of the product life cycle. It is interesting to note that this system is being developed by a major player within the industry, not by an IT company. The driver for this development is a perceived business need for concurrent engineering, utilising resources from around the world, and doing this successfully to gain a competitive advantage.

There are other complimentary approaches to encourage and facilitate this philosophy of concurrent engineering by providing an environment where a project model is represented within an intelligent (object-orientated, knowledge-based) integrated project database, acted upon by discrete yet integrated applications, across a transparent, distributed network. However, no complete system like this exists today. Elements of a system like this have been developed in prototype form within the European and UK funded research initiatives such as COMBINE, OSCON and SPACE.

Recently the very first object models were built using programs such as Reflex. These required very powerful (and hence very expensive) graphic workstations and servers. The performance provided by these machines is now available on high specification PCs at a fraction of the cost. Where once the extremely high cost of mini computers and Unix based workstations was a limiting factor in the development of the systems described above this is not the case today. Hardware performance or cost is therefore no longer seen as a limiting factor in implementing an integrated environment and hence we can expect to see soon the future development of these system concepts coming into everyday use.

4 NETWORKS AND COMMUNICATION

The development over the past few years has seen network technology and the speed at which data can be transferred, change out of all recognition. Within Binnie Black & Veatch for example, more than two hundred staff initially shared a 10-megabit Ethernet local area network (LAN) line. Less than two years on, many PCs now have their own dedicated 10 megabit connection to the LAN, which itself has a 200 megabit backbone.

One of the features of the integrated construction environments of the future is the premise that remote, distributed applications will operate on a central integrated project database. The aim of this distributed computing environment is to support better communication among client, design and build teams. This inevitably involves external communication via use of the Internet. Standards such as

CORBA are now available to facilitate the development of distributed object environments where objects can communicate with each other irrespective of their operating system, hardware or location. However, concern has already been raised regarding the rapid take up of Internet connections over the past few years and the system's ability to cope with the sheer volume of e-mail and Web traffic. Further improvements in network technology will need to be made and increased use of data compression, if the distributed project environment is to become a common place. Already at Binnie Black & Veatch we are making increased use of national and international, leased or purchased dedicated ISDN lines between team members in creating such project Intranets.

Having a large central integrated project database is often cited as one of the key components in an integrated environment, although systems are being developed relying on distributed rather than centralised servers. However, an integrated project database does not produce benefits by itself. It can only produce benefits when used in association with applications that develop, analyse, and communicate the design and project data contained within it. This assumption creates as many problems as it solves. For each application associated with the project a data model must be designed to ensure a smooth and efficient dedicated interface between that application and the database. Many of these applications may be from different software suppliers. If the structure of the central core is modified or the application is updated then the data model and associated interface has to be revised. The upkeep of these interfaces can be a time consuming and expensive operation.

5 STANDARDS

An important aspect of reducing the time, effort and cost spent on maintaining adequate links to each application is to use data exchange standards. More and more applications are complying with international standards such as STEP. The aim of STEP is to create an open standard for exchanging and sharing information. However, the take up of data exchange standards has been slow and historically they have failed to keep pace with the requirements and developments of the software technology.

The Industry Alliance for Interoperability (IAI) was established in 1995 to develop a common set of intelligent design objects for the construction industry as a basis for sharing information. It is hoped that as more and more IT vendors incorporate these into their applications it will facilitate application integration and communication. This will provide the environment needed to support an integrated project database.

6 CULTURE

The effective implementation of IT in an integrated environment will require a substantial degree of process re-engineering. As such the traditional methodology with the associated organisational and business structures that support it may have to be changed or reorganised.

Many of our organisations are based on "departments", such as "new works / operations", or "regions / head office" which accentuate the disciplines (and in some cases empires) into which we divide the work process. Our task-orientated approach encourages a narrow focus of each discipline on its own interest. Each time a boundary is crossed time, information, accuracy and Customer Focus is lost. Current thinking involves breaking down these artificial barriers and placing more emphasis on integrating staff expertise into project teams. These teams need not only involve the grouping of individuals into localised work areas. With the intelligent use of IT in the form of e-mail, video conferencing and integrated project databases, people can work in "virtual groups" and involve experts throughout the organisation and external people in the supply chain or customer base from around the world.

The adversarial nature of the construction industry over recent years has not leant itself to the open sharing of digital project information. Ignorance and misunderstanding in many cases has led to organisations insisting on the transfer of paper drawing and text as the only method of data exchange for fear of legal or contractual recriminations. As a result of partnering, PFI, and ECPM initiatives this is changing.

The power of the new tools within the Integrated Environment, and the benefits of small, high performance design/management teams will reduce traditional demarcations between each professional activity. The successful use of this new IT environment will depend on team members having a wide range of new skills in addition to their specific engineering disciplines. These skills will include:

- *Communication skills.* Communication and facilitation skills to improve the understanding of client requirements and to relate more effectively with other professions involved in the project life cycle.
- *Construction process understanding.* Integrated project members will require a knowledge and understanding of the whole construction process and of other disciplines. This is a requirement of concurrent engineering if the project organisation, processes and information flow is to deliver the required result and associated benefits.
- *IT Skills.* Working with the advanced IT tools that will support and enable this Integrated Environment it will be important for all team members be computer literate.
- *Design modelling skills.* Most design work is currently done in 2D CAD and, despite popular belief most engineers draw and think in two dimensions. The Integrated Environment enables engineers and designers to work directly in three dimensions. Advanced CAD and VR skills will be especially important.
- *Operation skills.* The new structure will require that the operational aspects of projects are integral to the design / construction and redevelopment planning
- *Project management.* Within the team there has to be a full appreciation of how the project is being managed and conversely there has to be an understanding of how to implement and manage the Integrated Environment in the most appropriate manner.

As with the development of any emerging technology the organisation must ensure that there is a culture of continuous improvement so that the benefits and problems encountered on previous projects are understood.

Computer automation of the mundane, repetitive aspects of engineering should free the engineer to apply his/her technical innovation and creativity, thus providing the mechanism to deliver economic, safe and environmentally sensitive solutions.

The only thing that has value throughout all time in the design/ construction/ operation process is the data, not any one of its many representations or visualisations. This being the case, proper use of the Integrated Environment with its true data-centric concepts can have enormous cost saving benefits, not only in the design stage but also during the construction and operation life cycle of the project. The potential improvement to the design process and resultant reduction in costs by implementing an Integrated Engineering Environment in the manner described above would necessitate a change in how a client awards contracts and judges cost effectiveness. The Integrated Environment provides the mechanism by which we can create value. The rewards from this should be shared among all parties making for a more profitable business and supply chain and hopefully a better served end user.

ACKNOWLEDGEMENTS

The author would like to thank Binnie Black & Veatch for permission to publish this paper and he also wishes to thank Ian Bush for his continual support in developing the IT strategy within Binnie Black & Veatch as discussed in this paper.

PART II

ANALYSIS SIMULATION AND DESIGN

A Loop Based Simulator and Implicit State-Estimator

J.H. Andersen *and* R.S. Powell

ABSTRACT

This paper presents a new formulation of the standard weighted least squares (WLS) state-estimation problem for water networks. The formulation follows naturally from the loop based network simulation method and inherits the advantages that such methods have over node based methods. The solution of the non-linear optimisation problem leads to an implicit iteration scheme which converges quickly due to careful avoidance of double iterations. This is achieved using a Lagrangian approach. Encouraging convergence results are presented for a small water network example.

1 INTRODUCTION

In network analysis packages, nodal demand values are assumed known and are used as the given variables from which the other hydraulic variables are derived. An initial estimation of these time-varying nodal demand values is obtained by multiplying the time-profile for each consumer type by the number of consumers of each type on a nodal basis. For a particular instant of time, the demand at each node is then found by adding up the consumption from each type.

In the context of state-estimation, these data are referred to as 'pseudo-measurements'. Engineers in the water industry are able to improve the accuracy of these demand estimates by carrying out statistical studies of measured demand patterns and reservoir outflows.

Real measurements may be gathered in the interest of improved monitoring for economical operation. This includes monitoring for the detection of abnormal consumption and possible leakage. Since real measurements are subject to random errors, statistical state-estimation could be used to find a coherent network state consistent with a hydraulic model. As a result a picture may be obtained which displays a somewhat uneven distribution of the estimation errors across the network.

The method of weighted least squares (WLS) state-estimation for water distribution networks is well known, e.g. Bargiela et al [2] and Powell et al [5]. The best fit of hydraulically consistent nodal heads is found by a series of normal projections. Although the numerical problem is well defined, it may lead to practical problems for large networks. In this paper a new implicit formulation of the WLS method is considered. The aim is to benefit from the smaller solution matrices which relate to the loop structure rather than the nodal structure of the network and to exploit the advantages that a looped solution provides over other methods when controlling elements are present, e.g. Andersen et al [1].

2 FORMULATION OF THE NETWORK EQUATIONS

As a basis for the mathematical derivations and for illustrating the resulting algorithms, a small piped network example is used. The example is hydraulically identical to the network used in Brdys and Ulanicki [2].

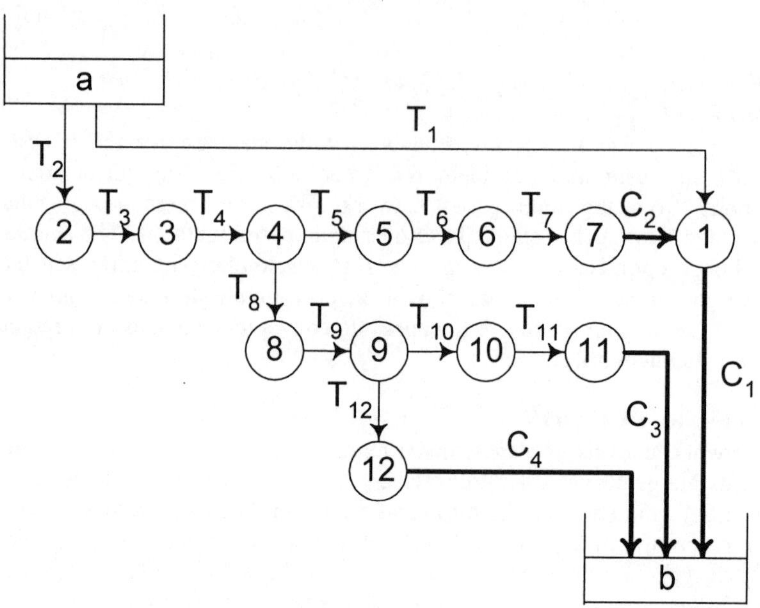

Figure 1. A Water Network Example

The network has $n = 12$ demand nodes and two fixed head datum nodes, reservoirs (a) and (b). A spanning tree is selected starting from reservoir (a) as the root node. Since all datum nodes in a sense belong to the same virtual node, reservoir (b) cannot be part of the spanning tree; the chord C_1 completes a pseudo-loop via the virtual datum node. For the general case it is assumed that the corresponding graph is connected.

Figure (1) shows the labelling and reference directions for the tree links (T) indicated by thin arrows and the chords (C) indicated by thick arrows. The chords are sometimes called co-tree links. The number of chords is equal to the number ℓ of fundamental loops. In this network we have $\ell = 4$. The network equations are as follows:

$$\mathbf{d} = \mathbf{U}\mathbf{q}_T + \mathbf{V}\mathbf{q}_C \tag{1}$$

$$\mathbf{q}_T = \Psi(-\mathbf{U}^T\mathbf{h} + \mathbf{a}) \tag{2}$$

$$\mathbf{q}_C = \Psi(-\mathbf{V}^T\mathbf{h} + \mathbf{b}) \tag{3}$$

Here **d** (n x 1) is the vector of nodal demands, $\mathbf{q}_T$ (n x 1) and $\mathbf{q}_C$ (ℓ x 1) are vectors of flows in the tree and chord links respectively and $\mathbf{h}$ (n x 1) is the vector of nodal heads. The matrix **U** (n x n) is the node-link incidence matrix for the tree, the network labelling is chosen such that **U** becomes upper triangular. The matrix **V** (n x ℓ) is the node-link incidence matrix for the chord flows. The vector $\mathbf{a}$ (n x 1) is the 'boundary' vector for the root node (fixed head node) and the vector $\mathbf{b}$ (ℓ x 1) incorporates the boundaries from all other datum nodes connected to the network graph. The diagonal vector function $\Psi(\delta\mathbf{h})$ is the non-linear hydraulic flow model for a given head loss vector $\delta\mathbf{h}$.

3 FORMULATION OF THE SIMULATOR PROBLEM

The first task is to formulate the simulation algorithm in the above notation. A simulation is defined as a solution of the network model for given nodal demands. Assuming that the flow function $\Psi(\delta\mathbf{h})$ is monotonic, its inverse function is then $\delta\mathbf{h} = \Phi(\mathbf{q})$. In the case of the Hazen-Williams flow model, the inverse is easily obtained. From equation (2), the nodal heads can be formally written:

$$\mathbf{h} = -(\mathbf{U}^T)^{-1}\left(\Phi(\mathbf{q}_T) - \mathbf{a}\right) \tag{4}$$

Substitution into (3) gives:

$$\mathbf{q}_C = \Psi\left(\mathbf{V}^T(\mathbf{U}^T)^{-1}\left(\Phi(\mathbf{q}_T) - \mathbf{a}\right) + \mathbf{b}\right) \tag{5}$$

Furthermore, substituting $\mathbf{q}_T$ via (1) leads to:

$$\mathbf{q}_C = \Psi\left(\mathbf{V}^T(\mathbf{U}^T)^{-1}\left(\Phi(\mathbf{U}^{-1}\mathbf{d} - \mathbf{U}^{-1}\mathbf{V}\mathbf{q}_C) - \mathbf{a}\right) + \mathbf{b}\right) \tag{6}$$

At this stage only simulation is considered, hence the only unknown variable present in (6) is the chord flow vector $\mathbf{q}_C$, though this vector appears on both left and right hand sides. Since the length of $\mathbf{q}_C$ is the same as the number of loops which may be considerably smaller than the number of nodes, there is scope for finding an efficient algorithm for solving (6). Once $\mathbf{q}_C$ is known, the network state follows readily via $\mathbf{q}_T$ from (1) and $\mathbf{h}$ from (2). Equation (6) is referred to as the '*chord-equation*'. A more familiar form of this equation may be obtained by operating the head loss function Φ on both sides of the equation. In this form the equation states that the head losses over the chords equal the head losses calculated the other way round the loops. This is the well known energy conservation statement.

Subtracting the right hand side of (6) from the left hand side casts the chord-equation into its residue form: $\mathbf{F}(\mathbf{q}_C) = \mathbf{0}$, this form is amenable to Newton-Raphson iterations:

$$\Delta \mathbf{q}_C = -\left(\frac{\partial \mathbf{F}}{\partial \mathbf{q}_C}\right)^{-1} \mathbf{F}(\mathbf{q}_C) \tag{7}$$

For the chord equation, the vectors $\mathbf{h}$ and $\mathbf{q}_T$ can now be considered as auxiliary dependent variables to assist the symbolic differentiation, hence:

$$\frac{\partial \mathbf{F}}{\partial \mathbf{q}_C} = \mathbf{I} + \frac{\partial \Psi}{\partial \delta \mathbf{h}} \mathbf{V}^T (\mathbf{U}^T)^{-1} \frac{\partial \Phi}{\partial \mathbf{q}} \mathbf{U}^{-1} \mathbf{V} \tag{8}$$

Here $\mathbf{I}$ is the (ℓ x ℓ) unit matrix and $\dfrac{\partial \Psi}{\partial \delta \mathbf{h}}$ and $\dfrac{\partial \Phi}{\partial \mathbf{q}}$ are diagonal matrices.

The derivatives are taken at $\delta \mathbf{h} = -\mathbf{V}^T \mathbf{h}(\mathbf{q}_C) + \mathbf{b}$ and $\mathbf{q} = \mathbf{q}_T(\mathbf{q}_C)$ respectively.

Since the above $\delta \mathbf{h}$ is an indirect calculation of the head losses over the chords, it was found in practice that a considerable speed-up in convergence could be obtained by using the following averaged argument for the first derivative in (8):

$$\overline{\delta \mathbf{h}} = \frac{1}{2}\left(\Phi(\mathbf{q}_C) - \mathbf{V}^T \mathbf{h}(\mathbf{q}_C) + \mathbf{b}\right) \tag{9}$$

The linear equation system (7) produces a compact and efficient iteration scheme. This is because $\mathbf{U}$ is upper triangular and hence easily solved, furthermore the Jacobian (8) is of order (ℓ x ℓ) only. The scheme with the enhancement (9) converges quickly for the network example, see Figure 2.

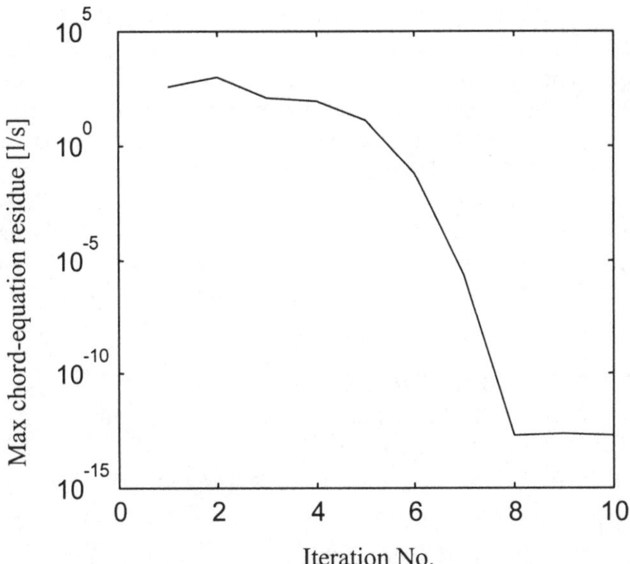

Figure 2. Convergence of the Simulator

A practical simulation scheme can exploit the network structure still further by recording the appropriate loops and their sparse couplings, e.g. see Andersen et al [1]. The chord-equation method was shown here as a prelude to the implicit state-estimation.

4 PRINCIPLES OF IMPLICIT STATE-ESTIMATION

Let the vector x (n x 1) be the unknown state variable. The WLS state-estimation minimises the objective function:

$$\text{Min: } \Omega = (\mathbf{z} - \mathbf{g}(\mathbf{x}))^T \mathbf{W}(\mathbf{z} - \mathbf{g}(\mathbf{x})) \tag{10}$$

where **z** (m x 1) is the measurement vector, **g**(**x**) represents the hydraulic model equations and **W** (m x m) is a diagonal matrix of measurement weights.

The corresponding Newton iteration step solves a normal equation problem using the Gauss-Newton approximation to the Hessian matrix:

$$\Delta\mathbf{x} = (\mathbf{J}^T\mathbf{W}\mathbf{J})^{-1}\mathbf{J}^T\mathbf{W}(\mathbf{z} - \mathbf{g}(\mathbf{x})) \tag{11}$$

where **J** is the Jacobian of **g**(**x**).

The choice of physical variables to represent the state vector **x** is yet unspecified. If the nodal heads represent the state vector (**x** = **h**) then **g**(**x**) may be readily calculated for head, flow and demand measurements. However, the solution matrix $\mathbf{J}^T\mathbf{WJ}$ may become large and difficult to solve efficiently. Alternatively, the state vector could be represented by the demand vector ($\mathbf{x} = \mathbf{d}$). This may have certain advantages, but the non-linear hydraulic model $\mathbf{g}(\mathbf{x})$ can no longer be readily calculated for the head and flow measurements. Intuitively, a secondary series of iterations may then be necessary. It will be shown that by using a Lagrangian approach such double-iteration can be avoided.

The term '*implicit*' is appropriate for this new method because of the apparent 'unavailablity' of the hydraulic model. The term '*Lagrangian*' normally implies a constraint, but no external constraint is imposed. Instead an internal constraint allows the introduction of an extended hydraulic model which can be calculated explicitly.

5 STATEMENT OF THE LAGRANGIAN

The Lagrangian method is often the method of choice when implicitly determined quantities appear in an optimisation problem.

It was shown above that for a given demand vector **d**, knowledge of the chord flows $\mathbf{q}_C$ solves the hydraulic model. In other words, if **x** is the demand vector and **y** is the chord flow vector (used here as independent variables) then in general:

$$\mathbf{g}(\mathbf{x}) = \widetilde{\mathbf{g}}(\mathbf{x}, \mathbf{y}) \text{ if and only if } \widetilde{\mathbf{F}}(\mathbf{x}, \mathbf{y}) = \mathbf{0} \qquad (12)$$

Here the function $\widetilde{\mathbf{g}}(\mathbf{x}, \mathbf{y})$ is the extended definition of the 'true' hydraulic model $\mathbf{g}(\mathbf{x})$. In the following, this extended function will be referred to as $\mathbf{g}(\mathbf{x}, \mathbf{y})$ in order to avoid confusion. Similarly, the residue function $\widetilde{\mathbf{F}}(\mathbf{x}, \mathbf{y})$ is the extended definition of $\mathbf{F}(\mathbf{y})$, which in turn is the chord-equation (6) in its residue form. Also the notation $\mathbf{F}(\mathbf{x}, \mathbf{y})$ will be used for $\widetilde{\mathbf{F}}(\mathbf{x}, \mathbf{y})$. Tables (1.)-(2.) summarise these definitions.

Table 1. The extended hydraulic model

Type	$\mathbf{g}(\mathbf{x}, \mathbf{y})$
Demand	$\mathbf{x}$
Head	$-(\mathbf{U}^T)^{-1}\left(\Phi(\mathbf{U}^{-1}(\mathbf{x} - \mathbf{V}\mathbf{y})) - \mathbf{a}\right)$
Tree flow	$\mathbf{U}^{-1}(\mathbf{x} - \mathbf{V}\mathbf{y})$
Chord flow	$\mathbf{y}$

Table 2. The extended residue function

	$\mathbf{F}(\mathbf{x},\mathbf{y})$
Chord residue	$\mathbf{y}-\Psi\left(\mathbf{V}^T(\mathbf{U}^T)^{-1}\left(\Phi(\mathbf{U}^{-1}(\mathbf{x}-\mathbf{V}\mathbf{y}))-\mathbf{a}\right)+\mathbf{b}\right)$

The state-estimation problem (9) can now be re-written as:

$$\text{Min: } \Omega = (\mathbf{z}-\mathbf{g}(\mathbf{x},\mathbf{y}))^T \mathbf{W}(\mathbf{z}-\mathbf{g}(\mathbf{x},\mathbf{y})) \quad \text{for } \mathbf{x}\in R^n \text{ and } \mathbf{y}\in R^\ell \tag{13}$$

subject to: $\mathbf{F}(\mathbf{x},\mathbf{y}) = \mathbf{0}$

Or the Lagrangian equivalent:

$$\text{Min: } L = (\mathbf{z}-\mathbf{g}(\mathbf{x},\mathbf{y}))^T \mathbf{W}(\mathbf{z}-\mathbf{g}(\mathbf{x},\mathbf{y})) + \boldsymbol{\lambda}^T \mathbf{F}(\mathbf{x},\mathbf{y}) \tag{14}$$

where $\boldsymbol{\lambda}\in R^\ell$ is the vector of Lagrangian multipliers.

The first order conditions are:

$$\left(\frac{\partial L}{\partial(\mathbf{x},\mathbf{y})}\right)^T = -2\,\mathbf{J}^T\mathbf{W}(\mathbf{z}-\mathbf{g}(\mathbf{x},\mathbf{y})) + \mathbf{K}^T\boldsymbol{\lambda} = \mathbf{0} \tag{15}$$

where J and K are Jacobians of g and F respectively.

Furthermore, the Jacobians can be partitioned as:

$$\mathbf{J} = \left[\mathbf{J}_\mathbf{x},\mathbf{J}_\mathbf{y}\right] = \left[\frac{\partial\mathbf{g}}{\partial\mathbf{x}},\frac{\partial\mathbf{g}}{\partial\mathbf{y}}\right] \quad \text{and} \quad \mathbf{K} = \left[\mathbf{K}_\mathbf{x},\mathbf{K}_\mathbf{y}\right] = \left[\frac{\partial\mathbf{F}}{\partial\mathbf{x}},\frac{\partial\mathbf{F}}{\partial\mathbf{y}}\right] \tag{16}$$

hence

$$\boldsymbol{\lambda} = 2\,(\mathbf{K}_\mathbf{y}^T)^{-1}\mathbf{J}_\mathbf{y}^T\mathbf{W}(\mathbf{z}-\mathbf{g}(\mathbf{x},\mathbf{y})) \tag{17}$$

which gives

$$\left(\mathbf{J}_\mathbf{x}-\mathbf{J}_\mathbf{y}(\mathbf{K}_\mathbf{y})^{-1}\mathbf{K}_\mathbf{x}\right)^T\mathbf{W}(\mathbf{z}-\mathbf{g}(\mathbf{x},\mathbf{y})) = \mathbf{0} \tag{18}$$

6 THE NEWTON STEP

Let $\mathbf{G}(\mathbf{x},\mathbf{y}) = 0$ comprise the total system of non-linear equations for which the solution is sought, that is

$$\mathbf{G}(\mathbf{x},\mathbf{y}) = \begin{bmatrix} (\mathbf{J}_x - \mathbf{J}_y(\mathbf{K}_y)^{-1}\mathbf{K}_x)^T \mathbf{W}(\mathbf{z} - \mathbf{g}(\mathbf{x},\mathbf{y})) \\ \mathbf{F}(\mathbf{x},\mathbf{y}) \end{bmatrix} = 0 \quad (19)$$

In the Gauss-Newton approximation, the Jacobian of $\mathbf{G}(\mathbf{x},\mathbf{y})$ becomes:

$$\frac{\partial \mathbf{G}(\mathbf{x},\mathbf{y})}{\partial(\mathbf{x},\mathbf{y})} = \begin{bmatrix} \left[-(\mathbf{J}_x - \mathbf{J}_y(\mathbf{K}_y)^{-1}\mathbf{K}_x)^T \mathbf{W}\mathbf{J}_x\right] & \left[-(\mathbf{J}_x - \mathbf{J}_y(\mathbf{K}_y)^{-1}\mathbf{K}_x)^T \mathbf{W}\mathbf{J}_y\right] \\ \mathbf{K}_x & \mathbf{K}_y \end{bmatrix} \quad (20)$$

The above Jacobian (20) divides into four blocks and can be expressed symbolically as:

$$\frac{\partial \mathbf{G}(\mathbf{x},\mathbf{y})}{\partial(\mathbf{x},\mathbf{y})} = \begin{bmatrix} \mathbf{A} & \mathbf{B} \\ \mathbf{C} & \mathbf{D} \end{bmatrix} \quad (21)$$

Inspection of the corresponding blocks of (20) and (21) suggests that the matrix **A** relates to the normal projection for a stand-alone WLS scheme if this exists. Similarly, the matrix **D** relates to a stand-alone chord-equation simulation scheme. The cross-terms **B** and **C** provide the glue which combines the two schemes into the implicit scheme. With this analysis, it is unlikely that **A** or **D** are singular, hence it is assumed that their inverse exists. The Newton iteration step can be written symbolically as:

$$\begin{bmatrix} \mathbf{A} & \mathbf{B} \\ \mathbf{C} & \mathbf{D} \end{bmatrix} \begin{bmatrix} \Delta\mathbf{x} \\ \Delta\mathbf{y} \end{bmatrix} = -\mathbf{G}(\mathbf{x},\mathbf{y}) \quad (22)$$

At this level of abstraction, the scheme (22) is of little practical use unless the promised efficiency in the solution of the linear system can be demonstrated.

7 REDUCTION OF THE LINEAR SYSTEM

The first reduction step consists of de-coupling the linear system (22). It was assumed that **A** and **D** were non-singular, hence:

$$\begin{bmatrix} \mathbf{A}^{-1} & \mathbf{0} \\ \mathbf{0} & \mathbf{D}^{-1} \end{bmatrix} \begin{bmatrix} \mathbf{A} & \mathbf{B} \\ \mathbf{C} & \mathbf{D} \end{bmatrix} = \begin{bmatrix} \mathbf{I}_x & \mathbf{A}^{-1}\mathbf{B} \\ \mathbf{D}^{-1}\mathbf{C} & \mathbf{I}_y \end{bmatrix} \quad (23)$$

which gives

$$\begin{bmatrix} \mathbf{I}_x & \mathbf{A}^{-1}\mathbf{B} \\ \mathbf{D}^{-1}\mathbf{C} & \mathbf{I}_y \end{bmatrix} \begin{bmatrix} \Delta\mathbf{x} \\ \Delta\mathbf{y} \end{bmatrix} = -\begin{bmatrix} \mathbf{A}^{-1}[\mathbf{G}]_x \\ \mathbf{D}^{-1}[\mathbf{G}]_y \end{bmatrix} \tag{24}$$

This form can be split up into two systems by means of substitution:

$$\left(\mathbf{I}_x - \mathbf{A}^{-1}\mathbf{B}\mathbf{D}^{-1}\mathbf{C}\right)\Delta\mathbf{x} = -\mathbf{A}^{-1}\left([\mathbf{G}]_x - \mathbf{B}\mathbf{D}^{-1}[\mathbf{G}]_y\right) \tag{25}$$

and

$$\left(\mathbf{I}_y - \mathbf{D}^{-1}\mathbf{C}\mathbf{A}^{-1}\mathbf{B}\right)\Delta\mathbf{y} = -\mathbf{D}^{-1}\left([\mathbf{G}]_y - \mathbf{C}\mathbf{A}^{-1}[\mathbf{G}]_x\right) \tag{26}$$

Equation (25) is a linear system of order n. However, it is easily seen that inversion of the matrix on the left-hand side of (25) can be reduced to a rank ℓ inverse update. Using the Sherman-Morrison-Woodbury formula [4] yields:

$$\left(\mathbf{I}_x - \mathbf{A}^{-1}\mathbf{B}\mathbf{D}^{-1}\mathbf{C}\right)^{-1} = \mathbf{I}_x + \mathbf{A}^{-1}\mathbf{B}\left(\mathbf{I}_y - \mathbf{D}^{-1}\mathbf{C}\mathbf{A}^{-1}\mathbf{B}\right)^{-1}\mathbf{D}^{-1}\mathbf{C} \tag{27}$$

As the matrix **D** already is of order ℓ, this leaves the above discussion to the inversion of the matrix **A** as this is the only non-trivial matrix of order n that could be potentially difficult to solve.

By the definition of **A**, see (20) and (21):

$$\mathbf{A} = -\mathbf{J}_x^T\mathbf{W}\mathbf{J}_x + \mathbf{K}_x^T(\mathbf{K}_y^T)^{-1}\mathbf{J}_y^T\mathbf{W}\mathbf{J}_x \tag{28}$$

Again, the Sherman-Morrison-Woodbury formula is employed; the result is given here for completeness:

$$\begin{aligned} \mathbf{A}^{-1} &= -(\mathbf{J}_x^T\mathbf{W}\mathbf{J}_x)^{-1} - (\mathbf{J}_x^T\mathbf{W}\mathbf{J}_x)^{-1}\mathbf{K}_x^T(\mathbf{K}_y^T)^{-1}\ldots \\ &\ldots\left(\mathbf{I}_y - \mathbf{J}_y^T\mathbf{W}\mathbf{J}_x(\mathbf{J}_x^T\mathbf{W}\mathbf{J}_x)^{-1}\mathbf{K}_x^T(\mathbf{K}_y^T)^{-1}\right)^{-1}\mathbf{J}_y^T\mathbf{W}\mathbf{J}_x(\mathbf{J}_x^T\mathbf{W}\mathbf{J}_x)^{-1} \end{aligned} \tag{29}$$

The outcome of this is that the inversion of the matrix **A** can be delegated to the inversion of the normal matrix $\mathbf{J}_x^T\mathbf{W}\mathbf{J}_x$ and matrices of the order (ℓ x ℓ)

Assuming that all free nodes have demand or pseudo-demand measurements, then the structure of $\mathbf{J}_x^T$ can be further detailed as:

$$\mathbf{J}_{\mathbf{x}}^{T} = \left[\mathbf{I}_{\mathbf{x}}, \mathbf{J}_{\mathbf{m,x}}^{T}\right] \tag{30}$$

where $\mathbf{J}_{\mathbf{m,x}} \in R^{m-n} \times R^{n}$ corresponds to the rows of $\mathbf{J}_{\mathbf{x}}$ for measurements other than demand measurements, hence $\mathbf{m}$ are the indices for non-demand measurements.

This leads to the structure of the normal matrix:

$$\mathbf{J}_{\mathbf{x}}^{T}\mathbf{W}\mathbf{J}_{\mathbf{x}} = \mathbf{W}_{\mathbf{x}} + \mathbf{J}_{\mathbf{m,x}}^{T}\mathbf{W}_{\mathbf{m}}\mathbf{J}_{\mathbf{m,x}} \tag{31}$$

Here the diagonal matrix $\mathbf{W}$ has been split into $\mathbf{W}_{\mathbf{x}}$ and $\mathbf{W}_{\mathbf{m}}$ corresponding to weights for demands and other measurements respectively. Hence, the inverse can be written as:

$$\left(\mathbf{J}_{\mathbf{x}}^{T}\mathbf{W}\mathbf{J}_{\mathbf{x}}\right)^{-1} = \mathbf{W}_{\mathbf{x}}^{-1} - \mathbf{W}_{\mathbf{x}}^{-1}\mathbf{J}_{\mathbf{m,x}}^{\mathbf{T}}\left(\mathbf{I}_{\mathbf{m}} + \mathbf{W}_{\mathbf{m}}\mathbf{J}_{\mathbf{m,x}}\mathbf{W}_{\mathbf{x}}^{-1}\mathbf{J}_{\mathbf{m,x}}^{\mathbf{T}}\right)^{-1}\mathbf{W}_{\mathbf{m}}\mathbf{J}_{\mathbf{m,x}}\mathbf{W}_{\mathbf{x}}^{-1} \tag{32}$$

Since the weight matrices are diagonal, this shows the inversion or solution of the normal matrix $\mathbf{J}_{\mathbf{x}}^{T}\mathbf{W}\mathbf{J}_{\mathbf{x}}$ can be reduced to the solution of a matrix of order (m-n) x (m-n), where m-n is the number of measurements other than demand measurements.

8 RESULTS

The main interest in testing of the scheme at this stage is the convergence properties compared to the standard WLS formulation as in (10). For the numerical experiment, artificial nodal measurements were generated comprising demands for all nodes plus head measurements for 5 selected nodes of the network in figure (1). The demand measurement data were derived from a set of nominal demands superimposed with Gaussian noise. The head measurement data were simulated to be hydraulically consistent with the nominal demands. Table (3) below shows the resulting data for the experiment.

Table 3. Generated measurement data and estimation results

Node	Head data	Nominal demands	Demand data	Estimated heads	Estimated demands
1	-	5.0	5.72	105.537	5.55
2	-	5.0	5.14	115.624	5.71
3	111.212	5.0	1.61	111.307	2.76
4	107.713	5.0	2.44	107.758	3.10
5	-	5.0	3.84	106.607	3.77
6	106.299	10.0	15.28	106.205	14.94
7	-	5.0	3.93	105.766	3.69
8	104.324	5.0	5.93	104.299	6.33
9		10.0	11.82	101.785	12.34
10	101.513	5.0	1.91	101.210	2.66
11	-	5.0	2.44	100.593	2.82
12	-	5.0	4.08	100.686	4.30
a	120	-	-	-	-
b	100	-	-	-	-

By definition the initial state $\mathbf{x}^{(0)}$ (heads) in the standard WLS is consistent with the hydraulic model $\mathbf{g}(\mathbf{x}^{(0)})$, hence a similar consistent initial condition was constructed for the implicit method. For the initial condition $\mathbf{x}^{(0)}, \mathbf{y}^{(0)}$ of the implicit method, $\mathbf{x}^{(0)}$ was set to the measured demand data and $\mathbf{y}^{(0)}$ was set to the chord flows derived from a simulation run. This initialisation is not strictly necessary for the implicit method, but simulation results are usually available prior to a state-estimation.

The enhancement as detailed in equation (9) was also applied to the implicit method. Figure 3 shows the convergence for different values of a common weight applied to all head measurements. These weights are relative to those for the demand measurements; in this experiment a common unit weight was assumed for the demands. (A weight $W_h = 400$ was used for the estimated values in Table 3.) The full curves show the maximum changes to the heads in the network. Since the implicit method is strictly not hydraulically consistent during the iterations, the dotted curves below the full curves show the maximum loop residue in the form of non-matching head losses.

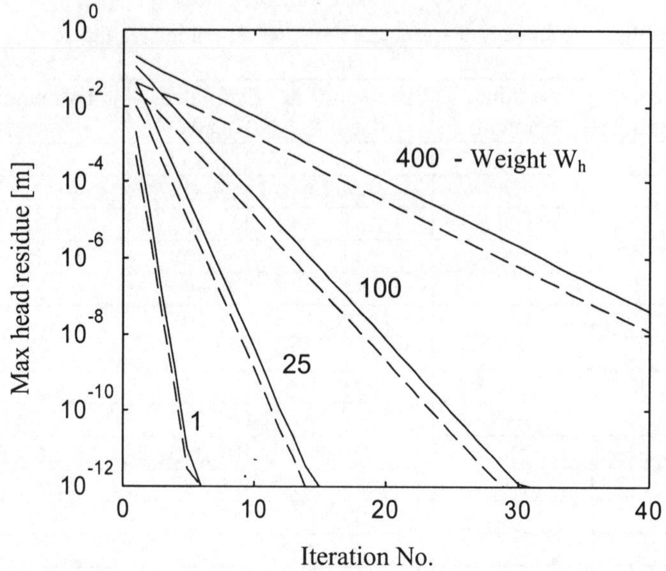

Figure 3. Convergence of the implicit WLS state-estimator

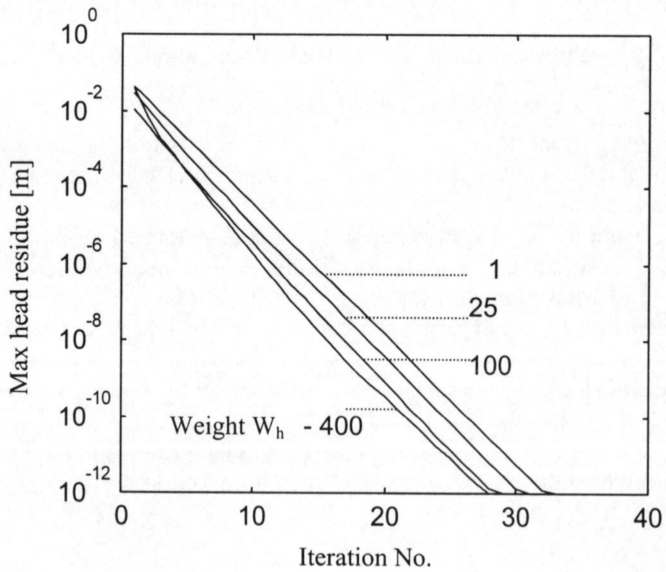

Figure 4. Convergence of the standard WLS state-estimator

The corresponding convergence curves for the standard WLS state-estimator are shown in figure 4. The curves are concentrated in a band suggesting less dependence on the weights in this case. The performance seems more favourable for large weights, but the workload in each iteration could in general be larger due to the large matrices compared to the implicit scheme. Also, in contrast to the implicit scheme, the convergence rate does not improve with reduced measurement count.

9 CONCLUSIONS

The implicit state-estimation method has, perhaps not surprisingly, been reduced to solving a system of normal equations. However, this method takes advantage of the particular measurement circumstances found in water networks. Normally network nodes are furnished with pseudo-measurements of demands. This 'background' information is substantiated with a much smaller number of real measurements, perhaps covering less than 10% of the network. Using the implicit method, the content of the normal matrix for such a system will consist of a diagonal matrix resulting from the demand pseudo-measurements plus a low–rank matrix contribution from the real measurements. This allows quick solutions of the normal equations via the inverse updating technique. Furthermore, the impact of a single measurement is solved almost as easily as a simulation run. Apart from any computational advantages, the main motivation for developing the implicit method was its natural relation to the loop based simulator and hence the convenient possibility of including controlling elements which may be in unknown states to be detected by the state-estimator.

REFERENCES

[1] Andersen, J.H. Powell, R.S. 'Simulation of Water Networks containing Controlling Elements'. Journal of Water Resources Planning and Management, ASCE May/June 1999, vol 125, Issue 3.

[2] Bargiela, A.B. 'On-line monitoring of water distribution networks' PhD Thesis, University of Durham, 1984

[3] Brdys, M.A., Ulanicki, B., 1994, *Operational Control of Water Systems: Structures, Algorithms and Applications*, (Prentice Hall, UK).

[4] Golub, G.H. Van Loan C.F. 1983 Matrix Computations. North Oxford Academic.

[5] Powell, R.S., Irving, M.R., Sterling, M.J.H., 1988, A comparison of three real-time state-estimation methods for on-line monitoring of water distribution systems. In B. Coulbeck (ed.) *Computer Applications in Water Supply, Vol. 1*, (Research Studies Press Ltd.), pp 333-348.

Dispersion of Mass in Intermittent Laminar Flow through a Pipe

S. G. Buchberger, Y. Lee, L. G. Bloom *and* B. W. Rolf

ABSTRACT

It is hypothesized that the mean effective axial hydrodynamic dispersion coefficient for random intermittent laminar pipe flow can be written,

$$E[K_I]=aE[K]^b,$$

where $a>1$ is a physically-based parameter, $0<b\leq1$ is an empirical regression exponent, and $K=r^2U^2/48D$ is Taylor's (1953) hydrodynamic dispersion coefficient for a soluble substance with molecular diffusion coefficient D, moving in steady laminar flow at mean velocity U through a pipe of radius r. The parameter $a=k_Sk_R$ is shown to arise from the product of two terms. The first term ($k_S\geq1$) accounts for the stagnation effects of intermittent laminar flow while the second term ($k_R\geq1$) accounts for the inherent random variability of the flow velocity. An experimental program is underway to corroborate the proposed expression for axial dispersion under random intermittent laminar flow. Preliminary data from a few laboratory tracer runs are presented.

1 INTRODUCTION

Peripheral zones of municipal water distribution systems often contain dead-end pipes. Dead-end pipes are supply lines that have only one connection to the looped portion of the water distribution system. Except for extraordinary circumstances (e.g., main break, fire flows, etc), water moves through a dead-end pipe in one direction only, along a path from the entrance to the point of withdrawal, in response to local downstream demands.

Even though the flow direction is known, water movement in dead-end mains is complex. Owing to the sporadic unpredictable nature of consumer demands, water in dead-end lines can be stagnant for long periods of time. When demands for water are made, the flow occurs as a burst of random duration and random intensity. Based on the nominal Reynolds number, the corresponding flow often is in the laminar regime.

Predicting water quality between the points of treatment and consumption in a distribution system is an important and difficult task (Clark et al., 1993). Not surprisingly, water quality predictions tend to be less reliable and water quality problems more prevalent in dead-end regions of the service zone. Here travel times are long, flow conditions are intermittent, disinfectant residual is often low, and bacteria counts can be high.

Certainly water quality complications arise from various biological-physical-chemical interactions among constituents mixed in the bulk water and attached to the inside pipe walls. Equally troubling are problems introduced by the random intermittent laminar flow that prevails in dead-end supply lines. These hydraulic conditions are very different from the simple steady-state plug flow assumptions invoked in current conventional approaches to model water quality in distribution systems. To improve water quality predictions in distribution systems, there is a need to investigate mass transport in pipes under stochastic sporadic laminar flow conditions.

2 OBJECTIVES

This paper has two objectives. First, it demonstrates that the combined effects of intermittent pipe flow and random water demands can significantly amplify the effective axial hydrodynamic dispersion coefficient for mass transported in laminar flow. Some preliminary data from a few experimental runs are presented to illustrate this amplification effect. Finally, there is a brief discussion on the implications of this finding for modeling water quality in municipal distribution systems, especially when applied to dead-end service lines.

3 ANALYTICAL METHOD

3.1 Steady Laminar Flow

In a classic paper, Taylor (1953) demonstrated that the concentration of soluble material injected into steady laminar flow eventually assumes a Gaussian distribution with distance along the pipe axis. The spreading of the tracer cloud with respect to the mean flow velocity U is characterized by an effective dispersion coefficient

$$K = \frac{r^2 U^2}{48D} \tag{1}$$

where r is the pipe radius and D is the molecular diffusion coefficient of the solute mass in water.

Dispersive transport in steady laminar flow arises primarily from the nonuniform (parabolic) velocity profile. Aris (1956) showed that the effects of molecular diffusion could be added to the expression in (1), but practically speaking, this contribution is dwarfed by spreading due to hydrodynamic dispersion. Analogous results showing a dependence between hydrodynamic dispersion and the square of the mean velocity have been tabulated for other instances of steady laminar flow (Fisher et al., 1979 p 93).

Following injection of the solute mass, a certain period of time t must elapse before the solute concentration attains a Gaussian distribution. In dimensionless terms, the Taylor initialization period can be expressed

$$\frac{Dt}{r^2} = T \tag{2}$$

Chatwin (1970) showed that the dispersing tracer cloud will be Gaussian provided T>1; Fisher et al (1979) indicate that the Gaussian distribution can be used with reasonable accuracy for T>0.40.

3.2 Periodic Intermittent Laminar Flow

Intermittent flow has periods of moving water punctuated by stagnant intervals of no flow. It is convenient to call periods of no flow "idle time" and periods with flow "busy time". Periodic intermittent flow means that the cycle of idle times and busy times follows a regular pattern.

For now it is assumed that all busy period flows are steady and equal. In the next section this assumption will be relaxed to allow intermittent flow that occurs as bursts of random duration and random intensity. Laminar flow implies the nominal Reynolds number ($\mathbf{R}=Ud/\nu$) is under 2,000 during the busy times.

Intermittent flow can be characterized several ways. It is especially important to make a distinction between *elapsed-time* and *busy-time* conditions. Elapsed time includes the complete sequence of all idle times and busy times. In contrast, busy-time refers only to periods when water is moving. If idle time does not occur, then elapsed-time and busy-time conditions are the same.

Any particular elapsed-time flow rate can correspond to many different busy-time scenarios. For example, an elapsed-time flow rate of 1 liter per minute can be generated by a sequence of busy-time pulses each 30 seconds in duration (with an intensity of 2 liter per minute) followed by 30 seconds of idle time, or by a sequence of pulses each 15 seconds in duration (with an intensity of 4 liter per minute) followed by 45 seconds of idle time. These three cases, each with identical elapsed-time mean flow rates, are illustrated in Figure 1.

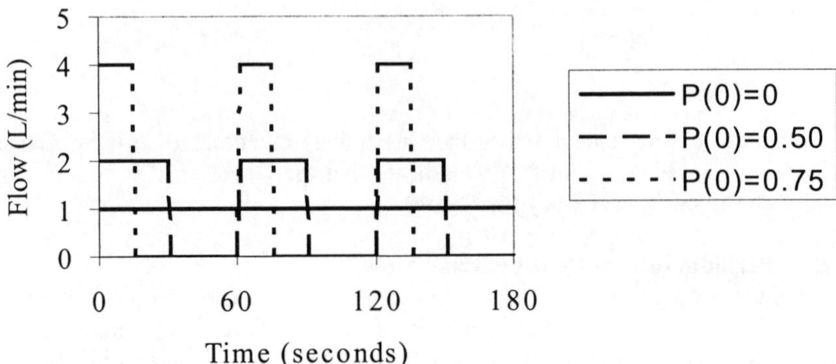

Figure 1. Three different busy-time flow conditions (two are periodic intermittent and one is steady) generate an identical elapsed-time mean flow of Q = 1 L/min.

To compare various busy-time scenarios (each having different degrees of intermittent flow), it is helpful to consider cases having identical elapsed-time flow rates. Let P(0) represent the percentage of time that the flow is stagnant. Then, for identical elapsed-time flow rates, the busy-time velocity U_B is related to the elapsed-time mean velocity U as follows

$$U_B = \frac{U}{1 - P(0)} \tag{3}$$

One effect of intermittent flow is to increase the busy time velocity by a factor of 1/[1-P(0)] compared to the elapsed-time steady flow. For example, if P(0)=50%, then U_B=2U. Hence, during a busy period, the intermittent discharge must have twice the velocity as the steady case to sweep the same volume through the same pipe during a given time interval.

During idle times, the velocity is zero and hydrodynamic dispersion does not occur. Molecular diffusion will continue, but its effect is negligible. Practically speaking then, the solute mass in intermittent flow can be considered frozen in place during idle times. During busy times, the average velocity is given by (3) and hydrodynamic dispersion resumes. It is conjectured, therefore, that a dispersion coefficient for intermittent laminar flow K_I can be written by analogy with equation (1)

$$K_I = \frac{r^2 U_B^2}{48D} = \frac{1}{[1 - P(0)]^2} \frac{r^2 U^2}{48D} = k_S^2 K \tag{4}$$

where K is given by (1) and k_S is a dimensionless coefficient accounting for the stagnation effects of intermittent flow on hydrodynamic dispersion,

$$k_S = [1 - P(0)]^{-1} \tag{5}$$

Since hydrodynamic dispersion in steady laminar flow varies with U^2, the anticipated effect of periodic intermittent flow is to amplify the mean flow dispersion by a factor of $1/[1-P(0)]^2$. *Therefore, for identical elapsed-time flow conditions, the spreading of the tracer cloud is expected to be greater under intermittent flow than under steady flow.*

3.3 Random Intermittent Laminar Flow

In dead-end regions of real water distribution systems, busy times and idle times do not follow a predictable cycle. Instead they display an irregular pattern driven by random demands imposed by residential consumers and other users. Considering the unpredictable nature of the busy-time pipe velocities, equation (4) should be expressed as the expected value of the hydrodynamic dispersion coefficient for random intermittent laminar pipe flow

$$E[K_I] = \frac{r^2 E[U_B^2]}{48D} = \frac{1}{1-P(0)} \frac{r^2 E[U^2]}{48D} = \frac{1+\Theta_U^2}{1-P(0)} \frac{r^2 E[U]^2}{48D} \tag{6}$$

In this expression, Θ_U is the coefficient of variation of the elapsed-time velocity and P(0) should be interpreted as the probability of zero flow in the supply line. Because virtually all field data are expressed in terms of elapsed-time conditions, it is convenient to define the expected value of the hydrodynamic dispersion coefficient based on an elapsed-time mean velocity

$$E[K] = \frac{r^2 E[U]^2}{48D} \tag{7}$$

Now (6) can be written

$$E[K_I] = \frac{1+\Theta_U^2}{1-P(0)} E[K] = k_R k_S E[K] \tag{8}$$

where k_S is defined in (5) and k_R is a dimensionless term accounting for the effects of random flow variability on the hydrodynamic dispersion coefficient

$$k_R = 1 + \Theta_U^2 \tag{9}$$

Equation (8) shows that intermittent and random conditions have a multiplicative effect on the hydrodynamic dispersion coefficient. Note that if the busy and idle cycles are periodic and the busy time flows are constant, then k_R=k_S and equation (8) reduces to (4).

To gain more insight into (8), suppose that residential water use behaves like a homogeneous Poisson Rectangular Pulse (PRP) process (Buchberger and Wu, 1995). Under the PRP hypothesis, water users at a typical home arrive at a rate λ and draw water for an average duration τ. Let $\rho=\lambda\tau$ and assume that water demands at a busy server have a coefficient of variation Θ_W. It can be shown that the coefficient of variation of the elapsed-time flow velocity is

$$\Theta_U^2 = -\frac{1+\Theta_W^2}{\ln[P(0)]} \tag{10}$$

and hence, under the PRP model for residential water use, equation (8) becomes

$$E[K_I] = \left(\frac{1}{1-P(0)}\right)\left(1-\frac{1+\Theta_W^2}{\ln[P(0)]}\right)E[K] \tag{11}$$

Dimensionless dispersion ratios, $K^*=E[K_I]/E[K]$, from equation (4) and equation (11) (with $\Theta_W=0.50$) are plotted in Figure 2 as a function of P(0). Even if $\Theta_W=0$, the value of K^* is always greater for random intermittent flow than for deterministic intermittent flow. The randomness in the PRP model is introduced by variability in water demands at a busy home *and* by variability in the number of busy homes along a dead-end line.

In dead-end zones of water distribution systems, it is not unusual for P(0) to exceed 0.90, especially along the supply line to the last few homes. If (4) or (11) are valid, then mass transport by hydrodynamic dispersion during intermittent flows in dead-end lines could be more than 100 times greater than that predicted by an elapsed-time mean flow analysis (see Figure 2). Indeed, Axworthy and Karney (1996) indicate that dispersive transport can be important in dead-ends mainlines, even when working only in terms of elapsed-time flow conditions.

The PRP model for residential water use leads to further simplifications in equation (11) under certain conditions. It can be shown that the probability of idle time for a dead-end main line supplying N homes is $P(0)=\exp(-N\rho)$. In remote service zones where the quantity $N\rho$ is small (say < 0.10), $1-P(0)=1-\exp(-N\rho)\approx N\rho$ and therefore (11) becomes

$$E[K_I \mid N\rho < 0.10] \approx \frac{1+\Theta_W^2}{N^2\rho^2}E[K] \tag{12}$$

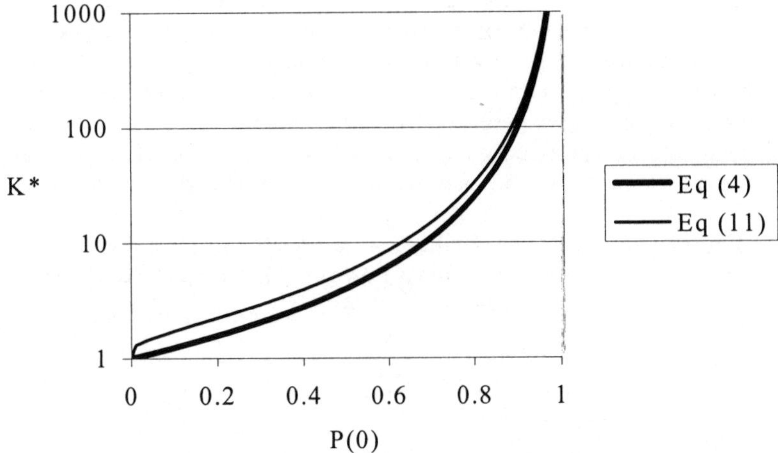

Figure 2. Variation in the effective axial hydrodynamic dispersion coefficient for intermittent pipe flow as a function of the percentage of idle time, P(0). K* is the ratio of the dispersion coefficient for intermittent conditions to the dispersion coefficient for steady conditions under identical mean elapsed-time flow rates.

Hence, it seems plausible that the effective hydrodynamic dispersion coefficient for random intermittent laminar pipe flow can be estimated from measurable statistical characteristics of residential water demands (ρ, Θ_W), the number of homes supplied by the dead-end mainline (N), and the mean elapsed-time estimate of hydrodynamic dispersion (E[K]).

4 EXPERIMENTAL METHOD

4.1 Experimental Facility

An experimental facility was constructed to provide controlled simulation of steady and intermittent flows through a dead-end pipe. The experimental facility has five components: (1) constant head tank containing distilled water, (2) long straight section of clear acrylic circular tubing (length, L=750 cm; radius, r=0.635 cm), (3) plane source tracer injection device, (4) flow control and measurement instrumentation (fluorometer, solenoid valves, differential pressure gage) and (5) computer controlled data acquisition and recording system (see Figure 3).

At the start of each run a known mass of Rhodamine WT tracer was injected as a plane source impulse into steady laminar flow moving through the acrylic tube. The concentration of the dispersing tracer cloud was measured by a Turner 111 fluorometer at a single downstream monitoring station. Water exited from the tube and was collected in an open vessel linked to a differential pressure gage, which was used to determine the flow rate. Tracer concentrations and flow rates were recorded versus time on a one second interval. During steady flow runs, the solenoid valves remained in the open position. For intermittent flows, the solenoid valves were programmed to open and close at a specified periodic cycle.

The experimental facility has the capability to replicate random demands resulting from typical intermittent indoor residential water use. At the time of this writing, however, the experimental program is still in the start-up phase. This paper, therefore, reports only on some preliminary findings from two runs made under steady laminar flow and three runs under periodic intermittent laminar flow.

Figure 3. Experimental facility to measure effective dispersion coefficient in random intermittent laminar flow (from Bloom and Rolf, 1999)

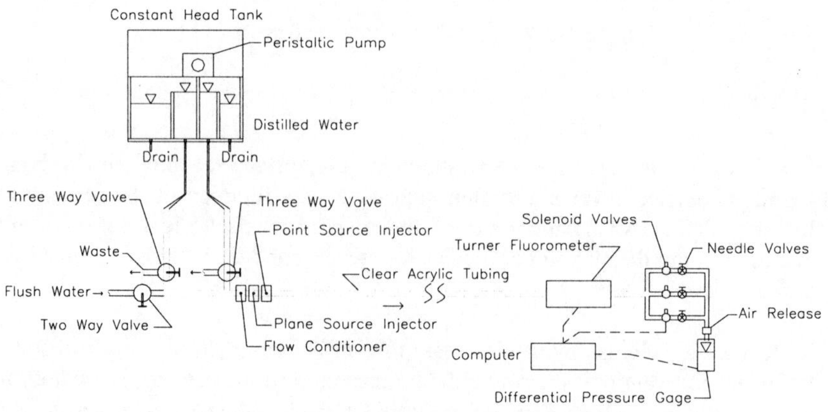

4.2 Experimental Analysis

The kth temporal moment of the tracer cloud is given by

$$m_k = \int_0^{\infty} t^k C(t)\, dt \tag{13}$$

where C(t) be the tracer concentration measured by the fluorometer at time t after the injection. The centroid of the tracer cloud (mean travel time) is

$$\mu = \frac{m_1}{m_0} \tag{14}$$

and the temporal variance of the tracer cloud is given by

$$\sigma^2 = \frac{m_2}{m_0} - \mu^2 \tag{15}$$

Estimation of the hydrodynamic dispersion coefficient is based on the assumption that K measures the rate of change of the variance of the tracer cloud (Fisher, 1968). One approximation recently used by Bello et al (1994) is based on work by Levenspiel and Smith (1957)

$$K_I = \frac{1}{4} U_B^2 \mu \left[-1 + \sqrt{1 + \left(\frac{2\sigma}{\mu} \right)^2} \right] \tag{16}$$

Equation (16) was applied to several experimental tracer clouds to estimate the hydrodynamic dispersion coefficient during steady and periodic intermittent flows.

5 RESULTS

Measured tracer cloud concentrations from two runs having identical mass injections and time-averaged discharges are plotted against time in Figure 4. The upper graph was obtained during a steady continuous run. The lower graph was obtained during an intermittent run with P(0)=50% and a busy pulse duration of τ=60 seconds. In the lower graph, only data obtained during the busy period are shown. The concentrations recorded during the idle times were omitted because these readings would otherwise *artificially* spread the tracer cloud shown in Figure 4 and, hence, confound the dispersion analysis.

Both tracer clouds show a characteristic steep leading edge and a long falling tail. Skewness is pronounced, but this is to be expected when measuring a dispersing cloud as it passes a single downstream station. Even though both tracer clouds correspond to identical elapsed-time mean flow rates, there is a marked difference between the steady flow and intermittent flow responses.

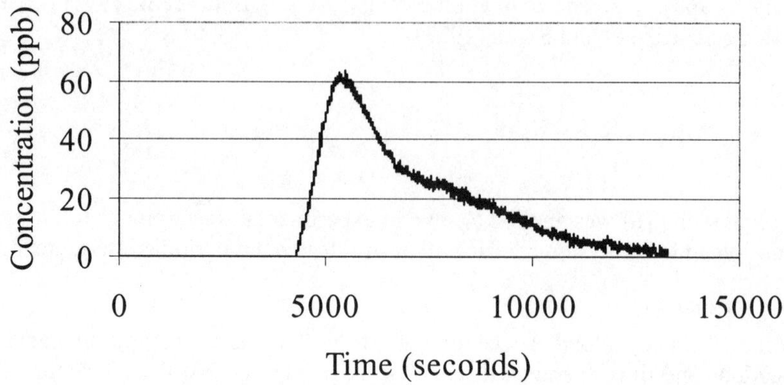

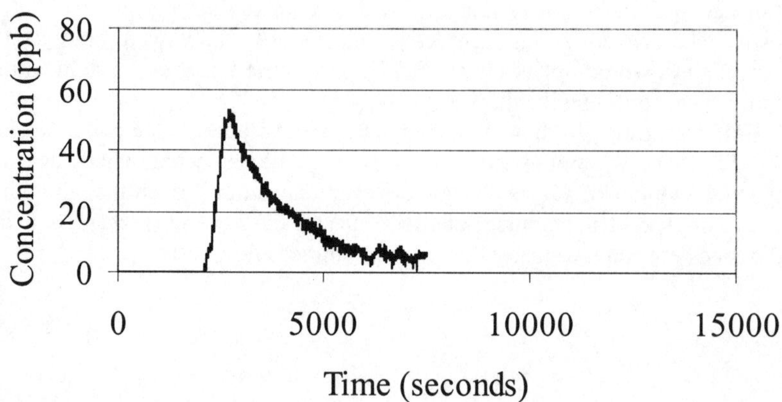

Figure 4. Busy-time tracer clouds for steady run S-1 (top) and intermittent run I-1 (bottom) from two experiments having identical time-averaged laminar flows.

Preliminary results from five experiments (two steady runs and three intermittent runs) are summarized in Table 1 for a mean discharge of Q=0.127 mL/sec and a Reynolds number, **R**=12.7. The duplicate runs (steady pair S-1 and S-2; intermittent pair I-1 and I-2) show good consistency. As expected, for a given mean elapsed-time flow rate, the estimated value of the dispersion coefficient increases with an increase in the percentage of idle time.

If equation (4) holds, then the intermittent flow case with P(0)=50% should lead to a 4-fold increase in dispersion over the steady flow case. Preliminary results in Table 1 indicate that the effective hydrodynamic dispersion coefficient increased by a factor ranging from K*=3.3 to 3.5. This range of values is within

85% of the conjectured increase. The lone case (run I-3) with P(0)=75% should lead to a 16-fold increase in dispersion. However, the estimate of the dispersion coefficient in Table 1 shows only a 6 to 7-fold increase over the steady flow result, considerably lower than expected.

Table 1. Preliminary estimates of dispersion coefficients from experimental tracer runs at a mean discharge, Q=0.127 mL/sec and Reynolds number, **R**=12.7.

(1)	(2)	(3)	(4)	(5)	(6)	(7)	(8)	(9)
Run No	P(0) (%)	τ (sec)	U_B (cm/sec)	μ (sec)	σ (sec)	T (--)	K_I (cm^2/sec)	K* (--)
S-1	0	S	0.10	6,996	1,883	0.87	2.37	1.00
S-2	0	S	0.10	7,128	1,890	0.88	2.35	1.00
I-1	50	60	0.20	3,851	1,277	0.48	7.70	3.27
I-2	50	60	0.20	3,947	1,338	0.49	8.22	3.48
I-3	75	600	0.40	1,918	631	0.24	15.13	6.41

Notes for Table 1:
(1) S is steady flow, I is intermittent flow.
(2) P(0) is the percentage of time the flow is idle.
(3) τ is the duration of busy-time water pulse.
(4) U_B is the busy-time flow velocity.
(5) μ is the centroid of the tracer cloud (from equation 14).
(6) σ is the standard deviation of the tracer cloud (from equation 15).
(7) T is the Taylor initial period ($T=D\mu/r^2$, with $D=5\times10^{-5}$ cm^2/sec).
(8) K_I is the effective dispersion coefficient (from equation 16).
(9) K* is the ratio of intermittent to steady flow dispersion coefficient, ($K^*=K_I/K$).

6 CONCLUSIONS

The preliminary results are encouraging, but it must be emphasized that the sample size is too small to draw meaningful conclusions. The discrepancy between predicted and computed dispersion coefficients in run I-3 can be attributed to many factors. Beyond examining sources of experimental bias (e.g., premature truncation of the tracer cloud and other issues), future laboratory work will focus on the potential transient side-effects of intermittent flow. Questions to address include: (a) Are the busy periods too short to allow full development of the parabolic laminar velocity profile (Bird et al., 1960, p 129)? (b) Does the stop and go flow behavior influence the length of the initial mixing period?

To account for intermittent flow effects, it may be necessary to generalize equation (8) as follows,

$$E[K_I] = aE[K]^b \tag{17}$$

where $a=k_S k_R$ is a physically-based dispersion amplification factor and $0<b\leq1$ is an empirical regression exponent. At present, even though experimental data are sparse, observed trends seem reasonable. For the same mean discharge, mass

dispersion in laminar flow is greater under intermittent conditions than under steady conditions. Lab runs are underway now to corroborate this notion. If confirmed, dispersive transport could be an important consideration in modeling water quality through dead-end zones of municipal distribution systems.

ACKNOWLEDGMENTS

Partial support for this work was provided through a seed grant from the University Research Council at the University of Cincinnati. The fluorometer was loaned by the Technical Support Center of the USEPA Breidenbach Environmental Research Center in Cincinnati, Ohio. We extend our sincere thanks to Robert Fitzpatrick and Robert Meunch for their assistance in constructing the experimental facility.

REFERENCES

Aris, R. (1956) *On the dispersion of a solute in a fluid flowing through a tube*, **Proceedings of the Royal Society of London Series A**, 235:67–77.

Axworthy, D.H. and B.W. Karney (1996) *Modeling low velocity/high dispersion flow in water distribution systems*, **ASCE Journal of Water Resources Planning and Management**, 122(3):218-221.

Bello, M.S., R. Rezzonico and P.G. Righetti (1994) *Use of Taylor-Aris dispersion for measurement of a solute diffusion coefficient in thin capillaries*, **Science**, Volume 226, 773-776.

Bird, R.B., W.E. Stewart and E.N. Lightfoot (1960) **Transport Phenomena**, John Wiley and Sons, New York, 780 pages.

Bloom, L.G. and B.W. Rolf (1999) *Experimental study of dispersion in intermittent laminar flow*, **Senior Design Project**, Department of Civil and Environmental Engineering, University of Cincinnati, Cincinnati, Ohio.

Buchberger, S.G. and L. Wu (1995) *Model for instantaneous residential water demands*, **ASCE Journal of Hydraulic Engineering**, 121(3):232-246.

Chatwin, P.C. (1970) *The approach to normality of concentration distribution of a solute in solvent flowing along a straight pipe*, **J Fluid Mechanics**, (43):321-352.

Clark, R.M., W.M. Grayman, R.M. Males, A.F. Hess (1993) *Modeling contaminant propagation in drinking water distribution systems*, **ASCE Journal of Environmental Engineering**, 114(4):454-466.

Fischer, H.B. (1968) *Dispersion predictions in natural streams*, **ASCE Journal of Sanitary Engineering**, 94(5):927-943.

Fischer, H.G., E.J. List, R.C.Y. Koh, J. Imberger and N.H. Brooks (1979) **Mixing in Inland and Coastal Waters**, Academic Press, San Diego, 483 pages.

Levenspiel, O. and W.K. Smith (1957) *Notes on the diffusion-type model for the longitudinal mixing of fluids in flow*, **Chemical Engineering Science**, (6):227-233.

Taylor, G.I. (1953) *Dispersion of soluble matter in solvent flowing slowly through a tube*, **Proceedings of the Royal Society of London Series A**, 219:186–203.

Taylor, G.I. (1954) *The dispersion of matter in turbulent flow through a pipe*, **Proceedings of the Royal Society of London Series A**, 223:446–468.

LIST OF SYMBOLS

a physically-based dispersion amplification parameter, $a=k_R k_S$
b empirical regression exponent
d pipe diameter (cm)
D molecular diffusion coefficient (cm^2/sec)
k_R parameter accounting for effect of random flows on dispersion coefficient
k_S parameter accounting for effect of stagnant time on dispersion coefficient
K dispersion coefficient for steady laminar flow (cm^2/sec)
K_I dispersion coefficient for intermittent laminar flow (cm^2/sec)
L pipe length (cm)
m_k k^{th} moment of the tracer cloud
N number of homes on dead-end stem
P(0) percentage of idle time
PRP Poisson rectangular pulse
Q mean flow rate (mL/sec)
r pipe radius (cm)
R Reynolds number, $\mathbf{R}=Ud/\nu$
t time (sec)
T initialization time for Taylor mixing, $T=Dt/r^2$
U elapsed-time mean flow velocity (cm/sec)
U_B busy-time flow velocity (cm/sec)

α average intensity of indoor water use at a busy server (mL/sec)
ρ utilization factor for indoor residential water use, $\rho=\lambda\tau$
λ average arrival rate of water user (sec^{-1})
μ centroid of tracer cloud (sec)
ν kinematic viscosity (cm^2 / sec)
σ^2 variance of the tracer cloud (sec^2)
τ mean duration water use pulse (sec)
Θ_U coefficient of variation of velocity in dead-end mainline
Θ_W coefficient of variation of indoor water use at a busy server

A Dynamic Hydraulic Model for Water Distribution Network Simulation

E. J. Dunlop

ABSTRACT

An overview is provided of the simulation engine contained in the new release of WADI (WADI-II), a program for dynamic (extended period) simulation of water distribution networks under development by the author. Features of WADI-II include: (i) Simulation of momentum effects for water in the pipes and (ii) De-coupled (or buffered) demand loading. Specifically, the differential equations of motion for water in the pipes of the network - ignoring compressibility - are solved. This is commonly known as "mass surge analysis". A modern Backward Differentiation Formula (BDF) implicit differential equation (i.d.e.) algorithm is applied to this problem.

The repeated formation, factorisation and application of the "Newton iteration matrix" constitutes a major component of the total computational effort required to run a BDF-based dynamic network simulation. A significant improvement in the efficiency of these operations will have a corresponding beneficial impact on the run-time of the full simulation process.

In WADI-II, the effective factorisation of the asymmetric Newton iteration matrix, which is $\geq (4.N)x(4.N)$ in size, is reduced to that of an NxN matrix which has a similar sparsity structure to the familiar nodal matrix and only slight asymmetry; the asymmetry is accommodated by application of Bennett's algorithm.

To account for the effect of storage at demand nodes, and the reality that network loading is a pressure driven forcing function, demand flows are modelled by de-coupling them from the network. "De-coupling" is accomplished by means of a pipe feeding a (notional) tank through a float valve - consumer demand being supplied from the tank. This demand model is particularly applicable to networks under stress, e.g. old overloaded networks, networks under maintenance (reduced capacity), fire demand loading conditions, etc. Representation of the pressure dependency and the re-phasing of demand is implicit in the de-coupled demand model. General details of the approach (problem formulation and solution method) are outlined in the paper.

1 INTRODUCTION

The objective of this paper is to give an overview of the simulation engine contained in the next release of WADI (WADI-II), a Fortran program for dynamic simulation of water distribution networks under development by the author. Features of WADI-II include:-

(i) Modelling of momentum effects for water in the pipes
(ii) De-coupled (or buffered) demand loading

A modern BDF implicit differential equation algorithm is applied to the problem of extended period simulation of the water network. Specifically, the differential equations of motion for water in the pipes of the network - ignoring compressibility – are solved. This is commonly known as "mass surge analysis", and it involves significantly less computational effort than a "water hammer analysis" which includes compressibility.

It is believed that, for the dynamic simulation of a typical urban network, which may involve several hundred kilometers of pipework with average diameter greater than 200 mm, momentum effects will not always be negligible. Such effects are ignored completely in conventional "extended period simulation" models, as these are implemented with repeated "steady-state" analyses – the so-called pseudo-dynamic approach. (Since pipe time constants, and hence response times, are large when flow tends to zero - typically 20 minutes, momentum effects may be particularly relevant in networks where flow reversal occurs within the network and when a quality model is to be built on top of the hydraulic model.)

Various methods for the mass surge analysis of networks have been described in the past (15,19). The advantages of such "dynamic" simulation have been pointed out by Cohen (20). However, the method has not become popular, presumably because of its significant or perceived computational overhead compared with conventional "extended period" simulation. In this paper, it will be shown that this overhead can be greatly reduced through the application of some native cunning.

De-coupled demands are used because although demand is <u>volume</u> oriented (e.g. "fill the cistern," "fill the bath" etc.), it is satisfied by a <u>pressure</u> loading on the distribution network - the consumer neither controls nor specifies flow rate. The specific pressure loading on the network is the datum level at the tap; this head is applied until the desired volume has been supplied and the tap is closed.

For the small region supplied by a network node, such demands can be "averaged" to make a demand flow profile, which is applied as a direct demand (or flow rate) loading. This is the conventional approach, but it does not represent the true physics of the situation.

To account for the effect of storage at demand nodes, and the reality that demand on a network is a pressure driven forcing function, demand flows can be modelled by de-coupling them from the network as shown in Figure 1.

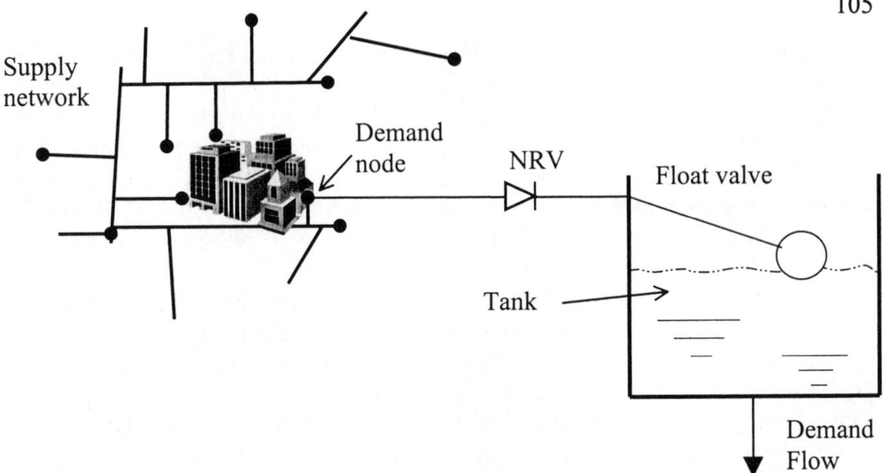

Fig. 1 De-coupled Demand Flow

If the demand flow exceeds the ability of the network to deliver that demand, the network will supply what it can, and the required flow will be "supplied" to the consumer from storage (or from notional storage - with an associated waiting time for the demand to be satisfied as the tank refills).

This demand model is particularly applicable to networks under stress, e.g.

- old overloaded networks
- networks with large industrial consumers
- networks under maintenance (reduced capacity)
- fire demand loading conditions
- rural networks with long pipe runs and large localised demands - probably fed from storage (e.g. milking parlours etc.)

The results will indicate realistic flows and pressures (e.g. waiting times for demand to be met instead of "negative" pressures; better modelling of the region around a heavily loaded node).

For networks that are able to supply the required demand easily, the float valve will balance the tank input and output flows, and the results will be similar to those of conventional models.

Representation of the pressure dependency and phasing of consumer demands is implicit in the de-coupled demand model; (i.e. demand will increase with pressure until a critical saturation point is reached, at which stage demand variation will cease – pressure dependency; also, water consumed from storage should normally be supplied from the network later - phasing). The modelling of other pressure dependent demands (e.g. leakage) will be described below.

Above all however, for a dynamic analysis which includes momentum effects, representation of the physics implied by de-coupled demand flows is essential if the procedure is to deal realistically with arbitrary demand flow profiles, (particularly "stepped" profiles).

The degree of realism of the truly "dynamic" model will be equal to or better than that of more conventional models, which use only "steady state" analysis repeated at, say, an hourly interval. This is because the full momentum effects are included in the dynamic model, and because the range of conditions for which the model is realistic is further extended as a benefit of the de-coupled demand arrangements. Also, the dynamic model provides the option of transient analysis subject to the limitations implicit in "mass surge analysis", and it can readily be extended to incorporate boundary conditions such as surge tanks. However, the objective in WADI-II is just to replicate and improve on the realism/functionality of normal "extended period simulation" software.

For a "gradually varied flow" regime, the effect of compressibility (ignored in the "mass surge analysis" model) is assumed to be negligible. This condition applies in normal circumstances with a well-run network. Calibration of the model is facilitated, since the degree of detail and accuracy of the model corresponds to that of simple instrumentation, which will monitor/log network data at, say, 1min/15min intervals. (Calibration for a "water hammer analysis" would require more extensive monitoring equipment to observe/record network behaviour, and the quantity of information provided from both the instrumentation and the model would be excessive for medium or large sized networks. Such analyses are most appropriate for small <u>sections</u> of a network where pump failure or other fault conditions could cause large transient problems.)

Ordinary differential equations arise from pipe dynamics (momentum) and reservoir/tank dynamics (filling/emptying); implicit differential equations, arise from the nodal flow-balance condition and direct nodal pressure dependent demand (e.g. leakage). The co-existence of tanks/float-valves, and of pipes with widely differing lengths and diameters in the same network, gives rise to very long and very short relative time constants (or response times) for different elements of the network. Furthermore, we normally require outputs only at, say, an hourly interval - which is likely to be very long relative to typical pipe time constants (say 5 seconds under normal flow conditions). Hence, the system of non-linear implicit differential equations representing the network dynamics will best be solved with a good "stiff" i.d.e. solver.

Using simple numerical integration schemes (e.g. straightforward Runge Kutta or Predictor-corrector) it will be necessary to use a time step which is small compared to the shortest time-constant in the network; otherwise, the procedure will be unstable and produce inaccurate results. However, a relatively short time step will mean a very lengthy sequence of iterations to fully represent the behaviour of those elements with long time constants, and to cover the required time interval, (say 24 hours). However, algorithms have been devised which can deal effectively with stiff i.d.e.'s. (24,25). Suitable i.d.e. solvers will most probably be implementations of the predictor-corrector algorithm using the BDF of variable order (usually 1 to 5); this is commonly known as Gear's method.

A very useful facility, built into some i.d.e. solvers, is the automatic location of the zeros of defined "root functions" which can correspond to, say

- flows about to go negative through an open NRV
- pressure drop about to go positive across a closed NRV

- upstream pressure about to drop below set head on an active Pressure Reducing Valve
- etc. etc.

i.e. points where network status changes occur that will modify the i.d.e.'s. If the chosen i.d.e. solver does not incorporate such a facility, it must be provided with some additional programming effort.

Essentially, BDF algorithms reduce computational effort by permitting large step sizes without loss of stability; the cost is significant additional complexity. The application of such algorithms requires the formation of certain vectors and matrices, and the solution of equations using these matrices.

For networks with N nodes and P pipes it is necessary repeatedly to form, factorise and solve equations using a large non-symmetric Newton iteration matrix – an unavoidable and essential part of the BDF procedure. N nodes implies N flow balance equations, N^+ de-coupled demand tanks with a corresponding N^+ pipes/float-valves connecting the demand tanks and P pipes which form the basic network; hence there will be N i.d.e.'s and $(2.N^++P)$ o.d.e.'s to be solved for our formulation. For instance, if two demand tanks (ground level and upper story) were supplied from each demand node, then $N^+ = 2.N$ and the Newton iteration matrix would be of order (5.N+P) x (5.N+P).

The repeated formation, factorisation and application of the Newton iteration matrix constitutes a major component of the total computational effort required to complete a BDF-based dynamic network simulation. A significant improvement in the efficiency of these operations will have a corresponding beneficial impact on the run-time of the full simulation process.

In WADI-II, a special strategy is adopted to process the large Newton iteration matrix with the greatest possible efficiency. This strategy will be described in detail.

In particular, the effective factorisation of the Newton iteration matrix is reduced to the factorisation of an NxN (only) matrix, which has a sparsity structure similar to the familiar NxN nodal matrix and only slight asymmetry.

We shall describe the WADI-II model as follows –

(i) Mathematical representation of main network elements, i.e. the differential equation for flow in a pipe.
(ii) The network incidence matrix, used to describe the behaviour of the network elements when they are connected as an integrated system.
(iii) Formation of the implicit differential equations for the network
(iv) The Newton iteration matrix [**W**], as required for Gear's method of solving implicit differential equations.
(v) An efficient procedure to solve the linear system

$$[\mathbf{W}]\,\{\mathbf{x}\} = \{\mathbf{b}\}$$

which takes advantage of the special structure of [**W**].

(vi) Determining consistent starting values for the i.d.e. variables and their derivatives

2 MATHEMATICAL REPRESENTATION OF NETWORK ELEMENTS

Only the differential equation for flow in a pipe and its time constant, and the modelling of leakage will be described in detail. The other network elements and modelling approaches, which are modelled in WADI-II but not described in this paper, are: minor losses, non-return valves, flow limiter valves, nodal head controllers (e.g. Pressure Reducing Valve's, Pressure Sustaining Valve's and Pressure Controlling Pump's), float valves, tanks, de-coupled demands.

2.1.1 The differential equation for flow in a pipe and its time constant

Consider the motion of fluid in a pipe of length L and X-sectional area A.

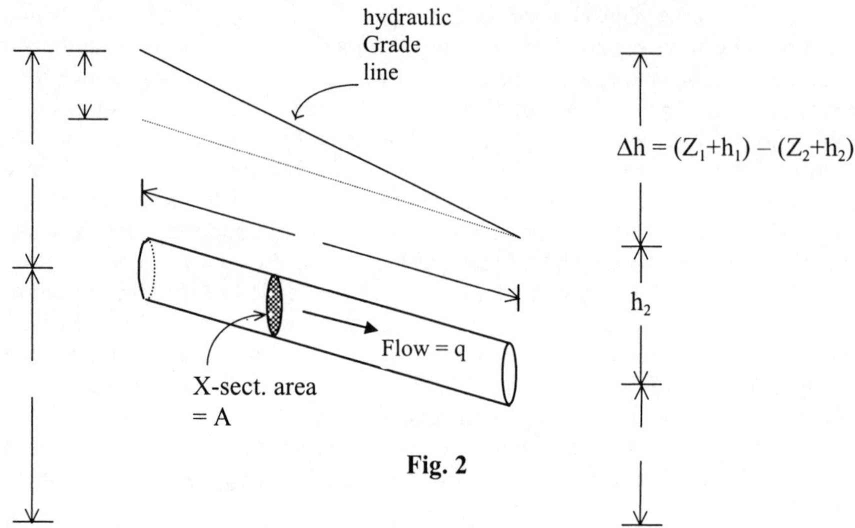

Fig. 2

Newton's second law:Force = Mass . Acceleration

Flow q gives rise to frictional head loss h_f,

so the head acting on the fluid = $\Delta h - h_f$.

Thus:

Force = $\rho.g.(\Delta h - h_f).A$

Mass = $\rho.A.L$

Acceleration = $dV/dt = (1/A).dq/dt$

Giving $\rho.g.(\Delta h - h_f).A = \rho.A.L.(1/A).dq/dt$

and $dq/dt = (A.g/L).(\Delta h - h_f) = \gamma.(\Delta h - h_f)$

This is a non-linear differential equation in q, since frictional head loss (h_f) is a non-linear function of flow (q).

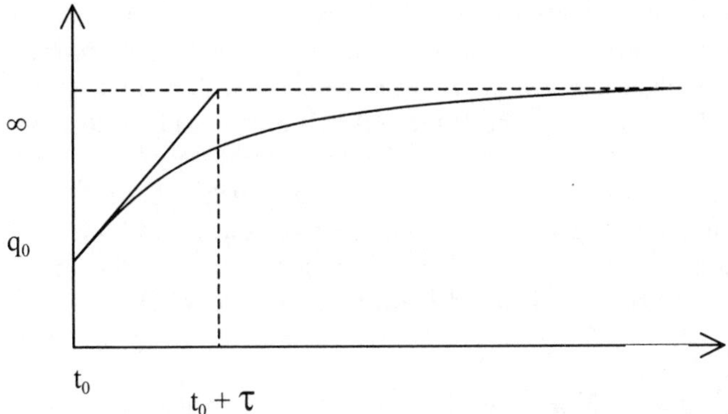

Fig. 3

It can be shown (15) that the time constant, τ, is given by

$$\tau = \left(\gamma \frac{dh_f}{dq}\right)^{-1} = \frac{L}{Ag} / \frac{dh_f}{dq}$$

$$= 0.5\pi D^3 / ((2f + q\frac{df}{dq}) \mid q \mid)$$

For fully turbulent flow (f = Constant), this reduces to

$$\tau = 0.25\pi D^3 / (f \mid q \mid)$$

$$= \sqrt{DL / (2gf \mid h_f \mid)}$$

and for laminar flow

$\tau = D^2/(32\nu)$

Thus as $|q|$ or $|h_f|$ gets small, τ gets large - subject to the limit under laminar flow conditions; e.g. under conditions of flow reversal, we must encounter laminar flow, and the time constant for a 300 mm pipe will peak at approx. 2500 seconds (or 42 mins). However, under more normal turbulent conditions ($| h_f |/L$ = 1‰, f = 0.03), the time constant will be only about 22 seconds for the same pipe.

2.1.2 Frictional head loss in pipes

In WADI-II the Darcy-Weisbach, (3,4) Colebrook-White (7,8,9) and Hagen-Poiseuille (1,2) formulae are used in the calculation of frictional head losses in pipes.

Friction factor behaviour in the critical flow region (2000<Re<4000) is modelled using cubic interpolation, which provides a smooth transition between laminar and turbulent flow. This is physically plausible, and assists numerical stability in our mathematical model because a gross discontinuity occurs if we switch from the turbulent to the laminar flow formula without the use of such a technique.

The code calculates both f and 2f+q(df/dq), which is needed in the network analysis calculations. It gives valid results for these quantities from Re=0 (laminar flow) upwards (fully turbulent flow). The compute time for this method is comparable with that of explicit approximations (11,12,13,14).

2.2 Leakage

It is assumed that at each demand node and reservoir the user will specify the leakage Q_L at a given head H_L, and its index of variation P; e.g. $Q_{LEAK} = Q_L(H/H_L)^P$ where the nodal head is H.

Alternatively, the loss per meter run Q_P for incident pipes can be specified, so a set of pipes of length L_j will contribute leakage $Q_{LEAK} = \left(\frac{1}{2}\Sigma Q_{pj}\,.\,L_j\right)\left(H/H_L\right)^P$

In general, we use

$$\begin{aligned} Q_{LEAK}(H) &= \phi(H) &&= K_{LEAK}.H^P && H>0 \\ & &&=0 && H\le 0 \\ \text{and } Q'_{LEAK}(H,H') &= \lambda(H).H' &&= K_{LEAK}.P.H^{P-1}.H' && H>0 \\ & &&=0 && H\le 0 \end{aligned}$$

Where K_{LEAK} for each node i given by

$$K_{leak} = \left(Q_L + \tfrac{1}{2}\sum Q_{pj}.L_j\right)/H_L^P$$

and the summation relates to pipes j incident on node i.

The index of leakage variation P will normally be between 0.5 and 1.5.

3 THE NETWORK INCIDENCE MATRIX

3.1 Definition of the incidence matrix

This well known matrix [**A**] has one row for each element of the network, and one column for each node. For element i (pipe, link or control device) we set the entries in the ith row of [**A**] to zero, except +1 for the column corresponding to the "start node" of element i and –1 for the column corresponding to the "end node" of element i.

We partition [**A**] into submatrices as shown below.

$$[A] = \begin{bmatrix} \mathbf{A}_{1,1} & \mathbf{A}_{1,2} & \mathbf{A}_{1,3} & \mathbf{A}_{1,4} & \mathbf{0} & \mathbf{0} & \mathbf{0} & \mathbf{0} & \mathbf{A}_{1,9} & \mathbf{A}_{1,10} \\ \mathbf{A}_{2,1} & \mathbf{A}_{2,2} & \mathbf{A}_{2,3} & \mathbf{A}_{2,4} & \mathbf{0} & \mathbf{0} & \mathbf{0} & \mathbf{0} & \mathbf{A}_{2,9} & \mathbf{A}_{2,10} \\ \mathbf{A}_{3,1} & \mathbf{A}_{3,2} & \mathbf{A}_{3,3} & \mathbf{A}_{3,4} & \mathbf{0} & \mathbf{0} & \mathbf{0} & \mathbf{0} & \mathbf{A}_{3,9} & \mathbf{A}_{3,10} \\ \mathbf{A}_{4,1} & \mathbf{A}_{4,2} & \mathbf{A}_{4,3} & \mathbf{A}_{4,4} & \mathbf{0} & -\mathbf{I} & \mathbf{0} & \mathbf{0} & \mathbf{A}_{4,9} & \mathbf{A}_{4,10} \\ \mathbf{0} & \mathbf{I} & \mathbf{0} & \mathbf{0} & \mathbf{0} & \mathbf{0} & -\mathbf{I} & \mathbf{0} & \mathbf{0} & \mathbf{0} \\ \mathbf{0} & \mathbf{0} & \mathbf{I} & \mathbf{0} & \mathbf{0} & \mathbf{0} & \mathbf{0} & -\mathbf{I} & \mathbf{0} & \mathbf{0} \\ \mathbf{0} & \mathbf{0} & \mathbf{0} & \mathbf{I} & -\mathbf{I} & \mathbf{0} & \mathbf{0} & \mathbf{0} & \mathbf{0} & \mathbf{0} \\ \mathbf{A}_{8,1} & \mathbf{A}_{8,2} & \mathbf{A}_{8,3} & \mathbf{A}_{8,4} & \mathbf{I} & \mathbf{0} & \mathbf{0} & \mathbf{0} & \mathbf{A}_{8,9} & \mathbf{A}_{8,10} \end{bmatrix}$$

The eight rows correspond to elements as follows:-

1 Pipes (general)
2 Float valve feeder links (tanks)
3 Float valve feeder links (reservoirs)
4 Float valve feeder links (demand tanks)
5 Links to tanks with float valves
6 Links to other tanks
7 Links to controlled nodes
8 Active control devices (PRV's, PSV's, PCP's)

The ten columns correspond to nodes as follows:-

1 General nodes
2 Associated nodes } Tanks with float valves
3 Other tanks
4 Controlled nodes
(1–4: Demand nodes)
5 Controlled nodes - {**H** ($\mathbf{q}_7$)}
6 Demand tanks- {**H** ($\mathbf{V}_6$)}
7 Tanks with float valves- {**H** ($\mathbf{V}_7$)}
8 Other tanks - {**H** ($\mathbf{V}_8$)}
9 Reservoirs with float valves- {**H**(t)}
10 Other reservoirs- {**H**(t)}

<u>Note</u> A "link" is a pipe whose length is short, i.e. approx. 1 to 10 times its diameter.

3.2 Application of the incidence matrix

3.2.1 [A] defines the continuity law at nodes

$$[\mathbf{A}]^{\mathrm{T}}\{\mathbf{q}\}+\{\mathbf{Q}_o\}=\{\mathbf{0}\}$$

element flows → $\{\mathbf{q}\}$; Nodal outflows → $\{\mathbf{Q}_o\}$

3.2.1.1 Demand node continuity

$$\begin{bmatrix} \mathbf{A}_{1,1} & \mathbf{A}_{1,2} & \mathbf{A}_{1,3} & \mathbf{A}_{1,4} \\ \mathbf{A}_{2,1} & \mathbf{A}_{2,2} & \mathbf{A}_{2,3} & \mathbf{A}_{2,4} \\ \mathbf{A}_{3,1} & \mathbf{A}_{3,2} & \mathbf{A}_{3,3} & \mathbf{A}_{3,4} \\ \mathbf{A}_{4,1} & \mathbf{A}_{4,2} & \mathbf{A}_{4,3} & \mathbf{A}_{4,4} \\ \mathbf{0} & \mathbf{I} & \mathbf{0} & \mathbf{0} \\ \mathbf{0} & \mathbf{0} & \mathbf{I} & \mathbf{0} \\ \mathbf{A}_{8,1} & \mathbf{A}_{8,2} & \mathbf{A}_{8,3} & \mathbf{I}+\mathbf{A}_{8,4} \end{bmatrix}^{\mathrm{T}} \begin{Bmatrix} \mathbf{q}_1 \\ \mathbf{q}_2 \\ \mathbf{q}_3 \\ \mathbf{q}_4 \\ \mathbf{q}_5 \\ \mathbf{q}_6 \\ \mathbf{q}_7 \end{Bmatrix} + \begin{Bmatrix} \mathbf{Q}_{\mathrm{LEAK}_1} \\ \mathbf{Q}_{\mathrm{LEAK}_2} \\ \mathbf{Q}_{\mathrm{LEAK}_3} \\ \mathbf{Q}_{\mathrm{LEAK}_4} \end{Bmatrix} = \{\mathbf{0}\}$$

or

$$[\mathbf{A}_1]^{\mathrm{T}}\{\mathbf{q}\}+\{\mathbf{Q}_{\mathrm{LEAK}}(\mathbf{H})\}=\{\mathbf{0}\}$$

Note: The summation of the last two columns of $[\mathbf{A}]^{\mathrm{T}}$ and the omission of $\{\mathbf{q}_8\}$ above follows from controlled node continuity (i.e. $\{\mathbf{q}_8\} = \{\mathbf{q}_7\}$).

3.2.1.2 Demand tank, tank and reservoir continuity

Again we sum the last two columns of $[\mathbf{A}]^{\mathrm{T}}$ to get

$$\begin{bmatrix} \mathbf{0} & \mathbf{0} & \mathbf{0} & \mathbf{A}_{1,9} & \mathbf{A}_{1,10} \\ \mathbf{0} & \mathbf{0} & \mathbf{0} & \mathbf{A}_{2,9} & \mathbf{A}_{2,10} \\ \mathbf{0} & \mathbf{0} & \mathbf{0} & \mathbf{A}_{3,9} & \mathbf{A}_{3,10} \\ -\mathbf{I} & \mathbf{0} & \mathbf{0} & \mathbf{A}_{4,9} & \mathbf{A}_{4,10} \\ \mathbf{0} & -\mathbf{I} & \mathbf{0} & \mathbf{0} & \mathbf{0} \\ \mathbf{0} & \mathbf{0} & -\mathbf{I} & \mathbf{0} & \mathbf{0} \\ \mathbf{0} & \mathbf{0} & \mathbf{0} & \mathbf{A}_{8,9} & \mathbf{A}_{8,10} \end{bmatrix}^{\mathrm{T}} \begin{Bmatrix} \mathbf{q}_1 \\ \mathbf{q}_2 \\ \mathbf{q}_3 \\ \mathbf{q}_4 \\ \mathbf{q}_5 \\ \mathbf{q}_6 \\ \mathbf{q}_7 \end{Bmatrix} + \begin{Bmatrix} \mathbf{Qo}_6 \\ \mathbf{Qo}_7 \\ \mathbf{Qo}_8 \\ \mathbf{Qo}_9 \\ \mathbf{Qo}_{10} \end{Bmatrix} + \begin{Bmatrix} \mathbf{0} \\ \mathbf{0} \\ \mathbf{0} \\ \mathbf{Q}_{\mathrm{LEAK}_9} \\ \mathbf{Q}_{\mathrm{LEAK}_{10}} \end{Bmatrix} = \{\mathbf{0}\}$$

also

$$\begin{Bmatrix} \mathbf{V}_6' \\ \mathbf{V}_7' \\ \mathbf{V}_8' \\ \mathbf{V}_9' \\ \mathbf{V}_{10}' \end{Bmatrix} = \begin{Bmatrix} \mathbf{Qo}_6 \\ \mathbf{Qo}_7 \\ \mathbf{Qo}_8 \\ \mathbf{Qo}_9 \\ \mathbf{Qo}_{10} \end{Bmatrix} - \begin{Bmatrix} \mathbf{Q_{D}}_6(\mathbf{t}) \\ \mathbf{0} \\ \mathbf{0} \\ \mathbf{0} \\ \mathbf{0} \end{Bmatrix}$$

Rate of change of volume in tanks/reservoirs — Outflows from network — Demands

$$\therefore \begin{Bmatrix} \mathbf{V}_6' \\ \mathbf{V}_7' \\ \mathbf{V}_8' \\ \mathbf{V}_9' \\ \mathbf{V}_{10}' \end{Bmatrix} = -\begin{bmatrix} \mathbf{0} & \mathbf{0} & \mathbf{0} & \mathbf{A}_{1,9} & \mathbf{A}_{1,10} \\ \mathbf{0} & \mathbf{0} & \mathbf{0} & \mathbf{A}_{2,9} & \mathbf{A}_{2,10} \\ \mathbf{0} & \mathbf{0} & \mathbf{0} & \mathbf{A}_{3,9} & \mathbf{A}_{3,10} \\ -\mathbf{I} & \mathbf{0} & \mathbf{0} & \mathbf{A}_{4,9} & \mathbf{A}_{4,10} \\ \mathbf{0} & -\mathbf{I} & \mathbf{0} & \mathbf{0} & \mathbf{0} \\ \mathbf{0} & \mathbf{0} & -\mathbf{I} & \mathbf{0} & \mathbf{0} \\ \mathbf{0} & \mathbf{0} & \mathbf{0} & \mathbf{A}_{8,9} & \mathbf{A}_{8,10} \end{bmatrix}^{T} \begin{Bmatrix} \mathbf{q}_1 \\ \mathbf{q}_2 \\ \mathbf{q}_3 \\ \mathbf{q}_4 \\ \mathbf{q}_5 \\ \mathbf{q}_6 \\ \mathbf{q}_7 \end{Bmatrix} - \begin{Bmatrix} \mathbf{0} \\ \mathbf{0} \\ \mathbf{0} \\ \mathbf{Q_{LEAK}}_9 \\ \mathbf{Q_{LEAK}}_{10} \end{Bmatrix} - \begin{Bmatrix} \mathbf{Q_{D}}_6(\mathbf{t}) \\ \mathbf{0} \\ \mathbf{0} \\ \mathbf{0} \\ \mathbf{0} \end{Bmatrix}$$

or

$$\left\{\mathbf{V}'\right\} = -\left[\bar{\mathbf{A}}_1\right]^{T} \{\mathbf{q}\} - \left\{\bar{\mathbf{Q}}_{\mathbf{LEAK}}\left(\bar{\mathbf{H}}\right)\right\} - \{\mathbf{Q_D}(\mathbf{t})\}$$

This defines the "tank/reservoir" dynamics.

3.2.2 [A] also defines the pressure drops across pipes and links

$$\{\Delta\mathbf{H}\} = [\mathbf{A}]\{\mathbf{Z} + \mathbf{H}\}$$

Nodal o.d. levels — Nodal heads (residual)

3.2.2.1 In our case

$$\{\Delta\mathbf{H}\} = \left[\begin{array}{cccc|cccccc} \mathbf{A}_{1,1} & \mathbf{A}_{1,2} & \mathbf{A}_{1,3} & \mathbf{A}_{1,4} & \mathbf{0} & \mathbf{0} & \mathbf{0} & \mathbf{0} & \mathbf{A}_{1,9} & \mathbf{A}_{1,10} \\ \mathbf{A}_{2,1} & \mathbf{A}_{2,2} & \mathbf{A}_{2,3} & \mathbf{A}_{2,4} & \mathbf{0} & \mathbf{0} & \mathbf{0} & \mathbf{0} & \mathbf{A}_{2,9} & \mathbf{A}_{2,10} \\ \mathbf{A}_{3,1} & \mathbf{A}_{3,2} & \mathbf{A}_{3,3} & \mathbf{A}_{3,4} & \mathbf{0} & \mathbf{0} & \mathbf{0} & \mathbf{0} & \mathbf{A}_{3,9} & \mathbf{A}_{3,10} \\ \mathbf{A}_{4,1} & \mathbf{A}_{4,2} & \mathbf{A}_{4,3} & \mathbf{A}_{4,4} & \mathbf{0} & -\mathbf{I} & \mathbf{0} & \mathbf{0} & \mathbf{A}_{4,9} & \mathbf{A}_{4,10} \\ \mathbf{0} & \mathbf{I} & \mathbf{0} & \mathbf{0} & \mathbf{0} & \mathbf{0} & -\mathbf{I} & \mathbf{0} & \mathbf{0} & \mathbf{0} \\ \mathbf{0} & \mathbf{0} & \mathbf{I} & \mathbf{0} & \mathbf{0} & \mathbf{0} & \mathbf{0} & -\mathbf{I} & \mathbf{0} & \mathbf{0} \\ \mathbf{A}_{7,1} & \mathbf{A}_{7,2} & \mathbf{A}_{7,3} & \mathbf{A}_{7,4} & \mathbf{A}_{7,5} & \mathbf{0} & \mathbf{0} & \mathbf{0} & \mathbf{A}_{7,9} & \mathbf{A}_{7,10} \end{array}\right] \left\{\begin{array}{c} \mathbf{Z}_1 + \mathbf{H}_1 \\ \mathbf{Z}_2 + \mathbf{H}_2 \\ \mathbf{Z}_3 + \mathbf{H}_3 \\ \mathbf{Z}_4 + \mathbf{H}_4 \\ \hline \mathbf{Z}_5 + \mathbf{H}_5(\mathbf{q}_7) \\ \mathbf{Z}_6 + \mathbf{H}_6(\mathbf{V}_6) \\ \mathbf{Z}_7 + \mathbf{H}_7(\mathbf{V}_7) \\ \mathbf{Z}_8 + \mathbf{H}_8(\mathbf{V}_8) \\ \mathbf{Z}_9 + \mathbf{H}_9(\mathrm{t}) \\ \mathbf{Z}_{10} + \mathbf{H}_{10}(\mathrm{t}) \end{array}\right\}$$

or

$$\{\Delta\mathbf{H}\} = \left[\begin{array}{c|c} \mathbf{A}_2 & \overline{\mathbf{A}}_2 \end{array}\right] \left\{\begin{array}{c} \mathbf{Z}+\mathbf{H} \\ \hline \overline{\mathbf{Z}}+\overline{\mathbf{H}} \end{array}\right\}$$

$$[\mathbf{A}_{7,1} \; \mathbf{A}_{7,2} \; \mathbf{A}_{7,3} \; \mathbf{A}_{7,4} \;|\; \mathbf{A}_{7,5} \; \mathbf{0} \; \mathbf{0} \; \mathbf{0} \; \mathbf{A}_{7,9} \; \mathbf{A}_{7,10}]$$

$$= [\mathbf{0} \; \mathbf{0} \; \mathbf{0} \; \mathbf{I} \;|\; \mathbf{-I} \; \mathbf{0} \; \mathbf{0} \; \mathbf{0} \; \mathbf{0} \; \mathbf{0}\,] \quad \text{(active head control devices)}$$

$$\text{or}= [\mathbf{A}_{8,1} \; \mathbf{A}_{8,2} \; \mathbf{A}_{8,3} \; \mathbf{I}+\mathbf{A}_{8,4} \;|\; \mathbf{0} \; \mathbf{0} \; \mathbf{0} \; \mathbf{0} \; \mathbf{A}_{8,9} \; \mathbf{A}_{8,10}] \quad \text{(non-active head control devices)}$$

To form $[\mathbf{A}_{7,1} \; \mathbf{A}_{7,2} \; \mathbf{A}_{7,3} \; \mathbf{A}_{7,4} \;|\; \mathbf{A}_{7,5} \; \mathbf{0} \; \mathbf{0} \; \mathbf{0} \; \mathbf{A}_{7,9} \; \mathbf{A}_{7,10}]$, we use appropriate rows of the above sub-matrices, depending on the active/non-active status of each control device. The A_{8i} entries indicate control device "end nodes". For the non-active case, links join control-associated demand nodes directly to the "end nodes" of the control devices; corresponding loss coefficients low/high can make the link open/closed as necessary.

4 FORMATION OF THE IMPLICIT DIFFERENTIAL EQUATIONS

4.1 The general initial value differential equation problem

We require $\{\mathbf{Y}(t)\}$ given :

(a) Initial conditions $\{\mathbf{Y}(0)\}$ and (ideally) $\{\mathbf{Y}'(0)\}$

(b) Differential equations $\{\mathbf{F}(\mathbf{Y}(t),\mathbf{Y}'(t))\} = \{0\}$

The Gear solution process for stiff/implicit differential equations requires frequent solution of linear equations of the form

$$[\mathbf{W}]\,\{\mathbf{x}\} = \{\mathbf{b}\}$$

where

$$[\mathbf{W}] = \left[\frac{\partial \mathbf{F}}{\partial \mathbf{Y}}\right] + c_j \left[\frac{\partial \mathbf{F}}{\partial \mathbf{Y'}}\right]$$

[**W**] is known as the Newton iteration matrix

We shall focus on a technique which will speed up the solution of these equations for the network mass surge analysis problem.

4.2 Dealing with [W]

In particular, we shall partition [**W**] so our problem is restated

Solve

$$\begin{bmatrix} \mathbf{W}_{1,1} & \mathbf{W}_{1,2} \\ \mathbf{W}_{2,1} & \mathbf{W}_{2,2} \end{bmatrix} \begin{Bmatrix} \mathbf{x}_1 \\ \mathbf{x}_2 \end{Bmatrix} = \begin{Bmatrix} \mathbf{b}_1 \\ \mathbf{b}_2 \end{Bmatrix}$$

Then, we obtain the solution in two steps

$$\left[\mathbf{W}_{1,1} - \mathbf{W}_{1,2}\mathbf{W}_{2,2}{}^{-1}\mathbf{W}_{2,1}\right] \{\mathbf{x}_1\} = \left\{\mathbf{b}_1 - \mathbf{W}_{1,2}\mathbf{W}_{2,2}{}^{-1}\mathbf{b}_2\right\}$$

Solve

and

$$\{\mathbf{x}_2\} = \mathbf{W}_{2,2}{}^{-1}\{\mathbf{b}_2 - \mathbf{W}_{2,1}\mathbf{x}_1\}$$

This is the key element of our approach. Great efficiency gains result because in our case

(i) [$\mathbf{W}_{2,2}$] is a relatively simple matrix, and it can be inverted analytically; also, [$\mathbf{W}_{1,2}\ \mathbf{W}_{2,2}{}^{-1}$] can easily be formed directly.

(ii) [$\mathbf{W}_{1,1} - \mathbf{W}_{1,2}\ \mathbf{W}_{2,2}{}^{-1}\ \mathbf{W}_{2,1}$] can also be formed directly; [$\mathbf{W}_{1,1}$] has the well known nodal sparsity structure, and no additional non-zeros are created by the subtraction of [$\mathbf{W}_{1,2}\ \mathbf{W}_{2,2}{}^{-1}\ \mathbf{W}_{2,1}$].

4.3 The Network i.d.e.'s

The system of i.d.e.'s for a network mass surge analysis (including leakage) is

$$\{\mathbf{F}(\mathbf{H}, \mathbf{q}, \mathbf{V}, t, V_{leak}, \mathbf{H}', \mathbf{q}', \mathbf{V}', t', V'_{leak})\} \quad = \quad \{\mathbf{0}\}$$

Or in more detail:

$$\left\{\begin{array}{c}[\lambda(\mathbf{H})]\{\mathbf{H}'\}+[\mathbf{A}_1]^T[\gamma]\left(\left[\mathbf{A}_2\ \overline{\mathbf{A}}_2\right]\left\{\begin{array}{c}\mathbf{Z}+\mathbf{H}\\ \overline{\mathbf{Z}}+\overline{\mathbf{H}}(\mathbf{q},\mathbf{V},t)\end{array}\right\}-\{\mathbf{h}(\mathbf{q},\mathbf{V},t)\}\right)\\ \{\mathbf{q}'\}-[\gamma]\left(\left[\mathbf{A}_2\ \overline{\mathbf{A}}_2\right]\left\{\begin{array}{c}\mathbf{Z}+\mathbf{H}\\ \overline{\mathbf{Z}}+\overline{\mathbf{H}}(\mathbf{q},\mathbf{V},t)\end{array}\right\}-\{\mathbf{h}(\mathbf{q},\mathbf{V},t)\}\right)\\ \{\mathbf{V}'\}+\left[\overline{\mathbf{A}}_1\right]^T\{\mathbf{q}\}+\left\{\varphi\left(\overline{\mathbf{H}}(t)\right)\right\}+\{\mathbf{Q_D}(t)\}\\ t'-1\\ V'_{leak}-\{1\quad 1\}^T\left\{\begin{array}{c}\varphi(\mathbf{H})\\ \varphi\left(\overline{\mathbf{H}}(t)\right)\end{array}\right\}\end{array}\right\}=\{\mathbf{0}\}$$

The system of ordinary differential equations is implicit (or differential algebraic if leakage is omitted). The five groups relate to:

Demand node continuity;
Pipe dynamics (momentum);
Tank/reservoir continuity;
Time;
Total cumulative leakage.

Where $[\mathbf{U}_1]=[\gamma]\left(\left[\overline{\mathbf{A}}_2\right]\left[\dfrac{\partial\overline{\mathbf{H}}}{\partial\mathbf{q}}\right]-\left[\dfrac{\partial\mathbf{h}}{\partial\mathbf{q}}\right]\right)$

$[\mathbf{U}_2]=[\gamma]\left(\left[\overline{\mathbf{A}}_2\right]\left[\dfrac{\partial\overline{\mathbf{H}}}{\partial\mathbf{V}}\right]-\left[\dfrac{\partial\mathbf{h}}{\partial\mathbf{V}}\right]\right)$

and $\{\mathbf{U}_3\}=[\gamma]\left(\left[\overline{\mathbf{A}}_2\right]\left\{\dfrac{\partial\overline{\mathbf{H}}}{\partial t}\right\}-\left\{\dfrac{\partial\mathbf{h}}{\partial t}\right\}\right)$

the Newton iteration matrix is given by:-

$$[\mathbf{W}]=\left[\begin{array}{c|cccc} [\mathbf{A}_1]^T[\gamma][\mathbf{A}_2]+c_j[\lambda(\mathbf{H})] & [\mathbf{A}_1]^T[\mathbf{U}_1] & [\mathbf{A}_1]^T[\mathbf{U}_2] & [\mathbf{A}_1]^T\{\mathbf{U}_3\} & \{\mathbf{0}\} \\ \hline -[\gamma][\mathbf{A}_2] & c_j[\mathbf{I}]-[\mathbf{U}_1] & -[\mathbf{U}_2] & -\{\mathbf{U}_3\} & \{\mathbf{0}\} \\ [\mathbf{0}] & \left[\bar{\mathbf{A}}_1\right]^T & c_j[\mathbf{I}] & \left[\lambda\left(\bar{\mathbf{H}}\right)\right]\left\{\frac{\partial \bar{\mathbf{H}}}{\partial \mathbf{t}}\right\}+\left\{\frac{\partial \mathbf{Q}_D}{\partial \mathbf{t}}\right\} & \{\mathbf{0}\} \\ \{\mathbf{0}\}^T & \{\mathbf{0}\}^T & \{\mathbf{0}\}^T & c_j & 0 \\ -\{\mathbf{1}\}^T[\lambda(\mathbf{H})] & \{\mathbf{0}\}^T & \{\mathbf{0}\}^T & -\{\mathbf{1}\}^T\left[\lambda\left(\bar{\mathbf{H}}\right)\right]\left\{\frac{\partial \bar{\mathbf{H}}}{\partial \mathbf{t}}\right\} & c_j \end{array}\right]$$

5 EFFICIENT FACTORISATION OF THE NEWTON ITERATION MATRIX

5.1 The inverse of $[\mathbf{W}_{2,2}]$

$[\mathbf{W}_{2,2}]$ can be inverted efficiently by partitioning (27).

Specifically,

$$[\mathbf{W}_{2,2}]^{-1}=\begin{bmatrix} \mathbf{T}_{1,1} & \mathbf{T}_{1,2} & \{\mathbf{T}_{1,3}\} & \{\mathbf{0}\} \\ \mathbf{T}_{2,1} & \mathbf{T}_{2,2} & \{\mathbf{T}_{2,3}\} & \{\mathbf{0}\} \\ \{\mathbf{0}\}^T & \{\mathbf{0}\}^T & c_j^{-1} & 0 \\ \{\mathbf{0}\}^T & \{\mathbf{0}\}^T & t_{4,3} & c_j^{-1} \end{bmatrix}$$

where

$$[\mathbf{T}_{1,1}]=\left[c_j[\mathbf{I}]\ -[\mathbf{U}_1]+c_j^{-1}[\mathbf{U}_2]\left[\bar{\mathbf{A}}_1\right]^T\right]^{-1}$$

$$[\mathbf{T}_{2,1}]=-c_j^{-1}\left[\bar{\mathbf{A}}_1\right]^T[\mathbf{T}_{1,1}]$$

$$[\mathbf{T}_{1,2}]=c_j^{-1}[\mathbf{T}_{1,1}]\,[\mathbf{U}_2]$$

$$[\mathbf{T}_{2,2}]=c_j^{-1}\left([\mathbf{I}]-\left[\bar{\mathbf{A}}_1\right]^T[\mathbf{T}_{1,2}]\right)$$

$$\{\mathbf{T}_{1,3}\} = c_j^{-1}[\mathbf{T}_{1,1}]\left(\{\mathbf{U}_3\} - c_j^{-1}[\mathbf{U}_2]\left(\left[\lambda\left(\bar{\mathbf{H}}\right)\right]\left\{\frac{\partial \bar{\mathbf{H}}}{\partial t}\right\} + \left\{\frac{\partial \mathbf{Q_D}}{\partial t}\right\}\right)\right)$$

$$\{\mathbf{T}_{2,3}\} = -c_j^{-1}\left(\left[\bar{\mathbf{A}}_1\right]^T\{\mathbf{T}_{1,3}\} + c_j^{-1}\left(\left[\lambda\left(\bar{\mathbf{H}}\right)\right]\left\{\frac{\partial \bar{\mathbf{H}}}{\partial t}\right\} + \left\{\frac{\partial \mathbf{Q_D}}{\partial t}\right\}\right)\right)$$

$$t_{4,3} = c_j^{-2}\{\mathbf{1}\}^T\left[\lambda\left(\bar{\mathbf{H}}\right)\right]\left\{\frac{\partial \bar{\mathbf{H}}}{\partial t}\right\}$$

5.2 Final statement of the algorithm to solve [W] {x} = {b}

We partition [**W**], and hence {**x**} and {**b**}.

Solve for $\{\mathbf{x}_1\}$ as follows

$$\left([\mathbf{A}_1]^T[\boldsymbol{\Gamma}][\mathbf{A}_2] + c_j[\lambda(\mathbf{H})]\right)\{\mathbf{x}_1\} = \{\mathbf{b}_1\} - [\mathbf{A}_1]^T\left[c_j\mathbf{T}_{1,1} - \mathbf{I} \quad c_j\mathbf{T}_{1,2} \quad c_j\{\mathbf{T}_{1,3}\} \quad \{\mathbf{0}\}\right]\{\mathbf{b}_2\}$$

where $[\boldsymbol{\Gamma}] = c_j[\mathbf{T}_{1,1}]\,[\gamma]$

then, we obtain $\{\mathbf{x}_2\}$

$$\{\mathbf{x}_2\} = \begin{bmatrix} \mathbf{T}_{1,1} & \mathbf{T}_{1,2} & \{\mathbf{T}_{1,3}\} & \{\mathbf{0}\} \\ \mathbf{T}_{2,1} & \mathbf{T}_{2,2} & \{\mathbf{T}_{2,3}\} & \{\mathbf{0}\} \\ \{\mathbf{0}\}^T & \{\mathbf{0}\}^T & c_j^{-1} & 0 \\ \{\mathbf{0}\}^T & \{\mathbf{0}\}^T & t_{4,3} & c_j^{-1} \end{bmatrix}\left(\{\mathbf{b}_2\} + \begin{bmatrix} [\gamma][\mathbf{A}_2] \\ [\mathbf{0}] \\ \{\mathbf{0}\}^T \\ \{\mathbf{1}\}^T[\lambda(\mathbf{H})] \end{bmatrix}\{\mathbf{x}_1\}\right)$$

It is obvious that the left hand side matrix for the solution of $\{\mathbf{x}_1\}$ has a simple (and very sparse) nodal structure. Hence, these equations can be solved quickly, using well tried methods (e.g. Gaussian elimination) that take advantage of sparsity and (near)-symmetry.

It can be demonstrated that $[\mathbf{T}_{1,1}]$, $[\mathbf{T}_{1,2}]$, $[\mathbf{T}_{2,1}]$ and $[\mathbf{T}_{2,2}]$ have simple structures and are very sparse. They are composed of sub-matrices that are either simple diagonals or subsets of the (very sparse) incidence matrix, [**A**] in structure. On account of this sparsity, $\{\mathbf{x}_2\}$ can be evaluated very efficiently.

5.3 The benefit of these results

Using the results described above, it follows that, for N nodes and P pipes, the bulk of the work in solving equations of the form [**W**]{**x**}={**b**} will involve a factorisation and solution with only an N x N sparse (near-)symmetric nodal matrix.

The straightforward approach would require similar operations with a much larger (3.N + P) x (3.N + P) non-symmetric matrix, because $[\mathbf{W}_{11}]$ is of order NxN and $[\mathbf{W}_{22}]$ is of order (2.N + P) x (2.N + P); – in fact, the size of $[\mathbf{W}_{22}]$ would be even greater if there were more than one demand tank per demand node (e.g. dual level consumer demand). Hence it is expected that the approach proposed above will produce a very respectable saving in computer time for the factorisation of the "Newton iteration" matrix.

For dense matrix calculations (work $\propto$ order^3), assuming that $P \approx N$, we will have an efficiency gain of about 100 times:- 4^3 due to matrix size, and further gain approaching 2 times because we can take advantage of the (near) symmetry of $[\mathbf{W}_{22} - \mathbf{W}_{21}\,\mathbf{W}_{11}^{-1}\,\mathbf{W}_{12}]$. For sparse matrix calculations, which are used in practice, the efficiency gain is less – but significant nevertheless.

Depending on data structures, the handling of sparsity etc., there could also be efficiency gains of up to 16 times (4^2) in the subsequent "solve" processes.

5.4 The application of Bennett's algorithm to factorise $[\mathbf{W}_{1,1} - \mathbf{W}_{1,2}\,\mathbf{W}_{2,2}^{-1}\,\mathbf{W}_{2,1}]$

We have seen that $[\mathbf{W}_{1,1} - \mathbf{W}_{1,2}\,\mathbf{W}_{2,2}^{-1}\,\mathbf{W}_{2,1}] = [\mathbf{A}_1]^T\,[\boldsymbol{\Gamma}]\,[\mathbf{A}_2]$. This matrix is nearly symmetric and we can apply Bennett's algorithm (21) to update the factorisation of its symmetric kernel, thus producing the factorisation for the complete matrix.

$[\boldsymbol{\Gamma}]$ has the sparsity structure of $[\mathbf{T}_{1,1}]$, i.e. it includes a diagonal and a small off-diagonal sub-matrix, $[\boldsymbol{\Gamma}_{2,5}]$. $[\boldsymbol{\Gamma}_{2,5}]$ has the same structure as $[\mathbf{A}_{2,2}]$ in the incidence matrix.

If there are no float valves feeding tanks, $[\boldsymbol{\Gamma}]$ will be purely diagonal. Furthermore, if there are no active nodal head control devices (PRV's etc), then $[\mathbf{A}_2] = [\mathbf{A}_1]$. If both of these conditions apply, then our matrix will be fully symmetric.

We define $[\boldsymbol{\Gamma}_D]$ as the diagonal part of $[\boldsymbol{\Gamma}]$.

$$
\begin{aligned}
\text{Then } [\mathbf{A}_1]^T\,[\boldsymbol{\Gamma}]\,[\mathbf{A}_2] &= [\mathbf{A}_2]^T\,[\boldsymbol{\Gamma}]\,[\mathbf{A}_2] + [\mathbf{A}_1 - \mathbf{A}_2]^T\,[\boldsymbol{\Gamma}]\,[\mathbf{A}_2] \\
&= [\mathbf{A}_2]^T\,[\boldsymbol{\Gamma}_D]\,[\mathbf{A}_2] + [\mathbf{A}_2]^T\,[\boldsymbol{\Gamma} - \boldsymbol{\Gamma}_D]\,[\mathbf{A}_2] + [\mathbf{A}_1 - \mathbf{A}_2]^T\,[\boldsymbol{\Gamma}_D][\mathbf{A}_2]
\end{aligned}
$$

~~$+ [\mathbf{A}_1 - \mathbf{A}_2]^T\,[\boldsymbol{\Gamma} - \boldsymbol{\Gamma}_D]\,[\mathbf{A}_2]$~~

The final term above is zero because of sparsity structure. Expressing the necessary updates in terms of the basic incidence matrix components, and omitting superfluous terms, we have:

$$[\mathbf{A}_1]^T[\Gamma]\,[\mathbf{A}_2] = [\mathbf{A}_2]^T[\Gamma_D]\,[\mathbf{A}_2] + [\mathbf{A}_{2,1}\ \mathbf{A}_{2,2}\ \mathbf{A}_{2,3}\ \mathbf{A}_{2,4}]^T[\Gamma_{2,3}]\,[\mathbf{0}\ \mathbf{I}\ \mathbf{0}\ \mathbf{0}]$$
$$+\left([\mathbf{A}_{8,1}\ \mathbf{A}_{8,2}\ \mathbf{A}_{8,3}\ \mathbf{I}+\mathbf{A}_{8,4}] - [\mathbf{A}_{7,1}\ \mathbf{A}_{7,2}\ \mathbf{A}_{7,3}\ \mathbf{A}_{7,4}]\right)^T[\Gamma_{7,7}]\,[\mathbf{A}_{7,1}\ \mathbf{A}_{7,2}\ \mathbf{A}_{7,3}\ \mathbf{A}_{7,4}]$$

Regarding the update $[\mathbf{A}_{2,1}\ \mathbf{A}_{2,2}\ \mathbf{A}_{2,3}\ \mathbf{A}_{2,4}]^T[\Gamma_{2,3}]\,[\mathbf{0}\ \mathbf{I}\ \mathbf{0}\ \mathbf{0}]$, note that column 2 of $[\mathbf{A}_2]^T[\Gamma_D]\,[\mathbf{A}_2]$ is the only column with a non-zero update.

Since $[\Gamma_{2,3}]$ has the same structure as $[\mathbf{A}_{2,2}]$, the structure of the update to column 2 is $[\mathbf{A}_{2,1}\ \mathbf{A}_{2,2}\ \mathbf{A}_{2,3}\ \mathbf{A}_{2,4}]^T[\Gamma_{2,3}] = [\mathbf{A}_{2,1}\ \mathbf{A}_{2,2}\ \mathbf{A}_{2,3}\ \mathbf{A}_{2,4}]^T[\mathbf{A}_{2,2}]$

Because $[\Gamma_{2,2}]$ is diagonal, this is obviously the same structure as column 2 of $[\mathbf{A}_2]\,[\Gamma_D]\,[\mathbf{A}_2]$, which is $[\mathbf{A}_{2,1}\ \mathbf{A}_{2,2}\ \mathbf{A}_{2,3}\ \mathbf{A}_{2,4}]^T[\Gamma_{2,2}][\mathbf{A}_{2,2}]$

Therefore this update introduces no additional terms in the original matrix. We do not obtain such an elegant result if it is permitted to connect more than one pipe to a supply tank outlet – hence the use of "links" and "associated nodes" in the model.

To factorise the matrix, we first factorise $[\mathbf{A}_2]^T[\Gamma_D]\,[\mathbf{A}_2]$ with an efficient sparse symmetric procedure. This will form the factors [**L**][**D**][**L**]^T, ([**L**]^T is not stored). We then allocate storage for upper triangle [**U**], and copy [**L**]^T into [**U**], so we have an [**L**][**D**][**U**] factorisation.

Bennett's algorithm can now be applied. This algorithm updates the [**L**][**D**][**U**] factorisation of a matrix [**A**], where [**A**] has been modified by the addition of [**X**][**C**][**Y**]^T (block matrix version) or c{**X**}{**Y**}^{T} (vector version, c scalar). For [**A**] of dimension n x n and [**C**] of dimension m x m, the algorithm will update a dense matrix in about $2mn^2$ add + multiply operations. Far fewer operations are required if we take advantage of sparsity, say by the use of linked lists to hold the non-zero elements of the [**L**][**D**][**U**] factorisation.

If we use the vector version,

No. of updates required = No. of active float-valves feeding non-demand tanks

+

No. of active head control devices (PRV's etc.)

It is possible to use the block matrix version and to perform the update in two steps (float-valves + head controllers). In this case, there will be infill in [**C**] and some additional storage will be required for this.

The "float-valve" update causes no additional infill in our [**L**][**D**][**U**] factorisation, just some asymmetry. The "head controller" update will result in a small asymmetric infill.

The bulk of the time to factorise a typical network matrix will be spent in factorisation of the symmetric part; the updates necessary to account for the few active control devices normally require an insignificant amount of time.

The process described in this section will factorise a nearly-symmetric matrix nearly as fast as it is possible to factorise a symmetric matrix.

Note that a suitable ordering algorithm (e.g approx. minimum degree) (23,26) should be used to re-number the equations so that the infill of additional elements during factorisations will be minimised.

WADI-II uses the sparsity structure of $[\mathbf{A}_1]^T[\Gamma]\,[\mathbf{A}_1]$, rather than that of $[\mathbf{A}_2]^T[\Gamma]\,[\mathbf{A}_2]$, for the ordering operation; this includes a few extra elements that can arise in the sparsity structure when nodal head controllers are not active (i.e. worst case).

An alternative to the approach suggested above is to use a "canned" equation solver which has been designed to work well with near-symmetric systems. A good example is the MA41 code from the Harwell Subroutine Library (22). The documentation for this routine mentions that the code can be suitably adapted for parallel processing, which could be relevant for some intensive modelling, optimisation or calibration applications.

6 DETERMINATION OF CONSISTENT STARTING VALUES

An implicit differential equation solver will require the specification of a consistent set of values for $\{\mathbf{H}\}$, $\{\mathbf{q}\}$, $\{\mathbf{V}\}$ ... and $\{\mathbf{H}'\}$, $\{\mathbf{q}'\}$, $\{\mathbf{V}'\}$.. etc. Salgado et alia (18) have shown how the Gradient Method can be extended to simulate pressure dependent leakage. This approach could be used to get initial values for $\{\mathbf{H}\}$, $\{\mathbf{q}\}$, $\{\mathbf{V}\}$ etc. and most i.d.e. solvers will have the facility of calculating a consistent set of derivatives.

The approach used in WADI-II is to use the dynamic simulation algorithm itself to obtain a consistent set of starting values for all variables.

The given "initial condition" data include the heads of non-demand tanks and reservoirs, pumping heads and valve settings.

When the initialisation phase is complete, the part of $\{\mathbf{V}\}$ corresponding to non-demand tanks and reservoirs and the value of t will be re-set to the appropriate initial values.

7 CONCLUSION

We have seen the motivation for the new method of analysis – including buffered demands, pressure dependent leakage, and modelling of network dynamics. The goal has been to replicate and to improve upon the functionality of the common "Extended Period Simulation" procedures, while not invoking massive additional compute time overheads.

We have seen how we model float valves and other elements, how we form the network relations with the incidence matrix, and how we put it all together to form the implicit differential equations for mass surge analysis.

We have then shown how we reduce the amount of computation in handling the factorisation and use of the Newton iteration matrix.

Proper attention to choice of an efficient equation solver and a suitable i.d.e. solver is essential for efficient implementation of the method described above. Ideally, the i.d.e. code should have a root finding capability. This will facilitate the detection of status changes in NRV's, PRV's etc., which must be reflected in the system matrices. The assessment of equation solvers for speed, memory requirements, ease of use, handling of sparsity and near-symmetry, etc., is best carried out with some "test bed" software. Such software would generate a large synthetic network matrix and corresponding right hand side, call the solver, and time the periods for optimal ordering, factorisation and solution of the linear system.

The model which has been described in this paper is a step closer to reality than the hydraulic model which underlies most current packages. It will be capable of providing the engineer with some useful additional insight into the behaviour of his/her distribution system.

REFERENCES

[1] Hagen (1839), Poiseuille (1840) – Experimental determination of Laminar flow formula

[2] Wiedemann, G., (1856) – Analytical derivation of Laminar flow formula

[3] Weisbach, Julius, "Die Experimental-Hydraulik," J.G. Engelhardt, Freiberg, Germany, 1855.

[4] Darcy, M.H. "Recherches experimentales relatives au mouvement de l'eau dans les tuyaux," Mallet-Bachelier, Paris, 1857

[5] Manning, Robert, "Flow of Water in Open Channels and Pipes," Trans. Inst. Civil Engrs. (Ireland), vol. 20, 1890

[6] Williams, G.S., and Hazen, A, Hydraulics Tables, 3rd ed., John Wiley and Son Inc., New York, N.Y., 1933.

[7] White, C.M. "The reduction of Carrying Capacity of Pipes with Age," J. Inst. Civil Engrs, (London), Vol. 7, p. 99 (1937-1938)

[8] Colebrook, C.F. "Turbulent Flow in Pipes, with Particular Reference to the Transition Region between Smooth and Rough Pipe Laws," J. Inst. Civil Engrs, (London), Vol. 11, pp. 133-156 (1938-1939)

[9] Colebrook, C.F. and White, C.M. "Experiments with fluid friction in roughened pipes," Proc. Roy. Soc., 161

[10] Moody, L.F. 1947. "An approximate formula for pipe friction factors". Mech. Engng., New York, 69.

[11] Wood, D.J. 1966. "An explicit friction factor relationship". Civil Engng., Am.Soc.Civ.Engrs., 36, No 12.

[12] Barr, D.I.H. Dec. 1975. "Two additional methods of direct solution of the Colebrook-White function", Proc.Instn.Civ.Engrs., 59.

[13] Haaland, S.E. 1988. "Simple and explicit formulas for the friction factor in turbulent pipe flow". J.Fluids Engng., ASME, New York, 105.

[14] Swamee & Jain. "Explicit approximation to the implicit Colebrook-White formula", from "Analysis of Flow in Water Distribution Networks" by P.R. Bhave (Technomic Publishing, 1991)

[15] Nahavandi, A.N. and Catanzaro G.V. "Matrix Method for Analysis of Hydraulic Networks", Jnl. Hyd. Div., ASCE, Jan. 1973

[16] Dunlop, E.J., 1976 "Dynamic Relaxation method for the analysis of hydraulic networks" (Same as gradient algorithm, but different derivation - described in original WADI documentation)

[17] Todini E. and Pilati S. 1988. "A gradient algorithm for the analysis of pipe networks",Computer Applications in Water Supply, Vol. 1, Systems Analysis and Simulation, edited by B. Coulbeck and C.H. Orr, Research Studies Press Ltd.

[18] Salgado, R., Rojo, J. and Zepeda, S. 1993. "Extended Gradient Method for Fully Non-linear Head and Flow Analysis in Pipe Networks", Integrated Computer Applications in Water Supply, Vol. 1, Methods and procedures for systems simulation and control, edited by Bryan Coulbeck, Research Studies Press Ltd.

[19] Chaudhry, M.H. "Transient analysis of water supply networks". Water Supply Systems, State of the Art and future trends, Ed. Cabrera, E. and Martinez, F., Computational Mechanics Publications, 1993.

[20] Cohen, J. "New trends in distribution research. Dynamic calculation and monitoring". Water Supply Systems, State of the Art and future trends, Ed. Cabrera, E. and Martinez, F., Computational Mechanics Publications, 1993.

[21] Bennett, John M. "Triangular Factors of Modified Matrices," Numerische Mathematik 7, 217-221, 1965

[22] MA27, MA37, MA41 "Symmetric and Near Symmetric linear equation solvers"; Harwell Subroutine Library, Release 12, Dec. 1995.

[23] MC47 "Minimum degree ordering for a Symmetric Matrix"; Harwell Subroutine Library, Release 12, Dec. 1995.

[24] Petzold, Linda R., "DDASRT stiff ODE/DAE/IDE solver with root finding (1982; rev. 1991)"; SLATEC library

[25] K. E. Brenan, S. L. Campbell, and L. R. Petzold, "Numerical Solution of Initial-Value Problems in Differential-Algebraic Equations"; Elsevier, New York, 1989.

[26] P.R. Amestoy, T.A. Davis, and I.S. Duff – An Approximate Minimum Degree Ordering Algorithm, ENSEEIHT Rapport Technique RT/APO/95/5 (ftp://ftp.enseeiht.fr/pub/numerique/REPORTS/RT_APO_95_5.ps.Z)

[27] Press W.H., Teukolsky S.A., Vetterling W.T., Flannery B.P. 1992. "Numerical Recipes in Fortran – The Art of Scientific Computing" Second Edition. Cambridge University Press.

Small-Scale Models for Validation of CFD in Water Applications

J.Hague, C.T.Ta *and* M.J.Biggs

ABSTRACT

The Computational Fluid Dynamics (CFD) technique has been applied in the water industry to obtain flow characteristics. Many treatment processes have been analysed for process understanding, troubleshooting, optimisation and scale up. However, for the technique to be accepted as an engineering design tool, experimental validations are required.

This paper discusses an experimental programme to carry out accurate measurements of flow velocity in models representing the flow conditions of various treatment processes. Laboratory-scale Perspex models have been employed for this purpose to allow the use of current advances in laser techniques for flow measurement. This is in addition to standard methods of flow measurement e.g. dye and chemical tracer testing.

The steady flow in a baffled tank is studied as in the dissolved air flotation tank (DAF) and the hydraulic flocculator. For single phase, the dye and tracer method is the standard procedure and, in addition, laser Doppler velocimetry (LDV) measurement has been carried out. The steady flow in a model of the rotating blade flocculator is measured using LDV technique to validate the use of the sliding mesh method in CFD calculation. The flow in the model of the service reservoir is measured using particle imaging velocimetry (PIV) to validate the use of the deforming mesh technique in simulating the fill/discharge flow condition. An appropriate model has also been built to demonstrate the optimised CFD results.

Experimental demonstration of CFD is particularly crucial in the case of multiphase flow, where intuitive understanding of the flow behaviour is limited. The applicability of both LDV and PIV techniques is assessed by making flow measurements on a Perspex model. In particular the measurement of the two-phase air/water flow in a DAF tank is currently studied. The unsteady flow due to top water variation is simulated using the volume of fluid method and flow measurements are made using the PIV technique.

1 COMPUTATIONAL FLUID DYNAMICS (CFD)

The technique of CFD is well established in many industrial fields as a powerful investigative tool. Traditionally the capital cost of CFD systems has been the major factor in limiting their use to within high-technology, high-added-value spheres of business such as aerospace and nuclear energy production. However the water treatment and supply industry is an emerging user of CFD technology. It is the adaptable nature of CFD models which allows the analysis of a wide range of water treatment and supply applications (Ta, 1999a). Hence CFD analysis within the water industry can offer broad-range solutions which justify the costs of the procedure.

The technique of CFD uses the computational power of modern CPU's to derive iterative solutions to the non-linear Navier-Stokes equations of fluid mechanics. The way that CFD tackles complex flow problems is to create a mathematical construct which corresponds to the flow space, and to define a network of nodes/cells within that space. The variables which define the flow field e.g. velocity, pressure, temperature, phase composition (for multiphase models), viscosity (non-Newtonian fluid) and density (for non-isothermal models and compressible fluids) are each assigned a value at each of these nodes. Certain source terms and boundary conditions are applied which characterise the system. A number of constitutive equations are defined which must be satisfied in order that mass and energy are conserved and that there is a momentum-force balance within the system. An iterative approach is used in order to converge to a solution under the specified conditions.

1.1 CFD model

The accuracy of CFD simulations is determined by three main factors. These are discussed below.

a) Grid cell size This defines the smallest individual feature that can be modelled within the grid-based network. It is important to tailor the CFD model to the process, based on understanding of the process hydrodynamics. In this way one can resolve the level of detail which is required.

b) Boundary conditions and closure relationships These define the framework of the CFD model. Processes of water treatment generally start and end with a hydraulic break which must be identified before setting up the model. In the calculation of flow distribution, the hydraulic breaks are used to set-up the pressure boundary condition at the outlet. Almost all water treatment processes are of the open-channel flow type. The upper boundary is normally specified as a frictionless wall. However, further complications arise when modelling a multiphase free surface. For example in the case of modelling bubbles in water, the top wall must act as a sink to remove air from the free surface.

c) Convergence The CFD iteration must converge for both mass and momentum. Typical convergence residuals are less than 0.001 (corresponding SI unit). The authors found further that the convergence curve must be smooth throughout, a condition which is particularly important in multiphase applications (Ta and Brignal, 1997).

1.2 Application

The attraction of CFD is that it not only can compute detailed flow maps of a particular process, but also that it can easily be adapted to take into account other factors. Examples are modified geometry, temperature fluctuations and countless other changes in the basic process under investigation.

In the utilities businesses, the CFD technique is used for trouble-shooting: to investigate the failure of certain processes to meet their target performance. An attraction of the technique is its non-intrusive nature, hence disruption to services is minimised. The technique is also used in the evaluation of remedial options in this respect, providing input at the planning stage before committing to major investment (Ta and Brignal, 1998. Salter et al, 1999).

For contractors, the CFD technique is used in flow distribution calculations where the objective is to deliver flow with the minimum of head loss. This application alone can easily justify the investment cost of a CFD system. In process design, CFD is employed at the preliminary stage to short-list potential designs and aid the decision-making process, hence reducing costs (Ta and Brignal, 1997). The technique of CFD is also employed for scale-up purposes of water-treatment processes.

2 EXPERIMENTS

For engineering applications, it is generally accepted that all mathematical models must be verified or at least demonstrated to be acceptable. Simulations by CFD models are no exception. This is particularly crucial in multiphase systems, where intuitive understanding of this flow behaviour is limited. Verification experiments are grouped into traditional and current approaches, and are discussed below.

2.1 Traditional approach

Results of CFD can be compared with measurements obtained from process performance studies, carried out on full-size or pilot-plant installations. The emphasis of these experiments is to obtain process performance data, such as removal efficiency and residence time distribution. However, the spatial details of the flow patterns within the process are not measured. It is difficult to produce detailed flow maps within full-size and/or pilot-plant installations. The reasons for this are that access to the flow field is restricted and the use of non-invasive laser techniques is not feasible. Chemical tracer testing and acoustic Doppler velocimetry (ADV) are discussed below.

2.1.1 Chemical tracer testing

Chemical tracer testing is a standard technique in the water industry. In waste-water treatment, radioactive tracers have also been used for the same purpose (Manuals, 1980). The tracer is a slug of a specific chemical in solution, and is added at a particular point in the fluid flow field. Samples of the fluid are taken at point(s) in space and time, and the concentration of the chemical is determined for each sample. The data that they provide can be used to determine process parameters such as average residence time and its variance.

If we consider a small volume element of the added chemical tracer, then the time taken for that volume element to flow through the system to the detector (the residence time) will depend on the route taken. The residence time distribution (RTD), deriving from multitudes of such small volume elements, therefore provides information about the flow patterns in the system. This is essential information for current engineering design. Furthermore from the RTD curve, the age profile of the exit water is calculated by summing the water volume for each time interval. This information is applied to quantify the effluent water quality (Ta, 1998. Salter et al., 1999).

2.1.2 Acoustic Doppler Velocimetry (ADV)

This equipment consists of an acoustic sensor (mounted onto the probe unit), a signal conditioning module and signal processor. The sensor has three acoustic receivers and a centrally mounted transmitter. The reliability of the data from ADV analysis is generally represented in terms of the correlation factor (Lohrmann et al., 1994). This refers to the correlation between the signals received by each of the three acoustic sensors. The focal point is designed to be constant, however, it is still possible to test at a range of known depths within the fluid. This is due to the fact that the probe unit is submersible.

The sample volume is a cylindrical volume within the flow field that surrounds the focal point of the acoustic signal. This sample volume is typically 5-6mm in diameter and 8-10mm in height, although the height of the cylinder can be reduced (to enhance resolution) by confining the extracted signal to within the central 1-2mm zone. The transmitter and receivers must all be in contact with the fluid that is being measured. This is essential in order to propagate the necessary acoustic signals.

This technique has been used in the measurement of flow velocity within full-size DAF tanks (Adlan et al., 1997. Eades and Beckley, 1998) and full-size flocculators (Ta et al., 1996).

2.1.3 Issues of scale

To achieve similarity between the experimental model and the full size installation on which it is based, both the geometry and the dominant forces must be taken into account. Common forces that are experienced within the flow field of water treatment processes are shown in Table 1, along with their associated dimensionless constants.

It is not feasible for any one given experimental arrangement to offer complete parity between the forces in the experimental model and those in the full-size installation. One given set of experiments may allow only one of the governing dimensionless numbers to be maintained constant between model and full-size tank.

Table 1 - Dimensionless constants and associated forces.

Inertial force	$u^2 A \rho$
Gravitational force	$g A h \rho$
Buoyancy force	$g V (\rho_i - \rho_j)$
Viscous force	$\frac{\mu u A}{y}$
Surface tension force	σd_b
Froude number (Fr) ratio of inertial / gravitational forces	$\frac{u^2}{gh}$
Reynolds number (Re) ratio of inertial/ viscous forces	$\frac{\rho u d}{\mu}$
Weber number (We) ratio of inertial / surface tension forces	$\frac{u^2 \rho_i d}{\sigma}$

where:
$\rho_{i,j}$ = density of phase i,j , u = velocity, A = cross-sectional area,
g = acceleration of free-fall, h = depth of liquid, μ = viscosity of liquid,
y = characteristic shear dimension, σ = surface tension of water,
d_b = bubble diameter (relevant to the DAF process)

The objective of the small-scale models on the other hand is not to achieve scaling of the process. Rather it is that the laboratory model is qualitatively comparable to the full-scale process, while allowing the experimental freedom to carry out rigorous verification of CFD models. The CFD model itself is then scaled-up in order that it is applicable to the full-scale process.

2.2 Current approach

The current approach puts emphasis on direct measurement of the velocity vectors within the flow field, so that it can be compared directly with CFD simulations. It involves the design of functional models of the process under investigation. The models are designed to be representative of the real process, while at the same time being of a laboratory-scale and allowing as much freedom as possible with regard to the types of flow measurement techniques which can be used.

The chemical tracer technique is still available for flow evaluation in the laboratory models. In addition optical techniques can be used, which rely on the flow field being penetrable by light. The optical techniques which can be used are dye tracer testing, laser Doppler velocimetry (LDV) and particle imaging velocimetry (PIV). The chemical tracer, dye tracer and LDV techniques are suitable for the measurement of steady-state flow conditions. The PIV technique is equally applicable to the measurement of time-dependent flow.

2.2.1 Dye tracer testing

This involves the simple injection of a visible dye into the flow field. The advantage of dye-marker tracing is that the flow patterns can thus be readily visualised and can be recorded with the use of camera or video equipment. However, the data derived from this technique may be difficult to quantify, as only the projection of the flow field onto a plane is recorded. This data is appropriate for systems which can be reduced to 2-D approximations because of symmetry in the third dimension.

Tracking the passage of the tracer material determines the instantaneous velocity field. The specific gravity of the tracer material is, however, likely to be greater than that of water. Therefore a significant component of the recorded velocity will be attributable to a gravitational body force rather than the velocity of the surrounding fluid. To render the tracer material neutrally-buoyant, either its density must be changed or the density of the fluid must be changed, e.g. by mixing an appropriate quantity of salt into the water. This is only feasible if the initial densities of the fluid and tracer material differ by less than approximately 3%. On the other hand, tracks of particles of known specific gravity can be compared directly with the CFD simulation.

2.2.2 Laser-Doppler Velocimetry (LDV)

This broad description comprises a host of techniques (Durst et al., 1976). The common factor between them is that they all rely on the measurable frequency-shift of an optical light source. Because the wavelength of light is orders of magnitude shorter than the wavelength of the acoustic waves used in ADV, the laser technique has a significantly higher resolution. Beams of coherent light are brought to a focus within the flowing fluid. The light is scattered from minute particles (10-30μm), "scattering centres" within the fluid. These may be naturally occurring (dust-particles, silt, etc.) or may have been deliberately added (known as 'seeding' the flow). The frequency of the scattered light is Doppler-shifted according to the velocity of the scattering particles, and it is thus that the particle velocity is measured.

In order to differentiate the very small Doppler shift that would be expected from flowing water systems ($v \ll c$ where v = velocity of fluid, c = velocity of light), the illuminating beam must be split into two identical beams which are brought to a focus within the fluid. Interference fringes are produced within the sample volume, which then interact with passing particles to produce a Doppler-shifted signal. The beat frequency of the Doppler signal is used to determine the fluid velocity.

Only one component of the fluid velocity can be measured by any one beam at any one time. The beam orientation is then changed in order to measure other components of the fluid velocity. Complex LDV systems do exist which can focus two or even three individual beam pairs onto a single measurement volume and in this way obtain simultaneous 2-D and 3-D velocity measurements. However, for steady flow situations it is often sufficient (and less expensive in terms of equipment cost) to carry out a number of separate, single-beam experiments.

The LDV technique has the advantage that it is non-invasive. It also has a very fast response time to changes in flow velocity, making it an ideal tool for the measurement of turbulent flow fields. These factors, coupled with its high resolution, make it a very useful technique in the verification of CFD models.

2.2.3 Particle Imaging Velocimetry (PIV)

In the late 1980s, PIV emerged as a powerful fluid dynamics measurement tool for obtaining instantaneous whole-field velocity data. The PIV technique was of sufficient maturity by the mid-1990s for commercial PIV systems to be developed. PIV offers a means of simultaneous evaluation of 2-D velocity fields. This is the main difference between PIV and point-measurement tools, such as LDV.

Another important difference is that PIV, as its name suggests, produces a live image of the seeding particles as they follow the flow. As in the case of LDV, the PIV technique relies on a laser optic probe. However, in the case of the PIV laser configuration, the beam is pulsed through a cylindrical lens at a very high frequency in order to produce a sheet of light within the sample volume. The measured response is that of particles traversing the light sheet. In doing so they give rise to a scattering of the light, which produces an image of the particle at right angles to the sheet. This image is then recorded by a high-speed, digital video camera, either on film or as data. Two 2-D images of the flow field are obtained, with a very short time lapse of known duration between them (a typical time step would be 1ms). These are used to calculate 2-D velocity vectors at each point where a particle has passed. After several hundred frames a full 2-D map of the velocity vectors is produced for that sample zone. The response time to fluctuations in velocity at particular points within the flow may be slower for PIV than for LDV, making it perhaps less suitable for detailed analysis of turbulence.

Seeding of the flow field is usually required in order to produce an abundance of scattering sites. In order that the particles can be successfully imaged, they do have to be of a certain size which is generally larger than is required by LDV. An appropriate seeding material for PIV analysis of flowing water would be powdered polyethylene or hollow, reflective, polymeric spheres. However, small bubbles within the water, as in the case of DAF, should also work well (Brücker, 1995).

A variation on the standard PIV system is the scanning PIV system (SPIV). This utilises an assembly of a rotating mirror and a cylindrical lens, positioned in front of the laser optics. An array of discrete, parallel laser sheets is produced, each separated from its neighbours by a small, fixed distance. Each of the individual laser sheets is produced exclusively and in a sequence that is determined by the rotating mirror. Likewise the scattering from each of the sheets is recorded independently by the high-speed camera. This arrangement allows the measurement of 3-D flow field.

3 CURRENT FOCUS

A number of water treatment processes are currently under investigation by CFD modelling. Small-scale models are playing and will continue to play a vital role in the verification of such CFD models. Details of these focus areas are given in the following sections.

3.1 Raw water reservoir

Water that is abstracted from rivers is often stored in reservoirs, prior to being fed into the water treatment works. The hydrodynamics of the process was studied using scale models that were mounted on a rotating table in order to simulate the Coriolis effect (Robinson, 1979). Models of Thames Valley reservoirs were built with various dimensions. The largest model measured 10m x 6m x 0.4m and the smallest 2m x 3m x 0.4m (length x width x depth).

Measurements were carried out using a dye tracer technique and photography of a buoyant object on the surface of the model. While the data may or may not apply to full-size reservoirs because of scaling issues, they are readily used for the validation of CFD models.

Raw water reservoirs represent a horizontal flow system. Studies are applicable to other processes including waste-water ponds (Salter et al., 1999) and disinfection tanks. Short-circuiting and dead zones are main concerns, and can be remedied using baffles or columns.

The aeration of reservoirs is a current focus of study. This is carried out by injecting air bubbles into a reservoir model. Measurements of the flow field induced by the bubbles are obtained using the PIV technique. Bubbles used for the aeration process are larger than 0.1mm. This bubbling system is similar to that of the ozonation process.

3.2 Hydraulic Flocculator

A hydraulic flocculator consists of a rectangular tank with flow under and over baffles, or flow around the baffles. This flow regime is also seen in ozone and dissolved air flotation tanks.

Figure 1 shows the Perspex model of a hydraulic flocculator, which was used for verification experiments where the LDV flow measurement technique was used. The dimensions of the model are 0.9m x 0.4m x 0.6m (length x width x height).

The important flow feature was identified as the vertical flow circulation. This causes short-circuiting in the tank. The vertical flow circulation was predicted by a CFD model where the free-surface was approximated by a frictionless fixed wall, and was compared with the results of LDV verification experiments (Figure 2). The agreement between the CFD simulation and the experimental results was found to be reasonable (within 10% throughout), justifying the approximation for the free-surface boundary.

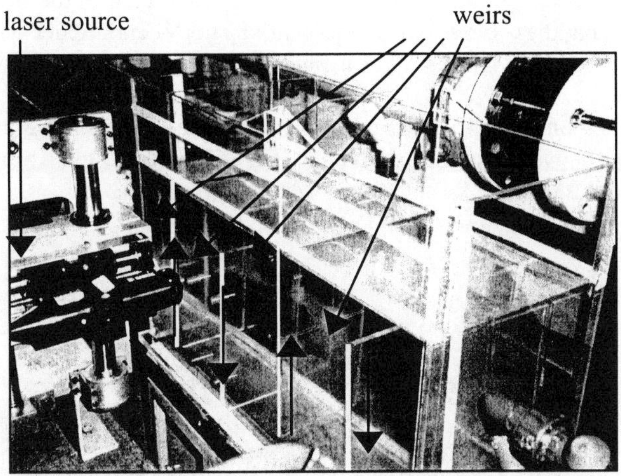

Figure 1 - Hydraulic flocculator model showing direction of flow. LDV apparatus is set up to the left of the picture.

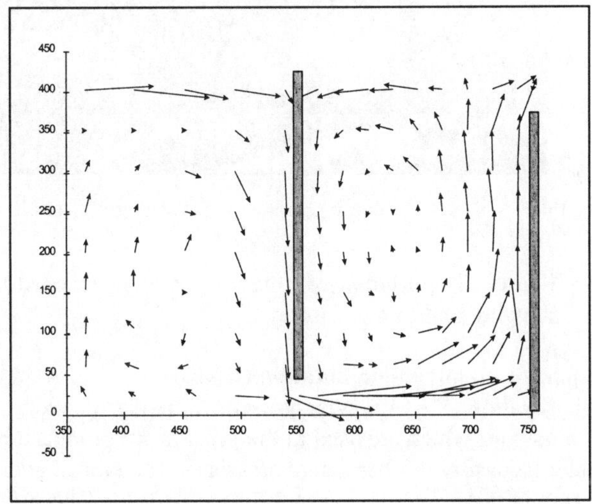

Figure 2 - LDV results showing the flow patterns within the hydraulic flocculator.

3.3 Rotating paddle flocculator

A Perspex model of a rotating paddle flocculator was designed and constructed, and is shown in Figure 3. The dimensions of the model are 0.7m x 0.4m x 0.55m (length x width x height). The inlet is shown at the bottom on one end of the tank. The water then flows from the inlet, over the first baffle and feeds into the main, central tank. In the central tank, shear stress is generated in the body of water by the rotating paddles. Flow out of the central tank occurs under a baffle, and the water then continues to flow over a final outlet weir which marks the end of the process.

The CFD simulation of this process makes use of the sliding mesh technique. Comparison between the measured and CFD simulated results is in progress. This CFD model is also used for mixing processes.

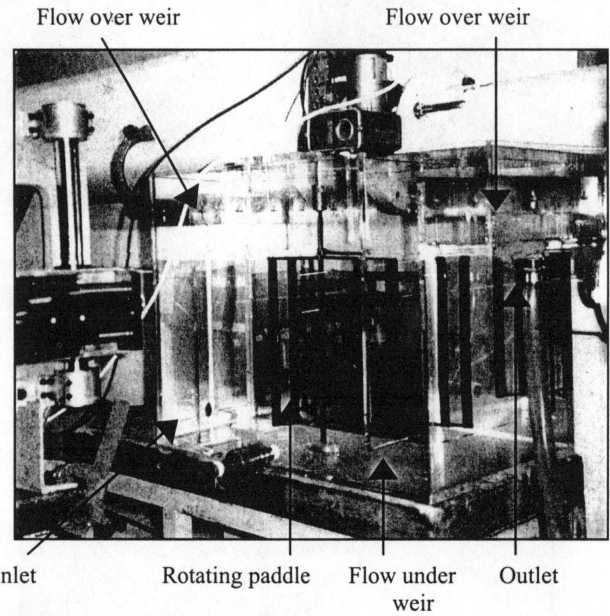

Figure 3 - Model of rotating paddle flocculator, showing main flow features

3.4 Co-current dissolved-air flotation (DAF)

Dissolved air flotation (DAF) is widely used in the clarification of potable water supplies. The bubbles which are used in the process are produced by means of air-saturated water (typically 4-7 bar gauge pressure). The typical average diameter of bubbles formed in this way is between 30μm and 70μm. These fine bubbles then attach to the surface of particles, lifting them out of the bulk fluid towards an upper, free surface where they can be collected.

A Perspex model of a DAF tank has been built. The dimensions of the model are 0.75m x 0.3m x 0.5m (length x width x height). For the single-phase model, no dissolved air is fed to the model, i.e. only water flows through the model. The flow regime is similar in some respects to that in the hydraulic flocculator. Figure 4 shows a dye tracer test in progress.

Figure 4 - Model DAF tank showing re-circulation currents, visualised by dye tracer testing

The next step is to investigate the flow in the two-phase (water/air) configuration. Figure 5 shows the cloud of air bubbles generated by the model during two-phase flow. The applicability of both LDV and PIV techniques is currently being assessed for two-phase flow measurement in the DAF model.

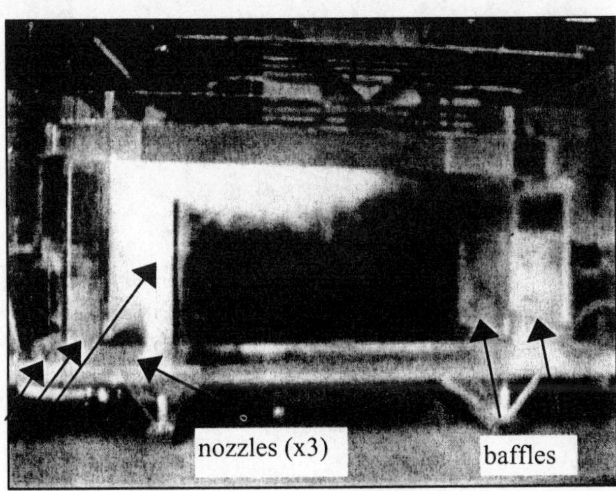

Figure 5 - Bubble cloud in DAF tank during two-phase flow experiment.

Computer modelling of DAF tank flow behaviour is limited. A number of computational fluid dynamics (CFD) models have been presented in the literature, either for single-phase flow (Ta and Brignall, 1997) or two-phase flow (Ta et al.,

1996. Fawcett, 1997). Some verification experiments in the form of dye visualisation experiments, underwater camera imaging and ADV measurements, were carried out by Ta and Brignall (1997) and Ta et al. (1996).

3.5 Service reservoir

Service reservoirs are used to store clean water in the distribution network. The aim is to maintain adequate pressure downstream and to maintain a supply for emergency use. The reservoir is normally filled during the night-time and emptied during the day-time. Because of the dynamic change of the water level, the flow is time-dependent. The effect of the variation is modelled using either the multiphase volume-of-fluid (VOF) method or the single phase deforming mesh technique (Ta, 1998). This model is also used for storm tanks (Ta, 1999b).

Figure 6 shows a still from a recording of dye tracer testing during the discharge cycle. The 2-D velocity vectors that were calculated from this work are shown in Figure 7. The PIV technique was also used to provide similar information on flow vectors around the outlet. The arrangement of the PIV equipment is shown in Figure 8, and a typical image of the seeding particles is shown in Figure 9.

Figure 6 - Dye tracer test in progress within service reservoir model during draining.

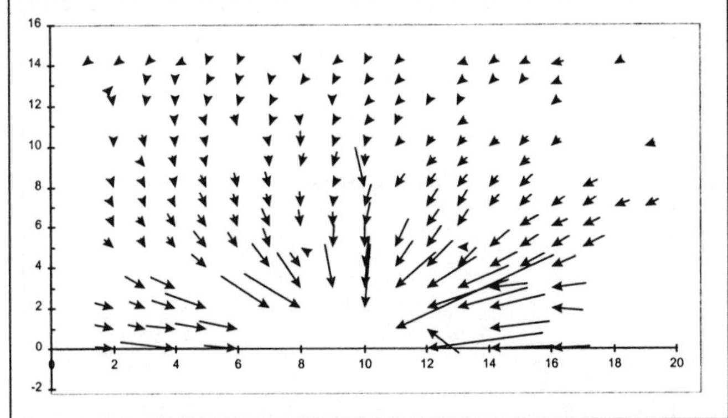

Figure 7 - Velocity vectors calculated from dye-tracer testing. NB: no measurements could be made in close proximity to the outlet (position 10 on x-axis) due to the response time of the camera.

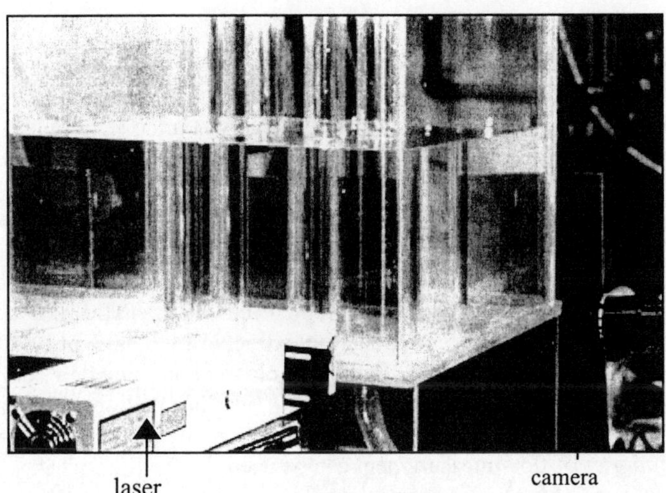

Figure 8 - Arrangement of laser and camera for PIV flow measurement on service reservoir model.

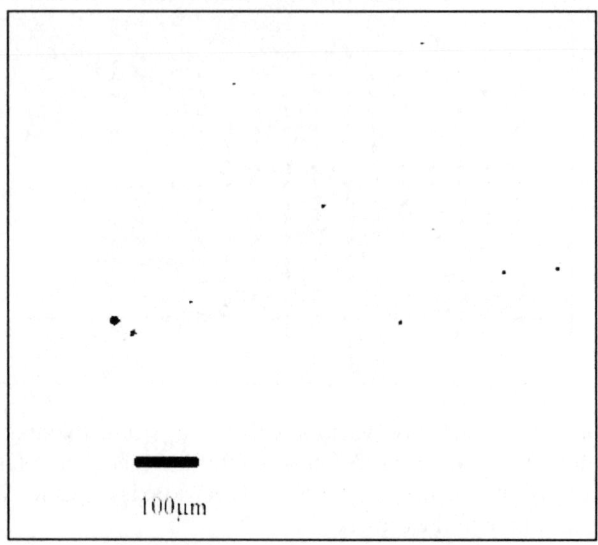

Figure 9 - A still taken from PIV measurements of seeding particles within the flow of the service reservoir model. The image has been inverted, showing the particles as dark spots.

4 CONCLUSIONS

The use of CFD as a design tool requires experimental validation in order to build confidence in the technique. This is especially true when the technique is breaking new ground, where the base of knowledge and experience is limited, as is the case in many water treatment and supply processes. The development of CFD models goes hand-in-hand therefore with detailed, accurate measurements of flow velocity within the process of interest.

The measurement of flow velocity fields is itself a huge challenge in most processes; each technique of flow measurement has its strengths and weaknesses and very rarely can all the required data be extracted by means of one technique alone. There are severe limitations, both technological and practical, on the range of flow measurements that can be made on full-size installations. The situation is less restrictive for transparent, laboratory models of treatment processes, where optical techniques of flow measurement can be used.

Specific verification experiments were given for five treatment processes: storage in a raw water reservoir, hydraulic flocculation, rotating paddle flocculation, dissolved air flotation (DAF) and service reservoir. These have been chosen to cover a number of CFD modelling techniques which are also applied to other processes. The three generic applications of CFD are steady flow (including horizontal and vertical flow as well as mixing), unsteady flow and two-phase bubbly flow.

REFERENCES

Adlan M.N., Elliott D.J., Noone G., Martin E.B. (1997). Investigation of velocity distribution in dissolved air flotation tank. In: *Dissolved Air Flotation, Proc. Int. Conf. Chartered Institute of Water and Environmental Management*, London

Brücker C.H. (1995). Digital PIV in a scanning light sheet: 3-D starting flow around a short cylinder. *Experiments in Fluids* **19** 255-263.

Durst F., Melling A. and Whitelaw J.H. (1976). *Principles and Practice of Laser Doppler Anemometry*. Academic Press. London.

Eades A., Beckley J. (1998). Private Communication.

Fawcett N.S.J. (1997) The hydraulics of flotation tanks : computational modelling. In: Dissolved Air Flotation, *Proc. Int. Conf. Chartered Institute of Water and Environmental Management*, London

Lohrmann A., Cabrera R., Kraus N.C. (1994). Acoustic doppler velocimeter (ADV) for laboratory use. In: *Proceedings : Fundamentals and Advancements in Hydraulic Measurements and Experimentation*, C.A. Pugh (ed.), Buffalo, New York, pp. 351-365.

Manuals of British practice in water pollution control (1980). Unit Process - Primary Sedimentation.*The Institute of Water Pollution Control.*

Robinson S.J. (1979). *Hydraulic Modelling of Circulation in Reservoirs.* M.Phil. Thesis. Faculty of Engineering, Imperial College of Science and Technology, London.

Salter H.E., Ta C.T., Ouki S.K., Williams S.C. (1999). Three-dimensional computational fluid dynamic modelling of a facultative lagoon to investigate a possible method for improving performance. *Proc. IAWQ 4th Specialist Int. Conf. On Waste Stabilisation Ponds : Technology and the Environment.* Marrakech, Morocco, April 20-23

Ta C.T. (1999a). Current CFD tool for water and waste water treatment processes. *Accepted for presentation - ASME PVP Conf.* Boston, U.S.A August 1-5

Ta C.T. (1999b). Computational fluid dynamic model of storm tank. *To be presented 8th Int.Conf. Urban Storm Drainage.* Sydney, Australia, August 30- Sept. 3

Ta C.T. (1998). Computational fluid dynamics tools for treated water reservoir mixing studies. *Proc.AWWA WQTC* San Diego, U.S.A Nov. 1-4, session 3a3

Ta C.T., Brignal W.J. (1998). Application of computational fluid dynamics technique to storage reservoir studies. *Water Science and Technology* **37** (2) 219-226.

Ta C.T., Brignal W.J. (1997). Application of single-phase computational fluid dynamics techniques to dissolved air flotation tank studies. In: Dissolved Air Flotation, *Proc. Int. Conf. Chartered Institute of Water and Environmental Management*, London

Ta C.T., Eades A., Rachwal A.J. (1996). Practical methods for validating a computational fluid dynamics-based dissolved air flotation model. *Proc. Int. Conf. AWWA WQTC*, Boston MA, USA

HIPERWATER: A High Performance Computing EPANET-Based Demonstrator for Water Network Simulation and Leakage Minimisation

V. Hernández, F. Martínez, A.M. Vidal, J.M. Alonso, F. Alvarruiz, D. Guerrero, P.A. Ruiz *and* J. Vercher

ABSTRACT

This paper presents the development of a demonstrator (HIPERWATER) based on parallel computing (or High Performance Computing) for the simulation of water networks, including leakage simulation and minimisation. EPANET package has been the starting point for the demonstrator. The paper discusses first the approach used in the application of parallel computing to the hydraulic and water quality simulations. Next, we outline a model that permits us to take into account the presence of leakage in water networks, and we propose a methodology, also using parallel computing, for the reduction of leakage by means of computing the optimal settings of Pressure Reducing Valves.

1 INTRODUCTION

EPANET (Rossman 1993a) is one of the most extended water network simulation packages in the world, due both to the good performance of its calculation module and to the support of the EPA (US Environment Protection Agency). As described in this paper, the HIPERWATER project (ESPRIT project 24003, http://hiperttn.upv.es/hiperwater/) (Hernández et al. 1999) was developed starting from the source code of EPANET, and considering two main objectives. Firstly, to introduce High Performance Computing in order to use the power of computer clusters to accelerate simulation tasks. Secondly, to introduce leakage modelling and develop a parallel optimisation module to reduce leakage by controlling network pressures.

In the near future, with the use of Geographical Information Systems (GIS), which are able to incorporate all the real elements of a network into a model, the number of such elements can dramatically increase. On the other hand, leakage models require iterative computations, which implies an increase in computation time. Finally, optimisation problems, based on repeatedly solving simulation processes, present high complexity.

These factors easily explain the need for reducing computation times for network analysis processes. This has been done in the HIPERWATER project, by introducing *High Performance Computing* (HPC).

The HPC technology, which is nowadays synonymous with parallel computing (several processors working together to solve the same problem) (Kumar et al. 1994), is a well-established technology. However, its use has been restricted until recently to large enterprises and universities. This was due to the high cost of HPC hardware, which has represented a barrier to its adoption by smaller companies. The advent of low cost multiprocessor systems offers now a powerful choice for industry at large. HPC makes possible, by means of the interconnection of PCs or Workstations, to reach a computational power similar to that of the supercomputers at lower costs.

The technical aspects covered by HIPERWATER are, firstly, simulation processes related to the hydraulic problem and water quality modelling. The hydraulic problem involves the solution of a non-linear system of equations, which is done by a parallel version of *Todini's gradient algorithm*. In its turn, water quality simulation is carried out by the *Discrete Volume Element Method* (DVEM), also with the necessary modifications so as to transform it into a parallel algorithm.

On the other hand, as an important added feature of the demonstrator with respect to *EPANET*, we present the incorporation of an integrated leakage model (Martínez et al. 1999), and the development of a methodology for leakage minimisation by finding the optimal pressure reducing valve settings. This optimisation problem, formulated as a non-linear programming problem, is solved by a parallel algorithm based on *Sequential Quadratic Programming*.

2 HYDRAULIC SIMULATION

The equations that model the water networks (and piping networks in general) are non-linear and therefore require an iterative solution, i.e. the problem is reconducted to the solution of a sequence of systems of linear equations. One of the most effective methods to perform this reconduction is the *Gradient Method*, proposed by Todini (Todini 1979; Todini and Pilati 1987). As shown in (Salgado et al. 1987), the fast and reliable convergence properties of the gradient algorithm make it preferable to other traditional methods, especially for design and operation optimisation purposes.

The standard version of the Gradient Method is represented by the following coupled equations at each iteration:

i) Heads update:

$$H^{k+1} = -\left(A_{21}G^{-1}A_{12}\right)^{-1}\left(A_{21}G^{-1}\left(A_{11}Q^k + A_{10}H_0\right) - \left(A_{21}Q^k - q\right)\right), \qquad (1)$$

ii) Flows update:

$$Q^{k+1} = \left(I - G^{-1}A_{11}\right)Q^k - G^{-1}\left(A_{12}H^{k+1} + A_{10}H_0\right), \qquad (2)$$

where:

k: Gradient Method iteration counter,
H: piezometric head, column vector of NN–NS elements,
NN: total number of nodes,
NS: number of source nodes (reservoirs or tanks),
H_0: piezometric head of source nodes, column vector of NS elements,
Q: flowrate, column vector of NP elements,
NP: number of network links (pipes, valves or pumps),
q: known nodal consumption, column vector of NN–NS elements,
A_{12}: ($NP \times NH$) edge-to-non-source-node connectivity matrix,
$NH = NN - NS$, number of non-source nodes,
A_{21}: transposed matrix of A_{12},
A_{10}: ($NP \times NS$) edge-to-source-node connectivity matrix,
$A_{11} = \mathrm{diag}(\alpha_i \,|\, Q_i^k \,|^{\,n-1} + \beta_i / Q_i^k)$, ($NP \times NP$) diagonal matrix,
α_i, β_i: known characteristic parameters of the link,
n: headloss-flow exponent,
$G = N\,A_{11}'$,
$N = \mathrm{diag}(n)$, ($NP \times NP$) diagonal matrix,
$A_{11}' = \mathrm{diag}\,(\alpha_i \,|\, Q_i^k \,|^{\,n-1})$, ($NP \times NP$) diagonal matrix,
I: identity matrix.

Equation (1) represents a linear system of NH equations in the unknown piezometric heads, and it can be expressed in the standard format of a linear system:

$$\left(A_{21}G^{-1}A_{12}\right)H^{k+1} = -A_{21}G^{-1}\left(A_{11}Q^k + A_{10}H_0\right) + \left(A_{21}Q^k - q\right). \qquad (3)$$

System (3) is a standard linear problem $A\ x = b$. Its coefficient matrix is symmetric, positive definite and sparse. These properties should be taken into account in an efficient implementation of the algorithm. For the solution of such systems of equations, two strategies are available: direct and iterative methods. A comparison of different linear solvers in the context of the Gradient Method is performed in (Salgado 1992), where it is shown that direct methods based on Cholesky factorisation are generally more efficient from the point of view of execution time.

The parallel implementation described in this section is based on the efficient implementation contained in EPANET, which uses the Gradient Method with a direct Cholesky linear solver for (3). During the simulation process, the solution of the hydraulic problem at each time step is obtained by iteratively applying expressions (3) and (2). Since the most computationally expensive task is the solution of the linear system of equations in (3), the major concern is to use an efficient parallel algorithm to solve sparse symmetric positive definite systems.

We have solved the given system by parallelising the following four consecutive stages: ordering, symbolic factorisation, Cholesky factorisation (Liu 1992; Kumar et al. 1994) and triangular systems solution.

In the case of hydraulic simulation, the ordering and symbolic factorisation stages need to be performed only once as an initialisation process prior to the simulation, because the matrix structure does not change in the different time steps. The ordering scheme applied is a key factor determining the efficiency of both sequential and parallel implementations. In the first case, the objective is to minimise the memory requirements and the number of operations during the factorisation of the coefficient matrix. In the parallel case, it also has to be taken into account that the degree of parallelism during the factorisation must be high.

In particular, the ordering method used is the *Multilevel Nested Dissection* (MND) algorithm (Bui and Jones 1993; Heath et al. 1991). This method has some advantages over other ones, such as *Minimum Degree* (MD, the method used by EPANET):

- It is usually faster than Minimum Degree and *Spectral Nested Dissection*, especially for large matrices.
- It leads to more concurrence and better load balance during the factorisation.
- Minimum Degree is serial in nature, but Multilevel Nested Dissection can be parallelised.

In fact, experiences have shown that not only is MND preferable to MD in parallel implementations, but also in the sequential case, especially when large matrices are encountered. MND reduces the number of non-zero elements in the matrix factorisation, accelerating the phases of factorisation and triangular systems solution.

3 WATER QUALITY SIMULATION

There are various methods that can be used for water quality simulation. The one that EPANET uses is the *Discrete Volume Element Method* (DVEM) (Rossman et al. 1993b). In this method, each hydraulic simulation period is divided into several quality simulation periods. DVEM partitions each pipe into a number of segments in which the volume remains constant while hydraulic conditions keep unchanged, and simulates the advance of the pollutant substance being measured along these segments. Three kinds of quality simulations are possible: substance concentrations, water age analysis and percentage of flow from a determined source.

In order to parallelise the quality simulation, the basic idea is to divide the water network in several parts, one for each processor in our system. The network partitioning seeks to minimise the number of connections between the different parts. Time spent on network partitioning is not a crucial issue, since this process is done only once, at the beginning of the computation.

The network can be considered as a graph where the vertices are given by the nodes and the edges of the graph are the pipes and valves of the network. Graph partitioning algorithms can be used to obtain a partition in which the number of edges (pipes) connecting different partitions is minimised. Therefore, the required communications during the simulation will be reduced.

In particular, given a graph $G = (V, E)$ with n vertices ($|V| = n$), we want to partition V into k subsets $V_1, V_2, ..., V_k$, such that the number of edges of E whose incident vertices belong to different subsets is minimised. In addition, all the subsets have roughly the same amount of nodes, and a node cannot belong to different subsets.

In particular, the algorithm used is known as *Multilevel Recursive Bisection* (Bui and Jones 1993; Hendrickson and Leland 1993). Since the partition of the network is not a time-consuming task, a sequential version of this algorithm has been used.

Once this partition of the network has been established, simulation can begin. Each processor performs in parallel the advance of the pollutant, with the need for communicating the concentration of frontier nodes.

For each hydraulic time step, quality simulation proceeds as follows:

- Hydraulic simulation results are read and distributed among the processors.
- Minimum travel time r is calculated and communicated. Each processor p calculates the minimum travel time of its local part of the network r_p, and then communications take place in order to find the minimum of r_p, where $0 \leq p < P$, P being the number of processors.
- Each processor divides its links into volume elements, according to the calculated minimum travel time r.
- At this point, quality simulation takes place at intervals of duration r. At each interval, each processor performs the DVEM algorithm into its local subnetwork. Communications take place in the phase of "transport into node".
- Results are sent to one of the processors, which writes them to disk.

4 OVERLAPPING HYDRAULIC AND WATER QUALITY SIMULATIONS

This section presents a parallel approach complementary to that presented in the previous two sections. The idea is to overlap hydraulic and water quality simulations, so that both take place simultaneously.

Hydraulic and water quality simulations are performed separately in EPANET, i.e. hydraulic simulation is first completely carried out and its results written to a file. Then quality simulation is performed, reading the hydraulic results from that file. This is shown in Figure 1 for a 24 hours simulation.

An alternative approach is to overlap both simulations. In this case, we have one or several processors in charge of the hydraulic simulation and another group of processors in charge of the quality simulation. As soon as the first group completes the simulation of a hydraulic step, the results are sent to the other group, which carries out the quality simulation. If the system is adequately balanced, which can be done by choosing the number of processors for the hydraulic and quality simulations, an effect of overlap takes place, as shown in Figure 2.

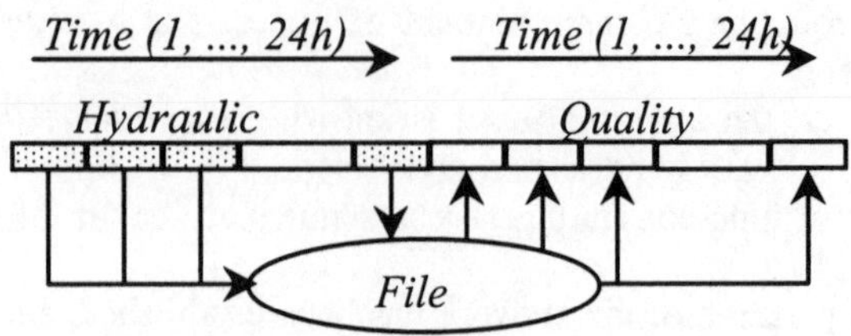

Figure 1: Hydraulic and water quality simulations in EPANET.

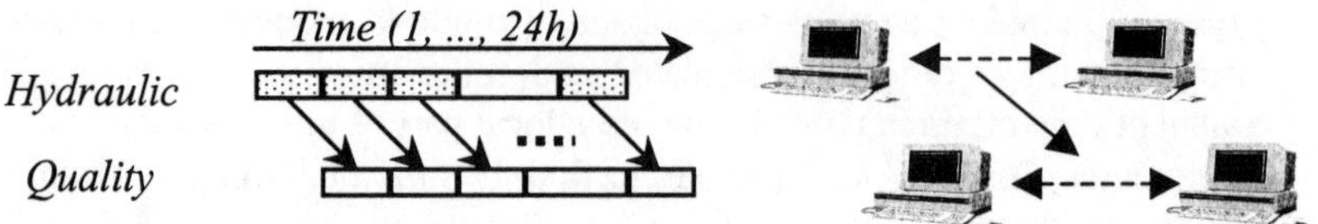

Figure 2: Overlapped hydraulic and water quality simulations in HIPERWATER.

5 LEAKAGE SIMULATION

While it is relatively easy to know the evolution hour by hour of the total flow injected into the network form different sources, it is generally impossible to obtain a similar evolution of each nodal demand. Normally all we know about nodal demands is their mean values. Despite this fact, classic models require that the user enters demand modulation curves as part of the input data.

On the other hand, the sum of demands for a given period will be less than the amount of water injected into the network for the same period, the difference being due to unregistered consumption, that we will consider leakage. From this difference we can compute the total leakage percentage with respect to the injected water.

Taking these facts into account, the main consideration for our leakage model is that modulation curves of nodal demands have to be obtained as a result of the application of the network equilibrium equations, considering the modulation of the water injected into the network, and certain variation laws imposed on leakage flows. Moreover, the global leakage percentage for the simulation period must be kept.

Different experiences have shown that leakage is very sensitive to network pressures. One of the equations used in the literature is based upon experimental studies and is presented as follows:

$$q_j = K_j p_j^{1.1}. \tag{4}$$

The variable q_j represents the leakage flow for a network node and p_j defines the pressure in such a node. The variable K_j represents a leakage coefficient to be determined for each node, which is supposed to stay constant for long periods of time. Its values can be obtained from the total leakage percentage observed in the whole network for district metering areas, under well-defined operating conditions (Martínez et al. 1999).

Taking leakage into account, the process of network analysis for a given time instant is carried out by the following iterative process:

1. Leakage flows are initially assumed to be zero at every node.
2. Temporary network pressures are determined by means of the standard Gradient Method described in section 2.
3. By means of the equation (4), an initial approximation of leakage for every node can be obtained.
4. Leakage flows are then added to real demands and new global consumption values are obtained.
5. Back to step 2. The iterative process finishes when leakage values at any node stay stable within a given desired accuracy.

6 LEAKAGE MINIMISATION

Total leakage is the sum of leakage at each simulation time step. At each of these time steps, we try to minimise leakage by controlling pressures with a number of Pressure Reducing Valves (PRV). We will make the assumption that we can minimise leakage at each time instant independently. Note that this independence is not strictly true, since PRV settings at an instant can alter the flows into/out of tanks, and therefore the network state at subsequent steps. However, this assumption can be considered a reasonable one, if we take into account that PRV usually control sectors with no tanks.

Germanopoulos (Germanopulos 1995) proposed an indirect formulation of the leakage minimisation problem, under the same assumption of independence between intervals. There, the objective is to minimise pressures in the network, with the constraint of keeping certain minimum pressures in some network critical points. That formulation permits the application of linear programming methods, although it does not strictly guarantee leakage minimisation.

For the HIPERWATER project, the problem of leakage minimisation at each time instant is formulated directly as a non-linear constrained minimisation problem

$$min\ f(x) = \sum_{j=1}^{NH} q_j(x)$$

$$\text{subject to: } p_j(x) \geq Pmin_j, \qquad j \in I_c$$

where:

x: PRV settings, a column vector of NV elements,
NV: Number of PRV valves that we can act on,
NH: Number of non-source nodes in the network,
$q_j(x)$: leakage flow at node j, expressed as a function of the PRV settings,
$p_j(x)$: pressure at node j, given by the formula $p_j(x) = H_j(x)-E_j$,
$H_j(x)$: piezometric head at node j,

E_j: elevation of node j (constant for each node),
$Pmin_j$: Minimum pressure imposed on node j,
I_c: subset of nodes where a minimum pressure is imposed.

An alternative, more standard way to express pressure constraints is:

$$g_j(x) = Pmin_j - p_j(x) \leq 0, \qquad j \in I_c,$$

where constraint functions $g_j(x)$ are introduced.

The optimisation problem stated is solved by means of the *Sequential Quadratic Programming Method* (SQP). We make use of the general-purpose implementation of the method in CFSQP (Lawrence et al. 1997). The SQP method is based on finding a step away from the current point by minimising a quadratic model of the problem. The elements that we have to provide for application of the SQP method are:

- The objective function $f(x)$ and the constraint functions $g_j(x)$. Given a value of x, we can compute the corresponding values $H_j(x)$ and $q_j(x)$ by solving the hydraulic analysis problem (with leakage) described in steps 1-5 of the previous sub-section. Then calculation of $f(x)$ and $g_j(x)$ is straightforward.
- Gradients of the previous functions with respect to the decision variables x: $\nabla f(x)$ and $\nabla g_j(x)$. These gradients are calculated by finite differences, although a deeper analysis could lead in the future to an analytical formulation of them.

In order to make a parallel implementation of this optimization process, we could think of applying parallelism to part of the hydraulic analysis, using the scheme outlined in section 2. However, an alternative approach is to take advantage of the time step sub-problems being independent of each other, thus assigning each of them to a different processor. This leads to a coarse-grain parallel implementation.

In order to make the implementation work efficiently independently of the number of processors and sub-problems, we have chosen a master-slave paradigm. The master has a pool of sub-problems to be solved, and initially gives one sub-problem to each slave. Then the master enters a loop where it waits for the solution of any sub-problem in any processor p; when this solution arrives it sends one of the sub-problems in the pool to the processor p. An outline of the algorithm of the master process follows:

```
    /* Nproc: number of processors
     * Nsub: number of subproblems
     */
Build pool of sub-problems
for p=0, 1,... min(Nproc, Nsub)-1
    Remove sub-problem sp from the pool
    Assign sp to processor p
end for
nrec = 0                /* number of solutions received */
while (nrec<Nsub)
    Wait for solution sol from any processor p
```

```
    Receive sol from p
    if (pool not empty)
          Remove sub-problem sp from the pool
          Send sp to processor p
    end if
end while
```

7 RESULTS

Results shown in this section have been obtained on a cluster of Pentium-Pro PCs linked with a Fast-Ethernet network. Operating system is Windows, and MPI (Message Passing Interface) (Dongarra et al. 1996) communication library is used. Each PC has 64 Mb RAM and runs at 200 MHz.

Three water networks (referred to as Test 1, 2 and 3) have been used for evaluating the performance of the parallel algorithms for hydraulic and water quality simulations. For the purpose of assessing the reduction in computing time that can be achieved by means of High Performance Computing, large water networks are used. These networks have been generated artificially, although they present features very close to those of real networks.

From a hydraulic point of view, these networks are very simple. They are fed by gravity pipes from one or more variable head tanks. Demand modulation is classified into five previously fixed patterns and randomly distributed among all the network nodes. Table 1 shows some summary information for the networks.

Table 1: water networks used for simulation.

	# Pipes	# Nodes	# Tanks
Test 1	4,901	2,501	1
Test 2	19,801	10,001	1
Test 3	34,516	32,404	4

Figure 3 shows the time spent on the simulation (both hydraulic and water quality simulations) by the HIPERWATER demonstrator, and also the time spent by EPANET. The demonstrator (with overlapping of hydraulic and water quality simulations) is run on 2, 3 and 4 processors (HW-2p, HW-3p and HW-4p respectively). It is easily appreciated that the computing time is notably reduced by the use of High Performance Computing in the HIPERWATER demonstrator.

On the other hand, tests 4, 5 and 6, corresponding to real water supply networks, have been used for the algorithms of leakage minimisation. Table 2 provides some summary information for them.

Figure 4 shows the computing time spent on the leakage minimisation process for the networks, considering from 1 to 5 processors. It can be observed the important reduction in computing time achieved by means of High Performance Computing.

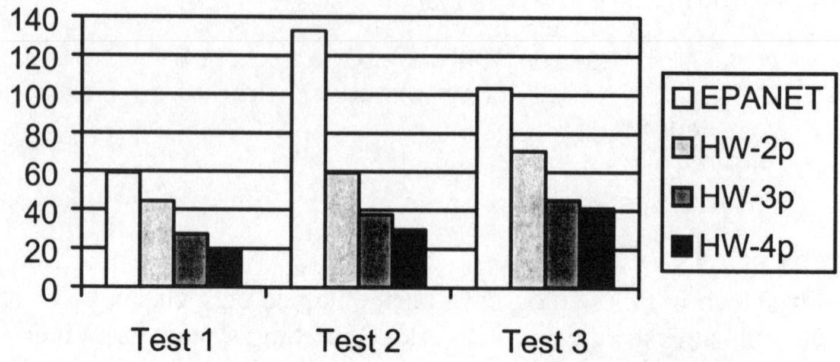

Figure 3: Simulation time (in seconds) of the HIPERWATER demonstrator, compared with EPANET.

Table 2: Water networks used for leakage minimisation

	# Pipes	# Nodes	# Tanks	# PRVs
Test 4	1,345	1,241	1	3
Test 5	805	770	4	3
Test 6	126	66	4	2

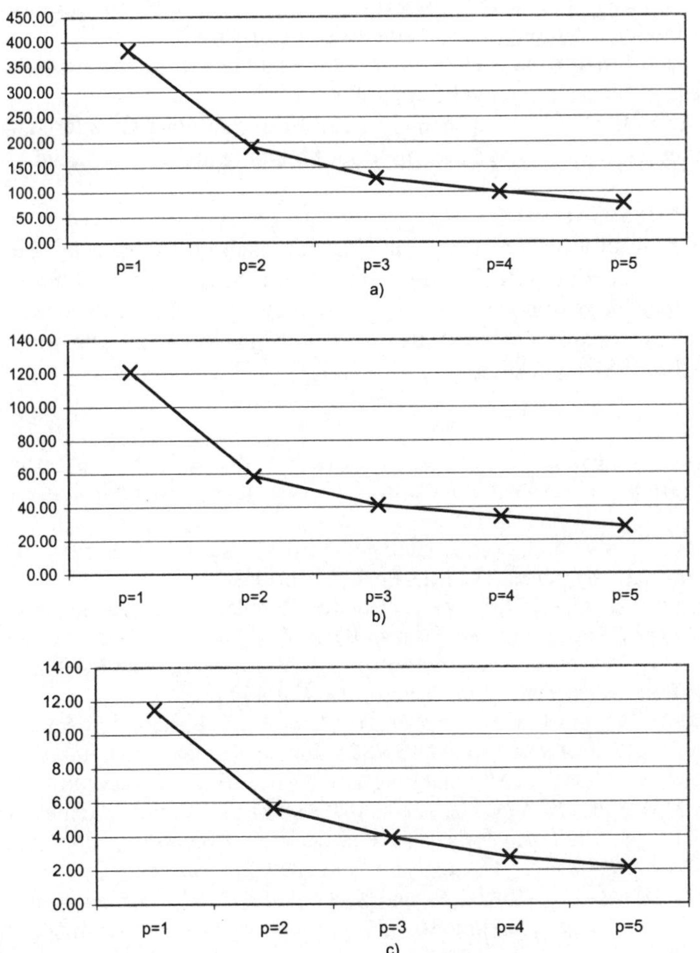

Figure 4: Computing time (in seconds) for leakage minimisation process as a function of the number of processors used: a) Test 4; b) Test 5; c) Test 6.

8 CONCLUSIONS

This paper shows the application of HPC techniques in water network analysis processes. Both hydraulic and water quality simulations are approached from the point of view of parallel computing, and results obtained show the interest in the application of parallel computing in processes of simulation of water networks.

On the other hand, a model for incorporating leakage in hydraulic simulation processes is briefly discussed. Based on that model, a methodology is presented for the determination of optimal pressure reducing valve settings minimising leakage. This process requires considerable computational power, which can be provided by means of the parallel implementation described in this paper.

Geographical Information Systems allow production of more complete network models, although their analysis is significantly more complex. A consequence of that is the need for more powerful computing resources, and hence the interest in the use of parallel computing.

Low-cost parallel systems, namely a cluster of Pentium PCs interconnected by a Fast-Ethernet network, have been used for obtaining the results shown.

ACKNOWLEDGMENTS

The authors wish to acknowledge the financial support provided by the ESPRIT programme of the European Commission, and also by research staff training grants from the Spanish government and the autonomous government of the Comunidad Valenciana, in Spain. We must also acknowledge André L. Tits for providing access to the CFSQP package.

REFERENCES

BUI, T., and JONES, C. (1993). "A Heuristic for Reducing Fill in Sparse Matrix Factorisation", *6th SIAM Conf. Parallel Processing for Scientific Computing*, 711-718.

DONGARRA, J., SNIR, M., OTTO, S., HUSS-LEDERMAN, S., and WALKER, D. (1996). *MPI: The Complete Reference*, The MIT Press Cambridge.

GERMANOPOULOS, G. (1995) "Valve Control Regulation for Reducing Leakage". *Improving Efficiency and Reliability in Water Distribution Systems,* 212-240.

HEATH, M. T., NG, E. G.-Y., and PEYTON, B. W. (1991). "Parallel Algorithms for Sparse Linear Systems", *SIAM Review*, 33: 420-460.

HENDRICKSON B., and LELAND R. (1993). "A Multilevel Algorithm for Partitioning Graphs". *Technical Report SAND93-1301*, Sandia National Laboratories.

HERNÁNDEZ, V., VIDAL, A.M., ALVARRUIZ, F., ALONSO, J. M., GUERRERO, D., RUIZ, P. A., MARTÍNEZ, F., VERCHER, J., and ULANICKI, B. (1999). "Parallel Computing in Water Network Analysis and Optimization Processes". *26th Ann. Water Resources Planning and Management Conf.*, ASCE, Arizona.

KUMAR, V., GRAMA, A., GUPTA, A., and KARYPIS, G. (1994). *Introduction to Parallel Computing: Design and Analysis of Algorithms*. Benjamin/Cummings, Redwood City, CA.

LAWRENCE, C. T., ZHOU, J. L., and TITS, A. L. (1997). "User's Guide for CFSQP Version 2.5: A C Code for Solving (Large Scale) Constrained Nonlinear (Minimax) Optimization Problems, Generating Iterates Satisfying All Inequality Constraints", *Technical Report TR-94-16r1*, Institute for Systems Research, University of Maryland.

LIU, J. W. H. (1992). "The Multifrontal Method for Sparse Matrix Solution: Theory and Practice", *SIAM Review*, vol. 34, no.1.

MARTÍNEZ, F., CONEJOS, P., and VERCHER, J. (1999). "Development of an Integrated Model for Water Distribution Systems Considering Both Distributed Leakage and Pressure-Dependent Demands". *26th Ann. Water Resources Planning and Management Conf.*, ASCE, Arizona.

ROSSMAN, L. A. (1993a). *EPANET User's Manual*, US Environmental Protection Agency.

ROSSMAN, L. A., BOULOS, P. F., and ALTMAN, T. (1993b). "Discrete Volume-Element Method for Network Water-Quality Models", *J. of Water Resources Planning and Management*, ASCE, 119(5), 505-517.

SALGADO, R., TODINI, E., and O'CONNELL P. E. (1987). "Comparison of the Gradient Method with some Traditional Methods for the Analysis of Water Supply Distribution Networks". *Computer Applications in Water Supply,* Volume 1, RSP, UK.

SALGADO, R. (1992). "Comparison between Linear Solvers for Sparse Systems in Steady State Pipe Network Analysis with the Gradient Method". *Numerical Methods in Engineering and Applied Sciences*, CIMNE, 297-306.

TODINI, E. (1979). "Un Metodo del Gradiente per la Verifica delle Reti Idrauliche", *Bolletino degli Ingegneri della Toscana*, No. 11, 11-14.

TODINI, E., and PILATI, S. (1987). "A Gradient Method for the Solution of Looped Pipe Networks", *Computer Applications in Water Supply,* Volume 1, RSP, UK.

ABOUT THE AUTHORS

For further information on HIPERWATER, please contact Prof. Vicente Hernández, coordinator of the project, Departamento de Sistemas Informáticos y Computación, Universidad Politécnica de Valencia, 46022, Valencia (Spain). Phone: +34 96 387 7356. Fax: +34 96 387 7359. E-mail: vhernand@dsic.upv.es, http://hiperttn.upv.es/hiperwater/.

For further information on the leakage model, please contact Prof. Fernando Martinez, Departamento de Ingeniería Hidraúlica, Universidad Politécnica de Valencia (Spain). Phone: +34 96 387 9610. Fax: +34 96 387 9619. Email: fmartine@hma.upv.es.

Construction and Use of a Dynamic Simulation Model for the Valencia Metropolitan Water Supply and Distribution Network

F. Martínez, M. Signes, R. Savall, M. Andrés, R. Ponz
and P. Conejos

ABSTRACT
This paper describes the experience of building a dynamic model for the Valencia water supply and distribution network, which provides water for 1,500,000 people through 1,500 km of pipe length. The network is supplied from two treatment plants and is regulated by pressure acting over a number of control valves. Moreover a nocturnal operating scheme is foreseen for saving energy. These factors have introduced a certain grade of complexity in order to build and calibrate a dynamic model. A quality model and a leakage model have been created using the dynamic model for the simulation stage. EPANET has been used as the base software for all such purposes.

1 INTRODUCTION

Building and calibrating the hydraulic model of a network in service has been carried out traditionally under static conditions for the maximum demand, when head losses are highest. However, the availability in recent years of numerous software packages performing dynamic simulation in extended period has motivated the building of dynamic models, which try to reproduce the time evolution of the most significant hydraulic variables of the systems (levels, flow-rates and pressures) in 24 hour periods. Calibration of such models is based on the registered values of chosen variables obtained from data-loggers or by telemetry. However, building and calibrating a dynamic model is far more complex than a static model. Difficulties often arise and surprisingly few papers have been written on experiences in building such models.

The authors have recently had the opportunity to build a dynamic model for the water supply system of the city of Valencia and its metropolitan area. The dynamic model is based on a previous static model, under a contract with Aguas de Valencia S.A., a private company in charge of the management of the network. The entire system is supplied from two treatment plants, Manises and Picassent, whose balanced outflows are determined by a set of remote controlled valves strategically situated throughout the network. These valves are continuously checked in order to maintain the pressure at various reference points. In addition, the Picassent plant is provided with a nocturnal operating schedule in order to save costs during the reduced tariff period, which significantly influence the variable registrations.

This experience has provided the opportunity to develop a 'modelling know-how', which is presented in this paper. The steps followed in building the model have been: volume balancing by sectors, nodal load allocation, establishment of pattern demands, screening of measurements, configuring the model to force some of the measurements and finally calibrating the system parameters (roughness, device control settings, etc). Field data available for such a purpose included: registers of injection flows, tank levels, flow-rates for selected mains and pressures at selected nodes, preferably upstream and downstream valves, as well as quarterly registered volumes at consumer meters and monthly registered volumes supplied to big consumers and peripheral demand nuclei. Model processing was been carried out using EPANET 1.1e.

The calibrated dynamic model was exploited for different goals: simulations under different operating conditions, establishing the water age according to control rules and checking a new leakage model developed by our team. Developments in the future will include extensions of the model as new application fields are foreseen to fulfil the service quality requirements.

2 DESCRIPTION OF THE NETWORK

Figure 1 illustrates the system to be modelled. Valencia is located on the eastern seaboard of Spain, with the terrain irrigated by the Turia River. *La Presa* treatment plant is located about 15 km upstream of the river estuary; it has supplied surface water to the city for over 100 years. Its treatment capacity, after several improvements, is currently 2.2 m^3/s. In the past, numerous boreholes complemented the water source but the majority of them have been shut down because of sanitary problems. In order to satisfy increasing demand during the 1980s, a new treatment plant was built at the south of the city, known as *Picassent*. Water is extracted from a transfer channel which transports water from the Jucar river located 40 km to the south of the city, to the Turia river. The current capacity of the new plant is 3 m^3/s, which added to the former plant is enough to supply the estimated consumption of the city for the next 20 years.

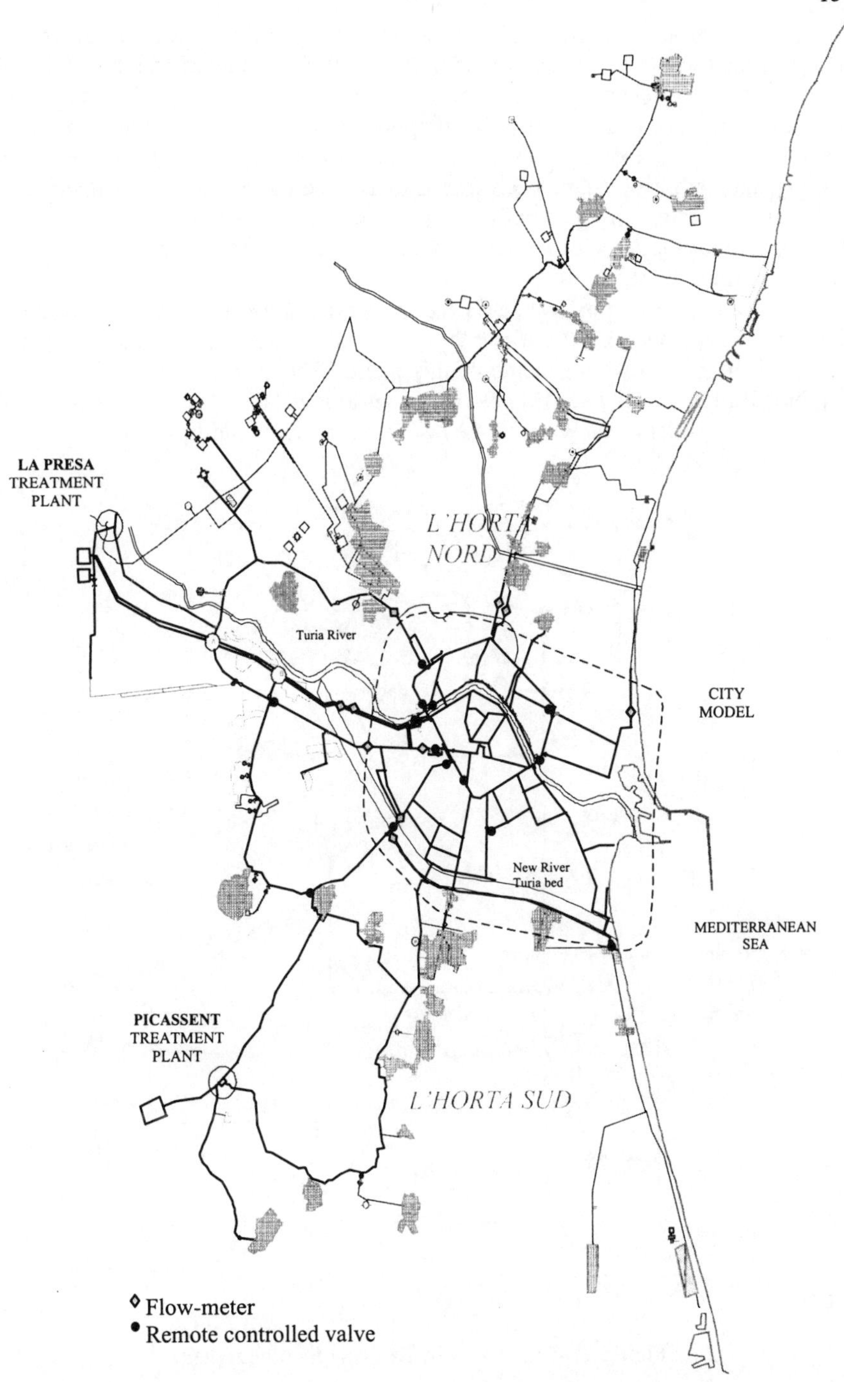

Figure 1. The metropolitan water supply and distribution system of Valencia area

The area around Valencia City is densely populated with 800,000 inhabitants living within the city and a total of 1.5 million within a radius of 15 km. In total 44 towns of different sizes constitute the metropolitan area of Valencia. In the past this area has been predominately supplied from boreholes, but due to the continuous increase of nitrates in the groundwater resulting from agricultural activity, the authorities have been forced to enlarge the current water distribution system of Valencia city to supply surface water to the whole metropolitan area. The capacity of the *La Presa* plant has recently been increased by 2 m^3/s to satisfy the new requirements.

The water distribution system is fed by gravity from the reservoirs located at the treatment plants. La Presa has 90,000 m^3 of storage capacity, at a height of 112m. Figure 2 shows a scheme of this plant. The Picassent reservoirs have a combined total capacity of 100,000 m^3 at an altitude of 92 m, with a total of 7 booster pumps raising the water to the reservoirs, as shown in figure 3.

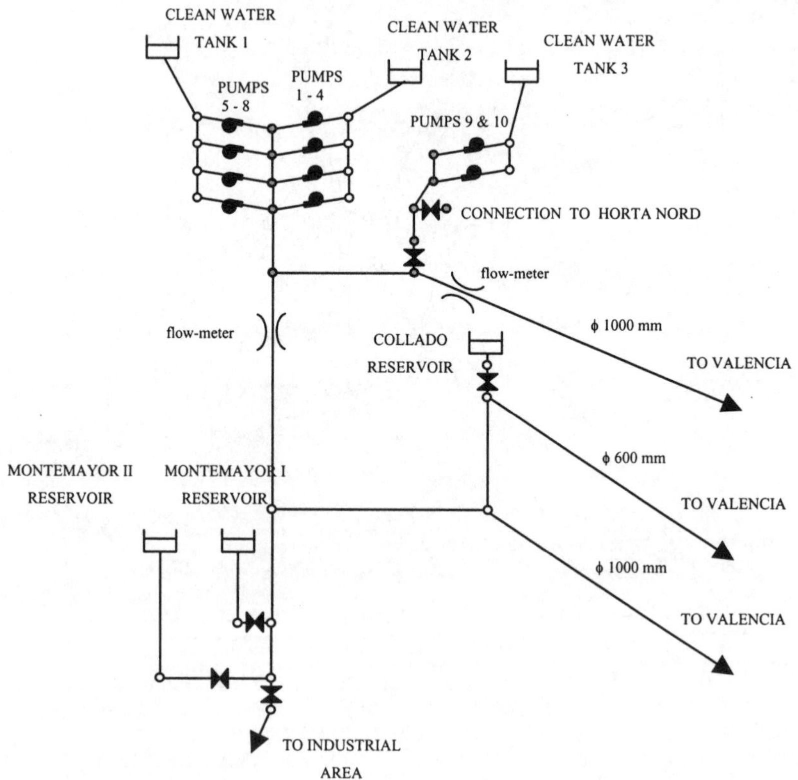

Figure 2. Scheme of the La Presa treatment plant

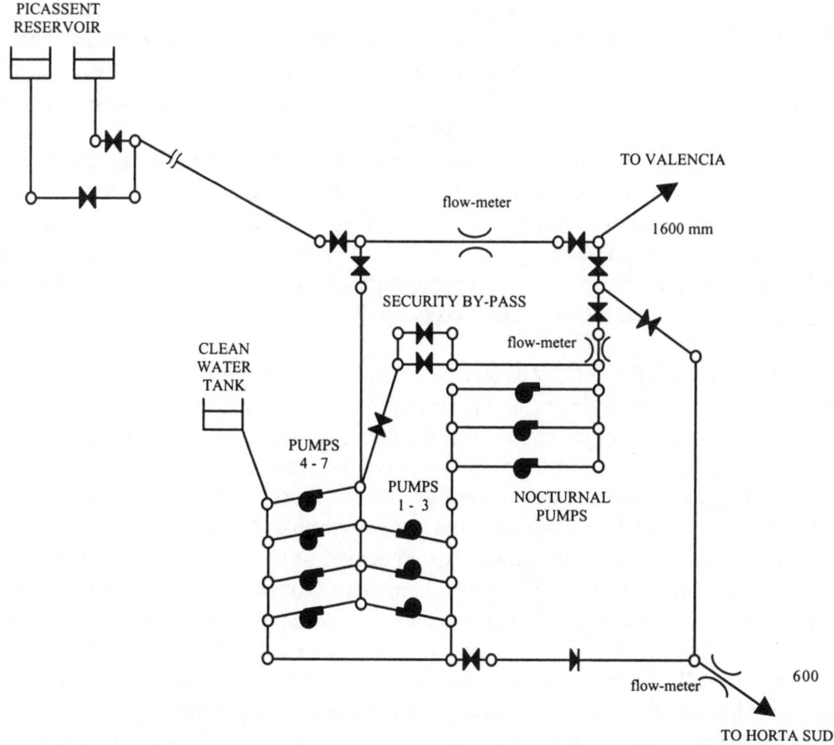

Figure 3. Scheme of the Picassent treatment plant, including the nocturnal pumps

Service pressures in the network must range between 30-60m, with the elevation at the east bound of the network near zero. Measures must be taken to control the pressures, particularly on a night when demand is low and consequently there are low head losses. For many years the system has been regulated by remote-controlled valves, the number now having increased to 30 valves (the most important are marked on figure 1 as big dots).

Valves are used to control the flow-rates injected from each plant. Depending on the demand, water availability and economic aspects, water is supplied from one or other of the two plants. In general, treated water coming from Picassent is cheaper than that coming from La Presa and is prefered. Moreover, the reservoirs of Picassent act as balancing reservoirs, in such a way that they are filled during the night when energy is cheaper and supply the stored water during the day when energy is more expensive. In order to fill the Picassent reservoirs during the night without increasing the service pressure at the entry point of the city, a valve isolates the reservoirs from the system, while three auxiliary pumps supply water directly to the city at low pressure. In brief, the whole system is fed from La Presa during the night and from Picassent in the morning, while in the afternoon both plants deliver water at approximately the same rate.

Five years ago the authors created a static model of the distribution network of the city. The calibration of the system was done for the peak demand period, focusing interest on system diagnosis. For this purpose pressures upstream and downstream of the valves were considered as data, with the valve opening maintained during the measuring period.

In order to accomplish more sophisticated tasks the model was enlarged. The modifications included: introduction of treatment plants, verify the new control rules required when the whole metropolitan system is in operation, check the efficiency of the regulation system, analyse the spatial and temporary leakage evolution, regulate demands acting on pressures during drought periods, develop a primary quality model, among many other applications.

3 LOADING THE NETWORK

Loading a dynamic model is not a simple matter. Demands and flows change continuously and although continuity for inlets and outlets at each node, as for the total network, is verified in the real system at each time instant, in practice, it is impossible to know the instantaneous flows everywhere. Injected flows are usually registered every five or ten minutes, but consumption is rarely registered with such accuracy. Usually, large consumers are provided with counter meters, which log the data daily or monthly. The same happens with domestic consumers, whose consumption is logged every one, two or three months, depending on the town size and the company policy. Furthermore, in this case, readings were not taken on the same day for all customers and the lag between the first and the last reading for a billing period can be of the same magnitude as the period itself.

Moreover, if we compare the total registered water in a given period with the total injected quantity, we will usually find an appreciable difference: the *non-accounted-for water*. Some of the items that cause this difference are: losses, under-registrations, consumers without meters, illegal connections and evaporation.

What can we do to load a dynamic model properly? Networks are sometimes provided with flow-meters in inner pipes that can contribute to a closer knowledge of how flow-rates span throughout the network and how the non-accounted-for water is distributed. However, when they are available, it is in a reduced number, not enough to make the loading problem deterministic. In the next section a practical solution is introduced.

3.1 Volume balance

In order to properly load a dynamic model we must proceed in such a way that continuity between produced and consumed water must be guaranteed on average values for monthly, daily and hourly periods. Firstly, a long term volume balance between registered consumption and produced water, based on counter meter registrations, is needed to obtain the base demand at nodes. Then a short-term balance based on flow-meters registrations will lead us to determine the final hourly demands at each node by using demand patterns.

In our case, a regular month was chosen to represent the long term demand and production, *October 1997*. Statistics confirmed that the total consumption in this month closely equalled the average annual demand of the network. Firstly the total production of treatment plants for this month was integrated from the flow-meters registrations and the trajectories of the different reservoirs levels. For this particular case several flow-meters were installed at the entry points of the distribution network of the city, in order to calibrate the former static model. Figure 1 indicates their locations. Today these flow-meters are integrated in the SCADA system of the control centre and registrations could be integrated as well in order to calculate the total volume injected into the city network during this month. The difference between the monthly volume produced and injected into the city represents the total volume consumed by the nucleus of population supplied from trunks between the plants and the city.

The connections to this nucleus were provided with counters providing the total demands for this month. By comparison, the unaccounted-for-water was determined and distributed between the different demand nuclei according to their demand figures, in order to fulfil continuity.

From billing corresponding between the period September-October, the total consumption for each customer during October was estimated. This data was used to allocate the base demands to the model nodes for the city area. A simplified method based on the population served by areas was used for this purpose. The network was divided into sectors and the customers served by each sector were identified according to their postal code. The difference between the total billed water prorated to October and the total volume injected into the city during this month was the unaccounted- for-water for the city area, which was apportioned between nodes using the allocation factors previously determined from billing. After this stage, average monthly demands were determined at each node of the model without any continuity violation.

3.2 Demand patterns

We are interested in determining the modulation of demands at each node. Registered flow-rates available for inlets and outlets constitute the basis for modulating the demands. From simple inspection of the water produced by the plants during October, an important difference in the modulation and the average demand was observed, from a working day to a weekend day. In particular, the ratio between the average demand for a working day and the monthly average demand was 1.035, and therefore all nodal loads were affected at first with this coefficient.

Next the modulated curve for the total injected flow from both plants was averaged for all the working days of the month. The same procedure was applied to the total inflow of the city, registered by the flow-meters shown in Figure 1, as well as to the outlets registered by the flow-meters available in some of the peripheral nuclei. The hour-by-hour difference between the produced water and the inflow to the city provided the demand curve pattern in the nuclei, supplied from trunks and not provided by the data-loggers. On the other hand the hour-by-hour difference between the inflow to the city and the outlets to the nuclei provided by the data-logger gave us the demand curve pattern for the city. In figure 4, patterns for the produced water and for the city demands are compared. Dotted curves represent the average factors plus and minus the standard deviation.

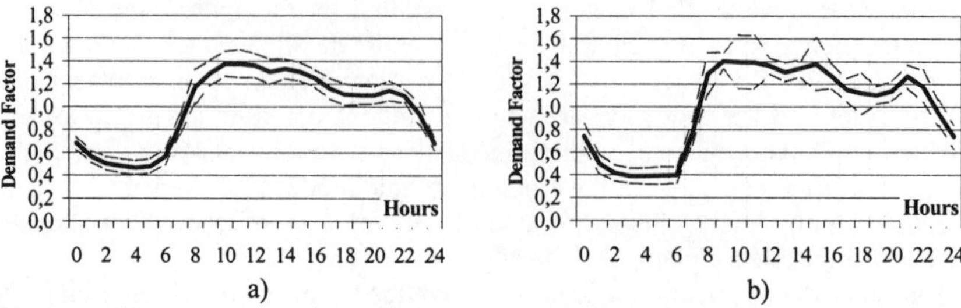

Figure 4. Demand patterns for a) produced water and b) city demand on a workable day

In total 6 demand patterns were identified: one for the city demand, four for the nuclei provided by data-loggers and one for the remaining outlets provided by counters. Figure 4 illustrates that there is a rapid increase in city demand from 6:30 to 8:30 a.m., just before the nocturnal pumping from Picassent is stopped. This abrupt change in demand created a lot of problems during the calibration stage.

4 BUILDING A DYNAMIC MODEL WITH EPANET

Building a dynamics EPS model required following these steps: defining the skeleton of the network on the basis of real pipes; adding control and regulation devices such as reservoirs, valves, pumps, etc; defining patterns for demands, levels, etc; introducing the control rules, and finally fixing the initial conditions to start the simulation. Other parameters like the total time simulation period, the hydraulic interval, the maximum number of iterations per step, etc, could be added depending on the features of the model. In the next section each of these steps are reviewed in detail.

4.1 Pipes skeleton and control devices

In order to build a dynamic model, little detail is required concerning the inner distribution network of the city. Pipe diameters less than 250 mm were at first disregarded, unless they closed an important loop. The skeleton for the distribution network was matched with the one used in the past to build the static model of the city, taking its diameters and roughness as data.

Starting with this reduced model, all mains and trunks up to the treatment plants were then added accurately, as well as all the control valves, whether operated by remote control or manually. All maintenance valves were disregarded for the sake of simplicity, taking into account that pipes can be closed directly in the model to simulate their action. Due to the large number of control valves, none of them is in reality provided with a local automation to maintain the upstream or downstream pressure, or the flow-rate, so most of them were modelled as throttle control valves (TCV). This case led to the introduction of these kind of valves in EPANET 1.1c. some years ago.

Treatment plants, reservoirs and pumping stations were modelled accurately as well, due to the important role they play in the dynamic model. Pumps can inject water directly into the network or they can feed the reservoirs, depending on the hour, the demand or the valve position. Pump intakes were modelled as reservoirs of fixed head and storage tanks as variable head. Figures 2 and 3 illustrate the detailed models of the plants, including the nocturnal pumps in the Picassent scheme, and all the valves involved in the plant regulation.

The complete scheme prepared to be run with EPANET 1.1e is represented in figure 5; it includes 604 pipes, 570 junctions, 9 reservoirs, 21 pumps and 74 valves, 30 of them remote controlled. Identifiers of pipes and nodes were assigned by zones, from 1 to 9,000. Valve links and their nodes were identified by 10,000's and for pumps by 20,000's.

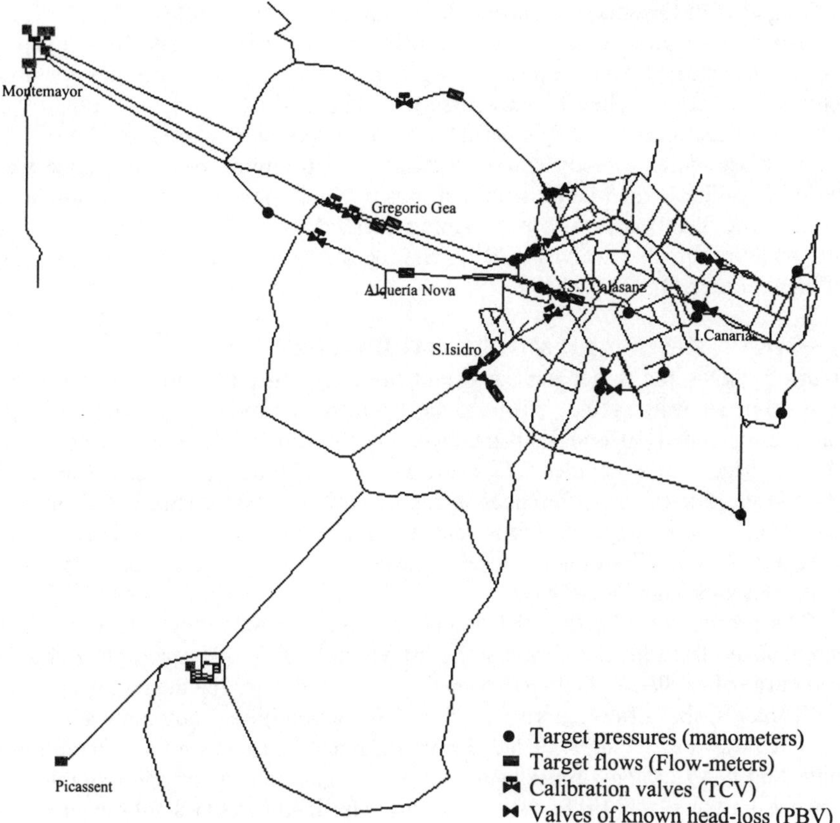

Figure 5. Scheme of the network prepared to be run with EPANET with the location of measuring points and regulating valves for calibration

4.2 Demands and patterns

In the [JUCTIONS] section of the input file just the base demands for the nodes of the city network were considered and the non-accounted–for water added using a MULTIPLY factor in section [DEMANDS]. Demands of the peripheral nucleus were declared directly in section [DEMANDS], augmented with the corresponding non-accounted-for water.

4.3 Control rules

The nocturnal operation of the auxiliary pumps in the Picassent plant was declared using OPEN and CLOSE options for valves in the section [CONTROLS] and the operation of the remaining pumps was based on time rules. Booster pumps do not operate by levels because storage capacity of the intake basins is low, this fact conditioning the pumping schedule for water production. A regular scheme for the working days of October was used to define the pump control rules.

There were some valves in the network whose opening schedules (open or closed) are repeated every day according to the demand pattern and the nocturnal pumping operation. Their control rules were declared based on time. Finally, 13 valves were regulated hour by hour to maintain the pressure references. Five of the valves, those which directly control the inflow distribution from the plants, were chosen to calibrate the model, with their position unknown during each simulation period. The head losses of the remaining 8 valves were forced, converting them into pressure breaker valves (PBV) and declaring time by time the average differential pressure for each.

5 MEASURINGS AND MODEL CALIBRATION

Figure 5 shows the target pressures and the target flows to be adjusted during calibration, as well as the valves to be operated, considered as TCV. Target pressures included 10 nodes upstream or downstream of PBV and 6 additional points. Target flows included 7 flow-meters. For calibration, Tuesday October 7th was chosen, due to the differences in the evolution of water supplied from each plant from day to day. Patterns and average demands were adjusted to this particular day for calibration purposes. The corresponding registrations were taken by telemetry through the SCADA.

The process of adjusting the model included the following. Firstly, plants were isolated from the network and the inflow curve from Picassent plant and the head curve of La Presa plant were forced. Due to the global balance of production and demands, the inflow curve from La Presa was adjusted automatically. The challenge was now to adjust the head from Picassent, pressures on the 16 reference points and flows on the 7 pipes acting on the 5 regulation valves and roughness of trunks from the plants to the city. Success was finally reached following a trial and error procedure. It is noticeable that trying to force pressures upstream and downstream of PBV by doubling them in a pressure reducing valve (PRV) plus a pressure sustaining valve (PSV) led to incompatibilities with the solver and thus a PBV model was chosen for these valves.

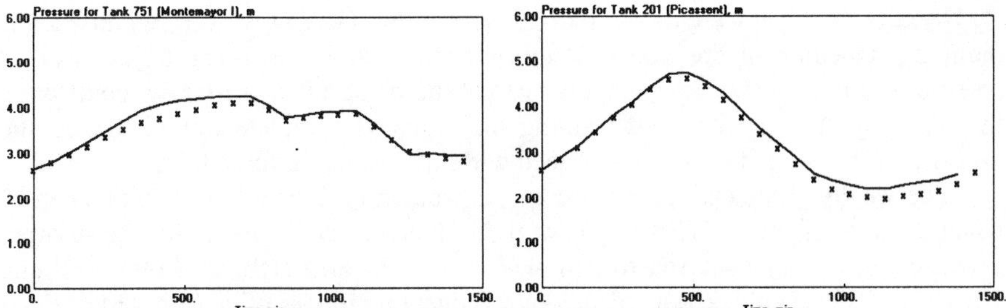

Figure 6. Calibration for levels of Montemayor and Picassent reservoirs

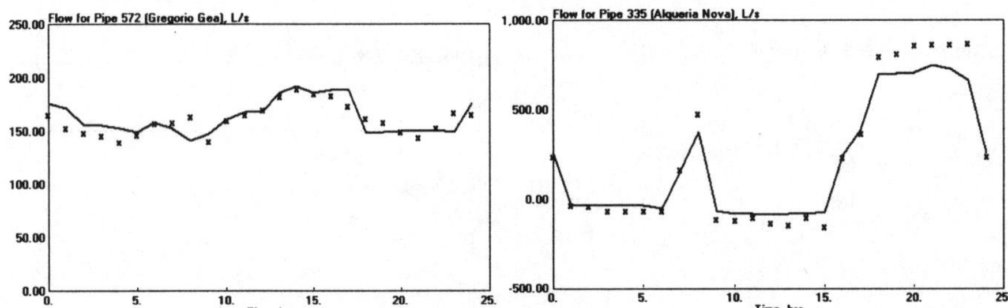

Figure 7. Calibration for inflows on the way of La Presa plant connection

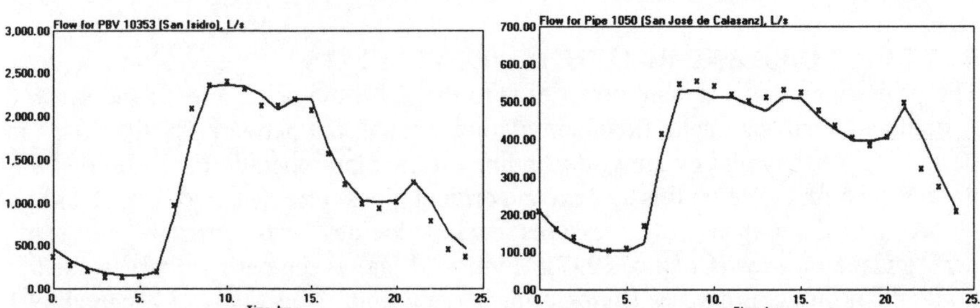

Figure 8. Calibration for flows in the way of Picassent plant and at an inner pipe

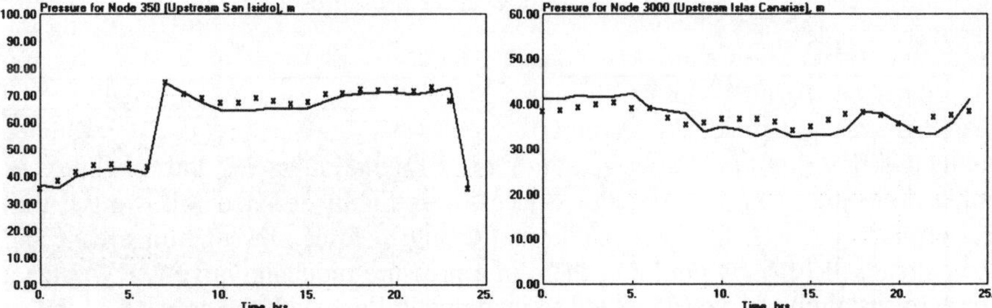

Figure 9. Calibration for pressures at the entry point from Picassent plant and at distant point from entries. The abrupt change on the first graph is due to the switch from nocturnal to diurnal pumping schedule.

Next both plants were calibrated independently. Given the outflow curves, the pumping schedules were adjusted in order to match the level trajectories of reservoirs and, as a consequence, the head at the plant outlets. Finally, connecting the plants to the network and running the adjusted complete model, heads and outflows of both plants were reproduced without needing to force them.

The results of the calibration process are shown in figures 6 to 9, which can be considered satisfactory. The graphs in figure 6 are given for time steps in minutes because instabilities occurred for simulations of 1 hr time step, as shown in figure 10. This is probably a bug of EPANET due to the use of a first order Euler integration scheme to compute levels when two identical reservoirs are connected in parallel.

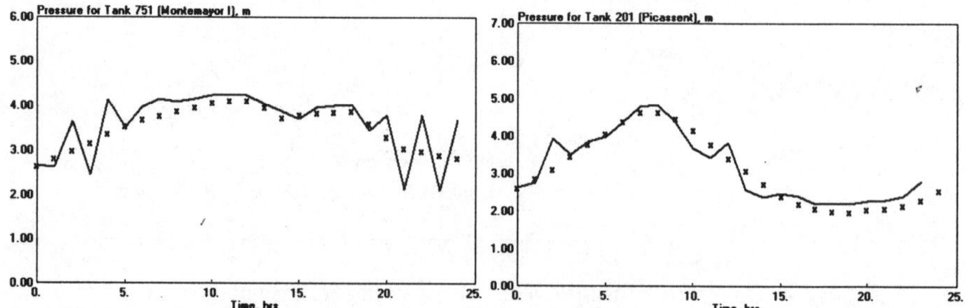

Figure 10. Evolution of reservoir levels when running EPANET with 1 hr time step

6 DIAGNOSIS AND FUTURE IMPROVEMENTS

The primary aim of a dynamic model is to diagnose the behaviour of the system. A complete analysis of the flows circulating through the network for the different periods: night, day and evening, depending on the plant operation scheduling, was done for the first time on this system and critical pipes were detected in each case.

A global analysis of pressures and velocities was carried out using performance indexes (Coelho, 1997). Figure 11 shows the penalty curves used to grade the absolute pressures (index 4 means optimum, 3 adequate, 2 acceptable, 1 to be improved and 0 unacceptable) and the next graph illustrates the performance diagram for 25% percentile bands. This shows the goodness of the current regulation scheduling, since 75 % of nodes have pressures above index 3, 3.5 being the average index.

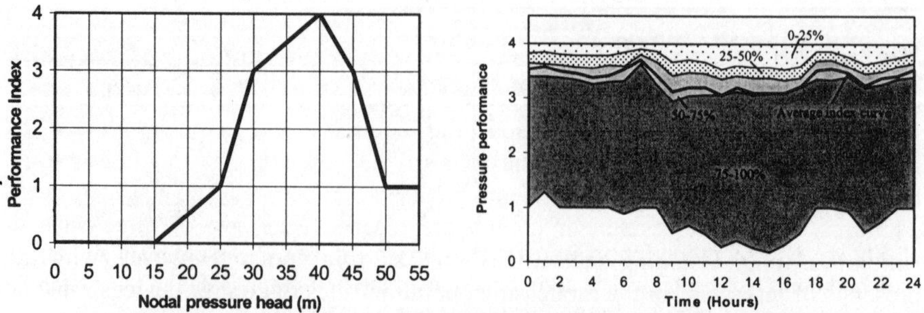

Figure 11. Penalty curve and performance diagram for pressures

Figure 12 shows the penalty curves used to grade the velocities and the corresponding performance diagram for 25 % percentile bands. In the nocturnal period most of the pipes show a poor index, below 1, while during the day only 25% of pipes are below 1. This means that the total pipe capacity is not used. This is usual in most networks and it will have important consequences for the quality model.

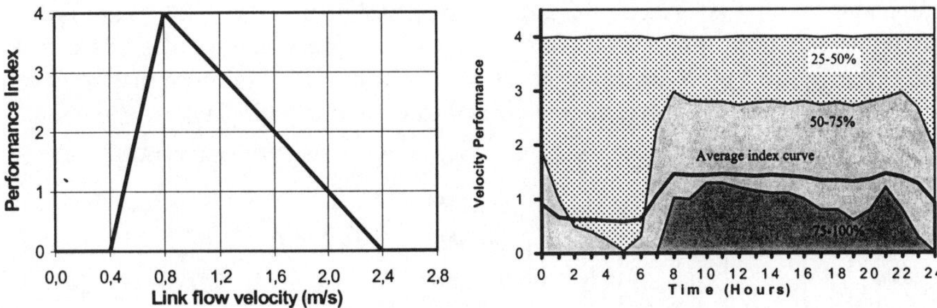

Figure 12. Penalty curve and performance diagram for velocities

Another benefit of the dynamic model is that sensitivity of the network variables to changes in the opening grade of the different valves can now be checked. For this purpose the valve coefficients are determined by comparing the measured differential pressure with the flow-rate given by the model. Using similarity laws and general curves from manufacturers, the real curve opening-valve coefficient for each valve can be determined by regression and applied to analyse future operating conditions.

7 APPLYING DIFFERENT QUALITY MODELS

Many dynamic models are built nowadays just to check water quality. Four types of quality analysis are regularly performed: water age, tracing from sources, mixing of conservative substances and evolution of non-conservative substances such as chlorine. All of them have been tried within the calibrated model.

Computing water age is the first step in the water quality analysis. Regulations sometimes refer just to water age. In general, water quality will deteriorate as water age increases. Limits of 24 hr or 48 hr are a good reference, but it depends on the quality at sources and other factors like pipe materials.

Two types of models were used for quality analysis. The first model substituted plants for the modulated injection flow and simultaneously forced the head in one of the entry points. The second model includes the plants in the model. The difference is the initial simulation time required to get stable results and the need for cycling the hydraulic behaviour of the network. Without reservoirs, hydraulic cycling is automatic if injection curves and demands are repeated every 24 hours, and stable results for quality are reached after two or three days of simulation time, assuming null values as a starting point. However, with the presence of reservoirs in the model, the operation rules of control devices must be adjusted in order to repeat the reservoir levels every 24 hours, in which case simulation times of five days or more are needed before steady state is reached.

Figure 13 shows the computed water age for two particular points using the model without plants and 0 as a starting age. The first point refers to Alqueria Nova at the entry from La Presa (see figure 5). Flow inversion occurs at this point due to the alternation of entries from both plants, which considerably affects the water age. The first graph illustrates two refreshments every 24 hours with a maximum retention time of 12 hours. The second graph refers to Islas Canarias point (see figure 5), 15 km from the plants. In spite of being more distant, water age is less, and more regular, than the previous point due to the lack of inversions.

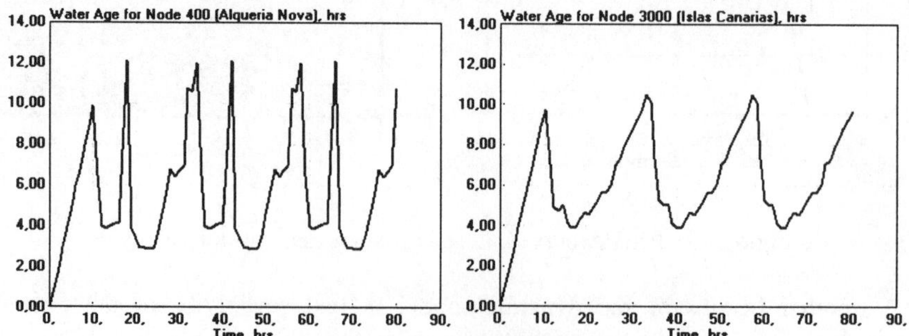

Figure 13. Water age for two particular points (not considering reservoirs)

Figure 14 shows the water age in reservoirs when considering just the plant models. In each case the outflow modulation from each plant was forced and the pumping scheduling adjusted to cycle the levels. Starting from age 0, values are stabilised after 7 days of simulation for La Presa plant, with water age for the Montemayor reservoir ranging from 45 to 60 hours. Regarding Picassent plant, stabilisation is reached after 5 days of simulation with water age for the reservoir ranging from 33 to 48 hours. Both models were run with 6 minutes (0.1 hour) of hydraulic time step due to the problem previously commented. The reservoirs increased the maximum water age for the points referenced in figure 13 up to 55 hrs and 35 hrs respectively. Water ages are not added directly because fresh water is pumped into the network during the day from direct booster pumps and a mixing process occurs.

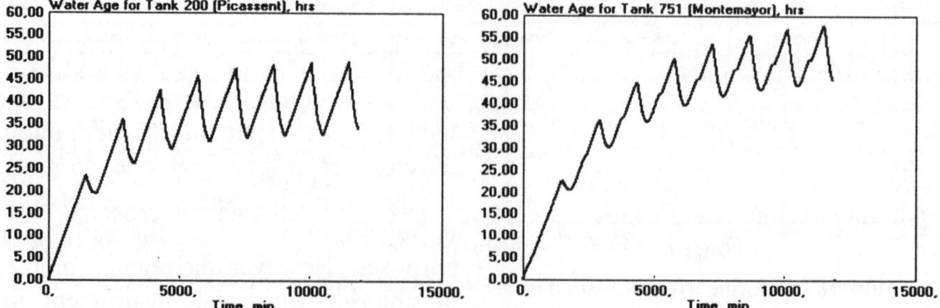

Figure 14. Water age for Montemayor and Picassent reservoirs

When the retention times are known, the next step is to run a mixing model of non-conservative substances. Nitrate was the waterborne substance chosen for this purpose, taking advantage of a failure in the Jucar-Turia transfer channel which occurred in November 1997. The channel was fed during one week with groundwater that contained an appreciable amount of nitrates, not treated by the Picassent plant, in substitution for water from Jucar River. Water coming from La Presa was surface water free of nitrates and a mixing occurred through the network. Regular samples taken in the network for Spanish health regulations permitted us to calibrate a mixing model. Mixing models of non-reactive contaminants are used today to calibrate those dynamic models referring to flow-rates, because concentrations depend on flows and it is far cheaper to measure the concentration of a substance than install a flow-meter. The mixing model ran without plants because concentration at the outlet plants was known, confirming our previously calibrated model by comparing concentrations at the sampling points.

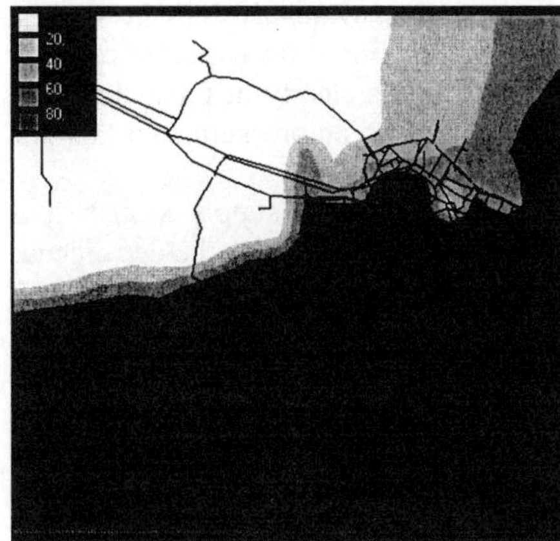

Figure 15. Source water tracing from Picassent at 12:30 p.m.

Once the model has verified flow-rates (or non-reactive pollutants) a tracing model can be applied to calculate the coverage of water coming from each plant, depending on the hour of day. Figure 15 shows the contours at equal stepped percentages for source water coming from Picassent at 12:30 p.m., after 5 days of simulation to achieve stability. The complete model was run in this case. Results indicate that water from Picassent reached the entire city during the morning for the actual scheduling, with a source of 100 % for south, centre and east bounds. Finally, a model of residual chlorine was developed from previous data.

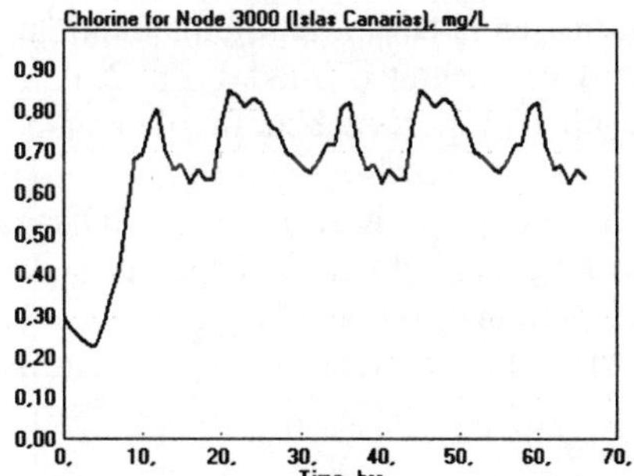

Figure 16. Chlorine evolution at Islas Canarias point Picassent at 12:30 p.m.

The bulk rate constant decay was determined by the bottle method, producing a value of $-0.72\ h^{-1}$ for the treated water (Vidal, 1996). An estimation of the wall rate constant was made, considering material, age and pipe diameters. In spite of the continuous detectors of residual chorine currently installed at twelve points in the network, its reliability was low for calibration purposes. However, the regular samples of chlorine were also insufficient for calibration purposes. A primary model is currently available, whose results for the distant point of Islas Canarias (figure 16) seem reasonable for the scarce data available. The model without plants was run for this case, considering chlorine registrations at the plant outlets as source data.

8 APPLYING A PRESSURE-DEPENDENT LEAKAGE AND DEMANDS INTEGRATED MODEL FOR SECTORISATION ANALYSIS

In section 2 we have emphasised the fulfilment of balance equations to determine demand patterns and the base average demands at nodes. According to this procedure, the non-accounted-for water was distributed proportionally to registered demands, spatially and temporally. However, in order to achieve a reliable dynamic model, such hypotheses cannot be sustained for the fraction of the non-accounted-for water corresponding to latent leakage. In effect, in a water distribution network, pressures vary during the day and leakage depends on pressures. In diurnal periods the highest demands and the lowest pressures are experienced, whereas in nocturnal periods the lowest demands and the highest pressure occur. Also, nodes with higher average pressures are expected to show higher percentage leakage. As a consequence, reality is clearly different from the conventional hypothesis. Furthermore, demands depend on pressure, and this fact is not taken into account by conventional dynamic models.

In order to build a reliable dynamic model of a water supply system it is necessary to both integrate leakage and to simultaneously consider the dependence of demands on pressure. In the literature few leakage models have been proposed, and most of them assume pre-defined coefficients to fix the dependence of leakage on pressure at each junction. In Martinez (1999) a complete formulation has been established to determine the leakage coefficients for all pipes and nodes considered in the model, using as data the overall values of efficiency for leakage observed for long time periods, which constitutes the only available data from water utilities. Secondly, once leakage has been characterised, the coefficients of the pressure-demand relationships can be determined as well.

Using the integrated pressure-dependent leakage and demand model, more reliable results can be obtained when simulations are carried out under new scenarios such as reduction of pressures for leakage control, failure of critical components of the network or restrictions during an extended drought. Important conclusions can be obtained from using the integrated model on leakage assessment and management policies, as another important benefit of dynamic models.

The proposed integrated model is currently being built in EPANET to enhance its performance. A partial model considering just global leakage at nodes proportional to global demands is ready at present and has been applied to assess new policies in order to reduce leakage on the Valencia water supply system by sectorization.

8.1 Fundamentals of the pressure-dependent leakage and demand model

The following general formula, experimentally verified, has been proposed by many authors to express the dependence of leakage on pressure in water distribution networks:

$$q = K\ p^{\beta} \tag{1}$$

where the exponent β takes values between 1.1 and 1.2. In our model, leakage in the modelled and non-modelled network are distinguished. In the former, leakage depends on the pipe characteristics according to the following reviewed formula:

$$q_1 = c_1 \cdot L \cdot D^d \cdot e^{a\,\tau} \cdot p_{av}^{\beta} \tag{2}$$

where L and D stand for the pipe length and diameter respectively, d is an exponent to be adjusted from field tests, k and a are coefficients depending on the pipe material, τ is the pipe age, and p_{av} the average pressure along the pipe. Finally c_1 is a coefficient unique to a whole sector, and is determined by forcing the total leakage volume for this kind of leak, summed over the whole simulation period and for all pipes in the sector, to be a percentage of the total demand. Once c_1 is identified, the leakage q_1 for each pipe is distributed in equal parts between its end nodes.

Also, according to experience, most leakage occurs on fittings not considered in the model, usually in small pipes and connections. This leakage is attached in our model directly to nodes using equation (1) in which the K coefficients must be identified for each node from the global percentage of leakage expected in it, or better still, using the following corrected expression:

$$q_2 = c_2\ Q_{c,T}\ p^{\beta} \tag{3}$$

which, once totally integrated, takes into account the proportionality of the total leakage with the total demand $Q_{c,T}$ and the average pressure at each node. In this case c_2 is unique for a sector and determined, as before, forcing a global leakage percentage in the sector for this kind of leakage.

Identification of any kind of leakage coefficient requires an iterative process because leakage depends on pressures, and pressures depend on total loads, including leakage. Once identified, subtracting each time the total leakage from the total injection, the total real demand at that time is obtained, which is then distributed between nodes according to its base demand derived from billing. Thus real demands can be identified for each node and time interval and then the demand coefficients can be obtained according to the following relationship proposed to express the variation of demands with pressures:

$$Q_{d,i}(k) = C_i(k)\left(p_i(k) - p_{m,i}\right)^{\gamma} \tag{4}$$

where $p_{min,i}$ is the minimum pressure required at node i to get any flow, γ an exponent to be determined from considerations of water use, either regulated by time or by volume, and $C_i(k)$ the wanted demand coefficient for this node at time k.

8.2 Application of the leakage model to the Valencia water supply system

The whole metropolitan system was divided into two sectors in order to apply the leakage model. The data provided by the flow meters, shown in figure 5, ensured that the global efficiency for the outer area could be differentiated from that of the city. According to Andrés (1995), in 1992 the non-accounted-for water for the whole system was approximately 30 %. Flow-meters indicated that 4 % corresponded to water taken in the outer area and the remaining 26 % to the distribution network. From results of leakage detection campaigns, approximately half of this percentage can be associated to latent leakage. In the outer area all leakage was assigned to the non-modelled network due to the large conveyance pipe diameters. Regarding the city, a quarter of the leakage was assigned to the modelled network and three-fourths to the non-modelled network, according to the field experiences and taking into account that diameters below 250 mm were not modelled.

Using equation (1) to characterise leakage in the non-modelled network with a common percentage to all nodes, and equation (2) for the modelled network, the different leakage coefficients were identified. Finally, demand coefficients for all nodes and time intervals were calculated using equation (4).

Results were assessed using the performance index shown on the left of figure 17 to produce the diagram shown on the right of the same figure. As expected, better results were obtained in the daytime hours. Results for the nocturnal hours were acceptable for just 50 % of the nodes.

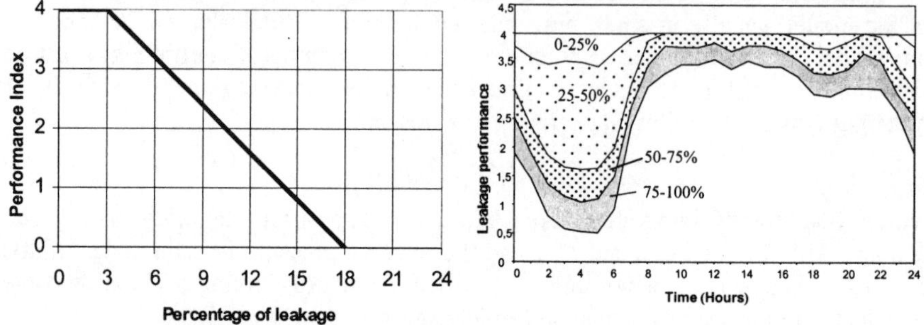

Figure 17. Penalty curve and performance diagram for leakage

In order to reduce leakage, a policy of pressure reduction was introduced, primarily for the whole city network. Acting on the entry valves, attempts were made to reduce the pressures to within the range of 25-30m for the entire network, with some success. Leakage in both the modelled and non-modelled city networks was 13 % down, while demand was reduced by just 3 %.

As an alternative, a better solution was proposed, isolating just the sectors with greater leakage percentages. Figure 18 shows the sectorisation proposed with five main sectors supplied from one or two entries maximum. At entry points, regulating valves were installed to reduce pressures to the range of 25-30 m for all internal nodes. Consequently, global leakage was reduced to 12 % for the modelled network and 8.5 % for the non-modelled one, with a demand decrease of 2.4 %, giving almost the same results as previously, but acting over just five sectors.

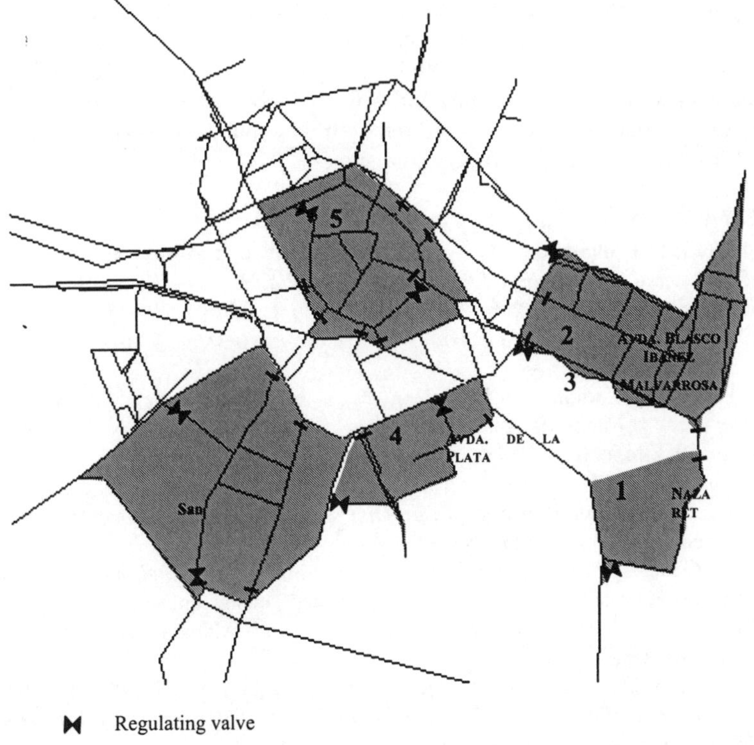

Regulating valve

Closed valve

Figure 18. Sectorisation proposed of city network to control and reduce leakage

In summary, sectorisation seems to be an effective way to control and reduce leakage. Its expected variation according to service pressure can be now evaluated due to integrating a leakage model into the dynamic model. Regarding the demands, reduction was not significant for this pressure range. However, minimum pressure was established for low flows of 10m to all nodes, important reductions in demand were obtained for service pressures below 20m, while leakage reduction in this range was not significant.

9 CONCLUSIONS

Extensive work on building and exploiting a dynamic model for the Valencia metropolitan water supply system, developed during the last two years, has been presented. Difficulties of loading and calibrating the model have been overcome successfully using the EPANET package. Aguas de Valencia is now intending to use the model to study new regulations in order to improve the quality service and to supply water to the new peripheral nucleus connected to the metropolitan network. In particular the north metropolitan area shown in figure 1 is close to completion and shall be added to the current model.

A dynamic model is much more productive than a static one. In the case of the Valencia water supply system, quality models and leakage-demand model simulations have been carried out as very important additions. A reduced leakage model has been built-into the EPANET 1.1e package to analyse the pressure-dependency of leakage and demands and a full version is being developed. The reduced leakage-demand model is currently in the public domain and can be requested from the authors (fmartine@hma.upv.es).

REFERENCES

Andrés, M. (1995) Leakage detection in practice. Application to the water distribution of Valencia, *Improving Efficiency and Reliability in Water Distribution Systems*, Kluwer Acad. Pub. Dordrecht, Boston, London, pp 97-105

Coelho, S.T. (1997) *Performance in Water Distribution. A systems approach*. Research Studies Press Ltd

Conejos, P. (1998) *Estudio de la Influencia de las Presiones en Red sobre las Demandas y Fugas de Agua Potable en determinados sectores de la Red de Valencia*. Graduation Project. ETSII. Valencia

Germanopoulos, G. (1995) Valve Control Regulation for Reducing Leakage, *Improving Efficiency and Reliability in Water Distribution Systems*, Kluwer Acad. Pub. Dordrecht, Boston, London, pp 165-188

Martinez, F., Conejos, P, Vercher, J. (1999) *Developing an integrated model for water* distribution system considering both distributed leakage and pressure-*dependent demands.* 26th Water Resources Planning and Management Conference. Tempe,Arizona.

Martínez, F., García-Serra, J.(1993) Mathematical modelling of water distribution systems in service. *Water Supply Systems. State of the art and future trends.* Comp. Mech. Pub. Southampton, Boston, pp 141-174

Ponz, R. (1998) *Incorporación de las Plantas Potabilizadoras al Modelo Dinámico de la Red de Suministro de Agua a Valencia, y verificación del Modelo Integrado. Aplicación al Diagnóstico del Régimen de Operación de la Red.* Graduation Project. ETSII. Valencia

Rossman, L.A. (1993) *EPANET User's Manual*. US Environmental Protection Agency

Savall, R. (1998) *Desarrollo de un Modelo de Calidad para la Red de Distribución de Agua Potable a la Ciudad de Valencia y Suministros en Alta*. Graduation Project. ETSII. Valencia

Signes, M. (1998) *Mejora del Control y Gestión de la Red Metropolitana de Abastecimiento de Agua a Valencia y su Entorno Urbano mediante la Confección y Explotación de un Modelo Dinámico*. Graduation Project. ETSII. Valencia

Vela, A., Espert, V., Fuertes, S., (1995) General overview of unaccounted for water in water distribution systems, *Improving Efficiency and Reliability in Water Distribution Systems*, Kluwer Acad. Pub. Dordrecht, Boston, London, pp 61-95

Vidal, R., (1996) *Implicaciones de los modelos de calidad en el diseño yoperación de las redes de distribución de agua potable.* Ph.D. ETSII. Valencia

Developing Real-Time Models of Water Distribution Systems

C. Orr, P. F. Boulos, C. T. Stern *and* P. E. P. Liu

ABSTRACT

This paper discusses some of the essential considerations in real-time hydraulic modeling of water supply and distribution systems. The paper outlines availability of real-time data from SCADA systems to their applications in on-line modeling and real-time control. An illustrative scheme is presented for an implementation of on-line modeling. The implemented scheme provides a basic platform for on-line model evaluation and real-time model simulation. To cater for more sophisticated control supported by many modern SCADA systems, comprehensive logical control rules have been developed and incorporated in the hydraulic modeling package to mimic the advanced control schemes. An information engineering approach is also discussed in the integration and data abstraction of various application requirements.

1 REAL-TIME HYDRAULIC MODELING

Hydraulic modeling has been widely used in recent years for a number of purposes, from planning, design, to operations and control in the management of water distribution systems. The modeling process is first performed by the construction of the structural system representation. The constructed topological network representation is then applied with operational data for full hydraulic analysis. The effectiveness of the overall models relies on the appropriate and accurate representation of the physical systems.

Water distribution system models are primarily represented by nodes and links. Network nodes may represent water sources, water storage, geographical locations, network junctions, and consumer demands; whereas network links may represent system piping, supply and booster pumps, and control valves. It is important to realize that network models are essentially mathematical abstractions of the physical systems. A modeling junction node may represent the aggregated demand characteristics for a number of consumer properties, whilst a modeling pipe may represent a number of physical piping components in the system. Similarly, a modeling tank may represent several storage tanks connected together, and modeling valves are used primarily to simulate the hydraulic properties of the system control valves.

In addition to the network components and their topological relationships, the hydraulic performance of the network modeling is further governed by the operational characteristics of the system model. Essential operational information ranges from spatial distribution and temporal variation of consumer demands, system hydraulic grade profiles, to the system control operating conditions. Such information is generally derived from measurement and system monitoring of consumption pattern and control decisions. To adequately model the physical system performance, it is imperative to provide sufficient and accurate monitoring information.

SCADA (Supervisory Control And Data Acquisition) systems provide two essential functions for: remote control of system operation and data acquisition from remote sensors. Collected information including operational status and hydraulic data measurements are typically managed and archived by the SCADA database. Analytical utilities, such as statistics and performance evaluation, have been introduced to provide additional information for many of existing system operational assessments. SCADA operations are performed on a real-time basis, typically on regular intervals and/or upon specific events.

Traditionally, the SCADA data archive has provided the hydraulic modeling applications with one of the essential sources to derive reliable operating condition information, such as system demand levels, demand variation patterns, reservoir water levels, and pump/valve operating modes. The availability of this operational information on a real-time basis provides further incentives to have automated on-line update of modeling operational data, which will then lead to the capability for real-time modeling.

On-line modeling has the added benefit of providing instant assessment of the system performance, especially under emergency operating conditions. Such on-line assessment can be used to support operational decisions in formulating effective operational control strategies. The recommended control policies can be further applied to SCADA systems to relay appropriate control status to the remote control station units. Fig.1 illustrates such an on-line model update and control assessment scheme. Several control objectives, such as pressure minimization, optimal pump scheduling policies under the normal operating conditions, have been considered. Such control objectives have been evaluated in conjunction with a range of modeling and optimization modules for the appropriate system control. The real-time SCADA data transfer has led to a number of pilot real-time control studies.

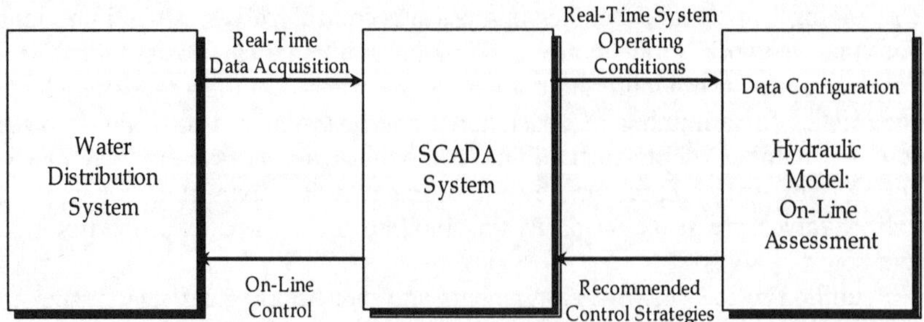

Fig. 1 On-line control scheme.

While the SCADA systems provide up-to-date information of the system operating conditions, it is important to realize that this information may not be sufficient in a number of areas and/or occasions. First, the SCADA system may not provide necessary measurements for every single system control element. Second, the remote sensors may fail to provide the expected measurement data at a given time, due to sensor or communication failure, resulting in erroneous and/or missing data. Third, the SCADA measurements may not have a direct correspondence in the modeling description. Finally, the SCADA system cannot provide demand information for all modeling demand junctions. Information derived from the SCADA system flow measurements can only be used to provide adjustment to the modeling demand conditions. In addition, the SCADA system status measurement may indicate a change in the modeling representation. Therefore, the extension of SCADA data to direct real-time modeling applications must be given careful consideration.

2 BASIC CONSIDERATIONS IN ON-LINE HYDRAULIC MODELING

The basic objective of an on-line modeling scheme is to update the relevant hydraulic modeling data description based on the current SCADA information available. Essentially, the modeling data subject to on-line consideration can be categorized into three sets: a control set, a boundary condition set, and a system flow set.

The control set X can be further sub-classed into:
Pump control set: $X_p|_i$, where $i = 1, \dots n_p$, where $n_p \leq$ total system pumps N_p
Valve control set: $X_v|_i$, where $i = 1, \dots n_v$, where $n_v \leq$ total system valves N_v
Pipe Link control set: $X_l|_i$, where $i = 1, \dots n_l$, where $n_l \leq$ total system pipe links N_l

The boundary condition set H can be sub-classed into:
Boundary reservoir set: $H_r|_i$, where $i = 1, \dots n_r$, where $n_r \leq$ total system reservoirs N_r
Storage tank set: $H_t|_i$, where $i = 1, \dots n_t$, where $n_t \leq$ total system tanks N_t

The system flow set Θ can be sub-classed into:
System inflow set: $\Theta_I|_i$, where $i = 1, \dots n_i$, where $n_i \leq$ total system inflows N_i
System outflow set: $\Theta_o|_i$, where $i = 1, \dots n_o$, where $n_i \leq$ total system outflows N_o
System storage set: $\Theta_s|_i$, where $i = 1, \dots n_s$, where $n_i \leq$ total system storage N_s
Demand flow set: $\Theta_d|_i$, where $i = 1, \dots n_d$, where $n_d \leq$ total demand junctions N_d

The pump control set X_p is described by the individual pump On/Off status as well as the speed settings for variable speed pumps. The pump On/Off status has been represented in a number of manners: from simple On/Off logic to a measurement of the given pump discharge flow. Each system pump can be modeled as an individual pump component. The pump speed can be given by the actual pump speed in rpm or given as a proportional measure within the applicable pump speed range, for which the minimum and maximum permissible pump speeds must be defined. The valve control set X_v is described in different manners based on individual modeling valve types, such as the downstream pressure for pressure reducing valve, valve opening position for throttling control valve, etc. An extension has been incorporated in the modeling system to relate the valve opening position to the throttling effect. The pipe link control set X_l is generally described by the link modeling status. The pipe link status can be applied to simulate a number of modeling scenarios, such as closing the tank outlet pipe to mimic a condition whereby the connecting tank is temporarily removed from the system for cleaning or maintenance, etc.

The boundary reservoir set H_r is described by the hydraulic grade values of individual reservoirs. Similarly, the storage tank set H_t is governed by the hydraulic grade values of respective storage tanks. The hydraulic grades at storage tanks will vary according to the storage conditions, which are generally measured as direct hydraulic grade values, storage water level fluctuations, storage volume changes, or percentages of stored water. Based on the geometric shape of the storage tanks, the storage and hydraulic grade relationship can be described by

$$H=f(V),$$

where H is the hydraulic grade value, and V is the measured storage volume. The functional tank volume storage relationship $f()$ may vary from the basic cylindrical form to any variable cross-sectional area tank characteristics. These reservoir and storage tank sets establish the reference conditions for the system hydraulic grade profile.

The system flow sets Θ_I *and* Θ_o can be used for two major purposes: to derive the system demand levels and/or to derive the storage condition, which can also be used to update the storage tank set H_t.

Although individual demand monitoring points may have been provided for selected major users, the majority of consumer demands are not monitored by the SCADA system. Demand allocation in hydraulic modeling remains an open area subject to a range of data updating techniques. However, the flow measurements available from the SCADA system often can be used to derive a demand update procedure amenable to the modeling purposes. In essence, system demands ΣQ_d are governed by total system inflow ΣQ_i, total system outflows ΣQ_o, and total system storage supply ΣQ_s as:

$$\Sigma Q_d = \Sigma Q_i - \Sigma Q_o + \Sigma Q_s$$

The majority of the system inflows and outflows are monitored by the SCADA system, thus providing the system in-out flow sets Θ_I and Θ_o. System storage supplies can often be derived from the storage tank inflow and outflow readings or from the changes in storage tank water levels, thus leading to the derivable storage supply flow set Θ_s. These available flow measurement sets Θ_I, Θ_o, and Θ_s, are often subsets of the total system inflows ΣQ_i, total system outflows ΣQ_o, and total storage supply ΣQ_s, respectively. Their net value is proportional to the overall system demand condition. Such proportional property can then be used directly to update the model demand definition scales.

It is noted that data measurements can be performed by different sensors from different manufacturers, resulting in a variety of data reading characteristics. Data measurement can be performed in different units and with different scales. All instruments are subject to their specific calibration and accuracy tolerances. Scaling factors and offset factors are necessary to convert the given data values into a format compatible with the modeling data definitions. Residual readings below given thresholds should be ignored. Readings may also be saturated due to instrumentation limitations. Filtering of these erroneous data conditions is essential.

On-line modeling can be performed on snap-shot or on extended time period time bases. Snap-shot time evaluation is performed based on the given SCADA data reading time. The associated hydraulic model controls, boundary conditions, and demand flows are updated with corresponding control set X, reservoir level set H, and system flow set Θ, respectively. The extended time period evaluation retains the storage characteristics at all system storage tanks from each SCADA data reading instance to all following SCADA time instances. In the absence of separate inflow and outflow readings for individual storage tanks, the changes in the storage tank levels can be used to derive the given system storage supplies ΣQ_s as:

$$Q_i = \Delta Storage_{ik} / \Delta\,(t_k - t_{(k-1)})$$

Where

$$\Delta Storage_{ik} = Volume_{i(k-1)} - Volume_{i(k)}$$

for individual tank i, with time advancing from stage $(k\text{-}1)$ to k. These storage supply changes are used in conjunction with the system inflow and outflow readings to adjust the system demand conditions. Similarly, where storage tanks are monitored only with inflow and outflow measurements, their value changes can be used to derive the corresponding storage tank volume change and hence the changes in the storage tank levels, as

$$\Delta H = f(tank\ inflow,\ \ tank\ outflow,\ \ \Delta T)$$

The SCADA data update can then be used to override existing modeling definitions, thus providing the more up-to-date modeling representation of the physical system. During the following hydraulic analysis, all updated information should maintain their respective values and status conditions and not to be further over-written by internal modeling operation.

3 EXTENDED LOGICAL CONTROL RULES

The system hydraulic performance is subject to the status of control set *X*. To effectively model the system performance, the hydraulic model should be capable of simulating a wide range of control scenarios. Essentially, control changes can be triggered by time events and/or by hydraulic conditions. Typical time events include: the actual time periods, and the time duration to reach to certain hydraulic conditions such as the time it takes to fill or drain a storage tank. Typical hydraulic conditions include: tank storage levels, junction pressures, pipe flow rates, etc., such as the set-point control in many existing utility systems. The hydraulic conditions can be further combined with the time factor to give the rate of change and the status of change, such as increasing or decreasing data parameter value. In addition, the control conditions may also be subject to the existing hydraulic status of selected system components, such as the On/Off status or speed setting of individual pumps.

Reliable and accurate metering sensors have already been able to provide detailed readings in a range of system quantities. Modern SCADA systems with integrated programmable logic controllers are now capable of providing sophisticated control implementation based on a combination of system conditions, such as storage status, off-peak tariff rate based pumping, pumping duration control, pumping volume control, etc. These systems will require control actions with corresponding hydraulic models far more sophisticated than the traditional set-point format. The control decisions will not only depend on single time or hydraulic conditions, but also on a logical combination of various event and conditional criteria. The capability to model these control rules with logical combinations of system conditions is particularly essential in SCADA based on-line modeling.

Comprehensive logic control rules have since been developed and applied to hydraulic models for systems controlled by these more sophisticated criteria. Essentially, the logic control rules are comprised of a set of logical conditional statements, each is associated with a set of corresponding control actions. The conditional statements may involve the assessment of a number of time event and/or hydraulic conditions. Each of these time and/or hydraulic conditions may be combined using a number of appropriate logical operators, such as *Logical And*, *Logical Or*, and *Logical Not*. All the logical precedence rules are observed. Logical conditions can further be nested to define appropriate priorities. When the given condition statement is assessed to be logically true, the resulting control actions are activated for the given set of control elements. If the given condition statement is assessed to be logically false, the assessment will be passed on to the following conditions. A typical set of logical control rule statements can be illustrated as below:

```
If (Logical Statement A)
 begin
   Control Action A-1
   ...
 end
Else if (Logical Statement B)
 begin
   Control Action B-1
   ...
 end
Else if (Logical Statement C)
...
Else
 begin
   Control Action Else-1
   ...
 end
Endif
```

Such compound control logic rules enable the users to operate their systems based on a combination of system variables, such as flow rate, tank water level, junction pressure, data rate of change, current status, etc. Fig. 2 below illustrates an example of logic control application.

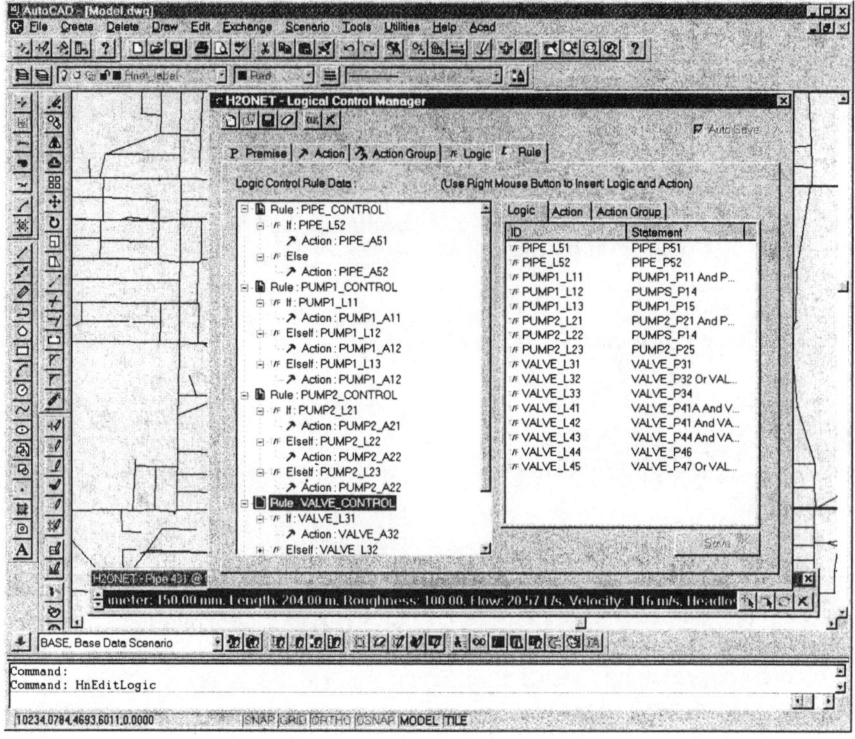

Fig. 2 Application of Logical Control in H_2ONET

4 SYSTEM INTEGRATION AND INFORMATION MANAGEMENT

For successful on-line modeling, SCADA data must be properly integrated with the hydraulic model. System integration comprises two main areas: data communication between the SCADA environment and the hydraulic model, and the common reference to the given information from both the SCADA and hydraulic model. These requirements are essentially the same and are applicable to all system integration applications. Within any given application domain, information is extracted from various sources of data. Management of a range of diversified data sources and extraction of the pertinent information contents is the key to the decision support for the given application.

Given two separate applications, such as SCADA and hydraulic model, the common means to exchange data is through a shared data repository. The most straightforward data exchange is made in the file exchange format in which data are compiled in a manner mutually recognizable by both exporting and importing applications. The exchange format may be defined from the export and/or import applications. For efficiency and for data security reasons, the data may be defined in an encrypted binary format. Alternatively, a standard ASCII format can be used to provide open data exchange. Data quantities are typically described with corresponding data headers and data tag identifiers, with any necessary and appropriate data delimiters. Data parsing techniques are used to encode and decode the corresponding data items. In a multiple application environment, a more general system analysis of the applications may lead to the adoption of databases as the data resource repository. Applications exchange data through the databases in the format as defined by the corresponding database systems. In both circumstances, computer network communication is allowing access from one application to the server system or to individually addressed machines.

The common reference of the provided information is also essential in the success of any system integration applications. The exporting and importing systems should configure their information exchange in such a format that the export data items have direct corresponding references in the importing applications. In general, such a configuration process is performed more easily with the modeling applications. For on-line hydraulic modeling, the configuration process includes two major steps for: configuration of the SCADA meter reference with the modeling element identification, and configuration of the SCADA data readings with the modeling parameters for the given modeling element. The first step involves matching the SCADA system meter names with the corresponding modeling elements, such as storage tanks, junction nodes, pipes, pumps or valves. The second step involves maintaining appropriate unit conversion, data type matching, such as node pressure level, tank water level, and data interpretation. The importing data from SCADA to hydraulic model are then filtered through this configuration process and converted into corresponding modeling values for on-line update, as illustrated in Fig.1. Similarly, any exporting data from hydraulic model to the SCADA system can also be converted back to the SCADA references for further processing within the SCADA environment.

Hydraulic solution is the basic modeling process in the hydraulic modeling. Extended period simulation (EPS) provides the quasi-dynamic capability in the hydraulic analysis. Some other commonly available processes include: water quality, fire-flow, and energy cost analyses. On-line modeling is another value-added process in the core hydraulic analysis. Within the modeling application, management of these process data and corresponding analysis results is posing a challenge to the modeling application development. Fundamentally, the modeling data and results can be categorized into a hierarchical tree type class structure. The basic network definition and hydraulic performance are the base class to all modeling applications. All other extended analyses inherit the base class characteristics while providing additional information on its own class' application. Fig. 3 depicts such a class structure of the modeling applications.

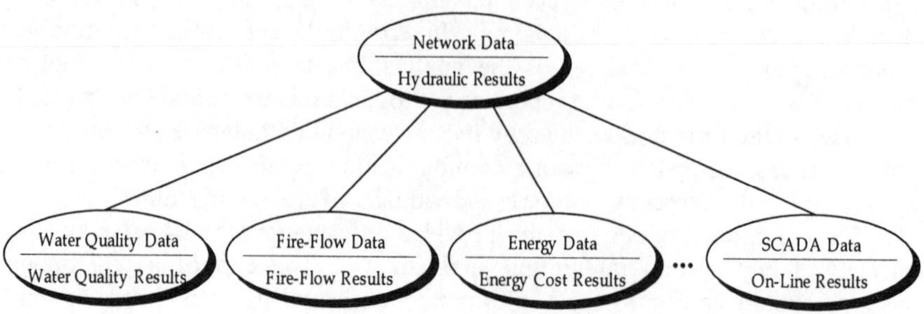

Fig. 3 Hierarchical data result structure in hydraulic modeling applications.

Data and information management in such an environment can also be organized in the same object-oriented manner. Data abstraction functions are provided at each class' level. The base class object provides the baseline data extraction functions for the network and hydraulic data. Additional data details can then be extracted from individual derived objects' own domains, subject to the given object type. All these data management objects are further derived from the basic data manipulation service object at the top level. The basic function of this top level object is to provide the fundamental data access services, such as file operator management and data type access. Some typical example service functions include: *FileOpen(FileName, FileMode)*, *IntegerRead(File_Pointer, Offset_Position, Data_Count)*, *VariableRead(File_Pointer, Offset_Position, Data_Count)*, *StringRead(File_Pointer, Offset_Position, Data_Count)*. All application level data accesses are then evaluated based on the appropriate offset counter. For instance, to assess the pressure at the nth node for the kth time period $P(n, k)$, the access function of

$$P(n, k) = VariableRead(*pFile, (b_{pk} + (n-1)*p_b), 1)$$

can be applied, where b_{pk} refers to the basic data offset to the pressure data section for the kth time period, and p_b refers to the basic pressure data spacing. Higher level data extraction functions can then be derived based on more abstract data information, such as the node junction identifier, which is then translated internally to the corresponding necessary offset, all transparent to the caller level. Data for the entire network can easily be retrieved with incremental data offset pointer n, and time series data can readily be organized based on the time stage offset pointer k, for all network elements $V(Network)(k)$ and all time stages $V(Time)(n)$, respectively.

The basic applications in on-line modeling include: system assessment, result comparison, and alarm status evaluation. These processes involve data access from: hydraulic modeling results, SCADA data, and configuration status references. The object-oriented approach as discussed above has been adopted to retrieve individual data items for assessment. Comparison can readily be made between the SCADA readings and modeling conditions, which can then be used for model calibration, etc. The full system hydraulic performance can be reviewed for the assessment of the system adequacy. Alarm conditions for tank water level, junction pressure, and pipe velocity violations can easily be highlighted. The corresponding hydraulic assessment can then be used to adjust any necessary control strategies to ensure the ongoing integrity of the given system operation.

5 APPLICATIONS AND CONCLUDING REMARKS

An implementation of the on-line modeling procedure as outlined in this paper has been directly incorporated within a widely used hydraulic modeling package as part of a standard modeling procedure[1]. The implementation can be used to perform snap-shot comparison. It can also be operated on a real-time basis with regular time intervals. Fig. 4 below illustrates an application of such an implementation scheme.

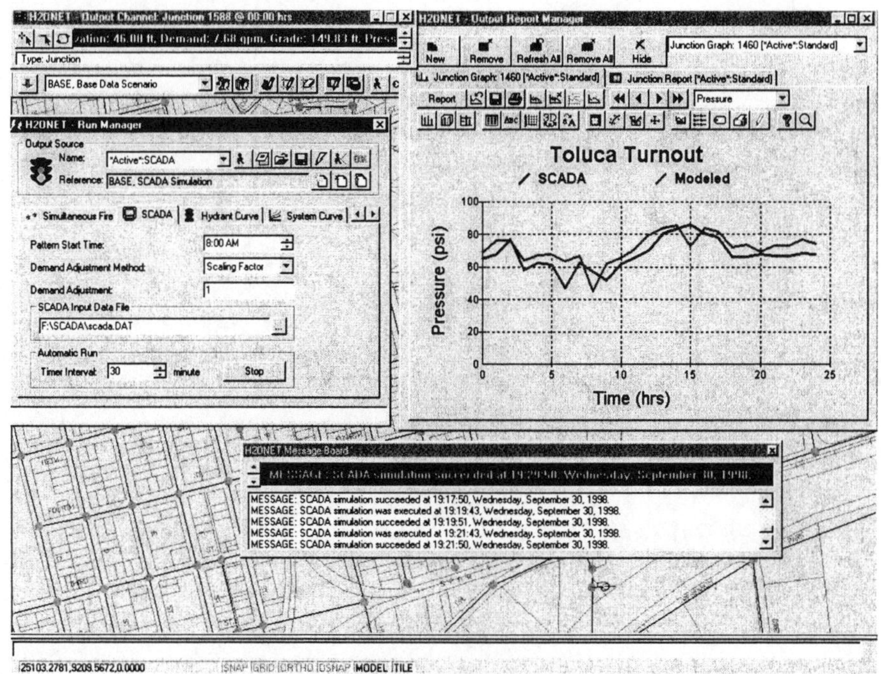

Fig. 4 On-line modeling with H_2ONET.

REFERENCES

1. *H_2ONET Users Guide – Graphical Water Distribution Modeling and Management Package*. MW Soft, Inc. 300 North Lake Avenue, Suite 1200, Pasadena, CA 91101 USA.

A Heuristic Approach to the Design of Looped Water Distribution Networks

E. Todini

ABSTRACT

Water distribution networks are generally designed using a complex minimisation of non-linear cost functions. For instance, minimum-cost, tree-shaped, irrigation-water distribution networks have been successfully designed using Dynamic Programming. Unfortunately, when dealing with looped water distribution networks, not only is there a dramatic increase in the number of alternatives, but most of all, the minimum-cost criterion loses its meaning as the sole criterion of economic evaluation.

The main reason for designing a distribution network as a "looped" network, is to guarantee the availability of water during pipe failures. The second objective: reliability, is opposed to the minimum-cost objective. The measure of the second objective is a "reliability index" defined as the ratio of the power (energy per unit time) not dissipated within the distribution network, to the total available power. Ignoring reliability inevitably leads to a spanning-tree design without loops. Consequently, the trade-off between cost and reliability is a major design-decision and it can be formulated as a vector optimisation problem with two objective functions: cost and reliability. The solution of a vector optimisation problem is the undominated or Pareto set in the space spanned by the objective functions. A heuristic technique is presented for the rapid approximation of the Pareto set. In addition, the proposed technique, which has been applied to the design of several real looped networks, shows that the required number of iterations is relatively small and that quasi-optimal solutions are easily obtained for each pre-set degree of reliability.

1 INTRODUCTION

The history of water distribution network design goes back to the beginning of the nineteenth century when industrialisation created the conditions for welfare in the cities. Initially the economical design was mainly oriented to minimise the cost of the main pipe bringing water to the city with respect to the cost of one or more reservoirs and the cost of the water distribution network.

Most of the nineteenth century systems have been designed in this way and in particular they were generally schematised as a main ring pipe of large diameter (for instance 1000 mm) with several smaller radial pipes (say 400 mm) reaching the centre of the ring in order to guarantee the delivery of water at any point, even if one of the pipes had a failure. Although the radial pipes were connected in the centre, because of computational difficulties they were considered disconnected during computation.

Under these conditions an analytical solution existed that allowed the engineer to provide a standardised design for the main pipes, while the very small ones (100-200 mm) were then introduced without real economical consideration.

Later, with the advent of digital computers, several approaches were taken into account for the minimum cost design of water distribution networks.

In the case of irrigation distribution systems, which are generally designed as trees, the problem can be easily solved by means of Dynamic Programming. Unfortunately this approach cannot be used for urban water distribution systems because in this case, owing to the need for delivering water to the public under failure conditions, a system is generally designed not as a tree but as a series of interconnected closed loops. In this case, given that the flow direction may vary as a function of the pipe diameters, to allow for the variables to be separable one must consider two possible flow conditions per pipe, which means that each pipe must be represented with two arcs with opposite flow direction, thus increasing enormously the computational effort.

Several direct optimisation approaches have been tried in the past, which were generally limited to networks with a few pipes, due to the computational effort required.

An interesting approach was given by Alperovits and Shamir (1977) who linearised the cost function, taking as the decision variables not the diameter of pipes but the length for which a given diameter had to be used. This approach reduced the original non-linear problem into a linear one for which linear programming could provide a solution with reasonable computer time requirements also in the case of very large water distribution systems.

Regardless of the optimisation approach taken for the optimal design of the looped water distribution systems, there is one basic question that rarely appeared in the literature: why are we designing a looped network topology ? The answer is obviously simple: the looped topology allows for redundancy, which means that there is enough capability in the system to overcome local failures and to guarantee the distribution of water to users.

Unfortunately all the "minimum cost" objective functions inevitably will lead to a spanning tree with no redundancies unless specific constraints are imposed (an interesting example was recently proposed by Stanic et al. (1998) who use an evolutionary algorithm for the selection of the most appropriate tree shaped network). If one does not believe this statement it is sufficient to look at table 5 reported in Abebe and Solomatine (1998) where several solutions are given for the two loop network. One can notice that the "best" solutions (Goulter et al., 1986, Kessler and Shamir, 1989, Eiger et al., 1994, Savic and Walters, 1997) are the ones that reduce to an allowed minimum the diameters of two of the pipes. Removing these two pipes, the looped network degenerates into the essential and minimum cost tree. But what happens if one of the remaining pipes fails: nodal demands cannot be met and this would also happen if the removed pipes have very small diameters as in the above mentioned optimal solutions.

This simple example shows that the choice of the "minimum cost function" is not one that allows for correctly designing a looped water distribution network because the objective function does not incorporate the concept of "reliability", which is the main reason for avoiding a tree shaped network.

Although the formalisation of the reliability concept could possibly be introduced into a more meaningful multi-objective function, in this paper a heuristic approach is introduced which is capable of describing the Pareto set, thus allowing the designer to choose reasonable solutions with limited computational time requirements.

2 THE CONCEPT OF RELIABILITY

In a water distribution system, the cost of power appears explicitly only if the water must be pumped: in this case one can notice that the optimisation problem may be posed by formulating a trade-off between the cost of pumping and the cost of the pipe network. If the pipes are too small the cost increases due to the necessity of pumping water. The installation and running costs may become too large to allow for an economical delivery in terms of demand of water and of sufficient head. When dealing with gravity driven water distribution systems, the cost which can be used in an optimisation scheme is generally limited to the cost of pipes and inevitably its minimisation produces a tree shaped network.

Nevertheless one can make the following considerations: if the water is delivered at each node, satisfying the demand in terms of flow and head exactly, whenever a pipe fails the water flow will change and the original network will be transformed into a new one with higher energy losses: this means that it will be impossible to deliver what is requested at all the nodes.

The immediate consequence is that in a looped network we would like to provide at each node more energy per unit time than that required, in order to have a sufficient surplus to be dissipated internally in case of failures.

If we denote with

$$P_{tot} = \gamma \sum_{k=1}^{n_r} Q_k H_k \tag{1}$$

the total available power at the entrance in the water distribution network, where γ is the specific weight of water Q_k and H_k are the discharge and the head relevant to each reservoir k, while n_k is the number of reservoirs, the following simple relationship exists:

$$P_{tot} = P_{int} + P_{ext} \tag{2}$$

where P_{int} is the power dissipated in the pipes while $P_{ext} = \gamma \sum_{i=1}^{n_n} q_i h_i$ is the power that is delivered to the users in terms of flow q_i and head h_i at each node i, with n_n the number of nodes.

It seems reasonable to define a reliability index I_r as:

$I_r = 1 - \frac{P_{int}^*}{P_{max}^*}$, where $P_{int}^* = P_{tot} - \gamma \sum_{i=1}^{n_n} q_i^* h_i$ is the amount of power dissipated in the network to satisfy the total demand and $P_{max}^* = P_{tot} - \gamma \sum_{i=1}^{n_n} q_i^* h_i^*$ the maximum power that could be dissipated internally in order to satisfy the constraints in terms of demand and head at the nodes.

After the appropriate substitutions, the reliability index can be written as:

$$I_r = \frac{\sum_{i=1}^{n_n} q_i^* \left(h_i - h_i^*\right)}{\sum_{k=1}^{n_r} Q_k H_k - \sum_{i=1}^{n_n} q_i^* h_i^*} \tag{3}$$

This is the most important index on which is based the proposed heuristic optimisation.

Another index to be accounted for is the failure index I_f defined as:

$$I_f = \frac{\sum_{i=1}^{n_n} I_{fi}}{\sum_{i=1}^{n_n} q_i^* h_i^*} \quad \text{where } I_{fi} \begin{cases} = 0 & \forall i : h_i \geq h_i^* \\ = q_i^* \left(h_i^* - h_i\right) & \forall i : h_i < h_i^* \end{cases} \tag{4}$$

This index is necessary both to identify infeasibilities during the optimisation process and to evaluate and compare the effect of pipe failures.

The third index is the available surplus of head into the most depressed node. This is an interesting quantity, that may give insight into the hydraulical behaviour of the network.

3 THE HEURISTIC APPROACH

Let us first note that the topology of an urban water distribution system is generally imposed by the structure of the urbanistic context: roads, buildings, industrial areas, hospitals, etc. and that a layout is generally available. Therefore the optimal design of the water distribution network to be solved assumes a predefined topology and a set of constraints in terms of amount of water to be delivered at the nodes, where the demand is concentrated, and of minimum head necessary to draw water from the tap.

The design demands are preliminarily established on the basis of population consumption and growth, industrial development, fire hydrants, etc. while the minimum head distribution is defined on the basis of the needs also in relation to the topographical elevation.

The available commercial diameters and the relevant costs (which would include additional costs such as for instance the excavation cost and the cost of special parts) complete the required information.

The proposed procedure starts from an initial set of diameters reasonably given by the designer on the basis of his experience. This is not essential, since any set of diameters can be used, nevertheless a reasonable initial solution can help in the search.

i) The first step is the network analysis, which is performed using the "global gradient" developed by Todini (1979), Todini and Pilati (1988) which has recently gained some credibility after its inclusion in EPANET (Rossman, 1993).

ii) From the results of an analysis, if the failure index is null the procedure starts reducing the pipe diameters, otherwise the pipe diameters must be increased.

iii) The reduction of diameters is performed according to the largest decrement in cost that will affect the generic pipe connecting nodes i, j:

$$\Delta C = C_{ij}^{k} - C_{ij}^{k-1} \tag{5}$$

for a reduction of diameters from class k to class $k-1$. It is obvious that the first pipes to be reduced in diameter are the longest and largest.

Before applying the reduction, three controls are made. The first is a velocity control: the maximum velocity in the modified pipe should not exceed a pre-determined value (generally set to 2 m/s). The second control is based upon an approximate estimate of the decrease in head $\delta\hat{h}_i$ of the most critical node, namely the node which head h_i resulted closer to the prescribed head h_i^* in the original network analysis, and the move is performed if the following constraint is satisfied: $h_i + \delta\hat{h}_i \geq h_i^*$. The third check is made on the basis of the reliability index as a function of an approximate estimate of the overall increase in internal power dissipation.

This procedure follows, without the need for a new network analysis at each diameter change, until no more moves are possible. At this point, given the new diameter configuration, a network analysis is performed, before starting again with point (ii).

iv) The increase in diameters proceeds according to the largest decrement of the internal power dissipation:

$$\Delta P_{int} = P_{int}^{k+1} - P_{int}^{k} \tag{6}$$

as a function of the change in the diameter $D_{ij}^{k} \rightarrow D_{ij}^{k+1}$ of the generic pipe connecting nodes i, j.

This will tend to increase the size not only of the very small and short pipes but also in particular, of the pipes where the flow is very low, in the kth configuration. As mentioned earlier, the method tries to mimic what a designer would do to improve, both in terms of cost and in terms of efficiency index, from the previous solution and the objective of the looped structure of the pipe networks is to try to distribute the flow more evenly among all the pipes, which is the opposite of concentrating the flow in a spanning tree.

The increase in diameters stops either according to a maximum number of moves or when the estimate of the reliability index is higher than the given value.

Again, at the end of the step, a new network analysis is performed and both the reliability and the failure indexes are computed before starting again from (ii).

After a number of steps, generally less than 50, it is possible to construct and analyse a reliability vs cost function that indicates the set of possible optimal solutions.

4 EXAMPLES OF APPLICATION

Two examples are reported in the paper in order to illustrate the concepts and the possibilities offered by the approach.

A first example can be derived from the Shamir problem reported in Abebe and Solomatine (1998). The network is an extremely simplified two loop network, the best economical solution of which is provided as in table 1, in terms of pipe diameters (which were kept in inches to be consistent with the previous papers), cost, reliability index and head surplus in the most depressed node.

The analysis based on the heuristic method, using the data provided in the above mentioned Abebe and Solomatine paper, allows to produce the edge of the Pareto set shown in figure 1, that gives better insight into the needs of tradeoffs between the two objectives, namely cost and reliability.

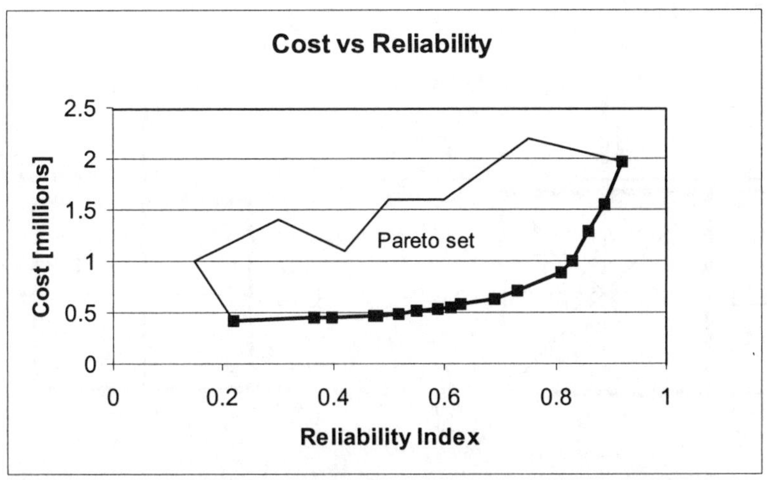

Figure 1. The limiting curve of the Pareto set of solutions in the two objectives space.

It is not difficult to notice that in the first part of the limiting curve one can more than double the reliability index with very small increases in cost. A set of four solutions is reported in table 1 in the range of .4 to .5 of the reliability index. It can be noticed that with an increase from 0.419 10^6 to 0.419 10^6 units of cost, the reliability index almost doubles as well as the head surplus (Sol. A). Moreover the designer can decide to spend a little more and retain solution D with a reliability index of .48 and a head surplus of 3.44 m.

Table 1. Comparison of the optimal solution with a set of alternatives.

Pipe n.	Cost. Opt.	Sol. A	Sol B	Sol. C	Sol. D
1	18	18	20	20	20
2	10	16	14	14	14
3	16	14	14	14	14
4	4	6	6	8	6
5	16	14	14	14	14
6	10	1	1	1	1
7	10	14	14	14	14
8	1	10	10	10	12
Rel. Ind.	0.22	0.41	0.47	0.48	0.48
Head Surp.	0.50 m	1.08 m	1.90 m	2.84 m	3.44 m
Cost	0.419 10^6	0.450 10^6	0.460 10^6	0.467 10^6	0.478 10^6

A second example is based upon the schematic looped main described in figure 2 using the same data of the previous example in terms of head losses function and parameters, diameter classes and cost.

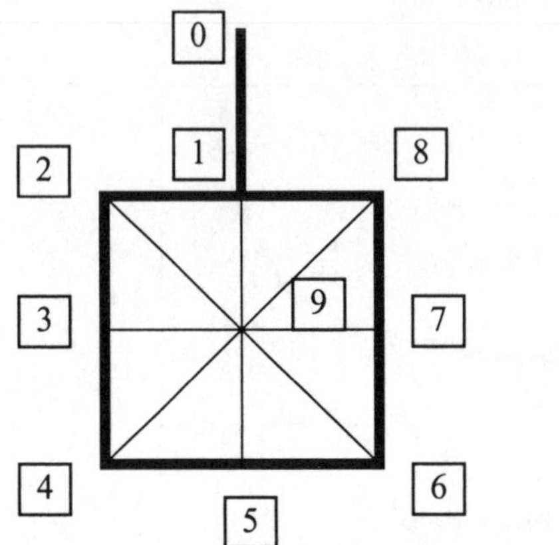

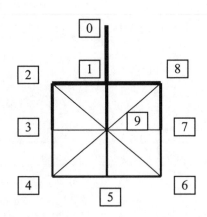

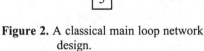

Figure 2. A classical main loop network design.

Figure 3. An improved design for the same system.

The diagrams shown in figures 2 and 3 are only topological. The pipe lengths are as follows: the pipe connecting nodes 0-1 is 2000 m long, the contour pipes are all 1000 m long and all the internal pipes connected to node 9 are 1210 m long. The following table 2 gives the head and the resulting flow in node 0 together with the prescribed minimum head and demand for nodes 1 to 9:

Table 2. Head and demand values for the second example.

Node	Head	Demand
0	200.	-180.00
1	197.	20.00
2	193.	20.00
3	192.	20.00
4	191.	20.00
5	191.	20.00
6	191.	20.00
7	192.	20.00
8	193.	20.00
9	189.	20.00

The results of this example are shown in table 3 where the classical main loop design is compared to what can be obtained using the heuristic approach described in this paper. The cost of the classical design is higher (.657 10^6) but produces a low reliability ($I_r = 0.20$) with only .7 m of surplus head available in the most depressed node. The new design is less expensive (.565 10^6), more reliable ($I_r =$ 0.52) and has a surplus head of more than 7 m in the most depressed node.

Table 3. Comparison between a classical main loop network design and a solution provided by the heuristic approach.

Pipe	Main Loop	Improved
0 - 1	16	18
1 - 2	12	14
2 - 3	12	10
3 - 4	12	8
4 - 5	12	1
5 - 6	12	1
6 - 7	12	8
7 - 8	12	10
8 - 1	12	14
1 - 9	3	10
2 - 9	3	1
3 - 9	3	1
4 - 9	3	1
5 - 9	3	8
6 - 9	3	1
7 - 9	3	1
8 - 9	3	1
Rel. Ind.	0.20	0.52
Head Surp.	0.70 m	7.08 m
Cost	$0.657\ 10^6$	$0.565\ 10^6$

5 CONCLUSIONS

The concept of minimum cost optimisation for looped municipal water distribution systems must be completely reviewed under the need for a more complex and multi-objective approach.

This paper has introduced a simple heuristic technique that tries to emulate the reasoning of an engineer aiming on the one hand to reduce the cost, while on the other hand to preserve a sufficient degree of reliability in the system to cope with possible failures.

After defining the problem in a two objective space, it was shown that the technique is capable of rapidly describing the optimal Pareto set, which allows one to find satisfactory solutions as a trade off between system reliability and cost.

From the two reported examples it clearly emerges that these solutions correspond neither to the minimum cost solutions nor to the classical main loop solutions, which are in general more reliable than the first ones, but can be further improved under a more systematic multi-objective analysis.

It is hoped that the considerations developed here will be retained and expanded by the scholars and designers that deal with optimisation of water distribution systems with the aim of producing both efficient and practical usable tools.

REFERENCES

Abebe A.J. and Solomatine D.P., (1998) - Application of global optimization to the design of pipe networks. In Hydroinformatics '98, Babovic and Larsen (eds), Balkema Rotterdam, pp. 989-996.

Alperovits E. and Shamir U., (1977) Design of optimal water distribution systems. Water Res. Research, 13(6):885-900.

Eiger G.U., Shamir U., Ben-Tal A., (1994) – Optimal Design of Water Distribution Networks. Water Res. Research, 30(9):2637-2646.

Goulter I.C., Lussier B.M., Morgan D.R., (1986) – Implications of head loss path choice in the optimisation of water distribution networks. Water Res. Research, 22(5):819-822.

Kessler A. and Shamir U., (1989) – Analysis of the linear programming gradient method for optimal design of water supply networks. Water Res. Research, 25(7):1469-1480.

Rossman L.A., (1993) EPANET, Users Manual. Risk Reduction Engg. Laboratory, Office of research & Devt., U.S. Env. Protection Agency, Cincinnaty, Ohio.

Savic D.A. and Walters G.A. (1997) – Genetic Algorithms for least-cost design of water distribution networks. Water Res. Planning and Management, 123(2):66-77.

Stanic M., Avakumovic D., Kapelan Z., (1998) – Evolutionary algorithm for determining optimal layout of water distribution networks. In Hydroinformatics '98, Babovic and Larsen (eds), Balkema Rotterdam, pp. 901-908.

Todini E., (1979) – Un metodo del gradiente per la verifica delle reti idrauliche(in Italian). Bollettino deglli Ingegneri della Toscana n.11.

Todini E. and Pilati S., (1988) – A gradient Algorithm for the Analysis of Pipe Networks. In B. Coulbeck and Chun-Hou Orr (eds) Computer Applications in Water Supply.Research Studies Press Ltd., pp. 1-20.

PART III

SIMULATION AND MODELLING

Adapting Models and Databases for Water Network Audits

C. Balmaseda, E. Cabrera, P. Iglesias,
E. Jr Cabrera *and* J.V. Ribelles.

ABSTRACT

Establishing the correct strategy for improving the hydraulic efficiency of a water distribution network is of paramount importance. The basic starting point for such a task is the knowledge of the weak points of the water distribution system. Such information can be found through an audit of the network. That balance implies subdividing the unaccounted for water in very clearly established and identified terms.

Based on the mathematical model of the network and a complete database of the demand characteristics of the system, this paper presents a general procedure for evaluating the different terms of that balance. The evaluation is done in an automatic and dynamic way since it can be performed systematically and periodically. Due to the methodology's long, complex and tedious calculations, the full method has been implemented as an auxiliary tool within the SARA workspace. SARA is a software package for network analysis developed by the authors among others. Using SARA the audit is feasible and can be systematically performed. Once it is completed, the efficiency of the management can be assessed by means of all the relevant indicators.

The paper outlines the basic principles of this new audit, and points out the basic hypotheses that support it. For a further understanding of the proposed methodology, an example is presented.

1 INTRODUCTION

Generally speaking with regard to leakage and losses in water supply systems, there has been a dramatic change of attitude in utilities of most of the developed countries in the past few years, especially since water management is focused on sustainable development (Lambert et al., 1998). It is nowadays quite usual to establish a target of a global hydraulic efficiency of 90% (AWWA,1996). As a matter of fact, many utilities around the world are reporting even higher values (Skarda, 1997; Van der Wilingen, 1997, Kawakita, 1998, Ng et al., 1998). Such figures were inconceivable just a few years ago.

On the other hand, it is important to outline that it is unanimously claimed that hydraulic efficiencies should be reported in relative terms, mainly for their use while benchmarking different utilities. Lambert et al. (1998) explain, with practical examples, the paradoxes that can arise due to the use of absolute efficiency ratios.

Losses should hence be expressed in relative indicators, for instance, m^3/km and h (Hirner, 1997). Utilities may then gain a deeper knowledge about the benefits obtained from a given strategy and action using such relative efficiencies rather than absolute values. Furthermore, if the system has been previously well defined, it is a simple task to obtain relative indicators from the absolute ones.

The change of attitude of the utilities with regard to water losses can be seen everywhere in the world. In developing countries, that change has been fostered by the international lending agencies (Yepes, 1995). The World Bank, just like other leading agencies, is able to lend money for urban infrastructures, but only if with the corresponding investment in the system improves its performance. In other countries, for instance those located in the arid or semi-arid areas, the pressure to minimise losses arose due to the need for a higher efficiency in the system foreseeing future long periods of shortages and droughts. In inefficient systems, rationing is implemented throughout service outage, this being a practical approach typical of desperate utilities and/or in relatively uncontrolled situations (Lund and Reed, 1995).

It is important to underline that even developed countries, with sufficient rates of fresh water per capita and per year, such as those in North-eastern Europe, also have important reasons for implementing active leakage controls and minimising unaccounted-for water. Partly because water conservation is not just important for saving water, but also as part of a long term strategy to provide safe and reliable drinking water supplies, whereby pollution is minimised (Beecher et al., 1998). Also last, but not least, because when applying full cost rates, unaccounted-for water increases lead to operation costs rises as well. As more severe standards regarding safe drinking water acts are being applied, costs are becoming higher, and so such policies make more sense as time goes by (Beecher et al., 1998).

Many activities of leakage management can be identified. Lambert et al. (1998) have organised the potential actions devoted to control losses into four different classes. Yet, strategies are not enough to achieve relevant hydraulic efficiency values, because it is clearly impossible to achieve good efficiencies with old pipe systems. Material, age and other risk factors for the pipes (damage arising from heavy traffic, poor workmanship in their laying, etc.) are questions, regarding leakage control, of paramount influence. For that reason a convenient rehabilitation and replacement frequency is necessary. With full cost recovery policies, including refund costs, old pipes are replaced in a systematic way and within a reasonable period. This fact explains why countries with political rates have less efficient water distribution systems than others with more fresh water resources per capita and per year. History, tradition and inertia play an important role in water policies rules in many countries. It is really difficult to change cultures widely developed throughout centuries (Cabrera and García Serra, 1997).

All in all, from an exclusively economic approach, it is well known that at a given rate, there is an optimum hydraulic efficiency value. For instance, in France and for the current prices (Villesot, 1997), the optimum leakage ratio is 20%. Values over 80% for hydraulic efficiency provide no return. Leakage management costs are higher than the savings obtained. For such reasons, some countries such as the UK by means of the OFWAT, link pricing rates policies closely to network efficiency - i.e. network investments - (OFWAT, 1994). Since improving the performance means higher costs and better quality, consumers must pay more per cubic meter, which is to be expected.

Sustainable policies from around the world have been widely accepted after the conclusions of the Brundtland Commission (Brundtland, 1987), that strongly recommend water conservation practices. Such practices imply a strict control of pollution and the use of full recovery costs. For that reason, experts (IWSA, 1995; EU, 1997) are following the same track. It is clear that the significant change of attitude of the utilities will consolidate this tendency, and that any research devoted to minimise water losses in a distribution system will gain, as time goes by, more and more relevance. That tendency is affecting water policies everywhere. Society is now in favour of water management rather than water development, as opposed to the tendency that has characterised the 20th century (Burgi, 1998).

In this paper a methodology for a complete and detailed audit of a water distribution system is presented. It is based on the mathematical model of the network and on a complete database which contains the more relevant characteristics of meters and consumers of the utility. The method has been implemented, as an auxiliary tool in SARA, a computer software package for network analysis.

2 GENERAL OVERVIEW

A standard terminology is the starting point for any analysis. Unfortunately (Lambert et al., 1998), the different terms involved in the balance of a water distribution system vary from one country to another. Some, like Germany (DVGW, 1986), France (AGHTM, 1990) or USA (AWWA, 1990) have their own glossary. In any case, the notation used in this paper has been chosen according to the needs of the audit herein presented. The notation fits very well the definitions given by Hirner (1998). Regarding the audit, all terms are expressed in volume per unit of time. Definitions are:

Q: Water supplied: represents all the input into the distribution system.

Q_r: Water consumed that has been registered by meters.

Q_u: Unaccounted for water (portion of the water supplied which has not been registered).

Q_{uc}: Unaccounted for consumed water (portion of the unaccounted for water delivered to the consumers that remains unregistered. It is not a real loss).

Q_{ul}: Unaccounted for water of the system due to real losses (leakage).

Q_{ucn}: Unaccounted for water consumed not metered (unmetered consumption, authorised or not).

Q_{uce} : Unaccounted for consumed water, due to meter errors (may be over metered, although usually under metered).

Q_{ucnl}: Unaccounted for consumed water, not metered, by legal consumer (consumption of an authorised user that does not have a meter device installed).

Q_{ucni}: Unaccounted for water consumed, not metered, by an illegal consumer (unauthorised and hence not metered consumption. Water that has been 'stolen').

It is important to underline that all the terms of the balance are positive except Q_{uce}. As will be later shown, in some few cases this flow can be negative (Hirner, 1998). However, most of the times, due to meter characteristics and flow rate percentages demand patterns, it is positive.

The summary balance is outlined by Figure 1, and it is clearly self-explanatory. As can be seen, any flow on the left side can be subdivided in two, giving full sense to the preceding definitions.

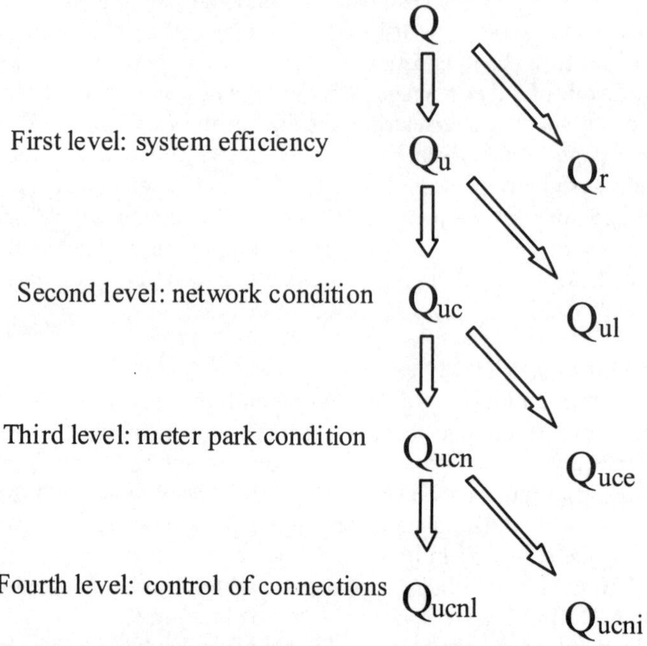

Fig. 1. Different levels of a water distribution network audit.

There is one significant flow rate that can be easily obtained from the defined ones, that is not included in the previous balance (Fig. 1). It is the water delivered to all the consumers, Q_d, equal to:

$$Q_d = Q_r + Q_{uc}$$

From the definitions, the following global water efficiencies can be defined:

- Global efficiency of the system, η_s:

$$\eta_s = \frac{Q_r}{Q}$$

- Network efficiency, η_n :

$$\eta_n = \frac{Q_d}{Q}$$

- Efficiency of the commercial management, η_m:

$$\eta_m = \frac{Q_r}{Q_d}$$

A relationship between the three parameters can easily be stated:

$$\eta_s = \eta_n x \eta_m$$

The full audit requires the evaluation of every flow rate in the balance. The procedure to be followed is: from Q and Q_r , both known from metering, Q_u may be determined ($Q_u = Q - Q_r$). The second term, which may be calculated according to the guidelines given by section 4, is Q_{uce}. Its value is obtained from the meter and consumer characteristics database. Finally, as outlined by section 5, Q_u is desegregated into Q_{uc} and Q_{ul}. That means that, up to the third level, all the flows considered by the audit shown on Fig. 1 are known. The last level, referred as control of connections, cannot be determined by analytical procedure due to the characteristics of Q_{ucn}. The only way to find the values for Q_{ucnl} and Q_{ucni} is by means of strict control of the connections to the different consumers, determining those who are stealing water.

3 REQUIREMENTS AND DATABASE NEEDED FOR THE AUDIT

The methodology developed needs:

- Meters at any point where water enters the system (delivery metering). This is necessary for evaluating Q, the volume, referred to a given unit of time, entering the system. No metering errors are considered in the evaluation of Q, because there are very few delivering meters and, due to their size, such meters have higher accuracy than consumer meters. Besides, delivery meters are often selected using the demand pattern vs. time of the system at the point where they are installed, resulting in more accurate measures. In any case, it is really easy to evaluate the accuracy and, if necessary, correct the measure accordingly.
- The highest possible number of demand consumer meters installed in their property. A smaller number of operating meters gives a smaller value for Q_r and, at the same time, higher Q_{uc} . This last value increases, as the previous one decreases at the same rate, giving more uncertainty to the audit.
- A well calibrated mathematical model of the network.

Any well-managed water distribution system accomplishes such requirements. Perhaps the last requirement is the most specific one, because all the networks are different, and for this reason the model must be tailored for each particular case. But, nowadays, it is very common for utilities to have their own model. According to a survey (Mc Elroy, 1985), in 1985 in the USA, 90% of the utilities with a demand level of 200 000 m^3/day use a mathematical model. AWWA promoted a similar study (Cesario, 1995) and the conclusion was that 86% of the utilities had, by then, a model regardless of their size. From those results it can be concluded that nowadays all well established utilities have their own mathematical model.

Regarding consumer data, which can be important, it is necessary to classify the data into different groups, attending to two criteria: type of consumer (commercial, industrial, domestic, etc.) and type and age of their meter. The first criterion defines the pattern of their consumption both vs. time and regarding flow rate percentages. For the second one, a characteristic curve for the meter (error against passing flow) will be assigned to the consumer.

In any case, and as far as the consumers database is concerned, for each consumer identified by an identification number (ID), the following data will be included:

- Type and age of the meter. As mentioned before, from these parameters, the meter is classified by the second criterion, with its corresponding characteristic curve.
- Type of consumer, for its allocation in one group by the first criterion. Every group has well defined:
    - Water demand pattern vs. time.
    - Flow rate percentages demand patterns.
- The node where the consumption of the consumer will be loaded on the model.
- The registered consumption during a given period of time (typically from one to three months).

Previously, the characteristic curves of significant samples of meters from each group have been determined in the laboratory. The average result is adopted as the representative curve of the group. As meters are manufactured taking into account the ISO 4064-1, the different classes defined by this standard can be adopted for defining the basic groups and, later, age can be adopted for a second level criterion (see Table 2, Section 6.3).

Water demand patterns can be obtained either from direct measurements of the consumers or patterns provided in the literature (Bowen et al., 1993; Yanov and Koch, 1987). In summary, average utilities with good management practices, can fulfil the necessary requirements. If not, it will be impossible to evaluate the different terms of the audit proposed. In that case, a more basic approach should be performed (AWWA, 1990).

We will proceed to explain now, in short, the methodology of the audit. There are basically two steps which are summarised in the two following sections.

4 STEP ONE: WATER METERS ERROR EVALUATION

Significant errors can arise during the metering process. Such errors will depend on the age of the meters, the characteristics of water (water may, or may not, generate incrustations in the meter channels) and, for sure, the flow rate percentages demand pattern with regard to the characteristic curve of the meters at the time of the measurement. For instance in Nuremberg (Hirner, 1998), after six years of metering operation, consumer meters have a tendency to over-register (about 1% - 2%). But the general tendency is to under register the real volumes consumed, due to excessive low flows of the patterns consumption. Such is the case in Zurich (Skarda, 1997) where, from a complete assessment, apparent losses of 4% were found, equal to the error of the metering park.

Matching for all the groups defined the real characteristics of the meter population with real flow rate percentages demand patterns, the total error of the water consumption can be determined and, then, the term Q_{uce} of the audit. As commented previously, in some cases, Q_{uce} can represent a significant value. Details of the procedure can be found in a recent investigation carried out in Valencia (Arregui, 1998).

Although being located in the third level of the audit, Q_{uce} is calculated prior to second level. This is because its value may be used, if significant, to modify Q_r accordingly and, later, desegregate its value into two terms Q_{uc} and Q_{ul}. With regard to the time scale of the problem all volumes in the audit are referred to in seconds or hours (flow rates), while consumed volumes are usually calculated per month. The scales of time are hence very different. In order to compare terms with the same units, a standardisation of the time scale is required. The variation of the consumed volume with time is also very important. In any case, this requirement is demanded by many other applications. For instance, it is a really basic point when adjusting the model.

5 REAL LOSES EVALUATION

The procedure for desegregating, through the network model, Q_u into two terms, Q_{ul} and Q_{uc} , at the second level of the audit, is briefly outlined. Details can be found in Almandoz et al. (1998). The procedure is based on the adjustment of the parameter $x = (Q_{ul} / Q_u)$, in a process very similar to the calibration of a model, but with just one parameter. For this purpose the network is analysed during 24 hours (extended period simulation) for different values of x assumed; x ranging from zero to one. Since all the terms of the audit have been loaded into the model, the water supplied to the network during simulation can be determined. The supplied flow has a different modulation for each value of the parameter x. Since the modulation of the water really supplied to the network is well known from the input meters, the parameter is adjusted, comparing the standard deviation of the solution with those corresponding to the different simulations.

As the load of the model includes all the terms for the distribution of Q_{ul} and Q_{uc} over the network, basic assumptions are necessary:

- Real losses, Q_{ul} , are assumed to be proportional to the length of the pipes, and distributed during the day, taking into account that leaks vary with the square root of the pressure of the nodes, that is variable with time during the extended period simulation. Particular characteristics of each network can modify the preceding basic spatial and temporal distribution in order to take into account other relevant circumstances (age, material, number of service connections, etc.), even different pressure exponents which appear in the literature (Germanopoulos, 1995; Bargiela, 1984; Lambert et al, 1998). This can be accomplished, if necessary, in a very easy way.

- The correct spatial and temporal distribution of the term Q_{uc} has more uncertainties. The basic assumption is that it is proportional, in space and in time, to Q_r. Nevertheless, depending on each particular network, Q_{uc} may take into account where and when that kind of consumption is spatially dominant (public buildings, own utility consumption, irrigation of parks, sewer and street cleaning, fire fighting and training, public fountains, freeze protection and, last but not least, water theft). If necessary, the flow rate Q_{uce} , previously determined, can be distributed correctly.

From a hydraulic point of view, and for current networks, as leakage is dominant during the night, higher values of x force a more uniform supply in the simulation done for that particular x value. For values of x tending to zero, Q_{uc} is dominant, producing the opposite effect and giving as a result a more modulated supply, that is to say, less uniform, in time. In such a case, there is a close relation between the values of x and the standard deviation of the water supplied.

In other words, this method based on the model benefits from the well-documented fact that leakage rates are very sensitive to pressure variation. And as far as the hypothesis concerns, leakage rates can be very consistent if the characteristics and distribution of non-metered demand are well known. In any case, they may be improved by trial and error. With all demands metered, if Q_{uc} is equal to Q_{uce} the analysis outlined in this paragraph should be dropped out; however, this is quite infrequent.

6 EXAMPLE

6.1 General information

Population: 90,000 habitants.
Total number of consumers: 28,464
Number of service connections (NoSC): 7631

6.2 Consumers characteristics

Total domestic consumers: 27,010

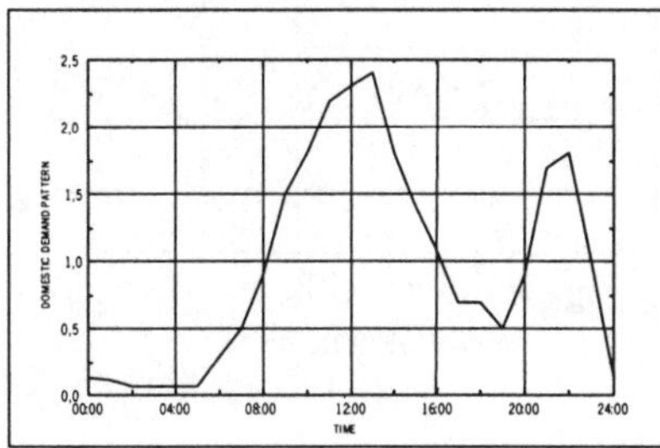

Fig 2. Domestic hourly demand pattern.

New building domestic consumers: 11,950

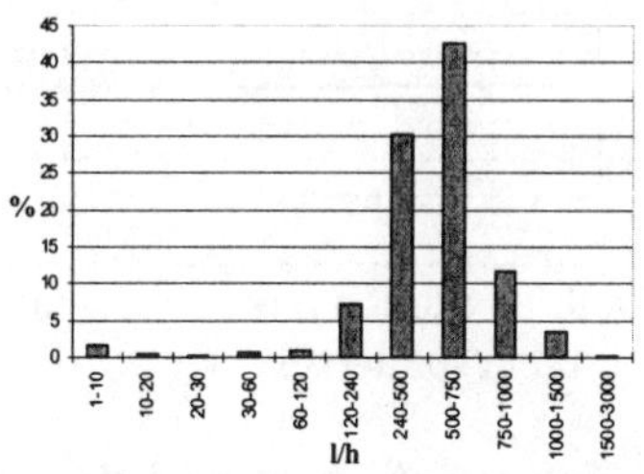

Fig. 3. Flow rate percentages demand pattern for domestic consumers in new buildings.

Old buildings dom. cons.: 10,683

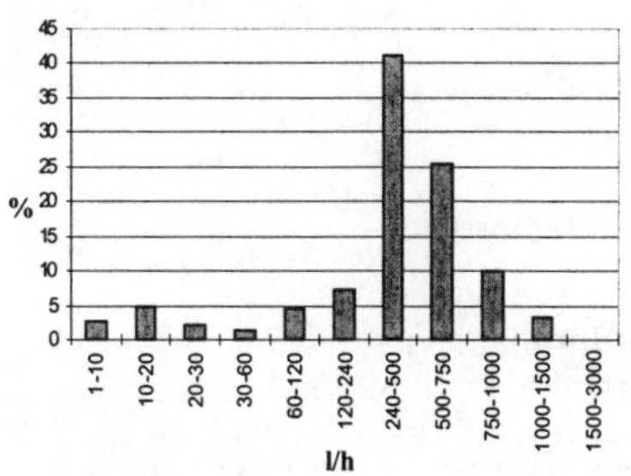

Fig. 4. Percent usage at flow rates for domestic consumers in old buildings (flow percentage demand pattern).

Residential areas dom. cons.: 4,377

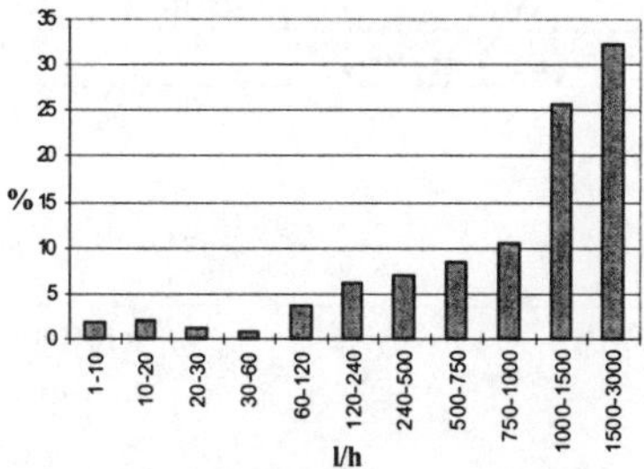

Fig. 5. Flow rate percentages demand pattern for domestic consumers in residential areas

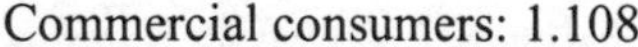

Commercial consumers: 1.108

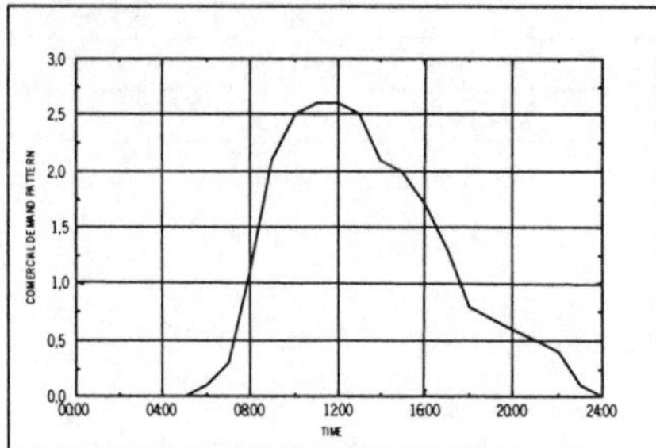

Fig. 6. Commercial hourly demand pattern.

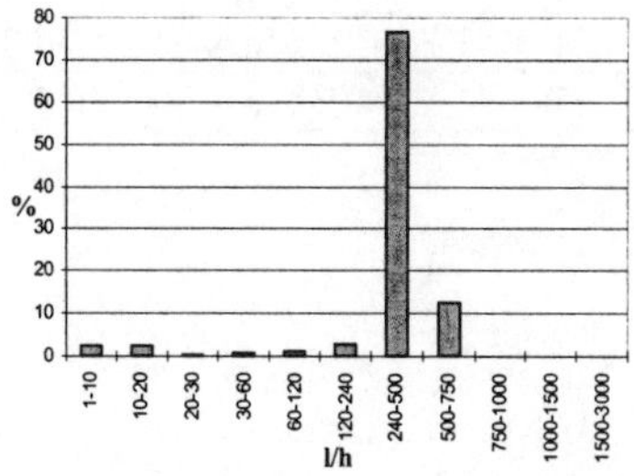

Fig. 7. Flow rate percentages demand patterns for commercial consumers

Institutional consumers: 62

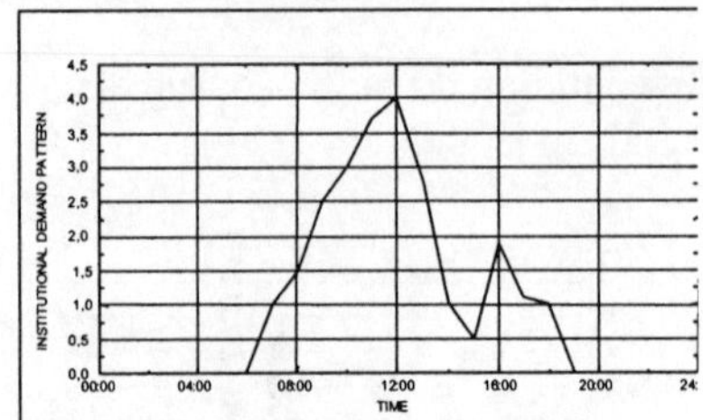

Fig. 8. Institutional hourly demand pattern.

Industrial consumers: 284

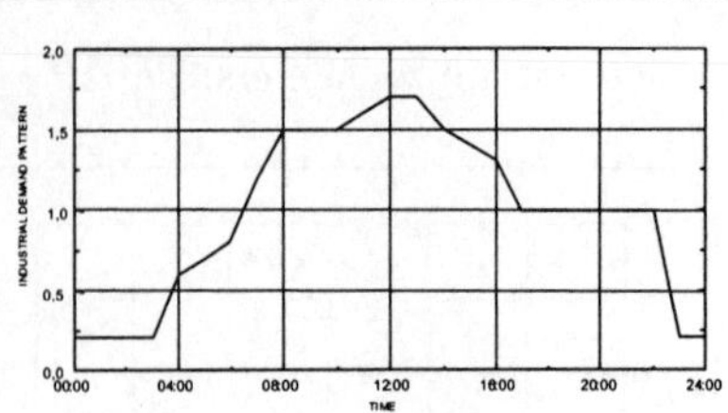

Fig. 9. Industrial hourly demand pattern.

6.3 Meter park characteristics

Table 1. Meter models

Type	1	2	3	4	5
Size	15 mm	15 mm	15 mm	25-65 mm	25-65 mm
Class (ISO 4064-1)	B	B	B	C	C

Table 2. Meter groups characteristics

Type	Age	Meter group	Minimum measurable flow (l/h)	Accuracy 50 l/h	Accuracy 750 l/h
1	1 – 5 years	1.1	15	101%	100.5%
1	6 – 10 years	1.2	25	100%	100.5%
2	1 – 5 years	2.1	20	102%	101%
2	6 – 10 years	2.2	30	99%	99%
3	6 – 10 years	3.2	30	95%	98%
3	> 10 years	3.3	40	90%	95%
4	1 – 5 years	4.1	Weighted accuracy 97%		
5	1 – 5 years	5.1	Weighted accuracy 96%		

Table 3. Meters groups percentage

Meter group	Domestic New build.	Domestic Old build.	Domestic Residential	Comm-ercial	Institutio nal	Indust.
1.1	20.9%	5.6%	9.2%			
1.2	15.8%		3.5%			
2.1		10.3%		1.7%		
2.2		14.2%		1.8%		
3.2	3%	3.4%				
3.3	2%	4.2%	2.8%	0.4%		
4.1						1%
5.1					0.2%	
Total	41.70%	37.7%	15.5%	3.9%	0.2%	1%

6.4 Network characteristics

Table 4. Pipe and node characteristics

Pipe	Diameter	Length		Node	Demand (l/s)	Elevation (m)
Pipe 1	600	600		2	5	40
Pipe 2	350	500		3	18	30
Pipe 3	350	300		4	15	25
Pipe 4	300	670		5	45	4
Pipe 5	100	600		6	20	4
Pipe 6	125	350		7	35	20
Pipe 7	300	100		8	15	25
Pipe 8	100	1140		9	22	40
Pipe 9	350	200		10	15	30
Pipe 10	100	400				
Pipe 11	100	600		*Tank* 1		85
Pipe 12	350	200				
Pipe 13	400	600				

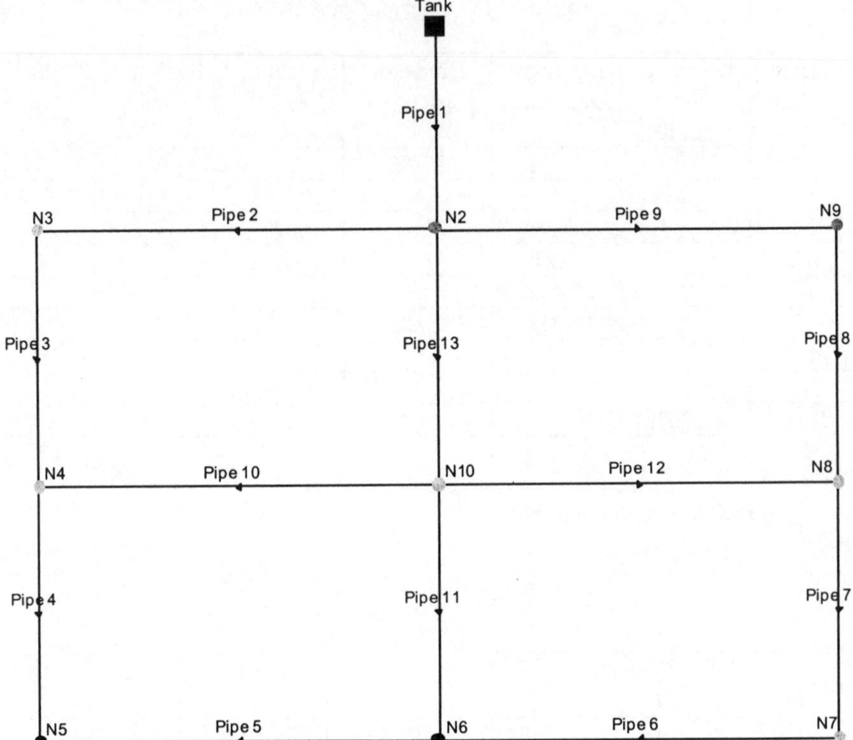

Fig. 10. Network layout

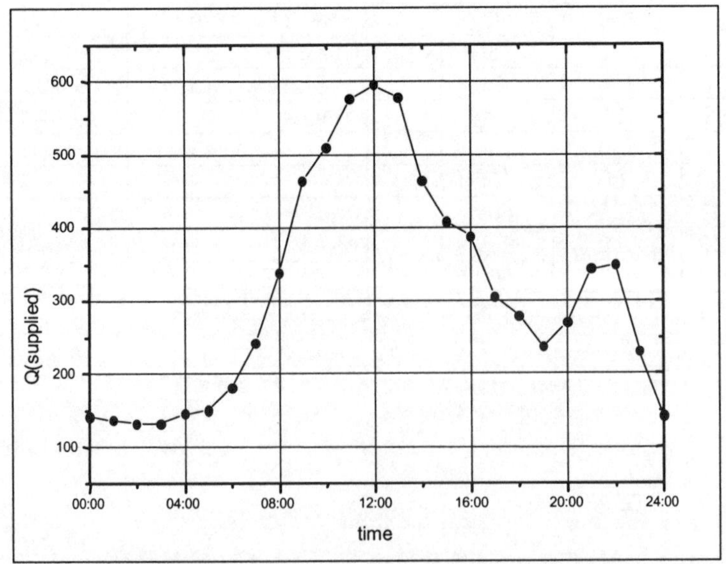

Fig. 11. Water supplied to the system (metered)

Standard deviation of the 24 hourly metered flows Sd(Q)=162.5 l/s
Average metered flow supplied Q = 307.3 l/s

7 RESULTS

a) Step 1: water meter's error evaluation V_{uce}=52.160 m^3
b) Step 2: Real losses evaluation. The standard deviation of supplied flow, calculated from different values of x, gives:

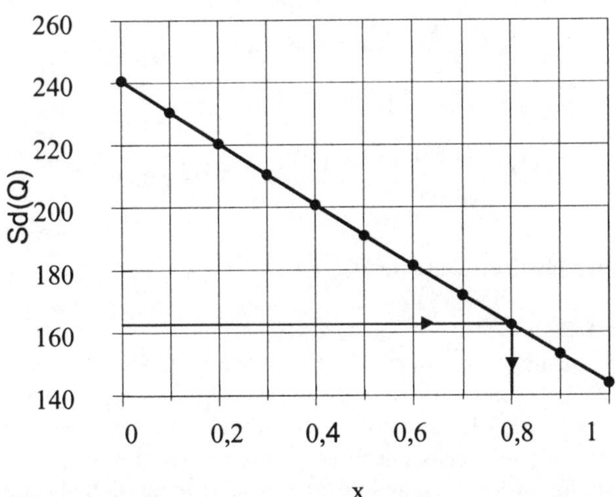

Fig. 12. Standard deviation of supplied flow for different simulations

Using the well known standard deviation Sd(Q)=162.5 l/s, one can calculate the x value

$$x = 0.8 \Rightarrow \begin{cases} Q_{ul} = x \cdot Q_i == 0.8 \cdot 117.25 = 93.85 \text{ l/s} \\ Q_{ucn} = 14.07 \text{ l/s} \end{cases}$$

Summary of flow results:

Q	Water supplied	307.25 l/s
Qr	Water consumed	190 l/s
Qu	Unaccounted for water of the system	117.25 l/s
Qul	Unaccounted for water of the system due to real losses	93.85 l/s
Quc	Unaccounted for water, consumed	23.5 l/s
Quce	Unaccounted for water consumed, due to meters errors	10.06 l/s
Qucn	Unaccounted for water consumed without metering	13.44 l/s

Absolute efficiency performance indicators:

Efficiency of the system	Efficiency of the network	Efficiency of the commercial management
$\eta_s = \frac{Q_r}{Q} = 61.8\%$	$\eta_r = \frac{Q_s}{Q} = 69.5\%$	$\eta_g = \frac{Q_r}{Q_s} = 89\%$

Relative leakage ratios

$$\frac{Q_{ul}\left(m^3/h\right)}{L\left(Km\right)} = \frac{93.85 \cdot 3.6}{200} = 1.69\ m^3 \big/ h \cdot Km$$

$$\frac{Q_{ul}\left(m^3/day\right)}{NoSC} = \frac{93.85 \cdot 3.6 \cdot 24}{7631} = 1.06\ m^3 \big/ day \cdot SC$$

both of which clearly equate to a poor network efficiency.

8 CONCLUSIONS

Water utilities can select and schedule different strategies devoted to minimise the unaccounted for water. All consumption connections can be controlled and registered, pipes can be repaired, rehabilitated or replaced, pressure can be regulated between reasonable levels, meters can be installed and/or repaired or replaced, etc. In order to schedule all these strategies, the economic approach criterion is the most reasonable one. However, the rapidly developing science of economic analyses is out of the scope of this paper. Nevertheless, the starting point of all the strategies weighted by such an analysis, is to gain full knowledge of the audit of the system.

This paper outlines a new structure of a water audit network, with a clear definition of the water balance components. The audit is supported by a complete consumer's characteristics database and, due to the necessary complex and tedious calculations, it has been integrated within SARA (software package).

The main message of this paper is that all water volumes should be metered, regardless of their use and/ or the characteristics of the consumer. By doing so, calculation of the water balance is made easier, although often reality is far from this situation. The general audit proposed in this paper may contribute to evaluate the different terms of the balance.

REFERENCES

AGHTM, 1990 *"Rendement des reseaux d'eau potable:definition des termes utilisés"* (In French) Association Générale des Hygiénistes et Techniciens Minicipaux. 4 815 90.

Almandoz J., Balmaseda C., Cabrera E., Iglesias P., 1998 *Análisis del agua no registrada: fugas y agua consumida no contablizada"* (In Spanish) Internal report. Grupo Mecánica de Fluidos. Universidad Politécnica de Valencia., Spain.

Arregui F. (1998) *"Propuesta de una metodología para el análisis y la gestión del parque de contadores de agua en un abastecimiento"* (In Spanish) PhD thesis. Grupo Mecánica de Fluidos. Universidad Politécnica de Valencia. Spain.

AWWA Leak Detection and Water Accountability Committee, 1996 *"Committee report: water accountability"* Journal of the American Water Works Association. July 1996., pp 108 – 111.

AWWA, 1990 *"Water audits and leak detection. M36. Manual of Water Supply practices"* American Water Works Association. Denver Colorado. USA. ISBN 0 89867 485 0.

Bargiela A., 1984 *"On line monitoring of water distribution systems"*. PhD thesis. Faculty of Science. University of Durham. UK.

Beecher J.A., Flowers J.E., Matzke C.S., 1998 *"Water conservation guidelines and the DWSRF"*. Journal of the AWWA. May, 1998, pp 60-67.

Bowen P.T., Harp J.F., Baxter J.W., Shull R.D., 1993 *" Residential water use patterns"* AWWA Research Foundation and AWWA. ISBN 0 89867 686 X. Denver. Colorado.

Brundtland, G.H. (1987) *"Our common future"* Oxford Academic Press. United Kingdom.

Burgi P.H., 1998. *"FORUM: Change in emphasis for hydraulic research at the Bureau of Reclamation"*. Journal of Hydraulic Engineering. ASCE. July, 1998 pp 658 – 661.

Cabrera E., García - Serra J. (1997) *"Problemática de los abastecimientos urbanos. Necesidad de su modernización"* (In Spanish) Grupo Mecánica Fluidos. Universidad Politécnica de Valencia. ISBN 84-89487-04-9

Cesario L. (1995) *Modeling, Analysis and Design of Water Distribution Systems"*. Edited by the AWWA. ISBN 0-89867-758-0. Denver USA.

DVGW, 1986 *"Water losses in water distribution systems: identification and assessment"*. Deutscher Verein des Gas und Wasserfaches (DVGW). Standard Regulations. Note - Paper W 391.

EU (European Union), 1997 *"Propuesta de Directiva del Consejo, por la que se establece un marco comunitario de actuación en el ámbito de la política de aguas"*. (In Spanish) Draft dated on 02.26.97. Brussels. Belgium.

Germanopoulos G., 1995 *"Valve control regulation for reducing leakage"*. E. Cabrera and A. Vela (eds.), Improving Efficiency and Reliability in Water Distribution Systems. pp 165 - 188. Kluwer Academic Publishers.

Hirner W. (1997) *"Maintenance and rehabilitation needs and strategies for water distribution netrworks"*. Water Management International, 1997, pp 221-231.

Hirner W. (1998) *"Definition of Performance indicators: the example of water loses"*. Proceedings of the Symposium on Application of Performance Indicators for Water and Sewerage Services in Europe. ENGREF. 4-5 June, 1998. Montpellier. France.

ISO (International Standard Organisation), 1993 *"ISO 4064-1: 1993. Measurement of water flow in closed conduits. Meters for cold potable water"* International Standard Organisation. Geneva. Switzerland.

IWSA, (International Water Supply Association), 1993 *"10 thesis for a drinking water tariff policy"*. Internal Document. IWSA, London. UK.

Kawakita K., 1998 *"Tokyo's modern water system celebrates its centenmial"* Journal of the AWWA. November 1998, pp 60-69.

Lambert A., Myers S., Trow S., 1998 *"Managing water leakage. Economic and technical issues"* Published by Financial Times Energy. London. UK.

Lund J., and Reed R.U., 1995 *"Drought water rationing and transferable rations"* Journal of Water Resources Planning and Management. ASCE. November 1995, pp 429-437.

Mc Elroy J.M., 1985 *"Water pipeline infrastructure: questions and answers."* Proceedings of the Distribution Systems Symposium. AWWA. Seattle, USA. 1985.

Ng K.H., Foo C.S., Chan Y.K., 1997 *"Unaccounted for water - Singapore's experience"* Journal Acqua. Vol 46, Nº 5., pp 242 – 251, 1997.

OFWAT, Office of Water Services, (1994) *"A better deal for metered customers. A report on tariff rebalancing and improvements in optional metering schemes"* OFWAT, Office of Water Services, March 1994. Birmingham. England.

Skarda, B.C. (1997) "The Swiss experience with performance indicators and special viewpoints on water networks" IWSA Workshop on performance indicators and distribution systems. LNEC. Lisbon. May 1997.

Villesot D. (1997) *"Methods and tools necessary to conduct a leakage survey"* Aquatech Europe Spring. Water Supply. pp 16 - 17.

Van der Willingen F.C. (1997) *"Dutch experience and viewpoints on performance indicators"* IWSA Workshop on performance indicators and distribution systems. LNEC. Lisbon. May 1997.

Yanov, D.A., and Koch R.N., 1987 *"A modern residential flow demand study"*.Session 27 of AWWA annual conference. Kansas City. USA.

Yepes, G (1995) *"Reduction of Unaccounted for Water. The job can be done"*. Water and Sanitation Division. World Bank, 1995

Evidence Supporting the Poisson Pulse Hypothesis for Residential Water Demands

S. G. Buchberger *and* Y. Lee

ABSTRACT

It is conjectured that indoor water demands at a single family home behave as a time dependent Poisson rectangular pulse (PRP) process. Water demands at 21 single family residences located along a common dead-end street in a small community near Cincinnati were monitored for a one-year period. Each home was instrumented with an automatic data logger that continuously recorded the total household water demand. Outdoor demands were screened and removed from the data base. Indoor water demands were converted to single equivalent rectangular pulses (SERPs) having a known start time, pulse duration, and flow intensity. The observed relative frequency of busy homes and the statistical properties of the flow rates along the dead-end main show good agreement with theoretical predictions from the PRP hypothesis for indoor residential water use.

1 INTRODUCTION

The Safe Drinking Water Act (SDWA) and its amendments require certain water quality standards to be maintained throughout municipal distribution systems in the US. Network models, now indispensable in sizing and operating water distribution systems to satisfy *hydraulic criteria*, are becoming essential tools for predicting the system's ability to comply with *quality standards*. Making accurate predictions of water quality between the points of treatment and consumption in a distribution system is difficult, however (Clark *et al.*, 1993; Rossman *et al.*, 1994).

Broadly speaking, the obstacles facing accurate modeling of water quality in distribution systems are twofold: uncertain flow rates and uncertain bio-chemistry. Flow patterns change continuously over time and across space in response to random demands imposed by many consumers dispersed throughout the service area. Complex biological-physical-chemical interactions often occur among constituents mixed in the water column or attached to the pipe walls or both.

The intertwined issues of random flows and uncertain reactions are especially acute in the peripheral areas of the network. Here travel times are long, stagnant conditions are common, chlorine residuals may be low and individual consumers measurably influence flow. Taken collectively, these factors make it difficult to achieve good water quality predictions in remote regions of a distribution system.

In view of increasingly stringent SDWA regulations, the next generation of network water quality models will adopt spatial and temporal resolutions having much finer scales than are customarily used today. This ultra-fine resolution will require a more detailed and accurate quantitative description of residential water use than is now available. The time dependent Poisson rectangular pulse (PRP) hypothesis is a promising new approach for modeling residential water demands on a fine scale.

2 OBJECTIVES

Using water demand data collected during a recent field monitoring study, this paper presents several analytical tests to check whether or not the PRP process is a reasonable description for residential water demands.

3 THE PRP PREMISE

Water use in a municipal distribution system may be amenable to analysis using queuing methods developed by Erlang (1917-18) over 80 years ago in his pioneering study of telephone demands for the communications industry. In the context of municipal distribution systems, the basic premise is that residential water demands at single family homes behave as a time dependent PRP process. This premise leads directly to expressions for the theoretical moments and probability distributions of busy homes, flow rates, travel times and disinfectant concentrations in dead-end pipes of municipal water distribution systems (Buchberger and Wu 1995, Wu 1996, Buchberger and Schade 1997, Buchberger *et al.*, 1998). A few key details of the PRP approach are sketched below.

3.1 Water Pulses

The frequency of residential water use is assumed to follow a Poisson arrival process with a time dependent rate parameter. When a water use occurs, it is represented as a single rectangular pulse of random duration and random intensity as illustrated in Figure 1. The rectangular pulse approximation has been verified in previous field studies (Buchberger and Wells, 1996).

It is unlikely that more than one pulse will start at the same instant. Owing to the finite duration of each water pulse, however, it is possible that two or more pulses with different starting times will overlap for a limited period. When this occurs, the total water use at the residence is the sum of the individual intensities from the coincident pulses.

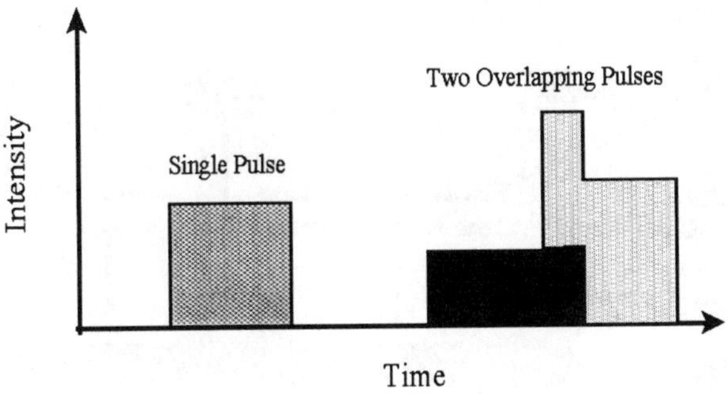

Figure 1. Poisson rectangular pulse process for water demands.

3.2 Dead-end Mains

Dead-end pipes are water supply lines that have only one connection to the main looped portion of the water distribution system (and, hence, have *at least* one terminal point). As shown in Figure 2, dead-end mains in residential neighborhoods can be represented as a sequence of pipe links that start and end with each service line connection. Dead-end mains are surprisingly common. Utility surveys indicate that dead-end mains often comprise 25 percent or more of the total infrastructure in a distribution system and tend to service a high percentage of the residential consumer base.

Except under extraordinary circumstances (e.g., main break, fire demand, etc), water moves through a dead-end pipe in one direction only, along a path from the entrance to the point of withdrawal, as dictated by local downstream use. Since water moves through a dead-end main in the downstream direction only, the principle of mass conservation alone is sufficient to estimate flow rates. Even though the flow direction is known, water movement in dead-end mains is complex. Owing to the sporadic unpredictable nature of consumer demands, water in dead-end lines can be stagnant for long periods of time. When demands for water are made, the flow occurs as a burst of random duration and random intensity.

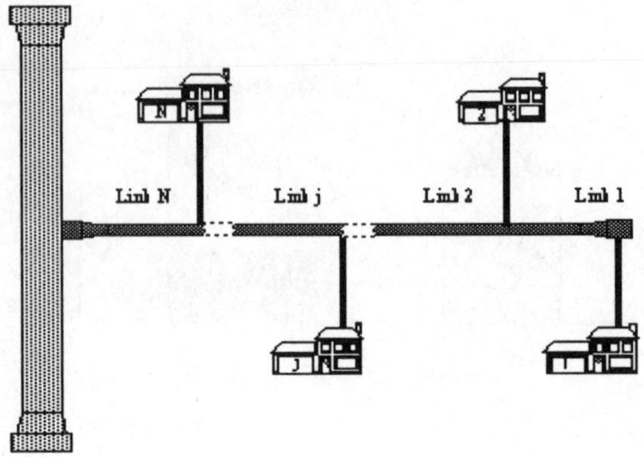

Figure 2. Definition sketch of a dead-end water main.

3.3 Busy Servers

Occupants of a home are considered "customers". Fixtures and appliances that draw water from the distribution system are called "servers". Suppose customers at residence j arrive according to a homogeneous Poisson process with parameter 8_j and receive service at rate μ_j. Let $K_j(t) = 0,1,2,\ldots$ be the number of busy water servers in residence j. The total number of busy servers in the block of N residences is

$$K_N^*(t) = \sum_{j=1}^{N} K_j(t) \tag{1}$$

For large values of t, the number of busy servers $K_N^*(t)$ forms a stationary stochastic process having a Poisson distribution

$$\text{Prob}[K_N^*(t) = k] = \frac{\left(\rho_N^*\right)^k \exp(-\rho_N^*)}{k!} \qquad k = 0, 1, 2, \ldots \tag{2}$$

with mean and variance given by the dimensionless parameter known in queueing theory as the "utilization factor"

$$\rho_N^* = \sum_{j=1}^{N} \frac{\lambda_j}{\mu_j} \qquad j = 0, 1, 2, \ldots \tag{3}$$

At a typical single family residence, the daily average utilization factor is about 5 percent (Buchberger and Wells 1996, Schade 1996). This means that indoor residential water demands tend to be short and infrequent. From Equation

(2), the probability of stagnant water in a dead-end main is $P[K_N^*(t)=0] = \exp(-\Delta_N^*)$; at a typical single family residence the stagnation probability is approximately $1-\Delta_1^*$.

If the service times are exponentially distributed, then it can be shown $K_N^*(t)$ has covariance function

$$C[K_N^*(t)] = \rho_N^* \exp\left[-\frac{\lambda_N^*}{\rho_N^*}|t|\right] \tag{4}$$

where; $\lambda_N^*=\Sigma\lambda_j$ (j=1,2,...,N). By definition, Equation (4) reduces to Equation (3) when t=0.

The number of busy servers $K_N^*(t)$ increases and decreases in response to customer arrivals and departures. During a time interval of duration D under steady-state conditions, the expected number of jumps up from k to k+1 busy servers ($J_{k,k+1}$) and jumps down from k+1 to k busy servers ($J_{k+1,k}$) are equal and given by

$$E[J_{k,k+1}] = E[J_{k+1,k}] = D\,\lambda_N^*\left(\text{Prob}[K_N^*(t) = k]\right) \tag{5}$$

The Poisson distribution given in Equation (2), the covariance function given in (4), and the expected number of jumps given in (5) each provide measurable theoretical predictions of busy server properties that can checked against field observations to test the PRP hypothesis for residential water use.

3.4 Flow Rates

The intensity of water use at a busy server in any residence is assumed to be an independent and identically distributed positive continuous random variable with mean α and variance β^2. The flow rate through link N (see Figure 2) resulting from water demands occurring downstream at K_N^* busy servers is:

$$Q_N^* = \sum_{j=1}^{N} Q_j \tag{6}$$

where Q_j is the flow into residence j. The mean and variance of Q_N^* are given by (Buchberger and Wu 1995)

$$E[Q_N^*] = \sum_{j=1}^{N} \rho_j \alpha_j \tag{7}$$

$$\text{Var}[Q_N^*] = \sum_{j=1}^{N} \rho_j\left(\alpha_j^2 + \beta_j^2\right) \tag{8}$$

where $\Delta_j = 8_j/\mu_j$. Equation (7) is a generalized version of the expression given by Linaweaver *et al* (1966) for the expected demand on a water system serving N homes. In Equation (8), it is assumed that the residences are using water independently of each other. Table 1 summarizes the mean and variance of the flow rate under the PRP hypothesis at several locations in the distribution system.

Table 1. Flow statistics in dead-end mainlines.

Point of Observation	Mean Flow Rate	Variance of Flow
One *busy* server	α	β^2
Service line into single home	$\rho\alpha$	$\rho(\alpha^2+\beta^2)$
Link N on dead-end mainline*	$\Sigma\ \rho_j\alpha_j$	$\Sigma\ \rho_j(\alpha_j^2+\beta_j^2)$

*the sum is taken over all links j = 1,2,…,N

4 DATA COLLECTION

An intensive field sampling program was implemented to monitor residential water use at 21 single family homes served by a common 15.2 cm cast iron dead-end main in a small community about 30 km east of Cincinnati. An instrument package was installed in each home to continuously record total water consumption through the service line to the residence. The instrumentation package included a Rustrak Ranger II datalogger, 4-29 mA impulse sensor, count and dump input pod, COM-504 modem connection and a high speed modem.

The impulse sensor transmitted a signal at one second intervals to the data logger where the readings were converted to a flow rate and stored. Using the dedicated modems, the water demands were downloaded nightly to a computer on campus. All residential demand data were archived twice according to two different schemes.

The first scheme simply tracked total flows through the service line to each residence. No attempt was made to identify coincident water use caused by two or more busy servers operating simultaneously in a home. The second scheme identified and separated coincident water uses and then converted each demand into a single equivalent rectangular pulse using an algorithm described by Buchberger and Wells (1996).

Indoor water demands were monitored for one year. Only data from February 1996 are presented here. About 57,850 water demands were recorded at 18 of the 21 homes during this 29 day monitoring period. Owing to family vacations and ownership changes, three of the homes were vacant.

5 RESULTS

The theoretical properties of busy servers and flow rates mentioned in Sections 3.3 and 3.4 offer several ways to test the PRP hypothesis. Two tests are considered here: (1) Probability distribution of busy servers (Equation 2) and (2) Statistical properties of flow rates (Table 1).

5.1 Distribution of Busy Homes

Two problems arise with the busy server distribution given in Equation (2). First, residential water use does not follow a homogeneous demand process. Most days have prominent diurnal cycles with periods of high and low water consumption. This problem is solved by treating the actual time dependent non-homogeneous water demands as a piecewise homogeneous process. To do this, the day is divided into 24 one-hour intervals each of which is considered homogeneous. The hourly neighborhood utilization factor $\Delta_N^*(t)$ in Equation (2) is estimated from the recorded water use data as the ratio of cumulative busy time to elapsed time. The computed Poisson probability distribution given in Equation (2) is then checked hour by hour against the observed relative frequency of busy servers.

The second problem occurs from a practical consideration. As yet, there is no affordable and convenient way to monitor all busy water servers in each home. Certainly, it is possible to analyze the signal of total flow into a home and from it extract demand signatures of individual water servers. This, however, is extremely cumbersome and not entirely objective.

Fortunately, for purposes of checking the PRP hypothesis, it is not necessary to secure data at the server level. Instead, the intractable problem is made manageable by shifting the "point of observation" from individual servers (as implied in Equation 2) to the entire household (as done with the monitoring program). In this manner, the problem changes from distribution of *busy servers* to distribution of *busy homes*. In a sense, each home is viewed as a single super server and effectively treated like a Bernoulli variable with two states: either on (busy) or off (idle). The water demand data archived under scheme number one are suitable for this purpose.

There is very good agreement between computed Poisson and observed hourly cumulative distribution functions of busy homes in February 1996 (Figure 3). For instance, during the hour starting at 7 AM, field observations show the chance of having at most 1 busy home is about 69% while the fitted Poisson distribution (series with hollow squares) gives $P[K \leq 1]=0.67$. The non-homogeneous (time dependent) nature of residential water use is clear with peak activity at breakfast and dinner and a stagnant period in the predawn hours.

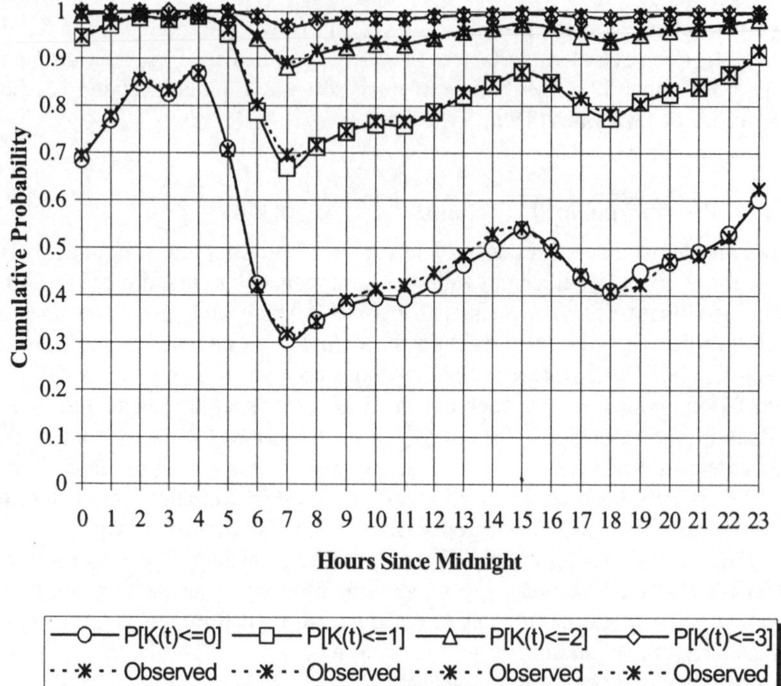

Figure 3. Comparison of computed and observed hourly cumulative distributions of busy homes derived from 57,850 water demands monitored along a dead-end mainline supplying 18 residences near Cincinnati during February 1996.

Table 2. Water demands recorded at study site in February 1996.

(1) home number	(2) pulse count (SERPs)[(b)]	(3) total demand volume (L)	(4) server busy time (min)	(5) customer arrival rate, λ (min^{-1})	(6) average demand volume, ϕ (L)	(7) average demand duration, τ (min)
1	1,398	1,044	231	0.0335	0.747	0.165
2	1,799	11,804	1,118	0.0431	6.561	0.621
3	3,271	7,670	978	0.0783	2.345	0.299
4	7,856	15,139	2,664	0.1881	1.927	0.339
5	5,986	25,556	3,217	0.1433	4.269	0.538
6[(a)]	--	--	--	--	--	--
7	2,983	14,882	1,612	0.0704	4.989	0.549
8	3,235	17,675	1,824	0.0755	5.464	0.564
9	3,918	9,421	1,073	0.0938	2.405	0.274
10	1,642	6,530	867	0.0393	3.977	0.528
11	2,083	10,733	1,352	0.0499	5.153	0.649
12[(a)]	--	--	--	--	--	--
13	3,150	15,632	2,288	0.0754	4.963	0.726
14	2,378	5,208	853	0.0569	2.190	0.359
15	2,912	17,429	2,486	0.0697	5.985	0.854
16	3,600	9,565	1,157	0.0862	2.657	0.321
17	2,909	9,584	1,257	0.0697	3.295	0.432
18	4,605	16,785	3,173	0.1103	3.645	0.689
19[(a)]	--	--	--	--	--	--
20	2,734	17,184	1,720	0.0655	6.285	0.629
21	4,421	25,448	4,179	0.1059	5.756	0.945
Total	60,835[(c)]	237,339	32,049	1.4568	72.704	9.481

(Note: Columns 8-14 and footnotes appear on the next page.)

5.2 Statistics of Water Demands

A second test of the PRP hypothesis is to examine the mean and variance of the water demands at individual homes and across the entire neighborhood. Conservation of mass (Equation 6) requires that the sum of water demands from all 18 households equals the total flow into the neighborhood. Further, by virtue of independent water uses, the PRP process requires that the flow variances be additive. Hence, the sum of the variances of demands from all 18 households should equal the variance of the total flow into the dead-end mainline.

Water demand data archived as SERPs under scheme number two (see Section 4.0) are suitable for this analysis. Residential demand statistics for February 1996 are summarized in Table 2. Equations (7) and (8) were used to estimate the mean and variance of the flow into individual homes. Resulting values are shown in columns 13 and 14, respectively, and then summed in the bottom row of Table 2.

Table 2. (cont'd) Water demands recorded at study site in February 1996.

(8) home number	(9) average of server use, α (L/min)	(10) variance of server use, β^2 (L/min)^2	(11) server (coef var)^2 $\Theta_W^2 = \beta^2/\alpha^2$	(12) residence utilization factor $\rho=\lambda\tau$	(13) avg home flow $\rho\alpha=\lambda\phi$ (L/min)	(14) var home flow $\rho(\alpha^2+\beta^2)$ (L/min)^2
1	4.536	21.33	1.036	0.0055	0.0249	0.230
2	10.57	42.55	0.381	0.0268	0.2831	4.132
3	7.838	40.26	0.655	0.0234	0.1833	2.380
4	5.687	17.23	0.533	0.0638	0.3629	3.162
5	7.940	37.55	0.596	0.0771	0.6120	7.756
6(a)	--	--	--	--	--	--
7	9.226	21.66	0.254	0.0386	0.3561	4.122
8	9.685	34.03	0.363	0.0437	0.4234	5.587
9	8.774	33.13	0.430	0.0257	0.2257	2.831
10	7.531	12.19	0.215	0.0208	0.1565	1.433
11	7.939	24.69	0.392	0.0324	0.2570	2.842
12(a)	--	--	--	--	--	--
13	6.857	13.85	0.295	0.0547	0.3750	3.329
14	6.096	10.53	0.283	0.0204	0.1244	0.973
15	7.007	21.68	0.442	0.0595	0.4169	4.211
16	8.278	23.89	0.349	0.0277	0.2294	2.560
17	7.630	21.16	0.363	0.0301	0.2298	2.389
18	5.289	11.42	0.408	0.0760	0.4018	2.993
19(a)	--	--	--	--	--	--
20	9.994	29.19	0.292	0.0412	0.4116	5.318
21	6.092	53.42	1.439	0.1001	0.6097	9.063
Total	136.96	469.77	8.726	0.7675	5.684	65.311

Notes:
(a) Homes 6, 12 and 19 were vacant during February 1996.
(b) SERPs are single equivalent rectangular pulses (Buchberger and Wells 1996).
(c) There were 2,985 coincident demands; hence, pulse count increased to 60,835.
(d) February 1996 had 29 days = 41,760 minutes.

Columns (1), (8) are assigned values; Columns (2), (3), (4) are measured values.
Col (5) = Col (2) ÷ 41,760
Col (6) = Col (3) ÷ Col (2)
Col (7) = Col (4) ÷ Col (2)
Col (9) = Col (3) ÷ Col (4) = Col (6) ÷ Col (7)
Col (10) = $[\Sigma d_i q_i^2/\text{Col (4)}] - [\text{Col (9)}]^2$
[where d_i is duration and q_i is intensity of ith water demand (SERP) at the home]
Col (11) = Col (10) ÷ $[\text{Col (9)}]^2$
Col (12) = Col (4) ÷ 41,760 = Col (5) × Col (7)
Col (13) = Col (12) × Col (9) = Col (5) × Col (6)
Col (14) = Col (12) × $[\text{Col (9)}]^2 \times [1 + \text{Col (11)}]$

Equation (6) was used to get $Q_N^*(t)$, the total flow into the dead-end main at time t, by summing the water demands measured each second at each individual home. Then the mean and the variance of the total flow $Q_N^*(t)$ were estimated with

$$E[Q_N^*] = \frac{1}{T}\sum_{t=1}^{T} Q_N^*(t) \tag{9}$$

$$Var[Q_N^*] = \frac{1}{T}\sum_{t=1}^{T} \left[Q_N^*(t)\right]^2 - \left(E[Q_N^*]\right)^2 \tag{10}$$

where $T=2.5\times10^6$ seconds is the duration of the February 1996 sampling period. Results from Equations (9) and (10) give $E[Q_N^*]=5.65$ L/min and $Var[Q_N^*]=67.68$ $(L/min)^2$.

Both compare closely with their counterparts given at the bottom of columns 13 and 14 in Table 2. The near perfect agreement (0.6%) between the estimates of the mean flow (0.6% difference) simply confirms that mass conservation is satisfied. The good agreement (3.6% difference) between the estimates of flow variability is taken as a strong sign that the variances are additive and the demands are independent, in agreement with the PRP hypothesis for residential water use.

6 SUMMARY

The premise that residential water use occurs as a non-homogeneous Poisson rectangular pulse process leads to a reasonable, rigorous and robust framework for solving contemporary problems that arise with water distribution systems. Several theoretical results from the PRP model for residential water use are outlined.

Water demand data were collected during an extensive field sampling exercise at 18 single family homes along a common dead-end main. These demand data confirm that discharge through a dead-end main is a random intermittent (laminar) flow process with long periods of stagnation (Buchberger *et al*, 1999). This picture is very different from the simple steady-state plug flow assumptions invoked in current conventional approaches to model water flow and quality in peripheral zones of municipal distribution systems.

Statistical properties of the busy homes and flow rates are extracted from these measured field data and used to corroborate the PRP hypothesis for residential water demands. The hourly distribution of busy homes shows excellent agreement with the Poisson probability model. The variance of individual residential flow rates appears to be additive as required by the PRP assumption of independent water users.

The PRP approach offers a stochastic complement to deterministic models which are indispensable for designing and operating municipal distribution systems to meet water demands and quality standards. The PRP model holds promise as a parsimonious and powerful tool to help network modelers improve predictions of water quality in municipal water distribution systems.

ACKNOWLEDGMENTS

This work was supported by grants from the National Science Foundation (NYI Award: BCS-9257608) and from the American Water Works Association Research Foundation (95-294).

REFERENCES

Buchberger, S.G. and L. Wu (1995) *Model for instantaneous residential water demands*, **ASCE Journal of Hydraulic Engineering**, 121(3): 232-246.

Buchberger, S.G. and G.J. Wells (1996) *Intensity, duration, and frequency of residential water demands*, **ASCE Journal of Water Resources Planning and Management**, 122(1): 11-19.

Buchberger, S.G. and T.G. Schade (1997) *Poisson pulse queuing model for residential water demands*, **Proceedings of 27th IAHR Congress**, San Francisco, CA, pp. 488-493.

Buchberger, S.G., J.T. Carter, and Y.H. Lee (1998) *Travel times in dead-end mains*, **Proceedings ASCE 25th Annual Conference Water Resources Planning and Management**, Chicago, IL, pp.260-265.

Buchberger, S.G., Y.H. Lee, L.G. Bloom and B.W. Rolf (1999) *Dispersion of mass in intermittent laminar flow through a pipe*, **Water Industry Systems: modelling and optimization applications, Volume 1,** RSP, UK.

Clark, R.M., W.M. Grayman, R.M. Males, A.F. Hess (1993) *Modeling contaminant propagation in drinking water distribution systems*, **ASCE Journal of Environmental Engineering**, 119(2):349-364.

Erlang, A.K. (1917-18) *Solutions of some problems in the theory of probabilities of significance in automatic telephone exchanges*, **Post Office Elec Engr J**, 10,189-197.

Linaweaver, F.P. Jr., J.C. Geyer and J.B. Wolff (1966) **Residential Water Use**, Johns Hopkins University, Baltimore, Maryland.

Rossman, L.A., R.M. Clark, and W.M. Grayman (1994) *Modeling chlorine residuals in drinking-water distribution systems,* **ASCE Journal of Environmental Engineering,** 120(4):803-820.

Schade, T.G. (1996) *Water demand and travel time in a residential dead-end loop*, **MS Thesis**, University of Cincinnati, 64 pages.

Wu, L. (1996) *Stochastic model of flow in the periphery of a water distribution system*, **PhD Dissertation**, University of Cincinnati, 207 pages.

LIST OF SYMBOLS

d pulse duration (min)
D time interval (sec)
j subscript for homes
$J_{u,v}$ number of jumps from u to v busy servers
$K(t)$ number of busy servers at time t
N number of homes on dead-end stem
$P(0)$ percentage of idle time
PRP Poisson rectangular pulse
q pulse intensity (L/min)
Q flow rate (L/sec)

SERP single equivalent rectangular pulse
t time (sec)
T duration of sampling period

C[.] covariance
E[.] expected value
Var[.] variance

* an asterisk indicates a quantity summed over the neighborhood.

α average intensity of indoor water use at a busy server (L/sec)
β^2 variance of intensity of indoor water use at a busy server $(L/sec)^2$
ρ utilization factor for indoor residential water use, $\rho=\lambda\tau=\lambda/\mu$
λ average arrival rate of water user (sec^{-1})
μ mean service rate (sec^{-1})
τ mean duration water use pulse (sec)
Θ_W coefficient of variation of indoor water use at a busy server, $\Theta_W=\beta/\alpha$

Application of Network Modelling for Operational Management of Water Distribution Systems

R. Burrows, G.S. Crowder *and* J. Zhang

ABSTRACT

Driven by the UK water industry's current desire for effective pressure management and the need to evaluate levels-of-service pressure systematically within its water distribution systems, simple means of pressure synthesis within suitably monitored District Metered Areas (DMAs) have been sought. A simplified implementation of hydraulic (flow/pressure) network modelling to a DMA's pipe network, accessed directly from Geographic Information System (GIS) records, is employed here to establish hydraulically robust estimates of supply pressure at property seed points. This enables production of thematic maps for pressure contouring or 'measle mapping' to highlight and report on levels-of-service performance and pressure failures. Such a procedure offers substantial improvement over pressure profiling based on the intuitive imposition of representative hydraulic grade.

The modelling capability so developed has potential as an integral element in routine performance evaluation by distribution system operators. This is enhanced by a complementary, regression-based approach which is shown to be capable of direct synthesis of internal pressures in the system, to adequate accuracies, as a function of the measured boundary conditions of head (pressure) and flow, such that 'real-time' performance monitoring of the DMA's function becomes possible.

1 NETWORK MODEL CONSTRUCTION

Seamless procedures have been developed in the NETBASE database [1] and Mapinfo-Professional (version 5) GIS [2] environments to interface between:- Corporate Geographic Information Systems (GIS), defining the pipe system and its assets; customer billing (Customer Account System, CAS) data for metered consumption; monitoring data of flow and pressure (and pump and valve statuses); and the chosen commercial network model for hydraulic flow synthesis (EPANET [3] being utilised in what follows herein). These enable rapid formulation and implementation of the network model for export of estimated pressures and flows back into the graphical environment (at the nodal level). The process is outlined schematically in Figure 1.

Automated routines, programmed in Borland Delphi (version 4) [4] translate the full GIS image of the pipe network into a reduced system of (geometrically accurate) pipe nodes and links. Customised algorithms rationalise detail and check pipe connectivity according to desire, ranging from production of 'all-mains' to 'trunk mains only' models. For the DMA level studies of interest here, property allocation to nodes is made via an initial assignment to the nearest pipe. Interactive procedures in the graphical (GIS) environment enable manual editing of the node/link system and property assignment, as well as enabling modification of the network configuration (i.e. adding/moving valves etc) in investigative planning or operational scenarios.

For hydraulic modelling at this level, total demand is allocated to nodes on a pro-rata basis to the number of properties assigned, large industrial users being treated individually. Net inflow to the network minus the identified large user flows, derived from monitoring (measured data) streams, defines this demand. Alternative, more comprehensive, means of demand specification can be made for more refined applications of the modelling procedure.

Specification of the 'source' head at inlet to the system enables a hydraulic analysis. Figure 2 displays a typical screen from a work exercise in the EPANET environment. EPANET facilitates model calibration based on simplified global pipe roughness factoring to tune the model predictions to available internal pressure data, at the chosen 'snapshot' time. Where absent from the GIS records, default pipe roughnesses are abstracted from database values according to age and material in order to initiate the investigation. In the NETBASE-Modelling routines [1] automated calibration procedures can be completed as illustrated by Figure 3 by running EPANET behind the scenes in 'DOS' mode. Manual intervention directly within EPANET (in the 'Windows' user interface environment) can be employed for more detailed calibration where pressure data justifies.

2 LEVELS-OF-SERVICE (DG2) REPORTING AND PRESSURE MANAGEMENT

Source data for the model scenario formulation and calibration can be sampled off incoming monitoring data from boundary meters and internal pressure logging points. This enables routine (retrospective) synthesis of flows and pressure through the DMA on a day-to-day basis. This opens up the possibility of 'real time' tracking of flow/pressure within the DMA as discussed shortly.

Chosen results output such as pressure, total head or demand can be exported for display in the Mapinfo-GIS environment as thematic 'measle' maps. These provide information down to the property level as shown in Figure 4. Associated histograms of pressures supplied in the DMA (Figure 4) and more

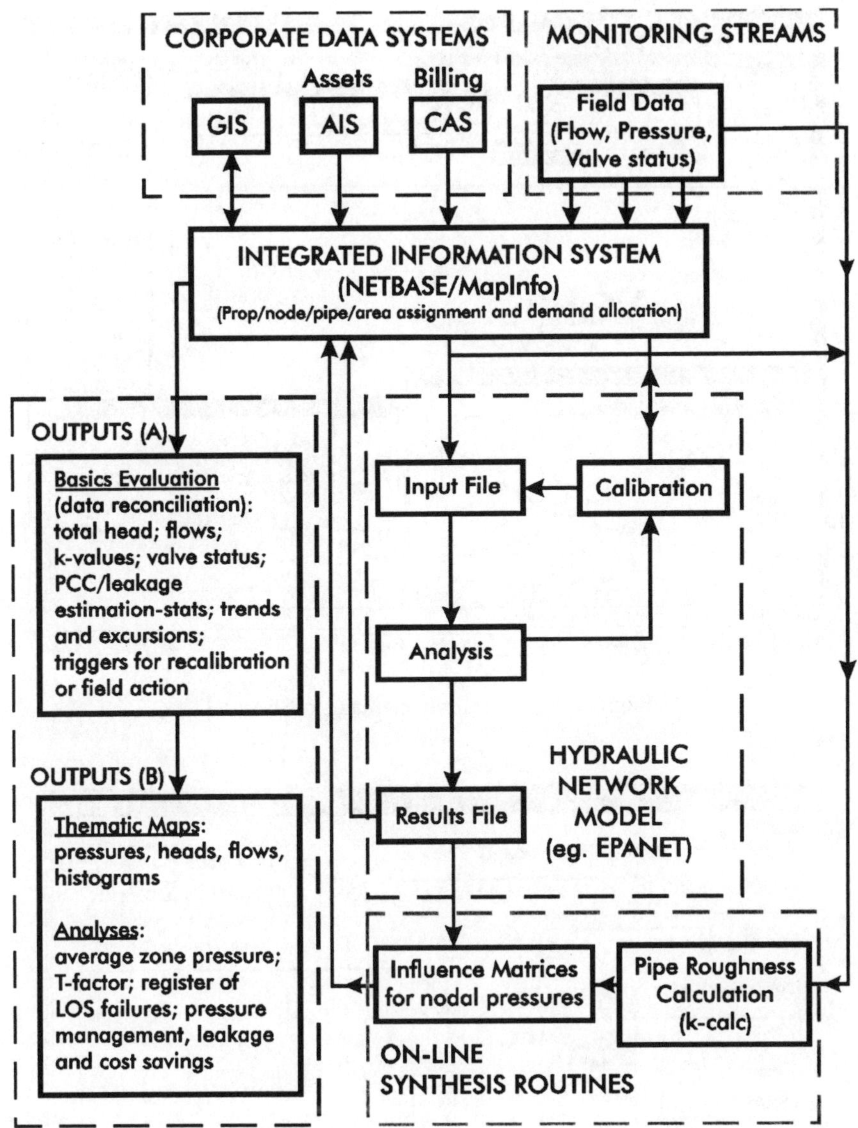

Fig. 1 A vision of distribution zone (DMA) diagnostics

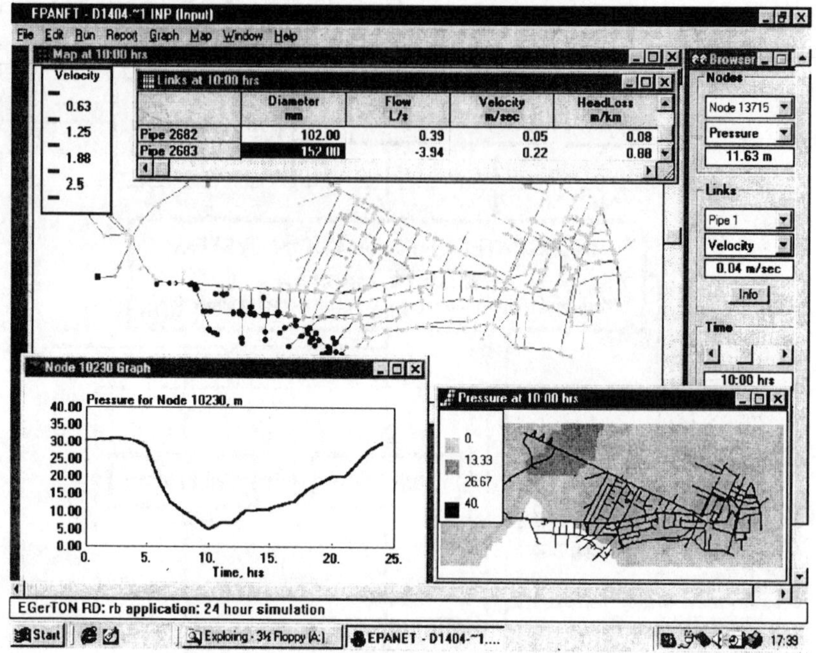

Figure 2 A typical screen image in EPANET

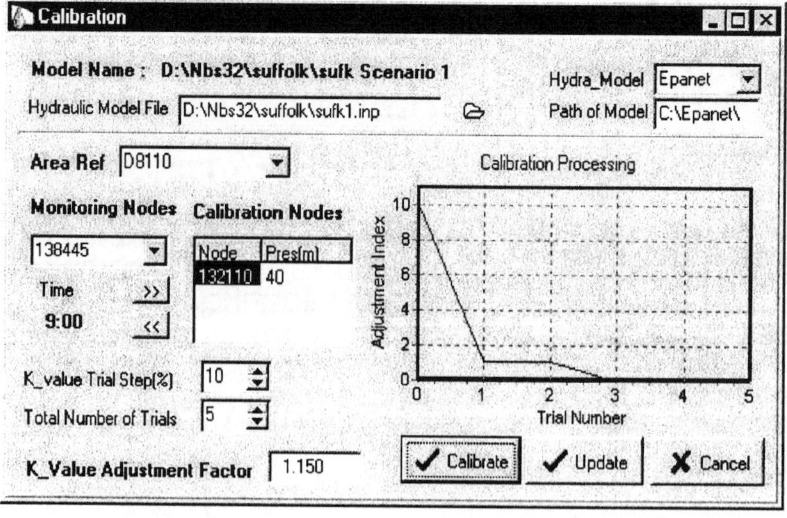

Fig. 3 User interface for a simple calibration study

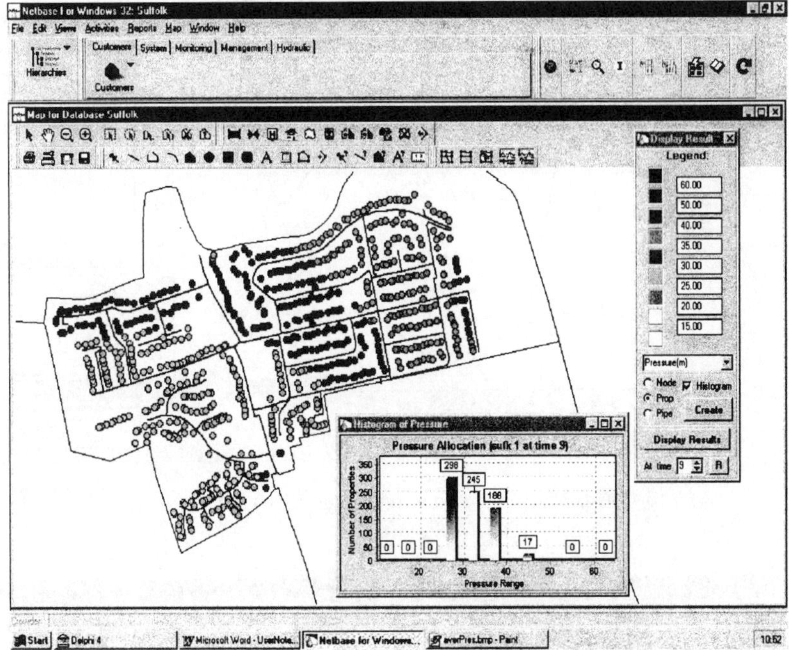

Fig. 4 Thematic mapping of pressures supplied

advanced hydraulic characteristics relating to the DMA's sensitivity to low pressure can be output for levels-of–service reporting to the water regulator.

Other hydraulic characteristics such as 'Average Zone Pressure' (AZP)-point and AZP-profile determination, Hour-Day (T)-Factor for daily leakage calculation (see Figure 5), and potential leakage reductions from optimal pressure management (Figure 6), can also be reported directly from any 24-hour simulation run.

3 AN INFLUENCE-MATRIX APPROACH TO NODAL PRESSURE SYNTHESIS

Where permanent retention of the EPANET models on the company's computer systems causes difficulty, a simplified procedure has recently been under evaluation [5,6,7]. This utilises the outputs of the network model (following calibration) to create influence functions between nodal (and hence property seed point) pressures and the instantaneous conditions monitored. These conditions might include total head and flow at the boundary inlets/outlets, and any large user consumption; as well as any pressures logged within the zone. With this approach, incoming monitored data can be converted directly, by simple algebraic relationships, into estimates of individual nodal pressure to enable standard reporting as implied by the flowchart in Figure 1.

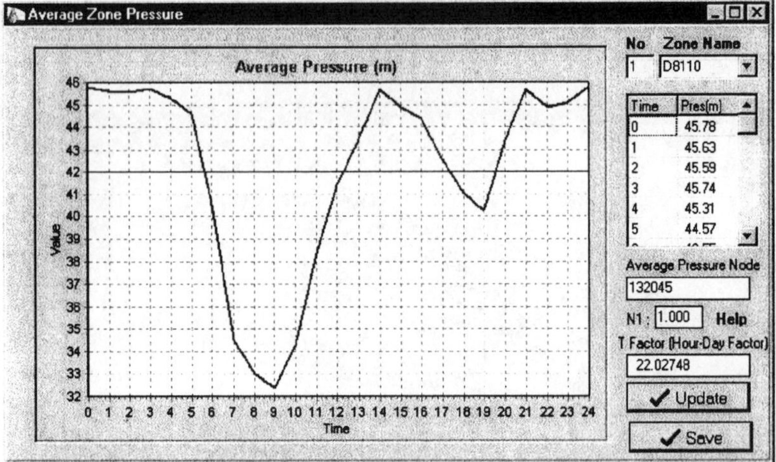

Fig. 5 Average zone pressure (AZP) determination

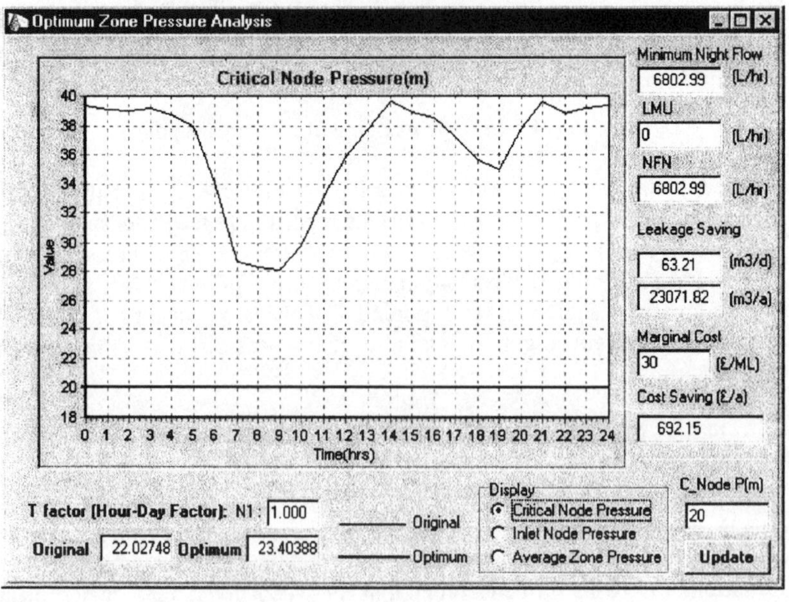

Fig. 6 Optimal pressure management assessment

The objective is to use an Influence-Matrix concept to synthesise total head (hence pressure) at chosen points (H_j) inside a DMA in terms of monitored boundary conditions of total head (H) or flow (Q) at inlet/outlet meters (but of necessity including also any very large users within the DMA).

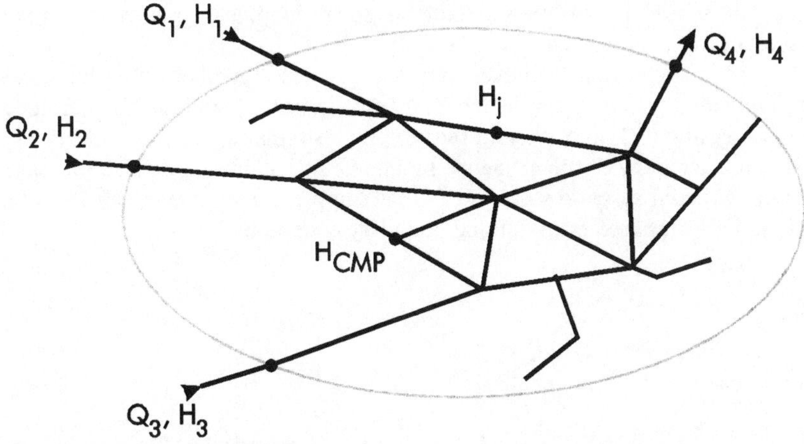

Fig. 7 Schematic of a District Metered Area

Referring to the schematised DMA shown in Figure 7, on conceptual grounds assume that the dependent variable (later designated - y) is expressible in terms of the (monitored) independent variables (later designated - x), as follows [6]:

$$\begin{aligned} H_j \quad &= A_1 H_1 + A_2 H_2 + A_3 H_3 + A_4 H_4 + A_5 H_{CMP} + B_1 Q_1^2 \\ &\quad + B_2 Q_2^2 + B_3 Q_3^2 + B_4 Q_4^2 + D_1 Q_1 Q_2 \\ &\quad + D_2 Q_2 Q_3 + D_3 Q_1 Q_3 + \ldots\ldots \\ &= \sum A_i H_i + \sum B_i Q_i^2 + \sum D_{ij} Q_i Q_j \qquad \text{Eq. (1)} \end{aligned}$$

where A_i, B_i, D_{ij} are influence coefficients to be obtained by regression analysis using either synthesised data from network modelling or the field data itself (as illustrated later). H_{CMP} here depicts the total head at the Critical Monitoring Point but can represent any internal monitoring point. Terms Q can include large metered users also.

The parameter groupings can be justified on the grounds of conceptual flow modelling. For dentritic (branched) systems, with nodal flow withdrawal well represented by a pro-rata allocation according to number of properties, this approach can be shown to be robust. In looped pipe systems where flow direction can vary from hour to hour, the approach becomes an approximation, and the sensitivity of its accuracy to such 'loopiness' has yet to be fully ascertained.

The regression is achieved by 'least squares' methods involving matrix manipulation presently implemented by spreadsheet operations but subsequently to be assimilated into customised coding. To maximise the robustness and potential accuracy of the influence functions the calibration data should actually be drawn from as wide a range of DMA supply conditions as can be enforced through the imposed or monitored boundary conditions.

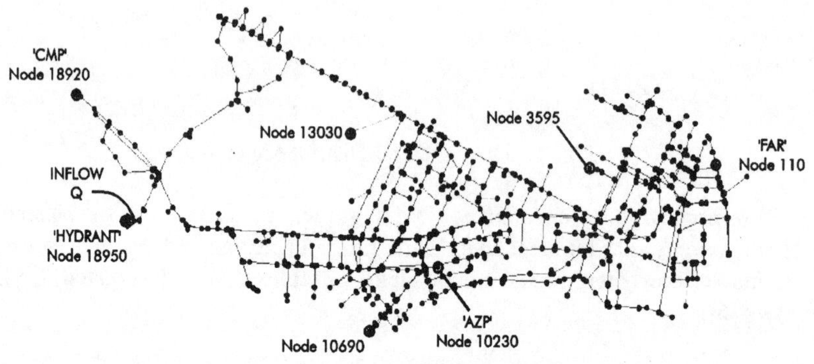

Fig. 8 The 'test' network

Promising preliminary results have been achieved by applying equation (1) to a multi-feed network [7]. By way of illustration here, however, a typical single feed multi-looped DMA in Birkenhead shown in Figure 8 has been considered [8]. As well as inflow (Q1) measurement, corresponding pressure data was available at entry (HYDRANT) and CMP, AZP and FAR points. The potential accuracy of the influence functions can be investigated by scatter plot and correlation statistics created from the calibrated EPANET model data arising.

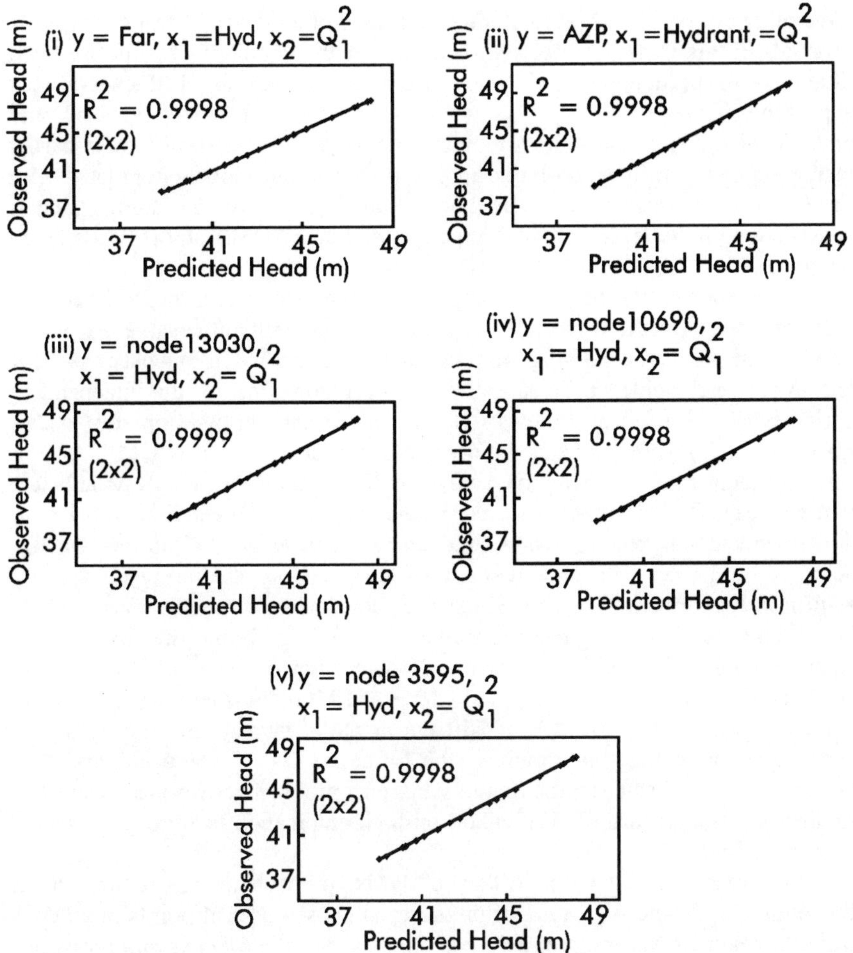

Fig. 9 Scatter plots using data from the network model

Figure 9 shows the 'Predicted Heads' output from the regression based Influence-Matrix approach based on only a single 24-hour data set of average-hourly (normal operational) conditions. These are compared against the calibrated EPANET results ('Observed Head') upon which they were based. It can be seen that the fit is almost exact, with correlation coefficients very close to 1.0. This applies both at the monitoring points (y = AZP and y = FAR) and at three other nodes selected at random. The degree of agreement implies that flow reversals in this DMA are not significant under the conditions synthesised and little loss of accuracy would follow from the use of the Influence-Matrix approach over the EPANET network modelling, whilst the physical and operational status of the system remains unchanged. These results offer partial validation of the Influence-Matrix approach to network models populated with nodal demands which all show identical diurnal profiles of variation which is consistent with the level of modelling addressed in the DMA studies considered herein.

To test the applicability of the approach to real conditions, the 24-hour field data collected from the DMA of Figure 8 have been input to the same regression process and predictions arising are presented in Figure 10. In this case the FAR and AZP head values (y) are estimated as regressions of the independent variables (x), selected as follows:- Hydrant head and Inflow (top row, a 2x2 matrix analysis); Hydrant head, Inflow and one internal head (i.e. CMP, AZP or FAR) (middle row, a 3x3 matrix analysis); and Hydrant head, Inflow and two internal heads (bottom row, a 4x4 matrix analysis) [8]. It is seen that with a 3- or 4-independent variable regression, prediction accuracies well within +/-1.0m are achieved, and even with only two variables predictions are rarely in error by significantly more than +/-1.0m. The general level of agreement is considered to be sufficient for most engineering studies. It should be borne in mind that the source data itself is subject to errors of inaccuracy (several of the pressure signals in this example reading only to +/-0.125m). Furthermore, the actual demands drawn from the system will not actually vary synchronously as the conceptual modelling (underpinning the Influence-Matrix approach) assumes. Given these circumstances, the findings are most promising, offering prospect of robustness, though further verification through application to other data sets is clearly required.

If ultimately validated, the Influence-Matrix approach might find application for routine 'real time' synthesis of pressures at chosen critical points in a DMA from the routine monitoring of only conditions such as inlet flows and pressures. In this case, a short field survey logging the pressures at these points whilst monitoring the normal data streams for the DMA (i.e. boundary head and flows) would enable production of the influence relationships, without the need for the network model.

It must be appreciated, however, that in both this case and application to the EPANET based prediction method, the influence functions will only remain valid whilst the physical status of the pipe system remains unchanged. Consequently, occasional change in internal valve statuses or operational state of the DMA will call for full re-calibration, necessitating the recovery from archive of the EPANET models, as depicted by Figure 1 (for the latter) or updated pressure logging of the critical points in the DMA (for the former).

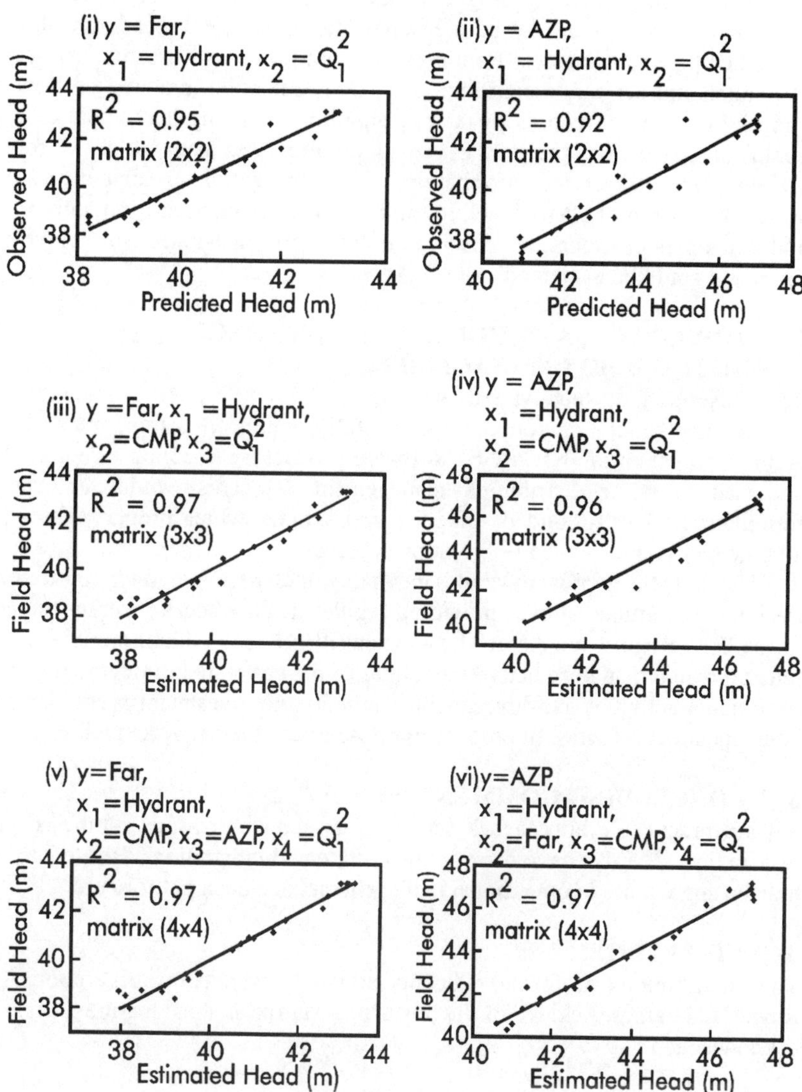

Fig. 10 Scatter plots using data from field study

4 DMA 'HEALTH-CHECK' MONITORING

The modelling approaches outlined above could potentially be operated automatically as 'snapshot' analyses drawing from successive download of monitored data streams, for evaluation of nodal (and thence property seed point) pressures. Outputs arising could then be integrated with other measures of performance depicted in Figure 1, to enable a comprehensive 'health-check' diagnosis of DMA status.

Such health checking might include [9] the continuous tracking of:- monitored inflows and large measured demands; total head profiles and computed pipe roughnesses, from pressure (and flow) data streams; and leakage evaluation and per capita consumption (PCC) estimation, drawing on flow data and billing records; records of mains condition, including bursts and customer complaints; as well as the synthesised supply pressures from the above routines yielding levels-of-service information. By interrogation of such measures of system behaviour at DMA level, in the form of the values and statistics arising, and any trends or excursions observed in time series generated, suitable triggers for action could be instigated.

5 TOWARDS A HOLISTIC APPROACH TO WATER DISTRIBUTION MANAGEMENT

The integration of such varied data streams and information databases with analysis and modelling routines can be fully embraced within open database architecture, as in the NETBASE software product, so enabling a more holistic approach to water distribution management [9]. This would facilitate the formulation of well-founded strategies and actions, which properly draw from all potentially inter-related information sources.

The holistic approach would inevitably include:- integrated performance tracking (as outlined above) providing regular health-checks meeting the needs of the U.K. Regulator's Asset Management Plan (AMP) submissions and 'July' returns; reactive and predictive modelling of networks in an integrated way, for operational activities and investment planning; and investment modelling of all interdependent activities in order to optimise asset renewal, risks and returns.

6 CONCLUDING COMMENT

Several practicable approaches for rigorous supply pressure synthesis within distribution networks have been outlined. Potential integration of these into more holistic approaches to operational management have been put forward.

ACKNOWLEDGEMENT

The contributions of former students at the University of Liverpool, Adam Rayne and James Ackley, to the Influence-Matrix studies are most gratefully acknowledged.

REFERENCES

1. Crowder & Co. Ltd, (1999),'NETBASE for Windows: Hydraulic Model Management', User Guide', 17-21 Price Street, Birkenhead, L41 6JN, UK, (e-mail crowders@compuserve.com).
2. MapInfo Corporation (1997) 'MapInfo Professional (5)' software documentation, MapInfo Corporation, Troy, New York, U.S.A.
3. Rossman L.A. (1994) 'EPANET-Users Manual', RREL, Office of Research and Development, U.S. Environmental Protection Agency, Cincinnati, OH 45268.

4. Borland International (1998) 'Delphi-4 for Windows 95 and Windows NT' Borland International Inc., P.O.Box 660001, Scotts Valley, CA 95067-0001, U.S.A.
5. Crowder & Co. Ltd, (1997) 'Pressure synthesis using Influence Matrices', unpublished exploratory study notes, August, (e-mail crowders@compuserve.com).
6. Burrows R. (1998) 'Synthesis of pressure inside a district metered area (DMA)', notes from an EPSRC/Industry short course on "Management of water distribution and leakage control", University of Liverpool, April, 4p. (e-mail r.burrows@liv.ac.uk)
7. Rayne A. (1998) 'Development and reliability study of an influence matrix method used for determining nodal pressure heads within a pipe network', B.Eng (Hons) Dissertation, Dept. Civil Engineering, University of Liverpool, UK, April, ~50p.
8. Ackley J.R.L. (1999) 'Reliability study of the application of the influence matrix concept to looped networks', B.Eng (Hons) Dissertation, Dept Civil Engineering, University of Liverpool, UK, April, ~80p.
9. Crowder G.S. and Burrows R., (1998) 'Developing a holistic approach to water distribution management', Proc. IQPC Water Industry Conference on "Modelling the Distribution System", IQPC London (http://www.iqpc.co.uk), July.

Linking SCADA to a Hydraulic and Water Quality Simulator at the 'Centre des Mouvements de l'Eau' (CME) in Paris, France

V. Tiburce, P. Chopard, J.Hamon, C. Elain *and* A. Green

ABSTRACT

The Centre des Mouvements de l'Eau (CME), operational since 1991, is a centralized Supervisory Control And Data Acquisition (SCADA) system for the water production and distribution network of the Syndicat des Eaux d'Ile de France (SEDIF) which serves approximately 4,000,000 inhabitants around Paris (France).

In 1994, the SEDIF and its operator , Générale des Eaux (VIVENDI group) initiated a project called the 'CME-phase II' project with the aim of linking the CME SCADA system with network analysis software (the Stoner Workstation Service) capable of simulating the hydraulic and water quality operation of the water network. This application has been operational since 1997 and provides the SEDIF with a new set of tools to better supervise the operation of the network, predict potential problems in the network and validate different operational strategies.

On-line simulations (i.e. in parallel to the real system operation) allow for anomaly detection and resolution: a new set of operational data is received from the SCADA system at a predefined time interval and is used to update the hydraulic model before executing a steady state simulation. Key points in the network can be closely controlled (without supplementary monitoring equipment) and potential discrepancies between calculated and observed data can be detected and analyzed. The entire process is fully automatic and does not require any operator intervention.

Off-line past simulations allow the operator to re-play an event that took place in the network using the operational data archived by the SCADA system with the objective of better understanding the hydraulic and water quality behavior of the network.

Off-line predictive simulations allow the operator to test various operational strategies. Using as a starting point the most recent set of operational data and defining the network demand and the facilities operation in the coming hours or days, the operator can initiate a variable simulation and assess the consequences of a specific operational strategy.

The CME phase II application is also extensively used to validate any extension or reinforcement project as it permits a quick extraction and validation of SCADA data to calibrate a hydraulic model before assessing the impact of a new extension or reinforcement scheme.

The paper will provide an overview of the CME phase II application, its operation and architecture and offer some example applications.

1 INTRODUCTION

One of the most prevalent analytical tools that can be layered upon a SCADA system is a network simulation. With the integration of these two technologies a wealth of applications can be realized from close to real-time control of systems through to operators training and more effective management and planning of the network.

This paper describes the work undertaken by the Syndicat des Eaux d'Ile de France (SEDIF) and its operator Générale des Eaux (GdE) at Paris (France) for developing a specific application which allows linking of the SEDIF telemetry system with a network simulator (the Stoner Workstation Service or SWS). The project involved several companies: Prolog Ingéniérie as leading consultant, GTIE as SCADA provider, Générale d'Infographie as GIS vendor and Stoner Associates as network analysis software provider.

2 BACKGROUND

2.1 The production and distribution systems of the SEDIF

The 'Syndicat des Eaux d'Ile de France' (SEDIF) gathers 144 municipalities in the inner and outer suburbs of Paris, France and supplies drinking water to a population of 4 million (see Figure 1). On a day-to-day basis, the SEDIF network is managed by the Générale des Eaux (GdE) which acts as an operator for the SEDIF.

The distribution system, consisting of 8,500 km of pipes, 45 pumping stations and 51 reservoirs for a total storage capacity of 700,000 m^3, supplies an average daily demand of 1 million m^3 to 500,000 domestic and industrial customers.

The system is divided into three general regions, each supplied by a water treatment plant located near one of the principal rivers in Paris:

- The plant at Choisy-Le-Roi treats water from the Seine (capacity of 800,000 m^3 per day) and feeds the Southern sector of the distribution network
- The plant at Mery-sur-Oise treats water from the Oise (capacity of 270,000 m^3 per day to be increased to 400,000 m^3 per day by the end of 1999) and feeds the Northern sector
- The plant at Neuilly-sur-Marne treats water from the Marne (capacity of 800,000 m^3 per day) and feeds the Eastern sector.

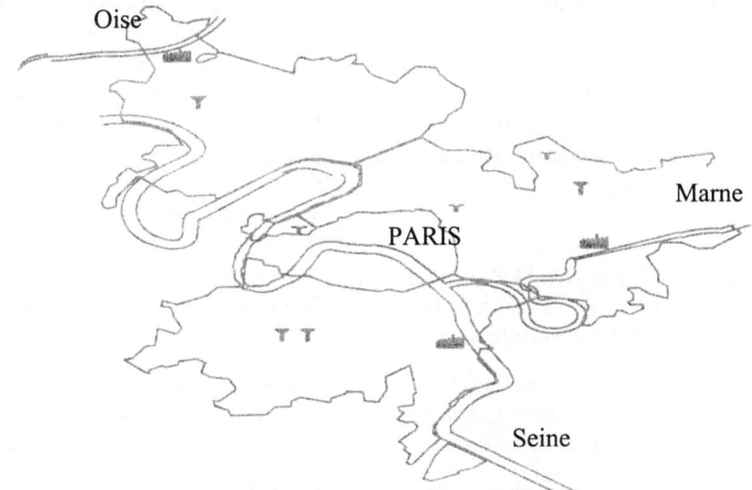

Figure 1 : Overview of the SEDIF system

The management of the system is conducted at three major treatment plants, which control:

- Operation of the treatment plant itself i.e. process and pumping stations feeding the distribution system;
- Operation of the facilities located in the distribution system supplied by the plant i.e. pumping stations, reservoirs, major valves, etc.

The capacity of each treatment plant is considerably larger than the actual daily production in order to ensure the safety of the supply e.g. in case of river pollution or incident at one of the other treatment plants. Large diameter pipes link the different sectors to allow the transfer of large flows.

2.2 The centralized supervision system : CME - Phase I

To facilitate the management, planning and operation of the entire SEDIF system, an operations co-ordination and monitoring system called the Water Movements Center (Centre des Mouvements de l'Eau or CME) was installed in 1991. This centralized system is connected in real time with the control command systems of the three treatment plants.

Furthermore, through a specific telephone application (Minitel), maintenance teams can manually provide the date and time for all the opening or closing of valves which are not connected to a monitoring system. The CME supervises in close to real time the position of approximately 5,000 key valves in the distribution system.

From these basic data, the CME calculates and subsequently archives complex variables, in particular the instantaneous water demand associated with the 40 small, self-sufficient networks which can be isolated and supplied from local storage facilities. A water forecast system for each of these 40 networks has also been developed.

The CME 's key role is to supervise and co-ordinate the actions taken by all three major regions in normal and crisis conditions. On a routine basis, the CME examines any significant maintenance work scheduled by any of the three regions and affecting the production or distribution systems. In periods of crisis, and in particular in the case of a river pollution alert, the CME acts as a crisis center collecting information and dispatching it to the treatment plants.

3 THE CME PHASE II PROJECT

The objective of this project is to link the telemetry system with a hydraulic and water quality simulator in order to:

1. Calculate hydraulic (pressures and flows) and water quality (chlorine residual and age of water) parameters throughout the distribution system and in this way complement the discrete measurement points of the telemetry system.
2. Simulate and predict the behavior of the network for various management scenarios.

The CME phase II application has been developed in order to meet four major operational objectives:

1. 'Real Time' Simulations
2. Past Simulations
3. Past Simulations
4. Predictive Simulations

3.1 'Real Time' Simulations

The objective is to perform automatically and continuously (i.e. at a time step specified by the operator) a steady-state simulation for the entire network in order to:

1. Control in real time the hydraulic behavior of the network i.e. a complete 'picture' of the system in pressure and flows is available every x minutes to help the operator assess the actual operation of the network.
2. Detect discrepancies between measured and calculated data that could signal potential problems in the system i.e. incorrect operation of an outstation, incorrect or missing information regarding the actual operation of the system such as a valve position, regulator set pressure, etc.

3.2 Past Simulations

This function allows the operator to replay a past event based on the data archived in the CME telemetry system. The objective here is to provide the operator with a training tool to better understand and analyze any incidents that occur in the distribution system and in this way gain experience in confronting various events.

3.3 Predictive Simulations

This function allows the operator to test various operating strategies and assess their future consequences on the behavior of the system. It requires several elements:

1. A demand forecast module to predict the water consumption for each system in the coming hours or days.
2. An operating strategy for the main facilities.
3. The latest network operational status extracted from the CME telemetry.

Taking all these elements from the separate sources, the application can create all the requested files to execute quasi-dynamic simulations. The objective here is to provide a day-to-day management tool to help the operator run the distribution system more effectively.

3.4 Planning Simulations

Typically, whenever any significant new planning project is assessed, an hydraulic model is built and calibrated in order to define precisely the characteristics of the future facility or pipe to determine its future hydraulic and water quality impact on the network. The CME telemetry system is sufficiently comprehensive so that in most cases the data it provides are sufficient to calibrate the model. However, until recently the process to extract data from the telemetry system, validate them and if necessary complement them with data issued from other sources was manual and obviously time consuming. The CME phase II application offers all the functions necessary to quickly and efficiently relate telemetry data and model parameters, control the quality of the archived operational data and complement them. Furthermore, the CME offers a large range of operational conditions against which the planning project can be tested.

4 MAIN PRINCIPLES AND ARCHITECTURE OF THE CME PHASE II APPLICATION.

The software 'Stoner Workstation Service' provided by Stoner Associates has been used by the SEDIF and its operator 'Générale des Eaux (GdE)' since 1992 and offers several key advantages:

1. The capability to simulate the behavior of large models.
2. Extensive functionality to take into account the complex rules which control the operation of the SEDIF facilities.
3. Water quality modeling capabilities.
4. Telemetry interface including a real-time module.

4.1 Mathematical models

In order to provide operators and modelers with as much flexibility as possible, about 10 models have been built which cover different zones and include various levels of detail:

- Models covering the three sectors but including only the strategic transmission system to analyze water transfer between sectors (in case of emergency) and use of the different water sources.
- Models limited to a specific sector but including only strategic mains to analyze storage and pumping management and assess the security of supply in normal and emergency situations.
- Models limited to one pressure zone within a sector.

Furthermore for each model built for the CME phase II application two versions have been created depending on the type of simulations to be executed:

- The models used for predictive simulations include a detailed description of the facilities and of the rules governing their operation;
- The models used for real-time and past simulations are simpler and do not explicitly include all the facilities. For example, a pumping station described with several pumps and valves in the predictive model will be represented as a simple source node, the flow out of the pumping station being provided by the telemetry system.

4.2 Telemetry data archived for the CME phase II application.

At the beginning of the project it was decided, mainly for system security, that the CME phase II application would not use the data directly from the CME master database but that a subset of operational data required for hydraulic and water quality data would be extracted and stored in a specific database following the principles described below:

- Every 10 minutes, about 1,000 continuous variables are stored in the phase II database;
- All logical variables (about 3,000 in total) are stored every time they change status;
- Data covering the last 14 months are stored on-line.

The continuous variables correspond to the following hydraulic and water quality parameters:

- 400 flow measurements (flow through individual pump and pumping station, water exchange with neighboring distributors, flow through regulators etc);
- 200 pressure measurements;

- 100 level measurements;
- 200 chlorine measurements (set point for chlorine booster stations or sources and measurements in the network);
- 80 demand values (actual and predicted demand for the 40 self sufficient systems)

The logical variables correspond to the following hydraulic parameters:

- 900 facility valve status;
- 1,800 network valve status;
- 150 pump status.

4.3 Setting up a simulation.

The objective that drove the entire project was to offer to the users of the CME phase II application a unique and robust simulation environment. The design of the application guarantees as much as possible that all data used to run a simulation are coherent and of sufficiently good quality to ensure the success of the simulation. A few simple principles were adhered to:

- The basic mathematical model is a true and tested representation of the physical network (topology, pipe and facility characteristics, etc).
- The link between telemetry data and model parameters is correct (extract the proper value for a reservoir level or a pump status).
- If telemetry data are missing or incorrect, replacement data are used.

The procedure eventually chosen is summarized in Figure 2 and relies on the following principles:

- To run a past, real-time or predictive simulation, the operator is limited to choose the mathematical model from a library of master models built, calibrated and tested by expert hydraulic users.
- These models include 'static' data which will not be modified when a simulation is performed i.e. (a) topology of the system, physical characteristics of the pipe and facilities, spatial distribution of the demand and (b) 'dynamic' data that will be modified when a simulation is prepared i.e. reservoir levels, network demand, equipment status, facility flows.
- When the operator sets up a simulation, the CME phase II application creates a new set of dynamic data and transfers it to the SWS together with a copy of the master model selected by the user;
- SWS updates only the dynamic data which guarantees (as far as possible) that the model used for the simulation is coherent.

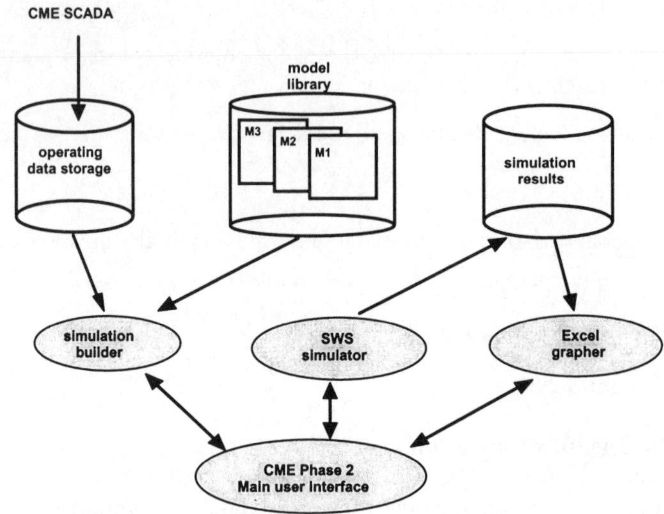

Figure 2 : Overview of the CME phase II application architecture

4.4 The CME phase II application interface.

There are two entry levels into the CME phase II application. Expert users have access to all the functions which allow them to link telemetry data and model parameters, to create libraries of complementary data and standard reports. Basic users have access only to the functions required to execute a simulation, i.e.

- Select the entry data for the simulation the user wishes to run i.e. the type of simulation, model selection, period of the simulation. Once all this information is provided by the user, a set of routines included in a simulation set-up module, creates and transfers all the necessary files to the SWS;
- The operator then executes the simulation in SWS and can analyze simulation results using all the graphical and tabular functions of SWS;
- The operator can analyze and compare results from different simulations in a spreadsheet environment e.g. MS Excel;
- The operator can transfer and analyze simulation results into a Geographical Information System (GIRIS provided by Générale d'Infographie) using the powerful query and display functions of a GIS.

5 IMPLEMENTATION OF THE CME PHASE II APPLICATION

5.1 Real-Time Simulations

On-line simulations are performed automatically at the CME continuously 24 hours a day, 7 days a week, for the entire SEDIF network. Every 10 minutes, a new text file which contains the latest operational and reference data from the telemetry system is created and placed in a specific file directory. An OnLine Module (OLM) which has been developed by Stoner Associates controls online operation of the SWS. Every 10 minutes the OLM initiates a new cycle which includes the following stages:

1. The OLM initiates the SWS.
2. It checks for the presence of a new operational text file in the appropriate directory.
3. If no text file is found, the OLM instructs the SWS to 'idle' until the next cycle.
4. If at least one text file is found, the OLM instructs the SWS to load the model file, load the operational text file and attempt to execute a steady state simulation.
5. If the simulation is successful, simulation results are written to a variety of text and binary files. If the simulation fails, a message is written to a log file and the process moves to step 6.
6. The OLM instructs the SWS to idle until the next cycle.

All results for Online simulations are kept for a sliding 24-hour period and can be analyzed by the operator in several ways:

1. A comparison file between calculated and observed data is always displayed to alert the operator for discrepancies which can signal monitor drift/failure or network problem;
2. The operator can analyze the last simulation results directly in the SWS;
3. The operator can analyze the last few simulations results through tendency curves displayed in an Excel environment.

Figure 3 represents an example of the different On-line simulation results displays available for the operator.

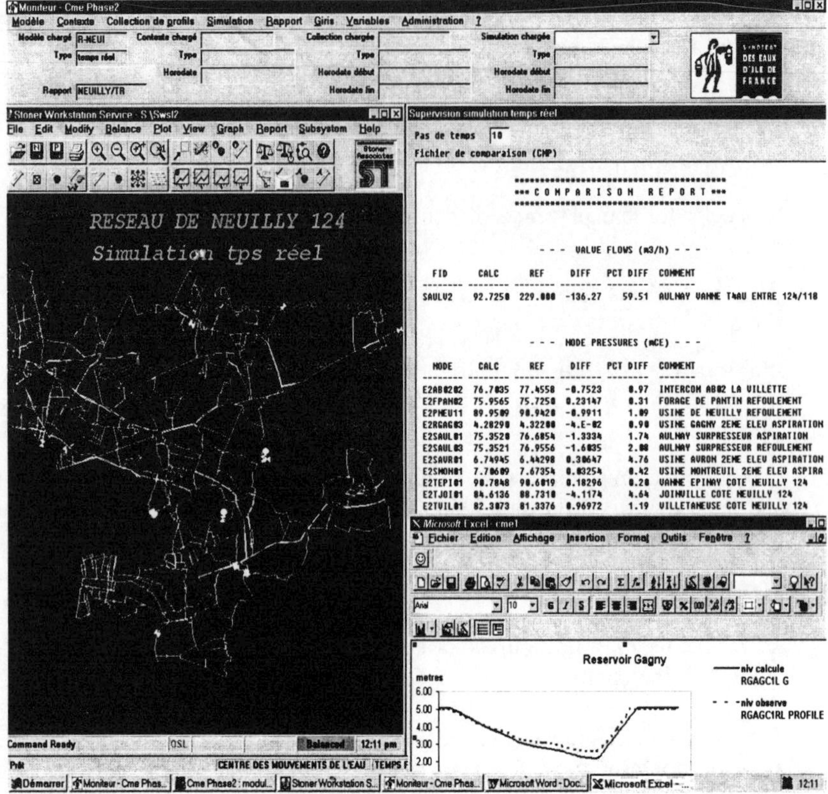

Figure 3: On-Line Simulation results display in the CME phase II application

5.2 Past simulations

This type of simulation allows the operator to replay a past event or incident based on the data archived by the telemetry system.

The operator indicates the type of simulation to be executed, selects a mathematical model and a starting and finishing date and time. The CME phase II application creates all the appropriate files and transfers them to the SWS. The operator can then initiate a quasi-dynamic simulation and analyze results directly in SWS or in Excel using pre-defined reports which typically include graphs, bar charts and tables.

Figure 4: Results for a past simulation run for a small distribution system

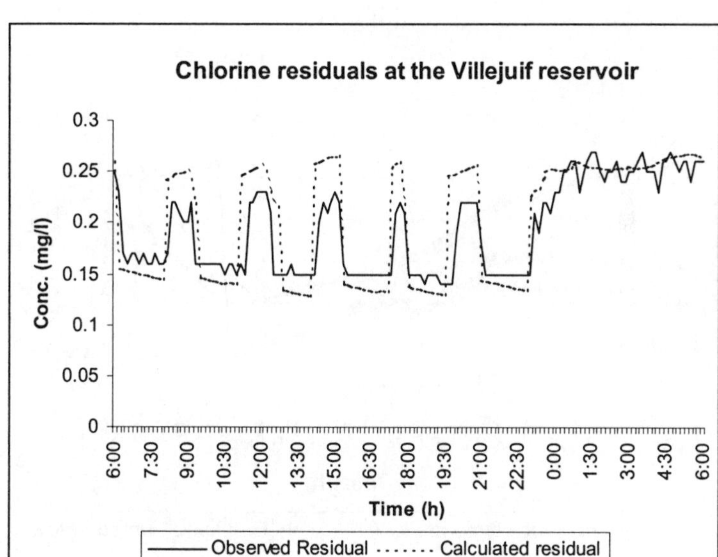

Figure 4 shows results for a past simulation run for a small distribution system called Villejuif. Observed and calculated chlorine residuals are represented. When the operator initiates a past simulation that includes water quality calculations, a minimum period of about 72 hours is selected to take into account the initialization period inherent to water quality calculation. This period corresponds to the time taken when the results are no longer affected by initial water quality estimates. Valid results are available only for the second part of the simulation (typically the last 24 hours).

5.3 Predictive Simulations.

This type of simulation allows prediction of the future behavior of the network for various operating strategies. Depending on the particular problem, the operator will choose a simple model from the model library which includes only one pressure zone or a more general model which includes several pressure zones. Some examples of this predictive type of simulation are shown below.

Example 1

At 6.00 pm the operator wishes to determine how the Montfermeil reservoir will operate during the night until 6.00 am the next morning. A predictive simulation is executed using the following information:

1. The latest operational conditions ie reservoir level and pump status at 6.00 p.m. extracted from the telemetry system.
2. Demand forecast for this network for the 12 hours to 6.00 am the next morning.
3. Rules controlling the operation of the pumping station.

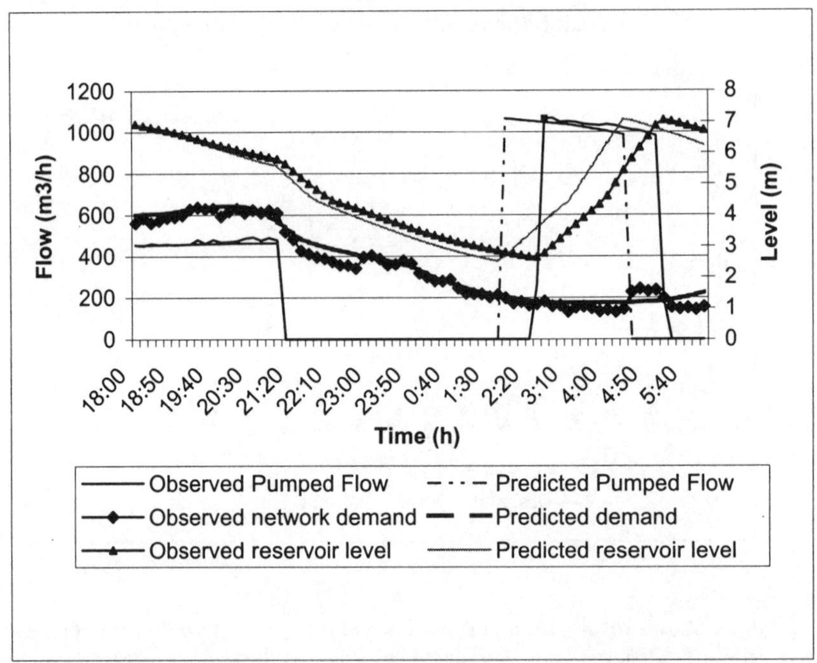

Figure 5 : Predictive simulation : comparison between simulation results and actual measurements

Figure 5 compares the results of the predictive simulation (reservoir level, pumping station flow and network demand) with the actual telemetry data.

Example 2

The operator wishes to determine the impact of the temporary closure (from 9.00 am until 5.00 p.m.) of an 800 mm diameter pipe located near the pumping station of Gagny.

Using the CME application the impact of this maintenance work can be simulated and a comparison made between the hydraulic behavior of the network with the 800mm diameter pipe in or out of operation (see Figure 6).

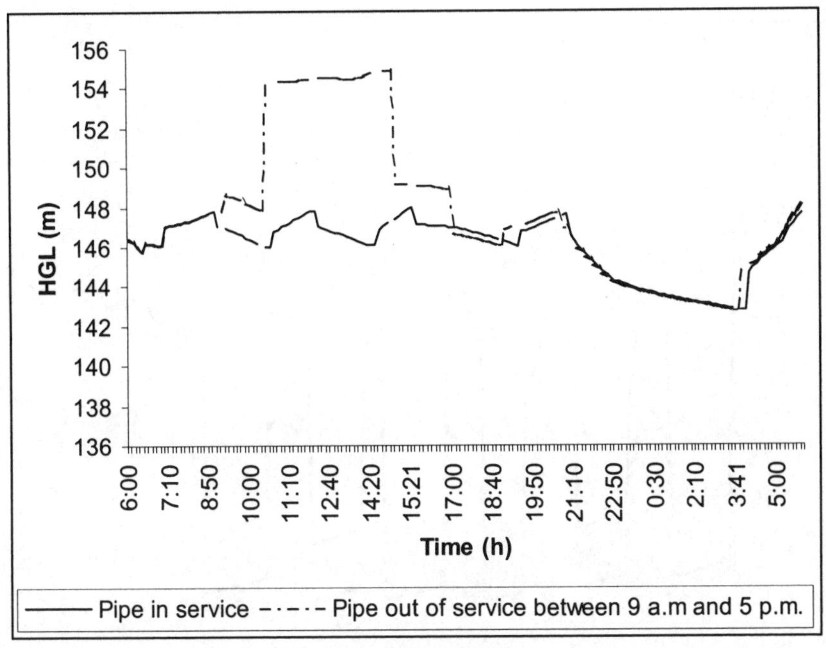

Figure 6 : Comparison of HGL for two operational scenarios

Example 3

A chlorine booster station is installed at the exit of the pumping station of Les Sorbiers which supplies the Villejuif reservoir. The usual chlorine concentration maintained by this booster station is 0.4 mg/l but the operator wishes to assess the impact of reducing this concentration to 0.3 mg/l for the Villejuif reservoir.

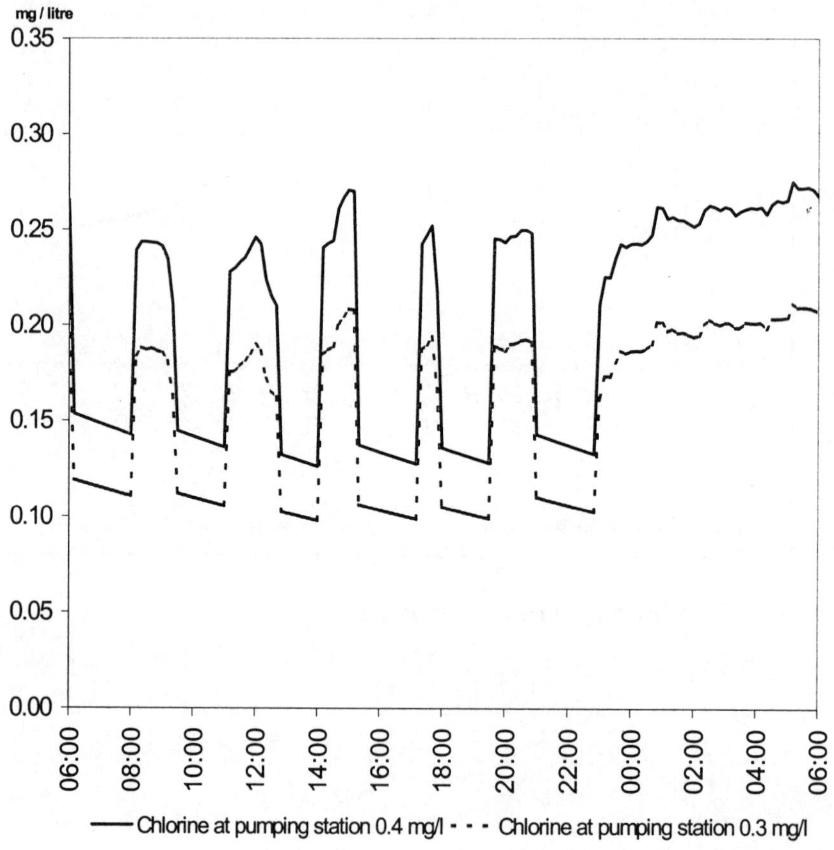

Figure 7 : Comparison of chlorine residuals for two operational scenarios

Using the CME phase II application, the operator can perform two quasi-dynamic simulations and compare the chlorine residuals in the Villejuif reservoir for the two situations (see Figure 7).

6 CONCLUSIONS

With the CME phase II application, the SEDIF has acquired a powerful tool to control and help manage and plan its distribution system to complement its existing SCADA system.

It will take several months to set up all the models and take full advantage of the capabilities of such an application but it has already provided useful information to validate the SCADA system (detect monitor drift), and better understand the hydraulic and water quality behavior of specific zones of the distribution network. Planning studies have already been completed in a comparatively short time as calibration data are quickly available for a number of operational situations. In time, real time simulations should also yield very useful information and provide an efficient and complementary service to the telemetry system.

Hydraulic Characteristics of Pressure Reducing Valves for Maximum Reduction of Leakage in Water Supply Networks

Luisa F. R. Reis *and* Fazal H. Chaudhry

ABSTRACT
A number of efforts have recently been reported in the literature seeking the minimisation of water losses and the optimal location of valves. As the demands and the reservoir levels may vary randomly, there will be a wide variety of openings necessary for the valves in the network. This paper studies the performance of pressure reducing valves in response to spatial variations of demands in the network. Fifty realisations of random geometric patterns of demand following a triangular probability density function are employed in the repeated minimisation of water leakage for different combinations of total network demand and reservoir levels to produce an equal number of responses in the network elements. These responses are analysed to identify the hydraulic characteristics of the valves necessary to ensure minimisation of leakage.

1 INTRODUCTION

A reduction of leakage in urban water distribution networks can be achieved through the control of excessive pressures in pipes by optimal location of pressure reduction valves (PRVs). The study reported in [2] expresses explicitly network leakage in the formulation of the problem of minimisation of water losses by means of PRVs whose opening can vary between fully open and closed positions. Since then, various studies have been conducted employing the formulation proposed by those authors. Some authors proposed alternative solution techniques in place of successive linearizations in order to circumvent convergence problems. Genetic algorithms were used by [1] and a procedure that considers a sequence of quadratic programming sub-problems for the determination of optimal valve openings was presented in [5]. Other studies considered the problem of optimal location of pressure reducing valves to obtain maximum reduction in water leakage while determining optimal valve openings for minimum leakage by genetic algorithms [1,3]. The valve location problem focusing the control of excessive pressures in the networks by isolating valves using genetic algorithms was also considered in [4].

While the studies on leakage minimisation report their results in terms of the discharge fractions as the control parameter, little is said about the relationship between these and the flow conditions in the pipe which could help visualise the hydraulic characteristics of the control valves. Such relationships are especially relevant in view of the normal variations in the valve operating conditions in the field. This paper presents the response of the pressure reducing valves for various simulated random spatial patterns of nodal demands and three distinct combinations of total demand and reservoir levels in an example network studied within the framework of the problem of leakage minimisation.

2 JOWITT-XU OPTIMIZATION PROBLEM

The problem of minimisation of system leakage can be posed as that of determining the optimal PRV opening $V(k)$ of kth valve, in order to minimise the sum of leakage in pipes spanning arbitrary nodes i and j, QS_{ij},

$$\underset{V(k)}{Min} \sum_{j \in R_i} QS_{ij} , \qquad i=1,2,...N \tag{1}$$

subject to:

1-The nodal mass balance in the form:

$$\sum_{j \in R_i} Q_{ij} + 0.5 \sum_{j \in R_i} QS_{ij} + C_i = 0; \quad i=1,2,...N \tag{2}$$

2- the minimum acceptable head for the nodes as:

$$H_i \geq H_i^*; \quad i=1,2,...NR \tag{3}$$

3- the upper and lower bounds for valve openings as:

$$V(k)^{min} \leq V(k) \leq V(k)^{max}; \quad k=1,2,...NV \tag{4}$$

where R_i is the set of nodes connected to node i, C_i the demand consumption at node i, N is the total number of nodes with unknown heads and NV represents the number of valves in the network.

The Hazen-Williams equation which relates flow head loss and pipe characteristics:

$$Q_{ij} = \frac{\alpha CHW_{ij} D_{ij}^{2.63} Sgn(H_i - H_j) \left| H_i - H_j \right|^{0.54}}{L_{ij}^{0.54}} \tag{5}$$

where: Q_{ij} represents the flow (L/s); D_{ij} pipe diameter (m); L_{ij} length of the pipe (m); H_i head at node i (m); CHW_{ij} Hazen-Williams coefficient for the pipe connecting nodes i and j; α, a constant whose value depends on the units used and $Sign(x)$ denotes sign of x.

In the case of a control valve located between nodes i and j, the flow-head loss relationship can be expressed as:

$$Q_{ij} = V(k) R_{ij} Sgn(H_i - H_j) \left| H_i - H_j \right|^{0.54} \tag{6}$$

where $V(k)$ is a parameter that represents the opening of the kth control valve. $V(k) = 1$ represents a fully open valve, which produces no head loss beyond that in the pipe element, and $V(k)=0$ a fully closed valve. R_{ij} incorporates the terms independent of head in the Hazen-Williams equation. The non-linear relationship between the leakage and average service pressure was approximated by the following function derived from experimental data:

$$QS_{ij} = CL_{ij} L_{ij} (P_{ij})^{1.18} = RS_{ij} (P_{ij})^{1.18} \quad (7)$$

In Eq. 7 QS_{ij} is water-leakage in the pipe element of length L_{ij} spanning nodes i and j; CL_{ij} is a coefficient that relates leakage per unit length of pipe to service pressure and depends on the system characteristics, such as age and deterioration of the pipe and the soil properties; RS_{ij} a composite parameter; P_{ij} is given by:

$$P_{ij} = 0.5\left[(H_i - G_i) + (H_j - G_j)\right] \quad (8)$$

where G_i is ground level at node i.

3 DEMAND PATTERNS

The system of equations defining the flows in a water distribution network is generally solved for a given demand pattern. The nodal demands are determined by dividing proportionally among them the demand of an urban area within a pipe loop. However, special demands can occur such as those caused by fire, breakage of pipes and other factors contributing to the instantaneous demand variations in the spatial demand patterns. As a consequence, the performance of such systems varies in time and assumes different values for different zones of the same network. Under such circumstances, the operation of control valves is subject to varying demand patterns, referred to as scenarios. This requires consideration of the stochastic nature of daily demand variations. Following [3], one considers nodal demands as random variables in order to evaluate the valve operation under different demand scenarios, assuming that the nodal demands follow a triangular probability distribution as follows:

(a) let $D_{max}(i)$ and $D_{min}(i)$ be extreme demands at node i;

(b) a vertex $h(i)$ of triangular probability distribution is thus given by

$$h(i) = \frac{2}{Q_{max}(i) - Q_{min}(i)} \quad (9)$$

The corresponding cumulative distribution function $F(i)$ defines the value of distribution function for design demand $DD(i)$ written as:

$$F0(i) = \frac{DD(i) - D_{min}(i)}{D_{max}(i) - D_{min}(i)} \quad (10)$$

Different scenarios can be produced by generation of random numbers $u(i)$ between zero and one for each node i. Two situations are possible. In the Case a, the random number $u(i)$ is less than or equal to $F(i)$, and the demand $D(i)$ can be expressed by:

$$D(i) = D_{min}(i) + \frac{u(i)}{F0(i)}\left[DD(i) - D_{min}(i)\right] \quad (11)$$

In Case b, where $u(i)$ is greater than $F(i)$, the demand $D(i)$ can be written as:

$$D(i) = DD(i) + \frac{u(i) - F0(i)}{1 - F0(i)}\left[D_{max}(i) - DD(i)\right] \quad (12)$$

The set of nodal demands obtained in this manner represents a simulation of the demand pattern (demand scenario) in the network. The spatial pattern of nodal demands *D(i),* thus generated, is adjusted proportionally so that the sum of these demands is the same for all the scenarios in the study sample in order to make comparisons between the results for various simulations.

4 STUDY EXAMPLE

The present study employs the hypothetical 25-node network (Fig. 1) with 37 pipes as in [1,2]. Valves are placed in pipes 1, 11 & 20. The network data differs slightly from those used by the various authors cited above. However, we adopt the data in [1], where the reservoir ground levels are set equal to 40m. Detailed information on nodal demands, the respective elevations, pipe lengths, diameters and the Hazen-Willians coefficient can be found in the latter references.

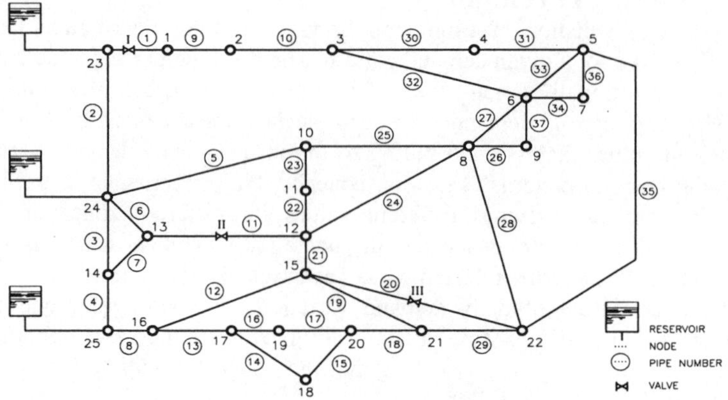

Fig.1 – Example Network [1]

For the hydraulic analysis of the network, the linear theory method was used as in the Jowitt-Xu algorithm based on Eqs. (1) to (8) for the problem of minimisation of pipe network leakage determining the optimal valve openings for a given valve combination. A constant value of 10^{-5} is assigned to the coefficients, Cl_{ij}, corresponding to pipe reach between nodes *i* and *j* in the leakage-pressure relationship for the network in Eq. (7).

The operational conditions in the network go through a continuous cyclic variation during a typical day with individual nodal demands varying in a random manner. However, in this study we summarise these conditions in three discrete combinations in terms of total network demand and reservoir levels as in Table 1.

Table 1 – Network Operational Conditions

Reservoir Nodes	Critical		Average		Below Average	
	Total Demand	Level (m)	Total Demand	Level (m)	Total Demand	Level (m)
23	90 L/s	55.900	150 L/s	55.200	210 L/s	54.500
24		55.500		55.000		54.500
25		55.600		54.825		54.050

These combinations are chosen so as to simulate the critical (high pressures and low demands), average and below average (low pressures and high demands) conditions to represent pipe pressures and discharges responsible for the amount of leakage in the network. Under each of these conditions, 50 random distributions of nodal demands were generated assuming $D_{min}(i)=0$ e $D_{max}(i)=2DD(i)$ in the description of the triangular probability distribution. The leakage minimisation model was run for each of these scenarios to obtain nodal heads, pipe discharges and optimal valve openings. In order to study the valve response, operation of the network with three valves operating simultaneously was considered. The minimum pressure at the nodes in the network was constrained at 25m.

5 HYDRAULIC CHARACTERISTICS OF VALVES

The Jowitt-Xu leakage minimisation model incorporates the operation of PRVs through a control parameter V_k defined as the ratio of discharge through the valve to a reference discharge that would occur in the presence of the same pressure differential with the valve fully open. This control parameter can not be related directly to the hydraulics of the valves because of the difficulty of defining the reference discharge in general. We analyse here the relationships between the

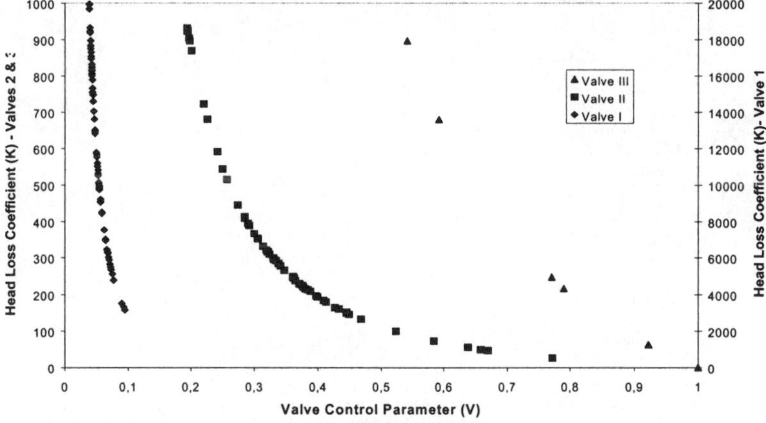

Fig. 2 – Head Loss Coefficient as Function of the Valve Control Parameter

hydraulic parameters such as pressure differentials, head loss coefficients, discharges and the flow control parameter V_k as indicated by the various simulations.

Head loss coefficients K (defined in $\Delta h = K.v^2 / 2g$) as a function of the valve control parameter for various valves are shown in Fig. 2. The three valves present distinct operational characteristics. Valve I, which is placed downstream from the reservoir at node 23, is heavily throttled in comparison to other two valves, with head loss coefficients of more than 3000 and the control parameter below 0.1. Valves II and III, which lie away from the reservoirs, act to control pressures for all the operating conditions, with the control parameter spanning its entire gamut of values. It should be observed that in the Jowitt-Xu model head loss through the valve is zero for the valve fully open. Head loss coefficients for Valve III are much higher than those for Valve II for a given value of control parameter. Only part of the simulated scenarios is included in Fig. 2 in terms of the head loss coefficient.

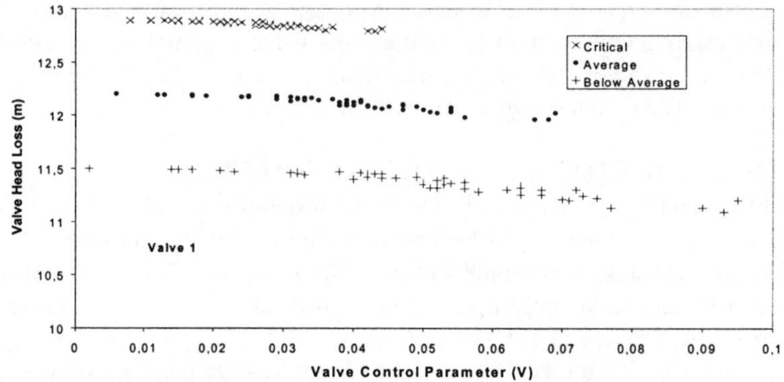

Fig. 3 – Valve head losses as function of control parameter – Valve I

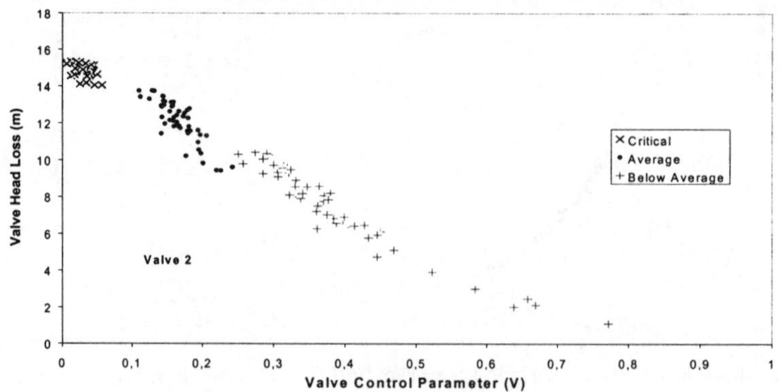

Fig. 4 – Valve head losses as function of control parameter - Valve II

Figures 3, 4 and 5 present all the simulation results on head loss for the three valves as functions of the control parameter. As expected, the valve head loss decreases with the control parameter. However, the patterns of variation of head loss with the control parameter are slightly different in their approach to zero at the fully open position in the case of Valves II and III. Head loss in Valve I (Fig. 3) depends directly on the level in the reservoirs, which is the reason for the shift in head loss variation from one operating condition to the other. In the case of Valve II (Fig. 4), the head losses under different conditions and nodal demand scenarios produce a continuous variation with the control parameter. The head loss behaviour of Valve III indicated by the simulations is different from the other valves in that, under critical conditions, the valve remains either fully open or closed.

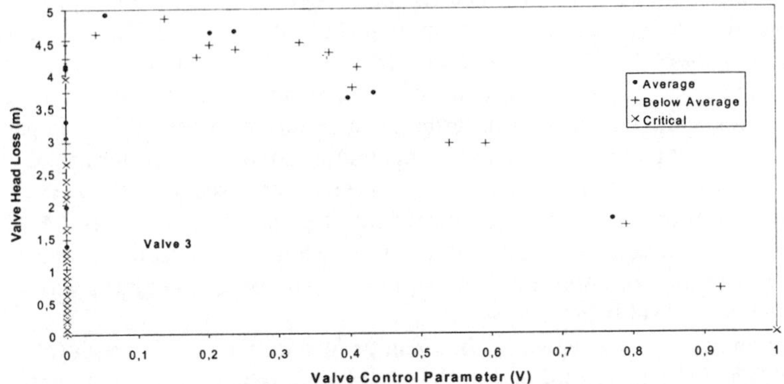

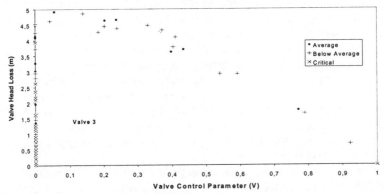

Fig. 5 – Valve head losses as function of control parameter - Valve III

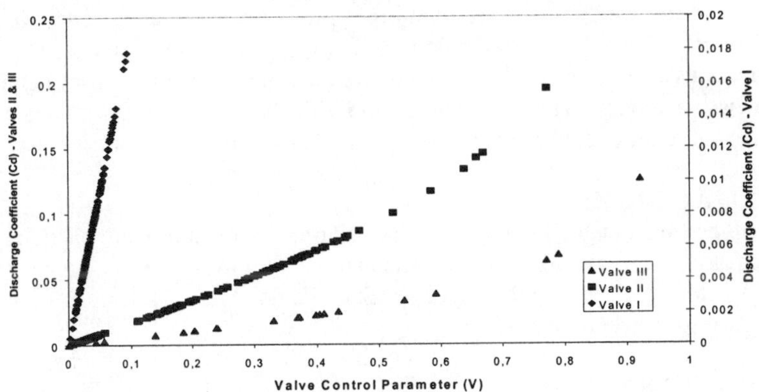

Fig. 6 – Discharge coefficient as function of valve control parameter

The discharge characteristics of the valves in the example network are presented in terms of discharge coefficient Cd (defined in $Q = Cd.A.\sqrt{2.g.\Delta h}$), based on the pipe cross sectional area A, in Fig. 6. As noted before, discharge coefficients for Valve I are very low, being highly throttled. Although these coefficients are indicated to be higher for Valves II and III, they are below 0.2. The variation of discharge coefficient with the control parameter in these valves follows the typical characteristic curves for V-notch and other valve types. Although Fig. 6 does not indicate this, it was observed from the simulation results that the discharge in pipe 20 with Valve III frequently oscillated in direction depending on the operating conditions in the network. Under critical leakage conditions, this valve was either closed or fully open.

6 CONCLUSIONS

The leakage minimisation studies reported in the literature generally incorporate a control valve parameter (V) defined as the ratio of flow through a valve and a reference discharge, considered as an indicator of valve opening as it varies between 0 (valve closed) and 1 (valve fully open). This description of control is rather incomplete given the difficulty of prescribing the reference discharge and thus requires some explanation in terms of hydraulic parameters. In this paper, a variety of combinations of network operating conditions in terms of nodal demands, total network demand and reservoir levels are used to evaluate the values of valve control parameter necessary for leakage minimisation. As these simulations of network operation also produce results on hydraulic parameters for each determination of control parameter, their relationships have been analysed to reveal the type of valve indicated.

It is shown that the valve in the vicinity of the reservoirs is highly throttled (K>2000) and the valves located further away may require lower but appreciable throttling with the head loss coefficient K in the order of thousands. It is further shown that the performance of the valves is highly differentiated depending upon the operating condition's potential for promoting leakage. The discharge coefficients indicated by the simulations show typical characteristics of pressure reducing valves in terms of the valve control parameters obtained from the leakage minimisation model. However, the relationship between this parameter and the valve travel requires further study to augment its usefulness in practice.

ACKNOWLEDGMENT

The authors are indebted to the Brazilian National Research Council (CNPq) for the research scholarships, to the Foundation for Investigations and Projects (FINEP) for project support and to São Paulo State Research Foundation (FAPESP) for travel grants.

REFERENCES

1. Gueli, R. & Pezzinga, G. (1998). “Algoritmi genetici per la regolazione di valvole ai fini della riduzione delle perdite”, *Atti*, XXVI Conv. di Idraulica e Costruzioni Idrauliche, Catania, Italy (in Italian).
2. Jowitt, P. W. and Xu, Ch. (1990). "Optimal valve control in water-distribution networks". *J. of Water Resources Planning and Management*, ASCE, vol. 116, n. 4, 455-472.

3. Reis, L. F. R., Porto, R. M. & Chaudhry, F. H. (1997). "Optimal location of control valves in pipe networks by genetic algorithm", *J. of Water Resources Planning and Management*, ASCE, vol. 123, n. 6, 317-326.
4. Savic and Walters (1995). "Integration of a model for hydraulic analysis of water distribution networks with an evolution program for pressure regulation", *Microcomputers in Civil Engineering*, vol. 10, n. 3, 219-229.
5. Vairavamoorthy, K. & Lumbers, J. (1998). "Leakage reduction in water distribution systems: optimal valve control", J. Hydraulic Engineering, ASCE, vol. 124,n. 11,1146-1154.

PART IV

MODEL CALIBRATION

Data Collection for Water Distribution Network Calibration

I. Ahmed, K. Lansey *and* J. Araujo

ABSTRACT

Calibration of a water distribution network is intended to develop a model that mimics field conditions under a range of demand distributions. Little work has been completed to identify the conditions under which useful data should be collected and locations where measurements should be taken. A heuristic three-step procedure is developed to assist in making these decisions. The basis for this procedure is a calibration assessment measure that estimates the error in predicted pressure heads throughout the system. The measure is computed by performing uncertainty analyses to predict the variance in the model predictions given the uncertainty of the field measurements. The trace of the covariance matrix of the predicted heads or single nodal variances can be used as measures of predictive uncertainty depending upon the objectives of the hydraulic model. Data collection experiments can be designed using a sensitivity based heuristic analysis by examining the change in uncertainty for system wide tests and critical pipe for individual pipe tests. An example system is analyzed to select calibration demand conditions.

1 INTRODUCTION

One of the most important problems concerning the use of mathematical hydraulic network simulators is to determine whether the model is capable of representing the physical system under study. Proper calibration of model parameters is not an easy task. Most calibration techniques have been deterministically based and little attention has been devoted to the impact of uncertainties in calibration efforts. Additionally, data collection is nearly universally overlooked because it requires great effort and is a costly process. However, this additional cost may be insignificant if compared to the consequences of making poor decisions based on inaccurate results.

Water distribution model calibration can be accomplished by adjusting network parameters so the model results match field measurements. The model should then be verified by comparing model results with a set of conditions that were not used to estimate the parameters. An alternate methodology for calibrating water distribution network models is discussed herein. A series of statistical procedures are linked to consider the effect of measurement uncertainty on the parameter estimates, to assess the impact of parameter uncertainty on model prediction and thus assist in defining data collection conditions based on model prediction uncertainty.

2 SOLUTION METHODS

2.1 Calibration Procedure

Network parameter estimation is performed by *ad hoc* calibration procedures, explicit methods, or optimization algorithms. The underlying idea is to adjust pipe roughness coefficients and consumer demands in order to force predictions of pressure heads and pipe flows to agree with measurements taken at a few locations in the system. Walski (1983, 1986), Bhave (1988) and Rahal et al (1980) have suggested improved *ad hoc* schemes. Explicit methods, also known as analytical or direct, solve a system of *n* nonlinear equations which describe the network hydraulics for a set of *n* unknowns which can include the desired parameters (Shamir and Howard 1968 and Ormsbee and Wood 1986).

These procedures, however, do not consider the quality of input values used for the calibration algorithm. They all attempt to find unique values for parameters without assessing their reliability. Araujo and Lansey (1991) evaluated the impact of input measurement uncertainty on the pressure heads predicted by a network simulation model.

A methodology for the calibration of water distribution systems considering uncertainties can be formulated from three basic steps (Araujo and Lansey 1991): Parameter Estimation, Calibration Assessment, and Data Collection. Lansey and Basnet (1991) applied a gradient-based optimization algorithm to estimate the unknown parameters in which the difference between the observed and computed values of pipe flows, nodal pressure heads and tank levels are minimized. The model assumes that the measurements are exact and focuses on minimizing the sum of the squares of the differences. Mathematically the *objective function* ***J*** is

$$J_{min} = \sum_{l=1}^{L} \sum_{i=1}^{I} (H^{o}_{i} - H^{p}_{i})^2 \qquad (1)$$

where J_{min} = objective function to be minimized, H^{o}_{i} = observed or measured head at node *i*, H^{p}_{i} = predicted pressure head at node *i*, *I* is the number of observation locations and *l* is the number of loads that field data has been collected under. The measurement locations may vary by measurement load.

Estimation alone does not provide the means to evaluate how well the model represents actual system behavior. Validation is usually accomplished by comparing model results with field measurements that were not used for parameter estimation. Assessing the impact of the parameter uncertainty on model predictions can improve on validation by determining the uncertainty in predicting performance for critical demand conditions. Assessment demands can be high demands that stress the system at critical nodes. These measures are based on the variance of the predicted pressure heads. The first order second moment (FOSM) analysis can efficiently and accurately estimate the covariance matrix of model output. The FOSM estimates are found by approximating a function with a Taylor series expansion around the mean value of the parameters and dropping higher order terms (Benjamin and Cornell, 1970).

The method is applied twice. First, it is used to approximate the covariance matrix of the roughness coefficient, Cov(C), (Yeh and Yoon 1981, Mallick and Lansey 1994, and Bush and Uber, 1998):

$$\mathrm{Cov(C)} = \sigma^2 \cdot [\frac{\partial H_M}{\partial C}] \cdot [\frac{\partial H_M}{\partial C}]^T \tag{2}$$

where σ^2 is the variance in measured head. This is also known as D-optimality. The matrix of the sensitivities of the pressure heads relative to the roughness coefficients is estimated analytically or by a numerical approximation as used here:

$$[\frac{\partial H_M}{\partial C}] = [\frac{H_M{}^L - H_M}{(C + \Delta C) - C}] \tag{3}$$

where $H_M{}^L$ is the head computed by simulation using the perturbed roughness values, H_M is the present best estimate of the computed pressure heads, C is the roughness coefficients, and ΔC is the error introduced in C. Since the purpose of a model is to determine the predicted head accurately, a second analysis of uncertainty is performed using Cov(C) to compute the covariance matrix of model predicted heads H_p. A first order approximation is also used to estimate Cov(H_P):

$$\mathrm{Cov}(H_p) = [\frac{\partial H_P}{\partial C}]^T \cdot \mathrm{Cov(C)} \cdot [\frac{\partial H_P}{\partial C}] \tag{4}$$

The trace of Cov(H_p), the sum of the absolute values of the diagonal terms, is used to represent overall model prediction uncertainty. The predicted heads are estimated for a demand condition selected by the modeler, to match the model's intended purpose.

2.2 Data Collection

If the reliability level based on FOSM analysis is unacceptable, it is necessary to decrease the uncertainty (variance) in model output by improving the knowledge of the estimated parameters. The problem is to determine the best (or at least worthwhile) field conditions (how much flow to induce and at which node) and to identify measurement type and locations that will provide the most useful information. If all nodes are not monitored, locations of accurate pressure head estimates should also be identified. Improving estimates of individual pipe roughnesses may also provide useful data but are beyond the scope of this paper. Heuristic procedures for designing field experiments for system wide tests are described below.

2.3 Global System Tests

A typical field test is to take pressure head measurements at selected points in the system under estimated demand conditions. A two step heuristic procedure is developed to identify useful field conditions under which to take measurements. The first step is to select a potential network loading pattern under which measurements should be taken. The second step is to identify the critical nodes where demands must be accurately estimated or measured. To do so, relative calibration improvement is measured by a reduction of the assessment measure (variance) after the proposed measurement load is used as available calibration information. Robustness is quantified by looking at the sensitivity of the assessment measure to small variations in the nodal demands during that load or nodal pressure head measurements.

To identify the measurement loading condition, the improvement of the calibration measure, M (i.e., the trace of the covariance matrix of the predicted pressure heads, $Tr(Cov(H_{pi}))$) is examined for l potential demand conditions through numerical simulation. The demand set can consist of different demands or the same demands with different measurement locations. A sensitivity vector $\underline{S}$ is formed:

$$\underline{S} = [\ \Delta M_1, \Delta M_2, \ldots, \Delta M_i, \ldots, \Delta M_l\ ] \tag{5}$$

$$\text{where } \Delta M_i = \Delta\text{trace } Cov(H_p)_i = \text{trace } Cov(H_a) - \text{trace } Cov(H_p)_i$$

The trace of $Cov(H_a)$ is the assessment measure at the current stage of the calibration. Each term, trace $Cov(H_p)_i$, corresponds to a prediction of the assessment measure that will result if the potential measurement load Q_{pi} is induced.

$Tr(Q_p)$ is estimated by calibrating the system with the previously measured conditions and the ith potential demand. The new condition is estimated numerically and included in the measurement set. The variance of the new condition for the measured nodes is the predicted measurement error based on the measurement devices. Clearly, this error should be less than the present model error.

Once the vector $\underline{S}$ is available, the potential load $\underline{Q_{pi}}$ corresponding to the largest ΔM_i is selected as the new measurement load. The load $\underline{Q_{pi}}$ is the desired field condition that should be induced when conducting the data collection. In the case when there is no potential load that contributes to improving the assessment measure, the modeler may try to improve estimates of C for individual pipes or accept the model as it is while recognizing the uncertainty in the model predictions. Data availability and accuracy will limit the accuracy of the model calibration.

A second sensitivity vector $\underline{S_q}$ can be determined after identification of several good measurement conditions. For a selected node j, this vector is expressed by:

$$\underline{S_q} = [\ \Delta M_{i1}, \ldots, \Delta M_{ij}, \ldots, \Delta M_{iN(J)}\ ] \tag{6}$$

where $\Delta M_{ij} = M - M_{p(ij)}$, M = improvement in the trace for the selected load, and M_{pj} = improvement in the trace for the selected load with small perturbation in the demand at node *j*. Each term in $\underline{S}_q$ corresponds to the difference of $Tr(Cov(H_a))$ and the trace obtained if the potential measurements load Q_p is used to estimate parameters *C* with the flow at node *j* increased by Δq_j. A similar vector can be developed for variation of the mean or variance of the pressure head. In this case, locations for precise (as is possible in water systems) measurements would be identified. When all nodes are not monitored, which is practically the case, the terms above can help identify which nodes should be monitored for pressure head.

3 APPLICATION OF DATA COLLECTION PROCEDURE

The pipe network shown in Figure 1 consists of 16 pipes, 12 nodes and a source reservoir with fixed grade at elevation 115.8m (includes pump energy in pipe 1). The total system demand is 0.267 cubic meters per second (cms) under normal demand conditions. Measurements under 4 loading conditions are considered to estimate the unknown pipe roughness coefficients *C* for all pipes. These loads are referred to as: Normal (N), Peak (P), Slack (S), and Fire at node 3 and 8. The peak and slack demand conditions are set by increasing and decreasing normal average daily demand by 40% and 60%, respectively. The fire fighting condition is produced at node 3 and 8 at 0.127cms with consumer withdrawal at all other nodes reduced by 80% of normal. $Tr(Cov(H_a))$ is obtained from calibration assessment with parameters estimated using 4 measurement loads.

In order to implement a system wide test, five potential network loadings have been determined as feasible conditions to be induced during field experiments. They form the array Q_p. Each load corresponds to a vector Q_{pi} that is composed of individual nodal demands q_{pij}, where *j* represents the node number. Table 2 shows the loading. The computed sensitivity vector ($\underline{S}$) shows that the potential load Q_{p2} is most sensitive with largest decrease in $Tr(Cov(H_p))$ i.e., the largest of ΔM_i. The potential load Q_{p2} is then perturbed by 0.057cms to predict the uncertainty measure (Table 3). An increased perturbation resulted in the largest decrease in Tr(Cov(C)). The two step heuristic procedure can be repeated with potential (sensitive) load Q_{p2} supplementing the original 4 measurement loads. Again, throughout the data collection process, estimated *C*'s are used to determine improvements in the assessment measure.

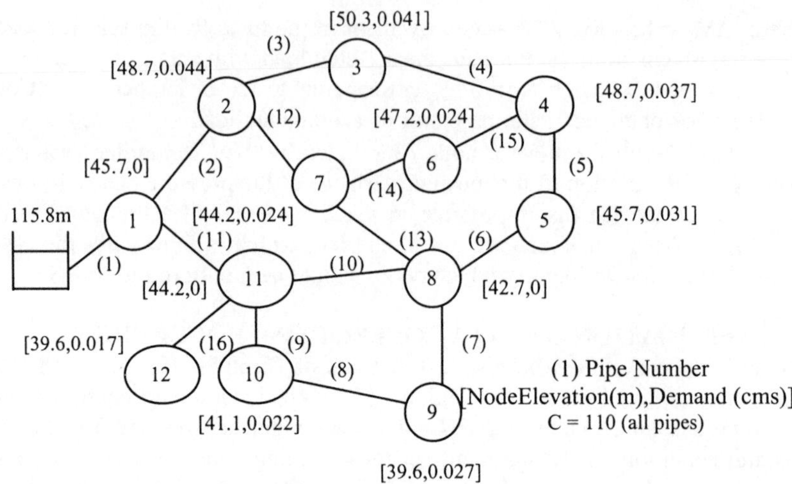

Fig. 1 Application Network

Table 1 Network Specification

Pipe Number	Length (m)	Diameter (m)	Pipe Number	Length (m)	Diameter (m)
1	3048.0	0.610	9	1066.8	0.305
2	1524.0	0.457	10	670.6	0.381
3	1524.0	0.406	11	1981.2	0.457
4	1676.4	0.356	12	1524.0	0.356
5	1066.8	0.305	13	1676.4	0.305
6	1676.4	0.356	14	914.4	0.356
7	1371.6	0.305	15	1219.2	0.305
8	762.0	0.152	16	1219.2	0.406

Table 2 Comparison of Assessment Measures for Potential Measurement Loads

Prediction of Uncertainty Measures				
Potential Load	Fire Node	Fire Demand (cms)	$Tr(Cov(H_p))$ (m^2)	Sensitivity Vector $\underline{S}$ (m^2)
Q_{p1}	1	0.142	33.3	46.2
Q_{p2}	7	0.142	23.6	55.8
Q_{p3}	4	0.142	53.9	25.6
Q_{p4}	4, 5	0.142, 0.085	50.7	28.8
Q_{p5}	2, 9	0.142, 0.071	52.9	26.6

$Tr(Cov(H_a)) = 79.5\ m^2$

Table 3 Adjustment of Nodal Demands for the Selected Measurement Load Q_{p2} (Demand Perturbed by +/- 0.057cms)

Prediction of Uncertainty Measures	
Node 7 Demand (cms)	Tr(Cov(C)) (m^2)
0.142	23.7
0.198	13.2
0.085	84.2

Second loading condition with largest decrease in Tr(Cov(C)) is induced.

4 CONCLUSIONS

A sensitivity based heuristic method is developed to study the uncertainty caused by measurement and estimation errors in water distribution network calibration. The procedure considers three components of the modeling process: parameter estimation, calibration assessment and data collection. The process accounts for the uncertainty in measurements, their impact on model parameters and the effect of these uncertainties on the outputs of the network simulator. Parameter uncertainty is transferred to model prediction uncertainty. A measure of calibration accuracy is defined based on the trace of the covariance matrix of the predicted nodal pressure heads. The methodology provides assistance in defining data collection strategies to improve the model predictive ability. Data collection experiments can be designed by examining the change in the assessment measure under different potential measurement conditions.

REFERENCES

Araujo, J., and Lansey, K.E. (1991). "Uncertainty Quantification in Water Distribution Parameter Estimation." Presented at the National Hydraulic Engineering Conference, ASCE, Nashville, TN.

Benjamin, J.R., and Cornell, C.A. (1970). *Probability, Statistics, and Decision for Civil Engineers*, McGraw-Hill, NY.

Bhave, P.R. (1988). "Calibrating Water Distribution Network Models." *J. of Environmental Engineering Division,* ASCE, 114(1), 120-136.

Bush, C.A., and Uber, J.G. (1998). "Sampling Design Methods for Water Distribution Model Calibration." *J. of Water Resour. Plng. and Mgmt.,* ASCE, 124(6), 334-344.

Lansey, K.E., and Basnet, C. (1991). "Parameter Estimation for Water Distribution Networks." *J. of Water Resour. Plng. and Mgmt.,* ASCE, 117(1), 126-144.

Mallick, K.N., and Lansey, K.E. (1994). "Determining Optimal Parameter Dimensions for Water Distribution Network Models." Presented at the ASCE Conference on Water Resources Planning and Management, Denver, CO.

Mallick, K.N., Ahmed, I., Tickle, K, and Lansey, K.E. (1999). "Determining Parameter Dimensions for Water Distribution Networks." (in review)

Ormsbee, L.E., and Wood, D.J. (1986). "Explicit Pipe Network Calibration." *J. of Water Resour. Plng. and Mgmt.,* ASCE, 112(2), 166-182.

Rahal, C.M., Sterling, M.J.H., and Coulbeck, B. (1980). "Parameter Tuning for Simulation Models of Water Distribution Networks." Proc. Instn. Civ. Engrs. Part 2, Sept. 751-762.

Shamir, U., and Howard, C.D.D. (1968). "Water Distribution Systems Analysis." *J. of Hydraulics Div.,* ASCE, 94(HY1), 219-234.

Walski, T.M. (1983). "Technique for Calibrating Network Models." *J. of Water Resour. Plng. and Mgmt.,* ASCE, 109(4), 361-372.

Walski, T.M. (1986). "Case Study: Pipe Network Model Calibration Issues." *J. of Water Resour. Plng. and Mgmt.,* ASCE, 112(2).

Yeh, W. W-G., and Yoon, Y. (1981). "Aquifer Parameter Identification with Optimal Parameter Dimension." *Water Res. Res.,* 17(3), 664-672.

A Clustering Technique for Parameter Estimation in Pipe Flow Network Models

A. Bascià, D. Termini *and* T. Tucciarelli

ABSTRACT

It is well-known that a low number of model parameters corresponds to a small sensitivity in their estimated values to the total error of the computed state variables (water pressures and flow rates) with respect to the measured ones. The clustering criterion normally used to group the unknown physical parameters into a few model parameters is highly subjective and often fails to provide computed values of the state variables that match the real ones within a given tolerance above and below the measured values. The proposed clustering technique is aimed at minimising the number of clusters, called zones, without any violation of fixed upper and lower limits of the computed state variables. The investigated physical parameters of the clustering optimisation problems are, in this first instance, restricted to the pipe resistances, defined as the ratio between the head loss per unit length and the square of the flow rate in each pipe. The methodology is here applied to synthetic examples of non-looped networks, where the effect of different numbers of measures and different tolerance values is investigated. The extension to looped networks is also discussed, but not documented.

1 INTRODUCTORY COMMENTS

Parameter estimation in a water distribution network plays an important role in correctly predicting the performance of the network under different loading conditions. The parameters (roughness and loss coefficients, internal diameters) cannot be measured directly, but have to be estimated on the basis of the measurements of pressure heads and water flow rates (called state variables), taken for a few nodes and pipes of the existing network.

Various techniques have been developed to solve the inverse problem of parameter identification. They can be divided into direct and indirect techniques [Neuman, 1973; Niranjan and Rao, 1996]. For the direct approach [Ormsbee and Wood, 1986; Boulos and Wood, 1990] the hydraulic parameters are considered as dependent variables, computed as the solution of a set of algebraic equations in order to exactly match the measured values. This implies the number of measures used is equal to the number of unknown parameters.

The indirect approach minimises an objective function, where the decision variables are the unknown model parameters. The objective function is usually a norm of the total error, that is the difference between measured and calculated state variables [Bhave, 1988; Ormsbee, 1989; Lansey and Basnet, 1991; Liggett and Chen, 1994, Tucciarelli *et al.*, 1999]. The indirect methods allow the use of a number of measurements larger than the number of parameters. In many calibration models [Hantush and Marino, 1997] the physical parameters are modelled as unknown coefficients of a dynamic system and estimated using Kalman filtering. Kalman filtering is very useful when the system is linear and different measures are available at different times.

It can be shown [Kendall and Stuart, 1973] that the use of least-square indirect methods or Kalman filtering provides an optimal parameter estimation, but this is true only when the measured state variables differ from the computed ones for an error that has a Gaussian normal spatial and temporal distribution. Moreover, it can be shown [Datta and Sridharan, 1994] that the sensitivity of the estimated parameters with respect to the total error strongly decreases along with the ratio between the number of measures and the number of parameters. When the physical parameters are discretized in a very large number of coefficients the information usually available from the measures is totally insufficient for a stable estimation of all the coefficients. This requires the "zonation" or "clustering" of the parameters, that is the choice of few unknown parameters related to all the coefficients with prescribed relationships. On the other hand, a small number of parameters provides a total error that is not consistent with the capability of the simulation model. Because of this, one of the main difficulties in solving the inverse problem is the choice of the model "zonation".

There are not many techniques to choose the "zonation" of a model. A simple one is based on clustering analysis [Kendall and Stuart, 1973], applied to the a-priori estimation of the parameters. Because this type of analysis does not have any relationship with the available measures, it is not possible to predict the effect of its results on the parameters' stability and on the simulation model accuracy. Moreover, there is no guarantee for the computed "zonation" to provide, at the measurement nodes, computed state variables that match the measured ones within a specified tolerance.

In the following, a new clustering technique is proposed. The main advantage of the proposed technique is that it guarantees the existence of at least one acceptable estimation of the unknown parameters, providing state variables that match the measured ones within a specified tolerance.

2 THE PROPOSED PROCEDURE

The main difficulty in the solution of the inverse problem is that real water pressures and flow rates differ from the computed ones for two different types of error. The first one is the measurement error, the second one is the model error.

Indirect methodologies for the parameter estimation assume a known distribution of the total error. Moreover, because one or a few measures are usually available in each measurement point, ergodicity assumption is also required for the distribution moments. In practice, Gaussian normal distribution is almost always assumed for the total error distribution. This assumption, which is questionable for the first type of error, is totally ungrounded for the second one.

Assuming a more pragmatic view of the problem, we propose a different approach where the solution of the inverse problem is found by means of the solution of a different minimisation problem. The physical parameters of the network are related to the decision variables of this minimisation problem with a known transformation. The objective function, the constraints and the decision variables of the problem are aimed to:

1) find a parameter set corresponding to computed water pressures and flow rates that are inside a fixed tolerance above and below the measured values.
2) find a parameter set with a small number of different values, corresponding to a small number of zones. This provides a small sensitivity of the parameters to the total error.

Assume in this first instance steady-state and totally turbulent flow in the network. Also assume that the only unknown parameters are the ratios between the head loss per unit length and the square of the flow rate in each pipe of the network, called resistances. If only water pressure measures are available, such a problem is of the following type:

$$\text{Minimise} \sum_i C_i\left(r_i^+ + r_i^-\right) \qquad (1)$$

$$\text{s.t} \qquad h_j^* - \sigma_j \le h_j(R) \le h_j^* + \sigma_j \qquad j=1,\dots,m \qquad (2)$$

$$0 \le R_j \qquad j=1,\dots,n \qquad (3)$$

$$R_i = \sum_j \left(r_j^+ - r_j^-\right) I_{ij} \qquad (4)$$

$$0 \le r_i^+, r_i^- \qquad i=1,\dots,n \qquad (5)$$

where m and n are respectively the number of measures and the number of pipes of the network, h_j^* is the jth measure, σ_j is the tolerance of the measure, that is the maximum expected total error, R_i is the resistance in each pipe of the network, C_i is any positive coefficient, r_i^+ and r_i^- are the decision variables and are two for each pipe of the network. The elements of the matrix I are zero or one. For each pipe i the element I_{ij} is equal to 1 only if pipe j is inside the path of a fixed tree (called resistance tree from now on) going from the root of the tree to pipe i. If the network is not looped, the root is the node with fixed water head and the tree is the network itself. The relationship h(R) is given by the simulation model.

If the network is not looped, problem (1)-(5) is linear and this guarantees convexity and uniqueness of the solution. Because, in each pipe, both the variables

r_i^+ and r_i^- have a cost but opposite effect on the constraint function, either the first or the second variable will be zero at the solution. Moreover, because the problem is linear, most of the optimal decision variables (all the non-basic and non-slack optimal variables) are equal to zero. Because of the transformation function (4), the number of different resistance values is equal to the number of nonzero optimal decision variables. This implies that the solution of the minimisation problem (1)-(5) corresponds to a small number zones, that increases with the number of measures and with the decrease of the tolerance value σ. When the network is looped the problem is no longer linear, but a similar behaviour can be obtained for the solution of optimisation problem (1)-(5), by choosing the resistance tree obtained from the real network by disconnecting, in each loop of the network, one pipe from its initial or final node. The best results, not documented here, are obtained by choosing the node that is likely to divide two pipes of the loop with different resistances.

3 APPLICATION TO A SYNTHETIC CASE OF A NON LOOPED NETWORK

The proposed procedure has been applied to the synthetic example shown in Figure 1. The network is composed of 31 pipes, one external and 31 internal nodes. The water demands and the topographic elevations of each node, the diameters and the lengths of each pipe are known. The resistance of each pipe has been estimated assuming the roughness coefficient given by a log-normal distribution with fixed mean and variance.

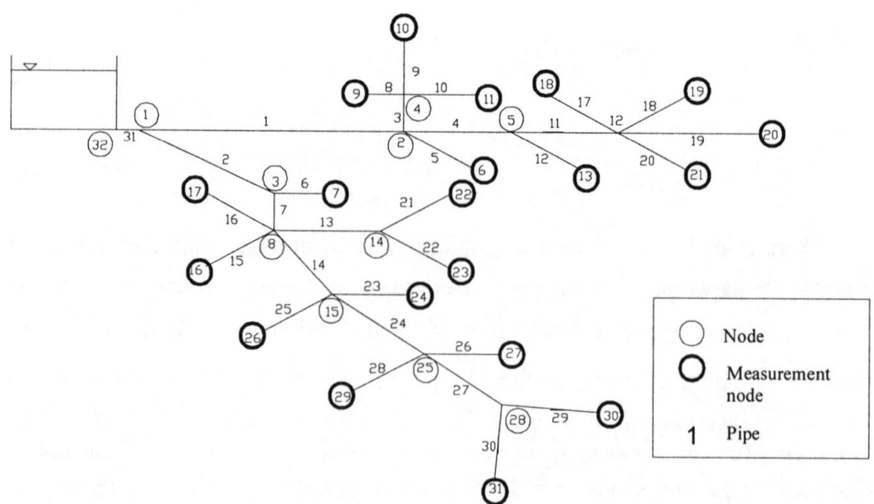

Fig. 1 Example Pipe Network

This assumption, along with the hypothesis of totally turbulent motion, allowed the Colebrook-White formula to be applied to determine the "real" resistance values R_j reported in Table 1. The pressure heads computed by the simulation model in each node of the network with the "real" resistances are reported in Table 2.

The procedure has been applied by using a different number of measures and a different tolerance constant for each node. In order to analyse the ability of the model to correctly estimate the state variables with a given number of measures and a decreasing level of tolerance, the pressure heads calculated by the model have been first compared with the pressure heads reported in Table 2.

Table 1 Resistance coefficients

Pipe	R ($m^{-6}s^2$)	Pipe	R ($m^{-6}s^2$)	Pipe	R ($m^{-6}s^2$)
1	2.99	11	421.47	21	3035.57
2	3.35	12	434.04	22	2829.42
3	413.72	13	3118.20	23	2855.69
4	451.32	14	2969.01	24	3050.76
5	438.14	15	3174.16	25	3169.18
6	3.10	16	2877.44	26	3039.65
7	3.34	17	436.00	27	3012.61
8	421.63	18	451.46	28	2931.61
9	435.04	19	430.22	29	3010.89
10	429.90	20	410.40	30	3174.73
				31	3.44

Table 2 Pressure heads

Node	Pressure head (m)	Node	Pressure head (m)	Node	Pressure head (m)
1	2.98	11	29.15	21	55.68
2	24.64	12	45.80	22	43.62
3	21.68	13	47.29	23	42.65
4	22.22	14	34.89	24	23.77
5	37.34	15	20.08	25	14.89
6	36.59	16	28.47	26	20.02
7	28.68	17	21.32	27	18.61
8	26.65	18	40.77	28	19.86
9	18.21	19	53.69	29	15.65
10	26.20	20	57.73	30	25.73
				31	22.59

Twenty measurement nodes (see Figure 1) were considered to solve the inverse problem. In Fig. 2 the average relative error between computed and 'real' pressure heads in the internal nodes not used for the measures is reported for each value of assigned tolerance. As shown in Figure 2, a precision of about 5% is achieved assuming a tolerance of 1 m. The precision strongly increases with decreasing value of tolerance.

In Fig. 3 the results of the procedure also show that the number of zones increases with the number and the precision of the measures. On the other hand, Fig. 4 shows the opposite behaviour for the stability of the estimated parameters. This stability has been associated with the average coefficient of variation (SVC) of each parameter R_j [Tucciarelli and Termini, 1998]. The SVC has in this case no statistical meaning, but it is a good measure of the stability because it is proportional to both the state variable sensitivity and the measure tolerance.

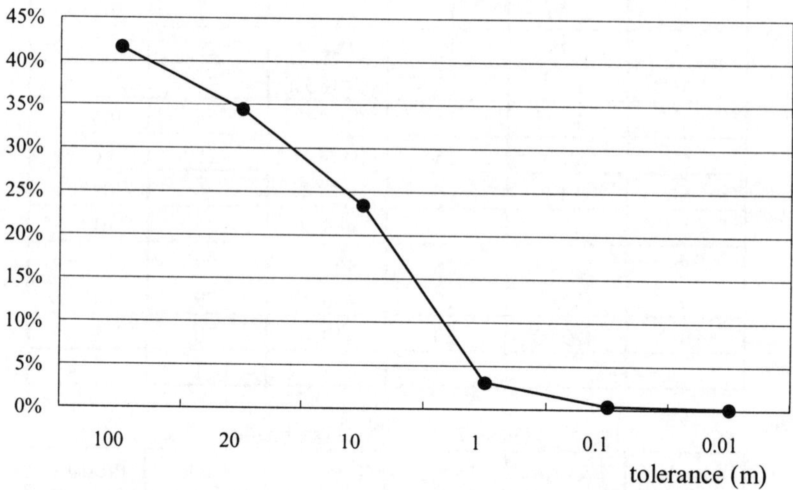

Fig. 2 Relative differences between computed and 'real' pressure heads

The behaviour of the SVC coefficient shows that matching the historical data is not always the best criterion to obtain a good model calibration. In this sense the measure tolerance can be thought of not only as the maximum expected measurement error, but also as the maximum inaccuracy that we can tolerate by the simulation model to guarantee stability. For this reason it can be useful, using a larger available number of measures, to increase the tolerance in order to obtain the same number of zones but a lower SVC. For example to the 'zonation' Z (see Fig. 3), obtained using 12 measured pressure heads and assigning a tolerance of 0.01m, corresponds the SVC coefficient shown in Fig. 4. If five new measures were available, in order to maintain the same value of SVC coefficient, it would be necessary to tolerate a greater measurement error, equal to 1 m.

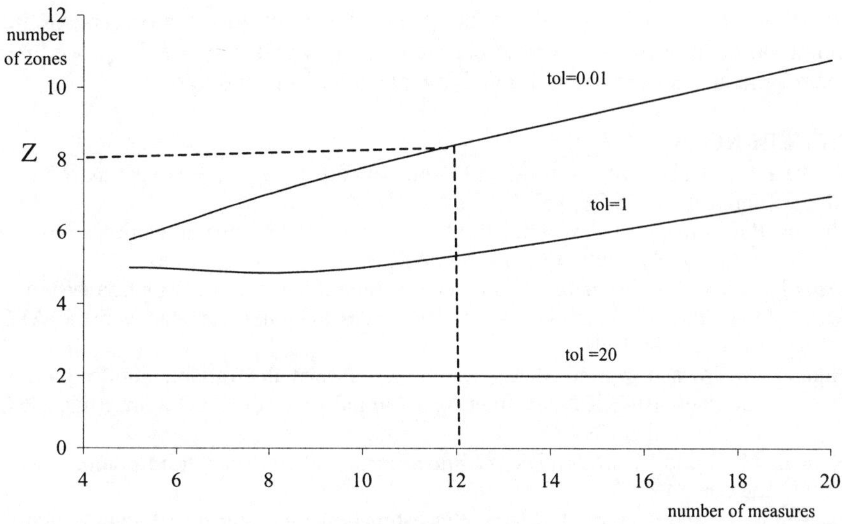

Fig. 3 Number of zones obtained varying the number of measures and the tolerance

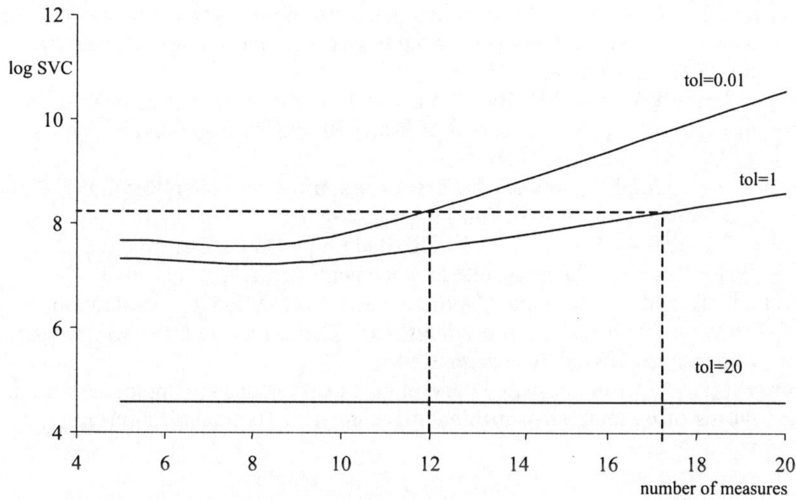

Fig. 4 Estimation error corresponding to different model 'zonations'

4 CONCLUSION

A new clustering technique that provides an optimal estimation of the model 'zonation' is proposed. It allows an estimation of the unknown parameters corresponding to computed state variables that match the measured ones within a specified tolerance. The procedure has been extensively applied only in the simple case of non-looped networks with only water pressure as measurements and pipe resistance as unknowns. Moreover, only one flow distribution and loading condition have been used for the calibration. The application of the procedure to more general synthetic examples and to real networks is in progress.

REFERENCES

Bhave P.R., "Calibrating water distribution network models", Journal Environmental Engineering, ASCE, vol. 114, n. 1, 1988.

Boulos P.F. and Don J. Wood, "Explicit calculation of pipe-network parameters", Journal of Hydraulic Engineering, ASCE, vol. 116, n. 11, 1990.

Datta R.S.N. and K. Sridharan, "Parameter estimation in water-distribution systems by least squares", Journal of Water Resources Planning and Management, ASCE, vol. 120, n. 4, 1994.

Hantush M. M. and Marino M.A., "Estimation of spatially variable aquifer hydraulic properties using Kalman filtering", Journal of Hydraulic Engineering, ASCE, vol. 123, n. 11, 1997.

Kendall M.G. and Stuart A., *The advanced theory of statistics*, third edition, Griffin, London, 1973.

Lansey K. E. and C. Basnet, "Parameter estimation for water distribution networks", Journal of Water Resources Planning and Management, ASCE, vol. 117, n. 1, 1991.

Liggett J.A. and L. Chen, " Inverse transient analysis in pipe networks", Journal of Hydraulic Engineering, ASCE, vol. 120, n. 8, 1994.

Neuman S. P., "Calibration of distributed parameter groundwater flow models viewed as a multiple-objective decision process under uncertainty", Water Resources Research, vol. 9, n. 4, 1973.

Niranjan Reddy P.V. and P.V. Rao, "WLS Method for parameter estimation in water distribution networks", Journal of Water Resources Planning and Management, ASCE, vol. 122, n. 3, 1996.

Ormsbee L.E., "Implicit network calibration", Journal of Water Resources Planning and Management, ASCE, vol. 115, n. 2, 1989.

Ormsbee L.E. and A.M. Don J. Wood, "Explicit pipe network calibration", Journal of Water Resources Planning and Management, ASCE, vol. 112, n. 2, 1986.

Tucciarelli T. and D. Termini, "Optimal valves regulation for calibration of pipe networks models", *Proceedings of Hydroinformatics '98 Conference*, Copenhagen, 24-26 September 1998.

Tucciarelli T., A. Criminisi and D. Termini, "Leak analysis in pipeline systems by means of optimal valve regulation", Journal of Hydraulic Engineering, ASCE, vol. 125, n. 3, 1999.

Leakage Detection in Single Pipelines Using Pressure Wave Behaviour

D. Covas *and* H. Ramos

ABSTRACT

Leakage reduction and control is an up to date issue that requires the attention of specialists and authorities. The analysis of hydraulic transients could be a useful tool for leak detection and location either in water conveyance systems or in water distribution networks. In this paper, different approaches to leakage detection and location in single pipelines, not yet used or implemented in practice, are presented.

Firstly, a mathematical formulation is presented, based on principles of mass and energy conservation, to locate one or more leaks in single pipelines under steady state conditions. Afterwards, transient analysis is used to detect leaks and sudden accidental bursts. The idea is to analyse how the transient regime is affected by the leak's presence and if its effect is sufficient to locate the leak both by the first wave reflected by the leak and by the free damped pressure oscillation generated in the transient regime.

Both approaches are recommended for leak location in single pipelines, nevertheless, its application in real and more complex systems, such as water distribution networks, is still a challenge for future generations.

1 INTRODUCTION

Leakage detection and location dates from the middle of the 20th century, with the exploitation of oil transportation pipelines. More and more, it is an up to date issue that requires the attention of specialists and authorities, in order to meet consumption growth as well as to assure future sustainable development. Leakage can vary between 10 to 40% of the total water volume distributed, which could be of great economic importance.

Nowadays, there are plenty of practical techniques and methods applied in real systems to detect and locate leaks, specially in water distribution systems, some of which have not shown feasible results. The idea of this paper is the presentation of different approaches to leakage detection and location in single pipelines, not yet used in practice.

The first approach based on mass and energy conservation principles, presents rapid formulae to locate leaks in single pipelines, as long as the physical characteristics of the hydraulic system and the steady conditions of flow and pressure in the extremities of the pipeline are known. The case of a pipeline with single and multiple leaks is studied.

Furthermore, hydraulic transient analysis could be a useful tool for leak detection and location in pipelines which could form a conveyance system for toxic or pollutant products with significant environmental impacts. The idea is to analyse how a transient regime, induced by a manoeuvre to close or open a valve, is affected by a leak's presence. The same conclusions could be applied to the occurrence of a sudden burst that causes a pressure variation, which travels upstream and downstream through the pipeline as a pressure wave. The detection and analysis of these transient pressures allows, in certain circumstances, the leak to be located in the hydraulic circuit.

2 LEAKAGE PHENOMENON

A leak, or a rupture, behaves like a small orifice with free discharge to the atmosphere or to the soil with a constant pressure that depends upon the type (e.g. coarse or fine particles) of the soil existing around the pipe and its compactness.

Neglecting the kinetic energy inside the conduit, the difference in elevation between the conduit axis and the leak, and the percolation velocity in the ground, the leak discharge Q_L (m^3/s) can be estimated by:

$$Q_L = C_D A \sqrt{2g\left(\frac{p_i}{\gamma} - \frac{p_o}{\gamma}\right)} \tag{1}$$

where C_D is the discharge coefficient, A is the orifice area (m^2), p_i and p_o are the inside and outside pressures (Pa), g is the gravitational acceleration (m/s^2) and γ the volumetric weight of the fluid (N/m^3). In Fig. 1, the main physical characteristics of the pipeline, the leak and the flow are presented.

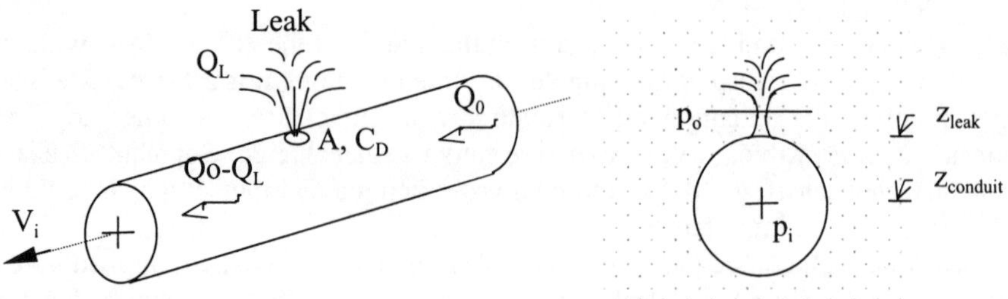

Fig. 1 Buried pipeline with a leak - main physical characteristics

3 LEAK LOCATION BY STEADY STATE FORMULAE

3.1 Introduction

In this section, a mathematical formulation is presented to determine the approximate location of one or more leaks in single pipelines, knowing the physical characteristics of the system (namely the inner diameter, the length and roughness of the pipe) and the steady state conditions of flow and pressure at the extremities of the pipeline. The formulation presented is based on the principles of mass and energy conservation applied to a control volume defined by a stretch of pipeline. The aim of this formulation is to deduce mathematical expressions that allow estimation of a leak or a rupture location in single pipelines. Nevertheless, it is not considered an alternative method for leak detection or location in water supply systems, but just another useful tool to help in this.

3.2 Isolated Leak

Based on the fundamental principles of mass and energy conservation, it is possible to deduce mathematical expressions that identify the leak's location in a single pipeline, under steady state conditions. The formulation is based on the control volume of a pipeline, delimited by upstream and downstream sections, $S_{①}$ and $S_{②}$. In Fig. 2, the fundamental variables used in this formulation are presented.

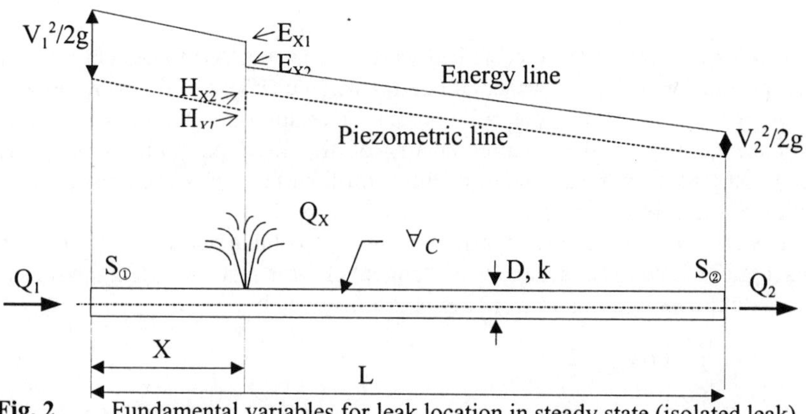

Fig. 2 Fundamental variables for leak location in steady state (isolated leak).

Mass conservation principle

Designating by M_S the total mass of the fluid inside the control volume, in the absence of flow sources, such as springs or wells, the time variation of the total mass is zero. Through Reynolds Theorem,

$$\frac{dM_S}{dt} = 0 \Rightarrow \int_{\forall_C} \frac{\partial \rho}{\partial t} d\forall + \int_{S_C} \rho \overline{V} \cdot \overline{n} \; dS = 0 \tag{2}$$

where ρ is the water density; $\overline{V}$ is the velocity vector in each control surface section; $\overline{n}$ is the external normal unit vector at the control volume surface. The integral volume, which represents the inside mass variation of the control volume per time unit, is zero, on the assumption that the fluid is incompressible. Hence, the expression is simplified to the surface integral, which stands for the mass flux through the boundary of the control volume per unit time

$$-\rho V_1 S_1 + \rho V_2 S_2 + \rho V_X S_X \quad \Rightarrow \quad Q_X = Q_1 - Q_2 \tag{3}$$

where Q_X is the leak's discharge; Q_1 and Q_2 are the flows at sections 1 and 2.

Energy Conservation Principle

Considering E_S the total energy of the fluid inside the control volume, for an inertial stationary reference linked to it, the time variation energy of the system is equal to the heat exchange, Q_c, with the exterior through the control surface per unit time, less the work, W, carried out by the non-conservative forces per unit time. Once again, through Reynolds Theorem

$$\begin{aligned} &\frac{dE_S}{dt} = \dot{Q}_c - \dot{W} \\ &\int_{\forall_C} \frac{\partial}{\partial t}\left(\rho\frac{V^2}{2} + \rho gh + \rho u\right) d\forall_C + \int_{S_C}\left(\rho\frac{V^2}{2} + \rho gz + \rho u + p\right)\left(\overline{V}\cdot\overline{n}\right) dS = \dot{Q}_c - \dot{W} \end{aligned} \tag{4}$$

where W is the sum of the mechanical work of concentrated forces (W_v), the work of tangential forces in the surface of control volume (W_T), and the work referenced to the movement of the control volume in relation to the inertial stationary reference (W_I); u is the internal energy of the fluid per unit mass; z is the topographic level of each section of the control surface; p is the pressure in each section of the control surface.

Under steady state conditions, the integral volume is zero. The only non-conservative force in the system is the tangential force that acts along the surface of the conduit, which results in work per unit time given by:

$$\dot{W}_T = \int_{S_C} \overline{\tau}\cdot\overline{V} dS \tag{5}$$

τ being the shear stress at the surface, defined, for Newtonian fluids, by:

$$\tau \propto \frac{dV}{dy} \tag{6}$$

Assuming uniform velocity distribution in each section of the conduit, the shear stress is zero, since the velocity does not vary over the cross-section. On the other hand, should a real diagram of velocity be considered, the velocity close to the surface is zero. Therefore, with both assumptions, the tangential force does not produce work.

The term $\dot{Q}_c$ is the flux of heat exchanged with the exterior of the pipe per unit time, namely the flux by the continuous and local head loss defined by:

$$\dot{Q}_c = \left[-\int_0^L J\gamma Q dx - \sum_i K_i \frac{V_k^2}{2g}\gamma Q_k \right] = -J_1 X\gamma Q_1 - J_2(L-X)\gamma Q_2 - \sum_i K_i \frac{V_k^2}{2g}\gamma Q_k \quad (7)$$

where J is unit head-loss; K_i is the coefficient associated with the local head-loss expressed in terms of kinetic energy; k is the index associated with sections $S_{①}$ and $S_{②}$.

The internal energy term of particles, u, is included in the unit head loss, J. In that way, the integral surface is given by:

$$\int_{S_C}\left(\rho\frac{V^2}{2}+\rho gz+p\right)\left(\vec{V}\cdot\vec{n}\right)dS = -\gamma Q_1\left(\frac{Q_1^2}{2gS_1^2}+H_1\right)+\gamma Q_2\left(\frac{Q_2^2}{2gS_2^2}+H_2\right)+\gamma Q_X\left(\frac{Q_X^2}{2gA_{ef}^2}+H_{Xext}\right) \quad (8)$$

where H_i is the piezometric head in the section i, $H=p/\gamma+z$; H_{Xext}, the piezometric head at the exterior of the leak, $H_{ext}=p_{ext}/\gamma+z$; A_{ef}, the leak's effective area ($A_{ef}=C_D.A_L$). Using equation (1) for the discharge law through an orifice:

$$A_{efL} = \frac{Q_X}{\sqrt{2g\left(E_{X1}-H_{Xext}\right)}} = \frac{Q_X}{\sqrt{2g\left(H_1+\frac{Q_1^2}{2gS_1^2}-J_1X-\sum_i K_i\frac{Q_1^2}{2gS_1^2}-H_{Xext}\right)}} \quad (9)$$

Replacing former equations into (4), and assuming constant pipe characteristics ($S_{①}=S_{②}=S$), the leak's location is defined by the distance X to the section $S_{①}$:

$$X = \frac{H_1 - H_2 + \frac{Q_1^2}{2gS^2} - \frac{Q_2^2}{2gS^2} - J_2L - \sum_i K_i\frac{Q_k^2}{2gS^2}}{J_1 - J_2} \quad (10)$$

In this way, knowing the physical characteristics of the conduit and steady state conditions of flow and pressure at upstream and downstream sections, it is possible to estimate the location of a leak based on equation (10) and to determine the effective leak area by equation (9).

3.3 Multiple Leaks

In the presence of more than one leak along the pipe, expressions (9) and (10) point to an equivalent leak, given by the equivalent effective area A_{Leq} and by the distance of the equivalent leak X_{eq}, which are related to the real leaks' areas A_{efLi} and distances X_i by the following expressions (Fig. 3)

$$A_{Leq} = \sum_i A_{efLi} \tag{11}$$

$$X_{eq} = \frac{\sum_i X_i \left(J_{Xi} - J_{Xi+1}\right) + L\left(J_{Xn+1} - J_{eq\,2}\right)}{J_{eq\,1} - J_{eq\,2}} \tag{12}$$

where J_{Xi} is the real unit head-loss upstream of leak i and J_{Xn+1} the real unit head-loss downstream of leak n. Nevertheless, these formulas do not allow determination of the exact location or the effective area of each leak since there are only two equations and $2n$ unknown variables (n being the number of leaks). To solve this indeterminacy, it is necessary to have measurements of discharge and pressure in n-1 pipe cross sections. Each leak must be located between two consecutive measurement sections, or the pipe length must be sufficiently short in order to be represented by an equivalent leak.

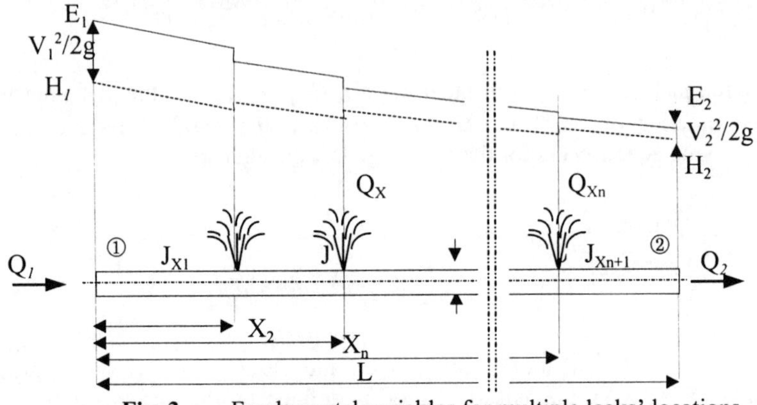

Fig. 3 Fundamental variables for multiple leaks' locations

The results from using expressions (9) and (10) in the presence of multiple leaks are ambiguous, requiring several accurate measurements in other sections of the pipeline, which would not be feasible in practice.

3.4 Final remarks

The rapid formulae presented allow the location and the estimation of the effective area of one or more leaks. Nevertheless, their application might present some practical difficulties, particularly in water distribution networks, due to the following factors:

- the accuracy in determining the roughness and the inner diameter of the conduits, particularly in old systems where the cross section is reduced due to encrustation phenomena;
- the measurement of flows in water distribution networks requires the installation of flowmeters;
- the distinction between real leaks and branch flows;
- ambiguous results in the presence of multiple leaks.

It must be emphasised that the use of a rapid formula related to the location of an isolated leak might have particular application to water or oil conveyance systems in which flows and pressures are well controlled in several sections. If a sudden rupture occurs, continuous monitoring of the system in conjunction with this formula might allow its rapid location.

4 LEAK LOCATION BY TRANSIENT ANALYSIS

4.1 Introduction

In the following sections, the possibility of detecting and locating leaks based on the transient pressures induced by a change in a valve setting is analysed and discussed. The idea is to analyse how the transient regime is affected by a leak both in the first reflected wave and in the general behaviour of the transient. The question is whether the leak is capable of altering the transient in such a way as to allow its location.

In order to solve this problem, two different approaches were followed, both involving the analysis of pressure (or piezometric head) time variation upstream of the valve. One approach focuses on the time analysis of the first wave reflected by the leak, allowing the location and the quantification of the leak. The other approach focuses on the frequency analysis of pressure variations, in order to identify, if possible, through the discrete frequency spectrum, the frequency associated with the leak.

Both approaches were followed in a theoretical way, having limitations and presenting good results under particular circumstances. Their practical application is still to be tested in the laboratory and in the field, other difficulties of implementation being expected.

4.2 Time analysis

Consider a stretch of pipeline with an upstream reservoir, a valve downstream and with a steady flow Q_o (Fig. 4). Let L be the length of the pipe, a the wave propagation celerity, S the cross-section of the pipe and T the closure time of the valve. *Rapid manoeuvres* are performed when the valve closure time T is less than the elastic reflection time T_{ER} of the hydraulic system (T_{ER}=$2L/a$), otherwise, they are called *slow manoeuvres*.

For rapid manoeuvres of total closure, neglecting the continuous head-loss and the kinetic energy variation, and assuming linear closure manoeuvre, the maximum surge head near the device generating the transient can be estimated by Frizell-Joukowsky formula (Fig. 4)

$$\Delta H = \frac{a}{g\,S} Q_0 \qquad (13)$$

Whenever there is a singularity in the conduit, be it a junction, a bend or a leak, it induces a reflected wave that interferes with the normal shape of the transient. Nevertheless, part of the incident wave is transmitted upstream (Fig. 5). Neglecting the energy head-loss along the conduit and in the singularity, the amplitude of the reflected wave added to the amplitude of the transmitted wave equals the amplitude of the incident wave (Fig. 5)

$$\Delta H=\Delta H_t+\Delta H_r \tag{14}$$

In the case of the singularity being a leak, during the occurrence of a transient regime, it undoubtedly reflects a wave that induces a sudden dampening in the upsurge next to the valve. For quasi-instantaneous manoeuvres, the location of the leak is calculated by the total time taken for the incident wave to arrive at the leak, be reflected and arrive back at the valve again, t^*. The distance of the leak from the valve, X, is given by

$$X = \frac{a\,t^*}{2} \tag{15}$$

If the closure time T is greater than time t^*, it might not be possible to identify in the shape of the transient pressure the instant of arrival from the first reflected wave. On the other hand, should the leak be far from the valve, its reflected damped wave might be attenuated due to continuous head-loss and the leak might not be identified.

Concerning the magnitude of the leak, quantified, for instance, by the relative leak flow $Q_{L0}/(Q_0-Q_{L0})$, this might be estimated by the amplitude of the reflected wave.

In order that the leak reflects a pressure wave, it is necessary that the leak's discharge suffers a variation. If the leaks' discharge was insensitive to the pressure/head variation during the transient regime, there would be no reflection and it would be impossible to determine the leak's position, either using time or frequency analysis. In fact, the leaks' discharge varies, in principle, like an orifice with constant opening, and, hence, the reflected wave would have the following theoretical value:

$$\Delta H_r = B\frac{Q_{L0}}{2}\left(1-\alpha\sqrt{1+\beta\frac{\Delta H_t}{H_0}}\right) \tag{16}$$

where $\alpha=1$ and $\beta=0$, when the leaks' discharge is constant and equal to Q_{L0}; $\alpha=1$ and $\beta=1$, when the leaks' discharge varies according to a quadratic law. When $\alpha=0$, the leak's discharge becomes zero with the arrival of the first reflected wave, increasing the upsurge ($\Delta H_r>0$).

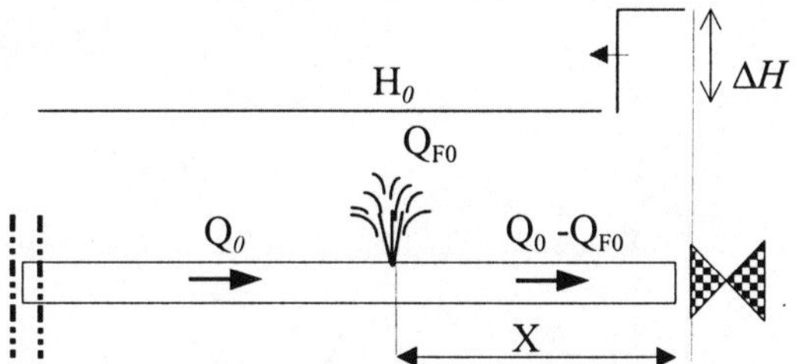

Fig. 4 First upsurge generated by instantaneous valve closure

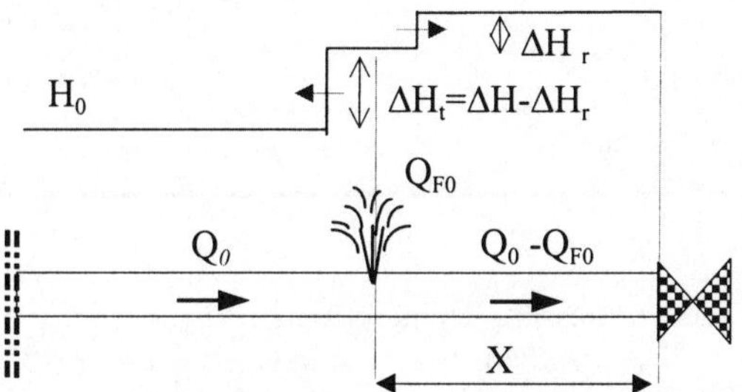

Fig. 5 Reflected and transmitted waves from the leak

The leak reflected wave ΔH_r, as it reaches the closed valve, incurs a 100% reflection, inverting its form completely. Hence, the sudden dampening ΔH_d observed in the pressure shape next to the valve is double the leak reflected wave and is related to flows Q_0 and Q_{L0} by the following expression (assuming $\alpha=1$, $\beta=1$)

$$\frac{Q_{L0}}{(Q_0 - Q_{L0})} = \frac{\Delta H_d}{\Delta H}\left[1 - \sqrt{1 + \frac{\Delta H + \frac{\Delta H_d}{2}}{H_0}}\right]^{-1} \tag{17}$$

In the presence of several leaks, each location and magnitude can be determined by the respective damped waves, as long as the closure time of the valve is less than time t^* of each leak (Fig. 6).

In Fig. 6, the pressure-time variation in the valve section is presented for an instantaneous valve closure manoeuvre and for two different leak magnitudes (5 and 9.5%) and locations (X/L=10 and 30%). The difference between the observed relative damping $\Delta H_d/\Delta H$ next to the valve, and the corresponding relative leak flow and real leak flow is due to the fact that in the development of formula (17) continuous head losses were neglected, and they become increasingly important the greater the distance of the leak from the valve and the smaller the leak's discharge.

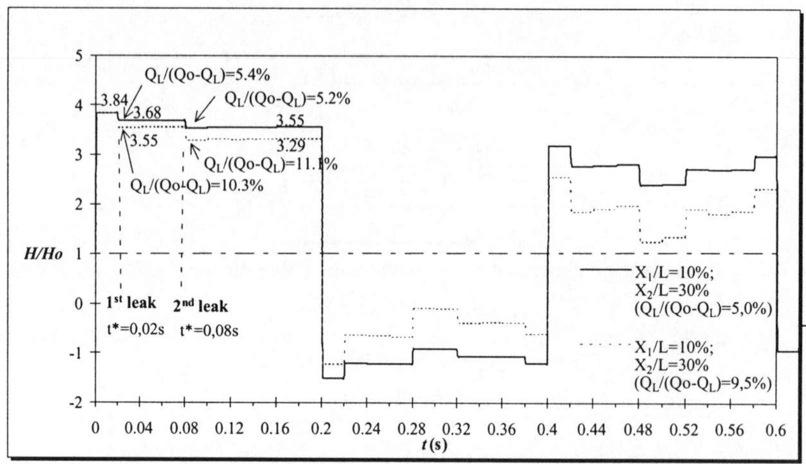

Fig. 6 Example of the sudden pressure damping induced by a leak (pressure-time variation obtain by mathematical simulation using Method of Characteristics, MoC)

The use of a quasi-instantaneous valve closure manoeuvre has proved to be a method theoretically exact and efficient, although it presents serious difficulties in practical implementation:

- The difficulty in performing quasi-instantaneous manoeuvres in valves.
- Even if quasi-instantaneous manoeuvres were performed, they would put at stake the security of the pipeline either by the occurrence of high or low pressures.
- Every singularity of the system, such as junctions, nodes and bends, reflects incident waves giving misleading information on the location of real leaks.
- In rigid conduits (made of steel, iron or concrete) the wave celerity is very high and, if the leak is near the measuring section, it might be difficult to identify the time instant t^*.

The use of this type of approach to detect leaks in real systems is only recommended for single pipelines, where all the physical characteristics are perfectly known. Even so, an instantaneous valve closure manoeuvre should always be preceded by the hydraulic simulation of the integrated system, in order to determine its rupture risk and to safeguard the security of the system.

4.3 Frequency analysis

Frequency analysis is based on the generation of a transient regime on the hydraulic system and the identification of dominant frequencies of the damped free oscillation by a Fourier analysis. This method is very sensitive to the topology of the system, being a potential future method for leak location in single linear pipelines, such as water conveyance systems.

In a single system with an upstream reservoir and a valve downstream, without leaks, the closure of a valve would generate a free damped oscillation with the main frequency associated with the position of the reservoir and given by

$$f = \frac{a}{4L} \tag{18}$$

This oscillation would correspond to the fundamental harmonic (Fig. 7(a)). The pressure-time response of the system next to the valve is presented in Fig. 7(b), as well as the discrete Fourier transform Fig. 4.4(c).

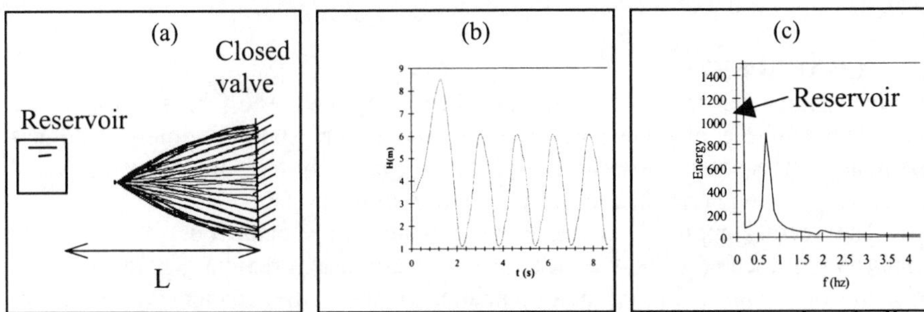

Fig. 7 Single pipeline without leaks: (a) pressure oscillation along the pipeline (fundamental harmonic); (b) pressure variation at downstream section; (c) Fast Fourier Transform - spectrum analysis

In the presence of a leak the same system, under the same perturbation (the valve closure), would have a response function, in terms of pressures close to the valve, with two main associated frequencies. The first frequency is related to the reservoir location, and a second one to the leak, given by expression (18) by making $L=X$. The second frequency has the third harmonic associated (Fig. 8) and can also be detected by Fourier analysis.

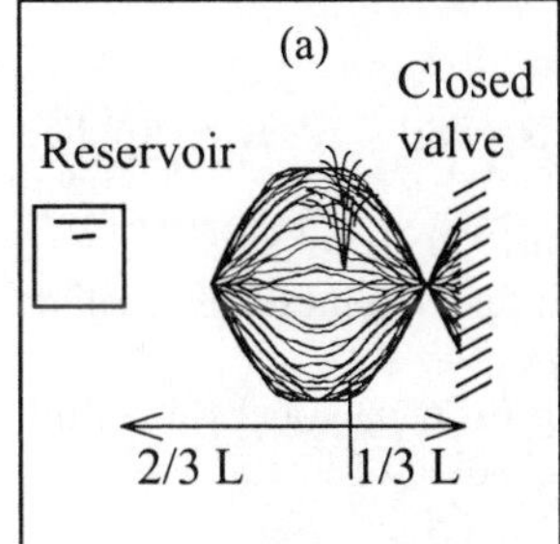

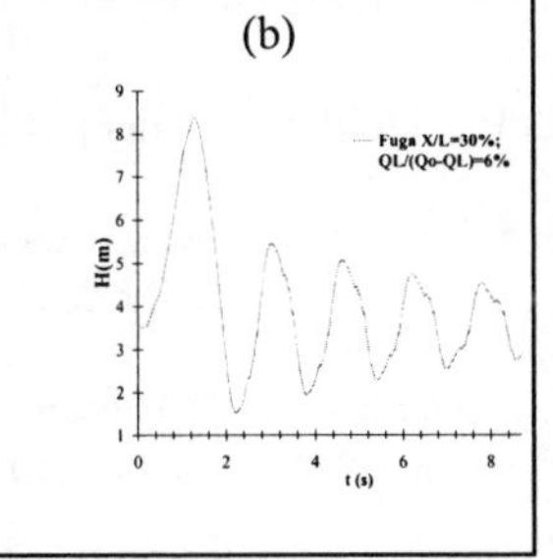

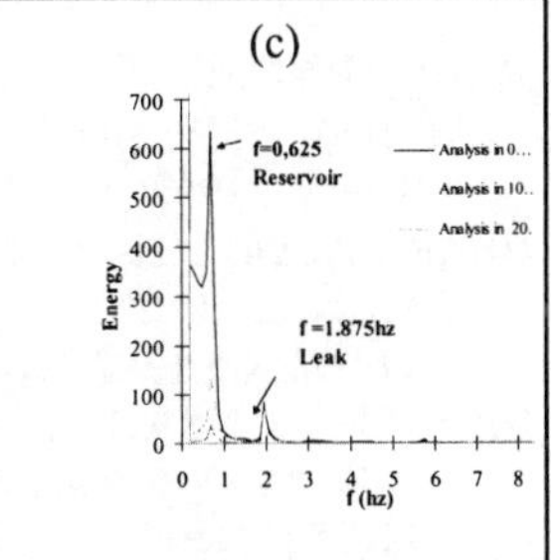

Fig. 8 Single pipeline with one leak located (X/L=30%): (a) pressure oscillation along the pipeline (3rd harmonic); (b) pressure variation in the downstream section; (c) Fast Fourier Transform - spectrum analysis

In this way, the leak's location would be estimated by the second dominant frequency f determined by Fourier analysis

$$X = \frac{a}{4f} \tag{19}$$

where a is the wave celerity and X the distance from the valve.

Although, theoretically, this method seems quite efficient, it has not yet shown good results in practice, as the frequency associated with the leak is dependent on the leak's position, the leak's magnitude, the presence of other singularities and the free damping of the system.

More research work is going to be done, in an EU Project SMT4-CT97-2188, based on experimental tests, in order to clarify the possible reasons for the failure of this method and under which circumstances it proves to have good results for leak detection and location.

5 CONCLUSION

Several approaches to leak location have been presented. The first is based on the knowledge of the pipeline characteristics and steady state conditions, and has particular relevance to those water or oil conveyance systems in which flows and pressures are well controlled in several sections of the systems.

The second approach is based on pressure wave behaviour. A theoretical analysis is presented of how a leak can affect a transient regime generated by the closure of a valve in a single pipeline. If the valve manoeuvre is quasi-instantaneous the first reflected wave from the leak allows the approximate location and the quantification of the leak. For other types of manoeuvre, the latter method is not applicable, and it is necessary to resort to frequency analysis of the spectrum of pressure energy applied to the section at the valve. Once again, the results of this method depend on the topology of the system and characteristics of the leak, and do not show good results in all circumstances. More research work is still to be done.

Both approaches are recommended for leak location in single pipelines, nevertheless, their application to real and more complex systems, such as water distribution networks, is still a challenge for future generations.

REFERENCES

ALMEIDA, A. B.; KOELLE, E. - Fluid Transients in Pipe Networks, Elsevier Applied Science, 1992

BEAR, JACOB - Hydraulic of Groundwater. MacGraw-Hill.1979.

BILLMAN, L.; ISERMANN, R..– Leak Detection Methods for Pipelines. Automatica, Vol 23,N.3 1987

CABRERA, ENRIQUE; VELA, ANTONIO F. - Improving Efficiency and Reliability in Water Distribution Systems. Water Science and Technology Library. Valencia, 21-25 November 1994.

CABRERA, ENRIQUE; MARTINEZ, F. - Water Supply Systems. State of Art and Future Trends. Valencia, 19-23 October 1992.

CHAUDHRY, M. H. - Applied Hydraulic Transients, Van Nostrand (2nd Edition), 1987.

COULBECK, BRYAN - Integrated Computer Applications in Water Supply - Volume I - Methods and procedures for systems simulation and control. Research Studies Press, 1993.

COVAS, D. - Leaks' detection and location in water distribution networks - Hydrodynamic Analysis's Method (M.Sc. thesis in Portuguese). Lisbon, July 1998.

WYLIE, E. B.; STREETER, V.L. - Fluid Transients, McGraw-Hill, 1978.

Optimal Logger Density in Water Distribution Network Calibration

W. de Schaetzen, M. Randall-Smith, D.A. Savic *and* G.A. Walters

ABSTRACT

If hydraulic models of water networks are to be used for predictive purposes with any degree of confidence, they generally need to be calibrated against field data. Model variables, often pipe roughness values, are adjusted until the predicted model values of flows and pressures at key points match those measured in the field.

Field data collection is a labour and equipment intensive task and is therefore expensive. The amount of field data required has always been determined by subjective judgement and it was therefore recognized as being beneficial to establish a rationale for determining the most appropriate levels.

The authors show in this paper using a case study from the UK that the problem can be addressed by taking the initially available field data set of the model and then selecting five reduced sets. Comparison of maximum and average deviations from model predictions, using the full data set, at selected reference nodes was made, allowing some initial conclusions to be drawn about the optimal level of field testing.

A Genetic Algorithm (GA) approach was used for the task of deriving a good calibrated network model for every case as it produces calibrations of a very consistent quality, removing much of the subjectivity that even the most gifted network modeller must employ.

Whilst further models would need to be similarly analysed to obtain statistically robust conclusions, the initial results suggest that, for the two models examined, there would be scope for substantially reducing the number of pressure monitoring points except where local detail and accuracy are highly critical.

1 INTRODUCTION

The number of monitoring points required to achieve a satisfactory quality of calibration is considered to be somewhat arbitrary and, given the high costs of field testing, a justifiable reduction in the monitoring density is highly desirable. One of the key strengths of using GA to calibrate models is the consistency of the results, with much of the subjective judgement associated with manual calibration being avoided. GA calibration therefore provides an ideal means of making sensible comparisons between results produced through calibrations arising from different numbers of monitoring points.

2 MODEL CALIBRATION USING GENETIC ALGORITHMS

To apply GA to the network calibration problem, the objective is to find the solution that has the minimum overall difference between field and model values of flow and pressure. This is achieved by performing a large number of runs of the network model using trial values of pipe roughnesses, which are adjusted throughout the process using the principles of natural evolution. The basics of the GA process are described in the following section. The decision variables are generally defined to be the roughness values for each pipe in the system. These can be expressed as the "k" value in the Colebrook-White formulation, or as the Hazen-Williams "C" coefficients. All pipes can have individually variable roughness values, or groups of pipes can be pre-selected to have a common variable roughness, the selection being based on diameter, age, material, location or a combination of these factors. Several snapshot simulations are used, generally at maximum and minimum demand times, and at an average time during a typical 24-hour cycle. At each measurement point and for each snapshot, the difference between simulated and observed data (head and/or flow) is calculated, and an overall error value for the whole network is minimised.

2.1 An Introduction to Genetic Algorithms

Evolution Programs, of which Genetic Algorithms are probably the best known type, are general artificial evolution search methods based on natural selection and the mechanisms of population genetics. They emulate nature's very effective optimisation techniques of evolution which are based on preferential survival of the fittest members of the population, the maintenance of a population with diverse members, the inheritance of genetic information from parents and the occasional mutation of genes. These algorithms are best suited to solving combinatorial optimisation problems that cannot be solved using more conventional operational research methods. Thus, they are often applied to large, complex problems that are non-linear with multiple local optima.

The analogy with nature is established by the creation in the computer of a set of solutions called a population. Each individual in a population is represented by a set of parameter values (e.g. a set of pipe roughness values) which completely describe a single solution to the defined problem. These are the encoded chromosomes.

The initial population of perhaps 100 solutions, which is usually chosen at random, is allowed to evolve over a number of generations. At each generation, a measure (fitness) of how good each chromosome is with respect to an objective function (e.g. the minimisation of the difference between field and model values of flow and pressure) is calculated. This is achieved by decoding the binary strings of the chromosome into parameter values which are then used to evaluate the objective function, (e.g. by running a network simulation model).

Then, based on their fitness values, individuals are selected from the population and recombined, producing offspring that make up the next generation. This is the recombination operation that is generally referred to as crossover because of the way genetic material crosses over from one chromosome to another.

The probability that a chromosome from the original population will be selected to produce offspring for the new generation is dependent on its fitness value. Fit individuals will have a higher probability than less fit ones resulting in the new generation having on average a higher fitness than the old population. This selection process parallels nature's processes of 'survival of the fittest'.

Mutation also plays a role in the reproduction phase, although it is not the dominant process. In GA each bit (or gene) is allowed a small probability to randomly mutate. If the probability of mutation is set too high, the search degenerates into a random process rather than the desired collective learning process.

3 PRESSURE MONITORING POINTS DENSITY ANALYSIS

The analysis is defined in two stages. The original model with the full calibration data set is firstly calibrated using a Genetic Algorithms approach at three different snapshots (i.e. representing the night, peak and average time conditions during the 24 hours calibration period). The calibrated model using the full calibration data set is defined as the "reference model". The "reference model" provides the total head values for the "references nodes". Those nodes are generally located at the model extremities where the greatest sensitivity to change might be expected.

Then the different reduced field data set models are calibrated using a Genetic Algorithms approach. The corresponding quality of each calibrated model with a reduced calibration data set is evaluated by comparing their predicted pressures at the selected references nodes with those obtained by the calibrated model using the full data set.

3.1 Case Study and Results

The numbers of pressure monitoring points or loggers used within the calibration of each reduced set and for two models are summarised in Table 1.

Table 1 Monitoring Point Reduced Sets

	Model 1 (2665 properties - 4 reference nodes)					
Reduced Sets (RS)	RS 1 [x]	RS 2	RS 3	RS 4	RS 5	RS 6 [y]
# loggers	26	15	10	5	3	0
# prop / logger	102.5	177.6	266.6	453.3	888.3	∝
	Model 2 (3822 properties - 16 reference nodes)					
Reduced Sets (RS)	RS 1 [x]	RS 2	RS 3	RS 4	RS 5	RS 6 [y]
# loggers	55	33	22	10	5	0
# prop / logger	69	115.8	173.7	382.2	764	∝

[x] original full set
[y] also called the "0" set

A "0"logger set model is included in the analysis to give an indication of the results which would be achieved with no calibration but with realistic Hazen Williams 'C' coefficients assigned to all the pipes depending on the material.

Key reference nodes are used for predicted pressure comparisons to assess the sensitivity of a calibrated model to reductions in the number of pressure measurement points used. Two criteria are taken into consideration when identifying the reference nodes:

- They should be nodes near the model extremities where pressure variations are greatest;
- They should be adequately distributed around the network.

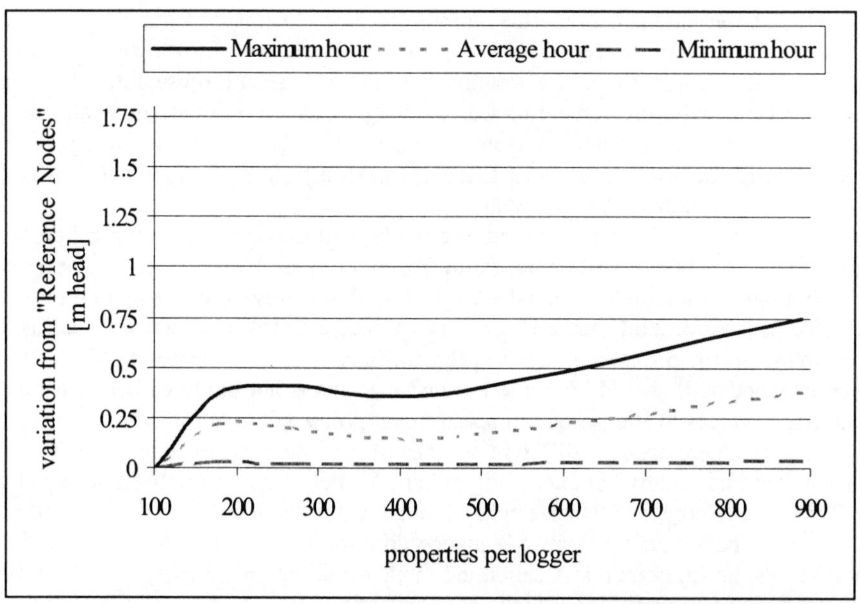

Fig. 1 Model 1 Results (Mean values at 4 Reference Nodes)

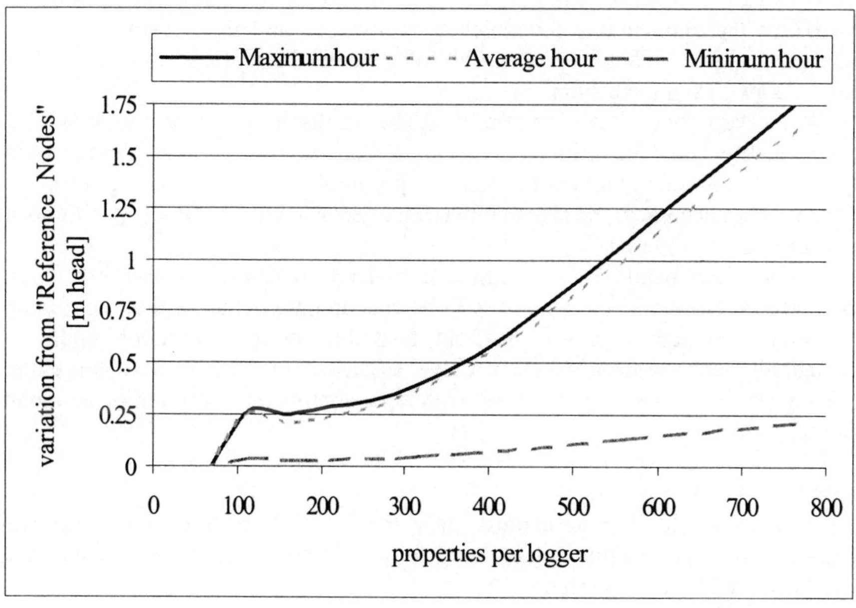

Fig. 2 Model 2 Results (Mean values at 16 Reference Nodes)

Two observations about the minimum and average demand condition snapshots, which are valid for both models, could be made from the results as illustrated in Figures 1 and 2. Minimum hour is not a good representation of how the error varies with the number of selected loggers, as the flows are generally very small at night and the analysis therefore insensitive. The errors at average hour have exactly the same profile as those at maximum hour except that they are smaller in magnitude by about 0.25m.

The degree of accuracy required of a model will to some extent depend on the applications to which it will be put following calibration. However, in carrying out a preliminary evaluation of the adequacy of model quality when calibrated on the reduced sets, a 'general purpose' measure of acceptability has been adopted as a maximum of 1m difference between the full and reduced data set results at the reference nodes. It should be borne in mind that this is not the error between field and model values at the pressure measurement points (field and model values are within 1m). There are two different sources of error to be considered which may have either a cumulative or cancelling effect. The performance criterion defined by WRc of 2m difference between field and model values is therefore still valid. [06]

In conclusion, adopting an acceptance criterion of 1m variation from the full data set results suggests that a calibrated model of acceptable quality could still be achieved with a property logger ratio in excess of 500.

4 SUMMARY AND CONCLUSIONS

The adequacy of a model calibrated with fewer and fewer monitoring points will depend on the applications for which it is intended and the accuracy which is therefore required. However, a number of general observations and conclusions may be drawn from the results.

Where very local detail is not critical, the results suggest that a high level of model quality could be achieved through a Genetic Algorithm calibration with significantly reduced numbers of monitoring points. In determining the adequacy of the model's accuracy, the risk that maximum errors could apply in specific local areas must be considered.

Further data would still be valuable to lend statistical credibility to these observations. However, there appears to be encouraging evidence that the number of pressure loggers deployed in field tests for model calibration could be substantially reduced from the current level assumed in this exercise (one pressure data logger for every 75 properties) without a significant compromise in model quality.

ACKNOWLEDGEMENT

The authors would like to express their thanks to Yorkshire Water Services Limited, and to Jay Naylor in particular, for providing the network models which have been used in the case study.

REFERENCES

[01] Ormsbee, L.E. and Lingireddy, S, Calibrating hydraulic network models, *journal AWWA*, Vol. 89, No. 2, pp. 42-50.

[02] Savic, D.A. and G.A. Walters, (1995), Genetic Algorithm Techniques for Calibrating Network Models, *Centre For Systems And Control Engineering*, Report No. 95/12, School of Engineering, University of Exeter, Exeter, United Kingdom, p.41.

[03] Savic, D.A. and G.A. Walters, (1995), Place of Evolution Programs in Pipe Network Optimization, *Integrated Water Resources Planning for the 21st Century*, M.F. Domenica (ed.), American Society of Civil Engineers, New York, USA, pp. 592-595.

[04] Savic D.A. and G.A. Walters, (1997), Evolving Sustainable Networks, *Hydrological Sciences Journal*, Vol. 42, No. 4, pp. 549-564.

[05] Walters G.A, Savic D.A, de Schaetzen W, Atkinson R.M., Morley M. (1998) Calibration of Water Distribution Network Models Using Genetic Algorithms, *in Hydraulic Engineering Software VII*, Computational Mechanics Publication, pp 131-140.

[06] Water Research Centre, 1989, Network Analysis - A Code of Practice, Water Research Centre, Swindon, England.

A Spatial Sampling Procedure for Physical Diagnosis in a Drinking Water Supply Network

O. Piller, B. Bremond *and* P. Morel

ABSTRACT

This article presents a tool enabling the physical diagnosis of a drinking water supply network operating in a quasi dynamic (extended period simulation mode). The method adopted pays particular attention to the positioning, nature and number of measurements.

The hydraulic models which solve the classic equations of mass and energy conservation pre-suppose good knowledge of the roughness of the pipes and of consumption. They enable calculation of the flows at each edge and the heads at each node.

Regarding the estimation of the parameters of the model, given the few measurements available, simplifications must be made to reduce the number of unknowns. One and the same roughness may be attributed to a class of edges of the same age and of the same material, and a mean demand may be attributed to a given type of consumer, for example a homeowner or an industrialist. The problem to be solved is a non linear least squares problem with constraints.

The solving of this inverse problem depends very much on the points of measurement and on the proper restitution of the hydraulic parameters. To overcome this difficulty, we propose selecting the set of measurements to minimise the influence of errors of measurement on the estimation of the state vector, while ensuring the observability of the network.

A solution is found using a greedy algorithm, which has the advantage of being in polynomial time.

1 INTRODUCTORY COMMENTS

The problem of identifying the network lies in finding the best estimate of the roughness and demand parameters from a set of measured values. The quality of the estimation of these parameters depends on the position, number and nature of the measurements. This choice must ensure the observability of the network (Cohen et al., 1987) and also prevent small errors in measurement resulting in an incorrect estimation of the parameters.

The purpose of this study, carried out at the Bordeaux regional centre of the CEMAGREF[1], is to propose an algorithm for the choice of measurements which satisfies these two conditions.

2 MATERIAL AND METHOD

2.1 The forward problem

After simplification of the distribution network and its representation by a set of edges and nodes (i.e. a graph), it is generally assumed that the average behaviour of a network is a succession of states of equilibrium in a steady state. The breathing of the tanks, the major transits, the major losses of head over the considerable lengths of pipes are all thus reproduced.

In what follows, time dependency is taken to be piecewise constant (i.e. a step-function). The diameters and lengths of the pipes are given. The water levels in the tanks, $h^f(t)$, are calculated on the basis of the previous time steps. $h^f(0)$ is given.

The forward problem consists of giving d(t), the vector of demand at the non-tank nodes, and Chw, the vector of the roughness coefficients. To determine the flow vector, q(t) and the vector h(t) of the total heads at the non-tank nodes, the system $(HQ)_t$ is solved :

$$(HQ)_t \qquad \begin{cases} A.q(t)+d(t)=0_n \\ \xi(t)-{}^tA.h(t)-{}^tA^f.h^f(t)=0_a \\ \xi(t)=\xi(Chw,q(t)) \end{cases}$$

where A is the incidence matrix reduced to non-tank nodes, A^f the incidence matrix reduced to tank nodes, and ξ the vector of the loss of head at the edges.

The first two linear equations express the conservation of mass and of energy. The last, or constitutive equation is an empirical law relating the flow and the roughness coefficient to the loss of head (e.g., the formula of Hazen-Williams). It will be supposed that the vector function $\xi \to \xi(Chw,q)$ was chosen to be continuously differentiable, strictly monotone in q, and to be the gradient of norm-coercive function: c in q (i.e.: c(q) tends to infinity as $\|q\| \to +\infty$).

Since Collins et al. (1978), we know how to set $(HQ)_t$ in an energy minimisation problem, which makes it possible, on the one hand, to evidence that $(HQ)_t$ has a unique solution q(t), h(t), and on the other hand, to apply descent algorithms which globally converge.

2.2 The inverse problem

It is now supposed that the roughness and demands are no longer known. The number of measurements of flow or head available is generally much smaller than the number of unknowns for a demand load (i.e., the number of edges plus the number of simple nodes).

---

[1] CEMAGREF, Agricultural and environmental engineering Research.

As demonstrated by Ormsbee (1989), an initial method for overcoming the lack of measurements consists of considering several loads. It is also worthwhile reducing the number of unknowns. The same roughness is often attributed to a class of edges of the same age and of the same material; consumers of the same type will be grouped together in a class of demand, for example the class of domestic consumers or the class of industrial consumers. This very frequently results in going from five hundred to fewer than ten unknowns.

If the roughness class vector is written as C, and the demand class vector as D^i (for $i=1,...,\tau$), the state vector x for τ periods is written as follows :

$$V(\delta x) = (S.J)^{+}.{}^{t}(S.J)^{+}$$

Solving the τ systems $(HQ)_t$ gives the vector y, of the flows and heads, as a function of x :

$$y = f(x) \tag{1}$$

The estimation of x, $\hat{x}$ is then sought, which minimises the sum of the square errors between the values observed, y^{obs}, and the values calculated by (1). The residuals are multiplied by a weight, w_{ij}, which takes account of the accuracy of the measurement apparatus and generally of the confidence in the result of the measurement. The least squares problem is solved:

$$\text{(CLS) to minimise } g(x) = \frac{1}{2}\left\|y^{obs} - S.f(x)\right\|_W^{\,2} = \sum_{i=1}^{\tau}\sum_{j\in M_i} w_{ij}.\left(y_{ij}^{meas} - f_{ij}(x)\right)^2$$
$$\text{subject to } x^{min} \le x \le x^{max}$$

where S is the selection matrix of the measurements, and M_i all the flows and heads measured for the period i.

Powell et al. (1987) showed that an automatic correction of the weights w_{ij} according to the weighted residues is more robust for the purposes of estimation. The resolution method that we have adopted is a slight modification of the algorithm of Levenberg-Marquardt (Nash, 1990, p. 211) to take account of all the constraints.

2.3 The problem of the choice of measurements

The measurement errors vector, δy, can now be introduced, and the non linear regression model written, as follows :

$$y^{obs} = S.f(x) + \delta y \text{ with } -\Delta y \le \delta y \le \Delta y \tag{2}$$

Bargiela et al (1989) addressed the problem of quantifying the impact of a measurement error δy on an estimation $\hat{x}$ in a drinking water supply network. The influence of measurement error on estimation is of the first order, δx, after linearisation of (2) around $\hat{x}$:

$$\delta x = \left({}^{t}J.{}^{t}S.S.J\right)^{-1}.{}^{t}J.{}^{t}S.\delta y = (S.J)^{+}.\delta y \tag{3}$$

where J is the Jacobian matrix of f in $\hat{x}$ and $(S.J)^{+}$ gives the pseudo-inverse of S.J within the meaning of Moore-Penrose. To write (3), it is supposed that S.J is of maximum rank (i.e. the network is observable). Also, $\hat{x}$ must be in the interior of constraints of (CLS).

To equilibrate the Jacobian of the model J, two changes of variables are made, the first in the space of the p parameters to be estimated, to homogenise them and the second in the space of the flows and heads, to obtain, by dividing by the a priori accuracy of the measurements, a measurement error vector, δy. All the components of δy are smaller than or equal to one. The influence of δx is thus increased by:

$$\sup_{\delta Y \in \overline{B}_{\infty}(0,1)} \delta x = \left\|(S.J)^{+}\right\|_{\infty}$$

Then, m measurements $m \geq p$ (therefore a selection matrix of the measurement S) are sought, where the influence of the measurement errors δx is minimal, subject to the constraint that the Jacobian of the problem (CLS) thus obtained is of maximum rank (algebraic observability condition). The problem of the choice of measurements is thus written:

(CM) to minimise $H(S) = \|S.J\|_{\infty}$, subject to $\text{rank}(S.J) = p$

Two algorithms were envisaged; an exhaustive algorithm, for small networks (where all the choices are explored) and a greedy algorithm, which has the advantage of being in polynomial time, for larger networks. A (k+1)-nth measurement is added to a choice of k measurements, that are not questioned, to minimise H, on the hypothesis of a maximum rank.

2.4 Simulations and measurement campaigns

We have chosen to illustrate our methodology taking the network of the town of Muret (Haute-Garonne, France). The diagram of the network is shown in Figure 1.

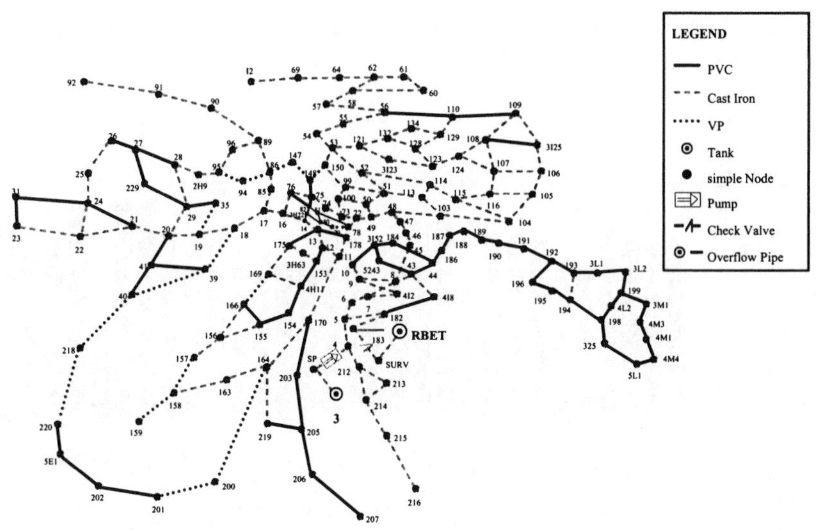

Fig. 1 Graph of the network of the town of Muret

Once in diagrammatic form, the Muret network is composed of 214 edges, 163 simple nodes and 2 tanks. It includes 50 loops and 1 elementary chain between tanks. The pipes are made of three materials, polyvinylchloride (PVC), cast iron, and cast iron with an internal coating (VP, Ville de Paris). Three classes of roughness are taken corresponding to the three materials. Similarly, it appears that two kinds of subscribers are sufficient to describe overall consumption behaviour. The number of parameters to be identified (adjusted) for this network is therefore equal to five, i.e $p = 5$ for a single load.

The automatic choice of measurements for prioritised head measurements is compared to the choice of the experts or to the systematic choice that consists in measuring the inflow and the pressures at the five antenna nodes.

Next we present the results of simulations with a random selection of errors repeated one thousand times according to a uniform law. These error vectors are added to the values calculated by the model, thus creating one thousand possible y^{obs}. CLS is then resolved for each of y^{obs}. The sampling distributions of the parameters obtained by this Monte Carlo method are then represented.

3 RESULTS

The accuracy attributed to the pressure measurements is 0.5 m of water (1 mH_2O = 3.3 ftH_20). The accuracy attributed to the flow measurements is deduced from the accuracy on velocity (of 0.01 m/s) multiplied by the cross section of the pipe. Figure 2 shows the sampling distribution of the head at node 207 in m. A chi-square goodness of fit test does not detect any significant departure from the uniform law, which means that the real head is just as likely to be 179.1 m as 180.1m.

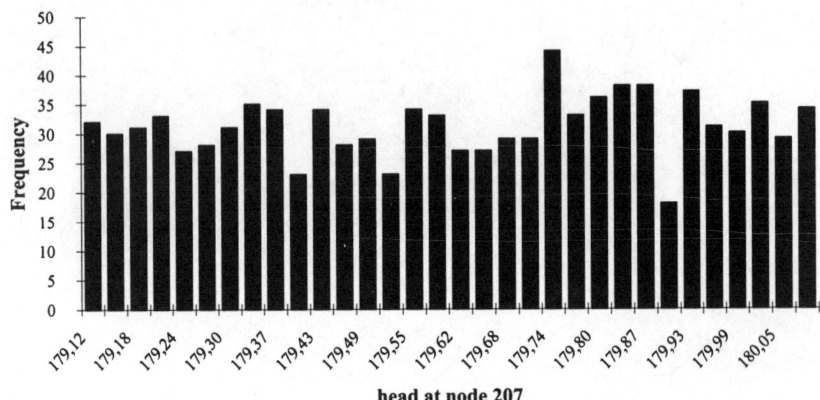

Fig. 2 Sampling distribution of a measurement

The true value of the roughness coefficient (corresponding to the Hazen-Williams formula) of the PVC pipes is $C_{PVC} = 141$ (i.e. approximately roughness of 0.05 mm). Figure 3 shows the sampling distributions of the estimation of C_{PVC}. For the expert choice, it can be seen that the estimations range from 70 to 295, whereas for the automatic choice (determined by CM) they are between 126 and 156.

To assess the efficiency of the two estimators resulting from the two-measurement choice, the mean square error is calculated as the sum of the bias squared and the variance. The scatter around the true value is thus measured, as it can be shown that:

$$\text{M.S.E.} = E(\hat{C}^{PVC} - 141)^2$$

It can be seen from Figure 3 that the most efficient estimator is that obtained with the automatic choice. The relative efficiency of the automatic choice, compared to the expert choice, is 1280/30 = 42.6.

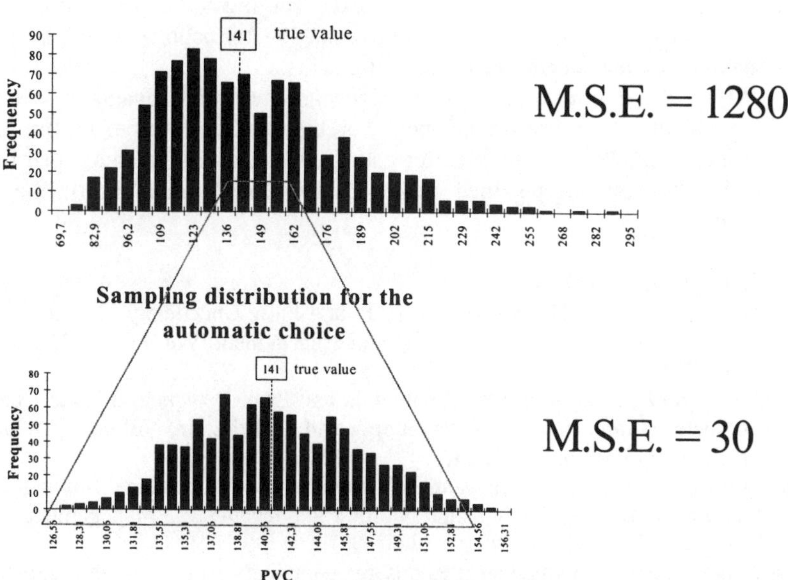

Fig. 3 Comparison of two sampling distributions for the estimation of the roughness of the PVC material

4 DISCUSSION

Special weight can be given to certain measurements in relation to others (e.g. pressure measurements) by attributing to them a smaller measurement error. The equilibration of J, the Jacobian matrix, according to the a priori accuracy of the measurements (cf. § 2.3.) reveals here its full importance.

For an optimal choice of measurement, the calibration of a drinking water supply network can prove to be difficult and time-consuming due to excessively flat piezometric landscape (no loss of head), valves closed when they were thought to be open, etc. For that reason, it is generally decided to open fire hydrants and to undertake sectorisations (closing of valves) to isolate certain zones or to create large transits.

We formed the hypothesis that the measurements errors were uniformly distributed. For measurement errors following a Gaussian law, this gives, for example: $E(\delta x) = 0_p$ and $V(\delta x) = (S.J)^{+}.^{t}(S.J)^{+}$ if $\delta y_i \in idN(0,1)$. The criterion H to be considered here could be the minimisation of the trace of $V(\delta x)$ (as the spectral radius or a matrix norm).

5 CONCLUSION

For the choice of the nature, number and positioning of the required measurements, we propose what is, to our knowledge, an innovative answer. The techniques implemented could be used in other contexts. The method consists of minimising the influence of measurement errors in the state vector estimation, whilst ensuring the maximum independence of the measurements.

Within this framework, a greedy algorithm, with the advantage of being in polynomial time, has been developed. For the resultant choice, the problem of identification of the network is better conditioned. It has been found that in small networks the solution obtained achieved the global minimum of the choice criterion.

REFERENCES

A. Bargiela and G. D. Hainsworth, Pressure and Flow Uncertainty in Water Systems, Journal of Water Resources Planning and Management, Vol. 115, n° 2, p. 212-229, March 1989.

G. Cohen and P. Carpentier, State Estimation and Leak Detection in Water Networks, Computer Applications in Water Supply and Distribution, Volume 1, Sept 1987, Research Studies Press Ltd., UK.

M. Collins, L. Cooper, R. Helgason, J. Kennington and L. Leblanc, Solving the pipe network analysis problem using optimisation techniques, Management Science, Vol. 24, n° 7, p. 747-760, March 1978.

J. C. Nash, Compact numerical methods for computers : linear algebra and function minimisation, Adam Hilger Ltd, Bristol, 1990.

L. E. Ormsbee, Implicit Network Calibration, Journal of Water Resources Planning and Management, Vol. 115, n° 2, p. 243-257, March 1989.

O. Piller, Modelling the behaviour of a network - hydraulic Analysis and a sampling procedure for estimating the parameters, thesis from the University of Bordeaux, defended on 03 February 1995, 288 pages (in French).

R. S. Powell, M. R. Irving, M. J. H. Sterling and A. Usman, A Comparison of Three real-time State Estimation Methods for on-line Monitoring of Water Distribution Systems, Computer Applications in Water Supply and Distribution, Volume 1, Sept 1987, Research Studies Press Ltd., UK.

Inverse Transient Calibration of Water Distribution Systems Using Genetic Algorithms

K. Tang, B. Karney, M. Pendlebury *and* F. Zhang

ABSTRACT

In this study, a new network calibration method was developed and applied using the currently popular idea of genetic algorithms combined with the state-of-the-art transient analysis model TransAM. The genetic approach has been used to represent many kinds of physical and engineering systems, from the modelling of eco-systems to the optimization of pipeline systems. The appropriateness of this approach for calibration lies in its ability to evolve a system of parameters to conform to a selected objective function. In the case of water distribution systems, the objective is to match as closely as possible a predicted response to a response measured in the field. The genetic approach seeks to adjust the physical parameters of the model (friction factors, demands, etc.) until they result in the convergence of the model's response to the behaviour of the actual system. Although traditional calibration approaches are inadequate and generally cannot work well in complex systems, the paper emphasizes that inverse transient approaches also have intrinsic challenges that must be carefully resolved.

1 INTRODUCTION

Almost every water distribution network comprises a unique arrangement of pipes, nodes, pumps, reservoirs, and valves. The peculiarities of geography, topography and history of an area tend to create distinctive arrangements of pipes, a history for the selection of various pipe materials, related challenges with respect to corrosion and scaling, and a unique pattern of demand. Also, all distribution systems undergo continual change, both in the short-term (e.g., as demands shift) and in the long-term (e.g., as hydraulic elements are added to or removed from the system). The challenge (and the strength) of such a system is that it behaves as a whole. Networks, with their complex combinations of pipe loops and redundant elements, produce an overall response in which the individual elements play a relatively minor role in the performance of the whole system. Although this is a highly desirable characteristic from a design point of view, it also makes it very difficult to isolate or extract the behaviour of individual components.

In this study, the recently-developed inverse transient calibration method is applied using the popular idea of genetic algorithms, combined with the state-of-the-art transient analysis model (TransAM). The appropriateness of this approach for calibration lies in its ability to evolve a collection of parameters to conform to a selected objective function. In the case of water distribution systems, the objective is to match as closely as possible a predicted response to a response measured in the field. The genetic approach seeks to adjust the physical parameters (e.g., pipe friction factors and system demands) until convergence is achieved between the model's response and the measured performance of the actual pipeline system.

Experience with the inverse calibration procedure has recently been obtained through studies of three water distribution systems in Ontario, Canada (i.e., at the cities of London and Thunder Bay and the Town of Ajax). The data required to perform the inverse calibration were obtained through a series of field tests conducted in 1997-99. In all cases, the collected data involve high speed pressure values measured during a transient event at strategic locations in the network.

2 GENETIC ALGORITHMS

Genetic algorithms (GAs) have been receiving increasing application in a variety of search and optimization problems. These efforts have been greatly aided by the existence of theory that explains what GAs are processing and how they are processing it. The theory largely rests on Holland's exposition of schemata, his fundamental theorem of genetic algorithms (Holland, 1975), and later work by several others (e.g., Goldberg, 1989).

Genetic algorithms require the natural parameter set of the optimization problem to be coded as a finite length string. Because GAs work directly with the underlying code, they are difficult to fool, since they are not dependent upon continuity of the parameter space and derivative existence. Genetic algorithms work iteration by iteration, successively generating and testing a population of strings. They work from a database of points simultaneously climbing many peaks in parallel, thus reducing the probability of finding a false peak.

Recently, genetic algorithms have received considerable attention regarding their potential as an optimization technique for complex problems and have been successfully applied to pipeline optimization, pump operation, as well as system reliability design. A relatively comprehensive approach for the use of genetic algorithms for steady state pipe network optimization has been developed over the last ten years (Goldberg and Kuo, 1987; Hadji and Murphy, 1990; Murphy and Simpson, 1992; Dandy et al., 1993; Simpson et al., 1993; Murphy et al., 1993; Simpson et al., 1994; Dandy et al., 1996; Halhal et al., 1997; Savic and Walters, 1997).

3 INVERSE TRANSIENT CALIBRATION

Calibrating a hydraulic model of a water distribution system can potentially result in a large cost saving to the utility. Water utilities spend a major portion of their operating budget on energy costs---often 50 percent or more---and typically between 90 to 95 percent of these energy costs are attributed to pumping treated water (Clingenpeel, 1983). Optimizing the operating policies could yield significant savings, and obtaining a calibrated hydraulic model is an important part of an optimized operating policy.

Calibrating a hydraulic model involves accurately simulating actual field pressures and flows for a given point in time or over a specified period of time. Traditional calibration procedures are labour intensive, tedious, and the resulting improvement in accuracy is sometimes questionable. For these methods, a distribution system is brought, or an attempt is made to bring it, to steady state, at which time either SCADA data or manually acquired field data is obtained. Manual changes are then made to the model in an attempt to match the actual steady state conditions. A model of a distribution system has thousands of variables, resulting in excessive work during calibration with often questionable returns. In fact, this type of approach may be quite impractical due to the size and complexity of the system involved. The calibration procedure used in the current analysis was developed by Hydratek Associates at the University of Toronto, building on previous work by Liggett (1994a, 1994b).

In summary, the current process differs from traditional techniques in that the calibration is performed on a non-steady state network. This feature allows for simple on-line collection of thousands of pieces of information, using a handful of high speed pressure transducers (hstp). These data can then be used to effectively solve or calibrate the model. The calibration is completed using a powerful solving technique known as a genetic algorithm.

3.1 Applying Genetic Algorithms

The inverse calibration approach borrows from nature, using its intelligence and wisdom in the simple form of genetics. Nature has been "calibrating" species and individuals to survive in the ecosystem for billions of years. The simple rule followed in nature is to create a large gene pool to allow the organisms or individuals to interact with each other and the environment to determine who are the most appropriate individuals to pass on their genes to the next generation. Through reproduction and occasional mutation, this process generally produces individuals that are better suited to survive in their environment: hence the expression, "survival of the fittest." In the current application, the individuals of a species are not used, but the analogy to genetics is still strong. In fact, a water supply system is viewed as an environment inhabited by "individuals" (data files) that have physical characters such as pipe friction factors, nodal demands, pipe diameters, wavespeeds, and pump and valve operating scenarios. Each artificial individual contains all parameters being calibrated as genes.

The genetic algorithm involves three major steps: gene typing, inverse calibration, and natural selection. The process requires a modelling program for pipe networks, and measured data of transient events in the actual system. Artificial individuals are used to work out the system parameters.

3.2 Step One: Gene Typing

The first step, gene typing, involves encoding all relevant physical values in an uncalibrated data file into binary numbers. The binary number system is a logical choice for gene encoding since a binary number consists of a sequence of bits, each of which takes on a value of either 1 or 0. A binary 1 is used to represent that a particular gene is switched on, while a binary 0 is used to represent that a particular gene is switched off. Consequently, an individual contains a number of genes, each stored as a binary number representing a physical characteristic. Since each characteristic can take on a number of values, the algorithm requires a lower and an upper bound for each parameter. For example, a 7-bit binary number with a possible numerical range of 0 to 127, can be used to represent the friction value of a particular pipe in the system. The following binary number (0010010) is equivalent to a Hazen-Williams friction factor of 111.3 if the upper and lower bounds are 180 and 100 respectively.

3.3 Step Two: Inverse Calibration

In the second step, inverse calibration, the individuals are converted to their actual physical parameter values. Since each individual contains all the physical characteristics of the system, the system must be modelled with a transient analysis computer model. The reason that a transient model is used instead of a steady state model is that a transient is very specific and short-lived, and only one particular set of parameters and/or events is likely to reproduce it in detail. By contrast, a number of different paths can be used to describe a steady state condition which are not necessarily physically unique given the limited amount of data that is usually available. Since a transient wave is specific to the actual system, measurements of the system in terms of pressure heads during a transient event can be used as a check on how well each individual mimics the actual transient. Therefore, in order to carry out an inverse calibration, we need accurate and frequent pressure measurements at various locations in the system.

All of the simulations carried out during the analysis were performed using the software program, TransAM (Transient Analysis Model). TransAM is a general purpose simulation model for calculating hydraulic conditions in pipeline systems. The inverse calibration process also requires the use of a genetic algorithm processor (GAP), which is used as a kind of supervisory program for the running of a wide variety of TransAM simulations. TransAM uses the method of characteristics, which is based on an established time increment, and calculations are continued for a specified time interval (Wylie and Streeter 1993; Chaudhry 1987). Initial conditions are defined by a steady state description supplied by the user, which may be obtained from a steady state network model or from the transient model itself. In addition, the standard solution by the method of characteristics has been improved to allow flexible friction term linearization (Karney and McInnis 1992).

3.4 Step Three: Natural Selection

In the third step, natural selection, all individuals of a generation are arranged in order by performance. This performance factor is determined by a least square error function, derived by comparing the measured pressure heads at the various locations for a test with the corresponding computer simulated heads during the transient event. The individual genes are deemed better performers if they have smaller error values.

Once the generation has been sorted, the better performers are given the chance to reproduce to form the next generation. The intention is that children of the current generation, being the combination of two parents, may perform better than either parent. Since not all of the characteristics of an individual may match the actual system, the combination of two individuals may produce new properties that better fit the desired attribute.

To enhance the chances of the new generation matching the actual physical system parameters, mutation is allowed to occur. Each new individual is allowed to have certain genes mutate randomly. However, the probability for mutation is set at a low value so as not to create an unstable or unrealistic calibration of the system.

When the new generation is complete, each individual is simulated and the whole process is repeated a specified number of generations or until a best individual is found. Once this best individual is found, its genes are converted to their corresponding values and used as the calibrated parameters in the model.

3.5 Summary

One of the major benefits of the inverse calibration approach is that any number of parameters can be calibrated. The procedure is simple, and can be efficiently implemented, with the computer doing most of the work in a timely fashion.

In addition to its flexibility and its ability to handle complex systems, inverse calibration has a number of advantages over traditional calibration procedures. It is an efficient and simple process to implement, and has some immunity to localization errors. Furthermore, the method can be applied using existing modelling programs.

4 FIELD TESTING

A critical requirement for the success of the inverse calibration approach is the availability of frequent and accurate pressure measurements during a transient event. The primary emphasis of the field tests is on routine or emergency shut-downs, or power-failure conditions at key pumping stations in the system, and the surge pressures created at various control locations due to these events. Prior to performing the field tests, the system must be assessed to develop a calibration approach. Every distribution network is unique, and so the approach to calibration must be tailored to meet the individual requirements of the system in question. Once a clear understanding of the system, and the approach to calibration, has been arrived at, the field tests can be performed.

4.1 Typical Field Test Procedure

In general, the tests should be performed when demand is low and relatively stable. Mid-morning is often ideal, but for practical reasons tests were carried out throughout the working day on two successive days. For each test, the specified pumps were operated until conditions in the system roughly stabilized, at which time one or more pumps were shut down and the consequent transient response of the system was recorded.

Although each test involves unique problems, a general procedure for the field tests can be described as follows:

- Install calibrated pressure transducers and verify their performance.

- Synchronize data loggers and run a confirmation test.
- Install data loggers at pressure transducer locations.
- Through the SCADA system, record reservoir levels, known flows, estimated demand, and system status information (such as other pumps running, valves opened or closed); record time and date.
- About 30 to 60 seconds prior to pump trip, initiate data loggers to record pressure data. Coordination and communication by telephone or radio is essential.
- Initiate the transient event.
- Observers at all locations record any "events of consequence" such as noise, hammering, bump, or check valve closure
- After about five or six minutes, stop data loggers and re-record any available flows, or reservoir levels
- Collect and compile data and observations in a central location for transfer to the system analysis/calibration stage.

4.2 Field Test Data

The immediate results of the field tests are a number of large data files containing high speed pressure measurements at each location for each test. Before conducting the calibration procedure it was necessary to format this data to make it more manageable. For example, it was necessary to filter out any noise picked up by the pressure sensors during the tests. Also, not all of the data for all of the tests is useful for the purposes of calibration.

5 INVERSE CALIBRATION USING THE FIELD DATA

This section contains a description of the calibration carried out on the City of Thunder Bay water supply system. Inverse calibration was carried out on the collected field data using a transient analysis program (TransAM) equipped with a genetic algorithm processor (GAP). Some of the primary parameters that influence the development and dissipation of transient pressures include acoustic wave speeds and friction factors of pipes, pump inertia, and nodal consumption. These parameters are given special attention in the model calibration.

5.1 Calibration Procedure

Before the inverse calibration is performed, the raw data obtained in the field must be conditioned to filter out unreliable, artificially induced signals that might have been picked up by the high speed pressure transducers and the data loggers. This is one of the reasons that a large sampling rate was required. Following the conditioning step, the data are formatted into one of the input streams for the Genetic Algorithm Processor (GAP). The processed field data is used to determine the fitness of each calibration member in each generation.

The GAP was programmed to run the transient simulator (TransAM) for populations between 50 and 100 members and for 30 to 50 generations. The calibration objects were nodal demands and pipe friction factors. For each generation, information about the simulated pressures at field measurement locations are recorded and analyzed by the GAP. At the end of each generation, the individuals are ranked and the best performer---with the closest pressure traces to the actual measured pressures---is recognized and allowed to continue into the next generation. The other members of the new generation are formed from the superior performers of the current generation through reproduction (i.e., gene swapping between parents) and a small probability of mutation. The process continues until a single individual is identified as the best performer.

5.2 Calibration Results

For the tests performed on Thunder Bay, the calibrated friction factors are lower than the friction factors in the original, uncalibrated model. A comparison of the calibrated and uncalibrated models can be shown using a plot of the measured pressure head, uncalibrated pressure head, and calibrated pressure head at the measurement nodes. Figure 1 and Figure 2 are typical plots of this nature.

Clearly, the inverse process is not magic. There are considerable challenges in selecting suitable transient events, in selecting sensor measurement locations and in choosing appropriate properties for the basic numerical model. If the source of the transient and the sensor locations are too remote, too little of the signal is transmitted to allow effective use of the method. The clear trade-off is between the amount of effort in collecting the data and in the level of detail furnished by the calibration.

Other issues also arise. If there are inconsistencies between the physical system and the basic capabilities of the numerical model, the optimization cannot be expected to work well. Many variations in the numerical approach are also possible and these have not been fully explored. For example, it may be preferable to use one of the sensor locations to drive the transient model, while the others are used to optimize the calibration process.

In comparing the change in friction factors to the change in nodal demands, it may occur that the nodal demands change more dramatically than the friction factors. This result occurred in the Thunder Bay analysis but was expected since the nodal demands were not known to a high degree of accuracy. Clearly the demands captured by the inverse transient method are specific to the time the field measurements took place, and thus may not be suitable for design.

6 CONCLUSIONS AND RECOMMENDATIONS

Particular examinations of the water supply system for several cities has provided fruitful information about the system behaviour to the daily operations of pump switching, shut-down and start-up as well as to some of the nodal demands and pipe friction factors. This analysis also provided important information on the feasibility of carrying out inverse calibration with genetic algorithms and the details and requirements of such tests. The overall performance of the transient calibration procedure is quite good, even though the potential of the inverse method has not yet been fully exploited. The low cost and overall effectiveness of this approach is quite evident considering the minimal field testing and human-power required for this analysis to be completed.

6.1 Future Work

Results of such a calibration would then be directly and automatically incorporated into the simulation model. In this way, an accurate assessment of the current state of the system, including the status of valves, the state of pumps and pipes, the current value of demand and the possibility of leaks could be accurately assessed. In addition, reliable predictions of the system's response could be made relating to operational and design decisions. For this hope to become a reality, it will be necessary to have rapid and reliable tranducers, data transfer lines, simulation software and optimization techniques. Although most of these components are available individually, to date the necessary connections of the various components has not taken place.

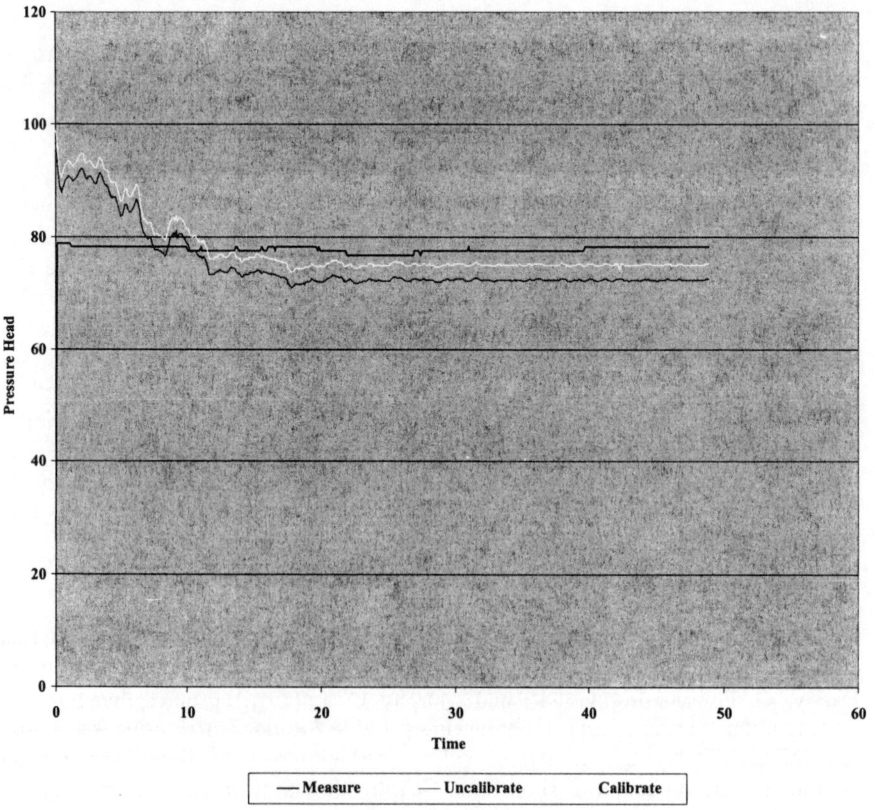

Figure 1 - Sample calibration results (North Ward Zone 3)

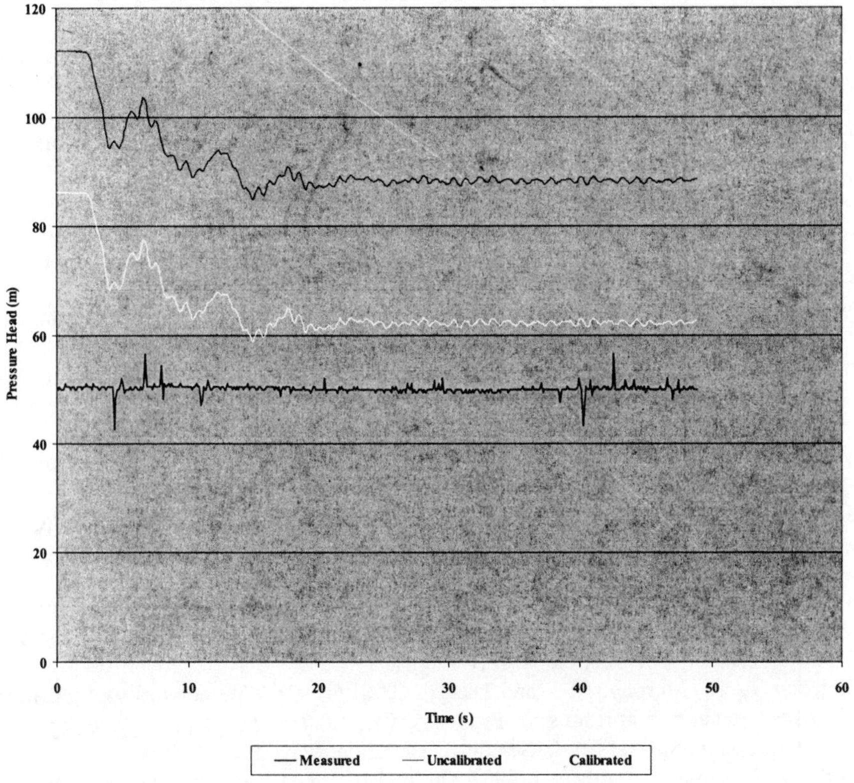

Figure 2 - Sample calibration results (North Ward Zone 1)

REFERENCES

Chaudhry, M.H. (1987). *Applied Hydraulic Transients*. Van Nostrand Reinhold, New York, N.Y.

Clingenpeel, W.H. (1983). Optimizing pump operating costs.: Management and Operations, *J. AWWA*, 502--509.

Dandy, G. C., Simpson, A. R. and Murphy L. J. (1996). An improved genetic algorithm for pipe network optimization. *Water Resources Research*, 32(2), 449-458.

Goldberg, D. E. and Kuo, C. H. (1987). Genetic algorithms in pipeline optimization. *J. Computing in Civ. Engrg.*, ASCE, 1(2), 128-141.

Halhal, D., Walters, G. A. and Savic, D. A. (1997). Water network rehabilitation with structured messy genetic algorithm. *J. of Water Resources Planning and Management*, 123(3), 137-146.

Holland, J. H. (1975). *Adaptation in Natural and Artificial Systems*. University of Michigan Press, Ann Arbor, Mich.

Hadji, G. and Murphy, L. J. (1990). Genetic algorithms for pipe network optimization. 4th Year Student Civil Engineering Research Report, University of Adelaide, Australia. pp. 134.

Karney, B.W., and McInnis, D. (1992). Efficient calculation of transient flow in simple pipe networks. *Journal of Hydraulic Engineering*, ASCE, Volume 118, No. 7, 1014--1030.

Liggett, J.A. and Chen, L-C. (1994a). Inverse transient Analysis in Pipe networks. *J. Hydr. Engrg.*, ASCE, 120(8), 934--955.

Liggett, J.A. and Chen, L-C. (1994b). Monitoring water distribution systems: the inverse method as a tool for calibration and leak detection. In: *Improving efficiency and reliability in water distribution systems*, E. Cabrera and Antonio F. Vela, editors, Valencia, Spain, Kluwer Academic Publishers, 107--134.

McInnis D., Karney, B. and Axworthy, D. (1997). TRANSAM Reference Manual. HydraTek Associates, (Ajax).

Murphy, L. J. and Simpson, A. R. (1992). Pipe optimization using genetic algorithms. Research Report No. 93, Department of Civil Engineering, University of Adelaide, Australia, June, pp. 95.

Murphy, L. J., Simpson, A. R. and Dandy, G. C. (1993). Design of a pipe network using genetic algorithms. Water, pp. 95.

Savic, D. A. and Walters, G. A. (1997). "Genetic algorithms for least-cost design of water distribution networks." J. of Water Resources Planning and Management, 123(2), 67-77.

Simpson A. R., G. C. Dandy and L. J.Murphy (1994). Genetic algorithms compared to other techniques for pipe optimization. *Journal of Water Resources Planning and Management*, Vol. 120, No. 4, July/August.

Simpson, A. R., Murphy, L. J. and Dandy, G. C. (1993). Pipe network optimization using genetic algorithms. Proc., ASCE, *Water Resources Planning and Management* Special Conf., Seattle, Washington, May, 392-395.

Walters, G. A. and T. Lohbeck (1993). Optimal layout of tree networks using genetic algorithms. *Engrg. Optim.*, 22, 47-48.

Wylie, E.B., and Streeter, V.L. (1993). Fluid transients in systems. Prentice-Hall, Inc., Englewood Cliffs, N.J.

Using a Kalman Filter Approach for Looped Water Distribution Networks Calibration

E. Todini

ABSTRACT

The paper presents an original technique to estimate unknown roughness coefficients in a looped water distribution network starting from an initial guess. The problem of estimating these values from the available measurements, known as the inverse problem, is generally carried out by minimising the square of differences of piezometric head or of nodal demands and leads to a more or less complex non-linear estimation problem. The formulation here presented converts the problem into a linear estimation problem, for which a Kalman Filter approach can be developed.

The observability conditions for this Kalman Filter hold provided that a sufficient number of independent sets of steady state observations are made, otherwise one obtains the solution closest to the initial guess in the least squares sense that satisfies the piezometric head constraints.

A simple example of application on an idealised network is used in order to show the convergence properties of the proposed method.

1 INTRODUCTION

When dealing with water distribution problems, one of the major uncertainties lies in the estimation of realistic values for the roughness coefficients to be used. In general the problem of estimating these values from the available measurements (known as the inverse problem) is carried out by minimising the square of differences of measured and computed piezometric heads and leads to a complex non-linear estimation problem, or by minimising the square of differences of measured and computed nodal demands, which may lead to a quadratic programming problem such as the one proposed by Greco et al. (1998).

In the formulation here presented, originally developed for the case of groundwater modelling (Ferraresi et al., 1996), the water consumption estimated as nodal demand is taken as the most uncertain quantity in the estimation problem, while the piezometric head, assumed perfectly known, is retained as a constraint. The formulation leads to a linear estimation problem, for which a Kalman Filter approach can be developed in order to estimate an a posteriori solution starting from a first guess evaluation of the roughness.

Following the formulation of a looped water distribution network given in Todini and Pilati (1988), which was used for the development of EPANET (Rossman, L.A., 1993), and by Todini (1999) to provide a unifying view of the looped network analysis algorithms, in very broad terms and, without considering - for the sake of simplicity though without compromising the substance of this paper - such special devices as pumps, valves and other dissipation elements, the head losses along the ith pipe of the network are represented by a monomial expression, such as:

$$f_i(Q_i) = \beta_i\, Q_i\, |Q_i|^{n_i - 1} \tag{1}$$

where: n_i is the exponent which takes account of both the flow regime (laminar or turbulent) and the dependence of the roughness coefficient on discharge, while β_i is a head loss coefficient which depends on the roughness coefficient c_i, on the diameter d_i and length of the pipe l_i and which, using the Gaukler-Strikler formula, can be expressed as a function of the roughness coefficient as follows:

$$\beta_i = \frac{\alpha\, c_i\, d_i^{8/3}}{l_i^{1/2}} \tag{2}$$

with $\alpha = 1000.\pi / 4^{5/3}$ when the flow is given in l/s.

A network comprising n_p pipes with unknown discharge, n_n nodes with unknown head and n_0 nodes with known head can be described in matrix form by the following system of equations:

$$\begin{bmatrix} A_{pp} & A_{pn} \\ A_{np} & 0 \end{bmatrix} \begin{bmatrix} Q \\ H \end{bmatrix} = \begin{bmatrix} -A_{p0}H_0 \\ q \end{bmatrix} \tag{3}$$

where:

$Q^T = \lfloor Q_1, Q_2, \cdots, Q_{n_p} \rfloor$ is the $\lfloor 1, n_p \rfloor$ vector of unknown pipe discharges

$H^T = [H_1, H_2, \cdots, H_{n_n}]$ is the $[1, n_n]$ vector of unknown nodal heads

$H_0^T = [H_{01}, H_{02}, \cdots, H_{0n_0}]$ is the $[1, n_0]$ vector of known nodal heads

$q^T = \lfloor q_1, q_2, \cdots, q_{n_n} \rfloor$ is the $[1, n_n]$ vector of known nodal demands

In equation (2) A_{pp} represents an (n_p, n_p) diagonal matrix whose elements are defined as $A_{pp}(i,i) = \beta_i |Q_i|^{n_i - 1}$ while $A_{pn} = A_{np}^T$ and $A_{p0} = A_{0p}^T$ are topological incidence sub-matrices, of size $\lfloor n_p, n_n \rfloor$ and $\lfloor n_p, n_0 \rfloor$ respectively, derived from the general topological matrix $\overline{A}_{pn} = \lfloor A_{pn} \vdots A_{p0} \rfloor$ of size $\lfloor n_p, n_n + n_0 \rfloor$ the definition of which is the following:

$$\overline{A}_{pn}(i,j) = \begin{cases} -1 \text{ if the flux of pipe } j \text{ leaves node } i \\ \ \ 0 \text{ if pipe } j \text{ is not connected with node } i \\ +1 \text{ if the flux of pipe } j \text{ enters in node } i \end{cases}$$

When dealing with the analysis of the looped water distribution networks, all the β_i appearing in equation (1) are assumed to be known together with the exponents n_i, all the nodal demands q and at least the head in one node H_0.

Although broader calibration problems can be found in the literature where the pipe roughness, the pipe flows, the nodal heads and the nodal demands are all quantities to be estimated (Cohen and Carpentier, 1988), all throughout this paper "calibration" means the estimation of the actual pipe roughness coefficients given the functional relationship of equation (1) with known coefficients n_i, known head at all nodes (namely H and H_0) and known nodal demands q. A discussion on the observability conditions will then treat the case of a reduced number of available measurements.

2 THE PROPOSED CALIBRATION METHOD

By taking advantage of the properties of the incidence matrices, equation (1) can be rewritten as:

$$\begin{cases} Q = -A_{pp}^{-1} \lfloor A_{pn} \vdots A_{p0} \rfloor \begin{bmatrix} H \\ \cdots \\ H_0 \end{bmatrix} = -A_{pp}^{-1} \overline{A}_{pn} \begin{bmatrix} H \\ \cdots \\ H_0 \end{bmatrix} = A_{pp}^{-1} [\Delta H] \\ A_{np} Q = q \end{cases} \tag{4}$$

Given that all the elements of vector ΔH are known and that matrix A_{pp}^{-1} is a diagonal matrix, equation (3) can be more conveniently re-written as:

$$\begin{cases} Q = D\,k \\ A_{np} Q = q \end{cases} \tag{5}$$

where D is the diagonal matrix of size $\lfloor n_p, n_p \rfloor$ representing the known head differences for all pipes and k the vector of size $\lfloor n_p \rfloor$ of unknown conveyances, namely $k(j) = \dfrac{1}{a_{pp}(j,j)}$.

When the network model is "calibrated" by substituting the first of equations (5) into the second, one expects:

$$A_{np} D k^* = q^* = q \tag{6}$$

Unfortunately this is not so since the conveyance first guess values generally differ from the real ones due to several causes, among which are the increased roughness and the narrowing of the pipe section due to ageing.

The formulation expressed by equation (5) is extremely useful because when D is known, the conveyance k is both linearly related to the flow, as in equation (5), and to the roughness coefficient through the following equation where all the terms, apart from c_i are given:

$$k_i = \frac{\alpha d_i^{8/3}}{l_i^{1/2} |\Delta H|_i^{1/2}} c_i \tag{7}$$

This allows for the application of the proposed technique by taking the conveyances k as the state vector to be estimated; the values for the roughness coefficient can then be computed once the estimation process is ended, without losing the properties of the estimator due to the linear relationship of equation (7).

Following what was done by Ferraresi et al. (1996) for the calibration of the groundwater models parameters, the following Kalman Filter can be formulated as follows:

$$k_t = k_{t-1} \qquad \text{system equation} \tag{8}$$

and

$$q_t = A_{np} D_t k_t + \varepsilon_t \qquad \text{measurement equation} \tag{9}$$

with t a given set of steady state observations and where ε_t represents a measurement error, which in this case can be viewed as an estimation error of the nodal demand q_t, and can reasonably be assumed Normally distributed, namely $\varepsilon_t \hat{=} N(\mu_t, R_t)$, with μ_t and R_t its bias and covariance matrix respectively. The $\lfloor n_n, n_p \rfloor$ matrix $A_{np} D_t$, called the measurement matrix, projects the state vector k_t into the measurement vector q_t. This is an important feature of the proposed approach, since it recognises the fact that the largest errors lie both in the estimation and in the assumption that the nodal demands represent the actual water consumption distributed along the pipes.

Starting from an initial guess value for k, namely $k_{0|0}$, and for the covariance matrix of errors of estimate, namely $P_{0|0}$, the following iterative scheme can be formulated, by substituting the extrapolation equations into the correction ones:

$$\begin{cases} \hat{k}_{t|t} = \hat{k}_{t-1|t-1} + P_{t-1|t-1} D_t A_{pn} \left(A_{np} D_t P_{t-1|t-1} D_t A_{pn} + R_t \right)^{-1} \left(q_t - \mu_t - A_{np} D_t \hat{k}_{t-1|t-1} \right) \\ \\ P_{t|t} = P_{t-1|t-1} - P_{t-1|t-1} D_t A_{pn} \left(A_{np} D_t P_{t-1|t-1} D_t A_{pn} + R_t \right)^{-1} A_{np} D_t P_{t-1|t-1} \end{cases} \quad (10)$$

For any given t, a limited number of iterations (5 – 6) on the same data are required to reach convergence.

3 THE OBSERVABILITY CONDITIONS

The observability conditions for such a scheme require the rank of the following matrix U to be equal to n_p, namely the dimension of the state vector:

$$U = \left\lfloor D_1 A_{pn} \;\vdots\; D_2 A_{pn} \;\vdots\; \cdots \;\vdots\; D_t A_{pn} \;\vdots\; \cdots \;\vdots\; D_{n_t} A_{pn} \right\rfloor \quad (11)$$

where n_t is the number of sets of observations. From this equation it appears evident that the state vector is always observable in the case of tree-shaped networks even when a unique set of measurements is available. In the case of looped systems, since the number of pipes is larger than the number of nodes, a unique set of steady state data is not sufficient in order to guarantee observability. Consequently, the proposed method, when used for estimating the roughness coefficients under steady state conditions using only one set of observations, loses its statistical properties, as will be seen from the example. The consequences are that, in the steady state case, two alternatives are available: either use several sets of steady state observations, or provide a meaningful estimate for the initial state, as for any Bayesian estimation procedure. Therefore, when one uses the proposed method in steady-state conditions with a single data set, the initial parameter estimates no longer represent a necessary step in the recursive procedure initialisation. Rather, they represent an indispensable and substantial component of the physical structure of the final solution which must be provided as the a priori Bayesian information, from which the final solution will deviate according to the optimal correction factor given by the Kalman gain times the innovation. Nevertheless, the topology of the connections, expressed by the incidence matrix which appears in the observability matrix, requires the use of several sets of observations in order to guarantee that the flow of information reaches all the pipes in the network.

As mentioned earlier, the proposed technique requires the demand and the head to be measured at all the nodes. This is not possible in real world cases and the following considerations can be made: if not all the q are measured, the observability conditions will only require a larger number of observation sets; the problem is more complex for the case of missing H. In this case it is necessary to proceed from coarser to finer meshes by first identifying the main pipes roughness using a very coarse mesh, and proceeding then to the analysis of finer meshes with different sets of measurements.

In any case, due to the fact that the observability conditions are mainly dominated by topology, it is not difficult to define an experiment, even without the real measurements, that shows the pipes where the roughness coefficient estimation will be more difficult, as will be seen in the following example.

4 AN EXAMPLE OF APPLICATION

The algorithm described by equations (9) was applied to the simple example of figure 1:

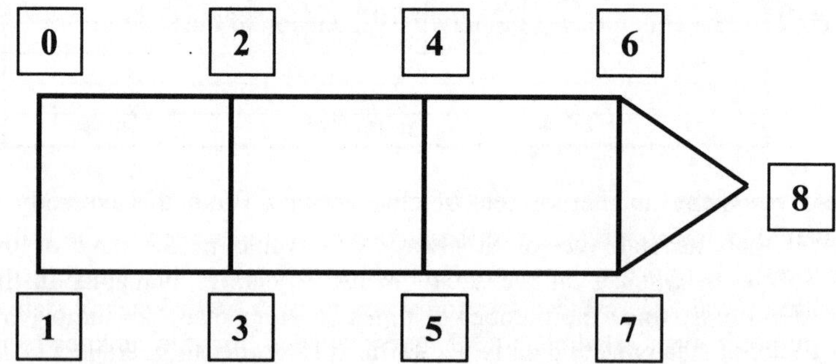

Figure 1. A schematic representation of the simple looped network

Table 1 provides the pipe data in terms of diameter, length and roughness coefficient, while table 2 shows the values of the head and the demand measurements.

Table 1. Pipe related quantities in the proposed example of figure 1

Pipe	d_i [mm]	l_i [m]	c_i .
0 - 1	400	603.876	100
0 - 2	400	373.398	100
1 - 3	400	812.080	100
2 - 3	300	466.556	100
2 - 4	400	400.788	100
3 - 5	400	845.086	100
4 - 5	300	315.987	100
4 - 6	400	769.652	100
5 - 7	300	509.562	100
6 - 7	100	269.182	100
6 - 8	400	855.814	100
7 - 8	300	305.436	100

Table 2. Node related quantities in the proposed example of figure 1

Node	Head [m]	Demand [l/s]
0	115.000	-680
1	107.637	181
2	107.603	64
3	106.096	63
4	104.304	95
5	104.254	67
6	102.394	75
7	102.434	25
8	101.964	110

As mentioned in the previous section, when several sets of steady state data are available, the convergence of the algorithm is guaranteed by the Kalman Filter properties. Therefore it was felt interesting to analyse the convergence of the algorithm toward the correct solution when using a unique steady state case. For this purpose, one hundred initial guess values for the unknown roughness coefficients (all equal to 100) were independently generated with a uniform distribution between 80 and 120. The theoretical variance of the generated values is 133.33, which expresses our uncertainty. The experiment consisted in verifying if the proposed method was able to reduce it.

The algorithm described by equation (9) was simplified in that the matrix $P_{t-1|t-1}$ was taken as the identity matrix and was not updated. The reason for doing so relates to the fact that the updating is meaningful when new sets of data, showing different distributions of heads and demands, are used.

Table 3 shows that the method is unbiased and that it noticeably improves in a number of pipes, while in other pipes no substantial modifications can be observed. Nevertheless the overall variance is reduced by 62% as one can see from the last column of table 3 where the gain is defined as:

$$\text{Gain} = 1 - \frac{\text{Final Variance}}{\text{Initial Variance}} \tag{12}$$

Figure 2 shows the initial and final values for all pipes; in particular one can see that pipe 6-7 shows the worst performances, pipe 6-8 shows a good improvement, and pipe 3-5 an extremely large improvement in the estimation variance.

The experiment confirmed that the algorithm is convergent, but, as mentioned earlier, the observability is not guaranteed for all the pipes when a unique set of steady state data is available, unless a very good initial guess is provided, in particular for the most penalised pipes. The sensitivity analysis performed shows which pipes can hardly be affected by the corrections, and for these pipes one can either provide better initial guesses or special experiments could be conducted by partialising the network, as proposed by Tucciarelli and Termini (1998).

Table 3. Statistical analysis of the roughness coefficients before and after the estimation phase.

Pipe	Initial Mean	Final Mean	Initial Variance	Final Variance	Gain
0-1	100.62	100.20	145.87	6.38	0.96
0-2	98.67	99.86	149.74	3.98	0.97
1-3	100.86	100.51	122.36	40.47	0.67
2-3	99.82	98.88	158.71	113.86	0.28
2-4	101.60	99.85	134.79	13.56	0.90
3-5	99.72	100.09	148.75	3.60	0.98
4-5	99.25	99.22	128.98	129.18	0.00
4-6	101.82	100.20	149.14	10.08	0.93
5-7	101.22	99.59	118.19	33.56	0.72
6-7	99.35	99.42	131.42	142.08	-0.08
6-8	100.63	100.48	131.25	51.91	0.60
7-8	99.26	99.40	128.42	79.70	0.38
Overall			**137.30**	**52.36**	**0.62**

5 CONCLUSIONS

The major advantage of the proposed formulation is its linearity, thus allowing for the use of a Kalman Filter. This does not solve all the issues relevant to the general problem of estimating all the roughness coefficients in a looped pipe network. Nevertheless it is the first time that a linear estimator is derived and its full possibilities of application in practice are still to be assessed.

Another interesting possibility given by the Kalman Filter formulation is relevant to the assessment of which pipes are scarcely affected by a given set of measurements by performing a simulation similar to the one shown in the previous section: this can help in locating which flow measurements are necessary in order to improve both the efficiency of the estimation and reducing at the same time cost of measurement campaigns.

It is evident that further work has still to be done and it is the intention of the author to perform extensive testing of the proposed algorithm on several existing water distribution networks in order to assess all of its potentially practical uses.

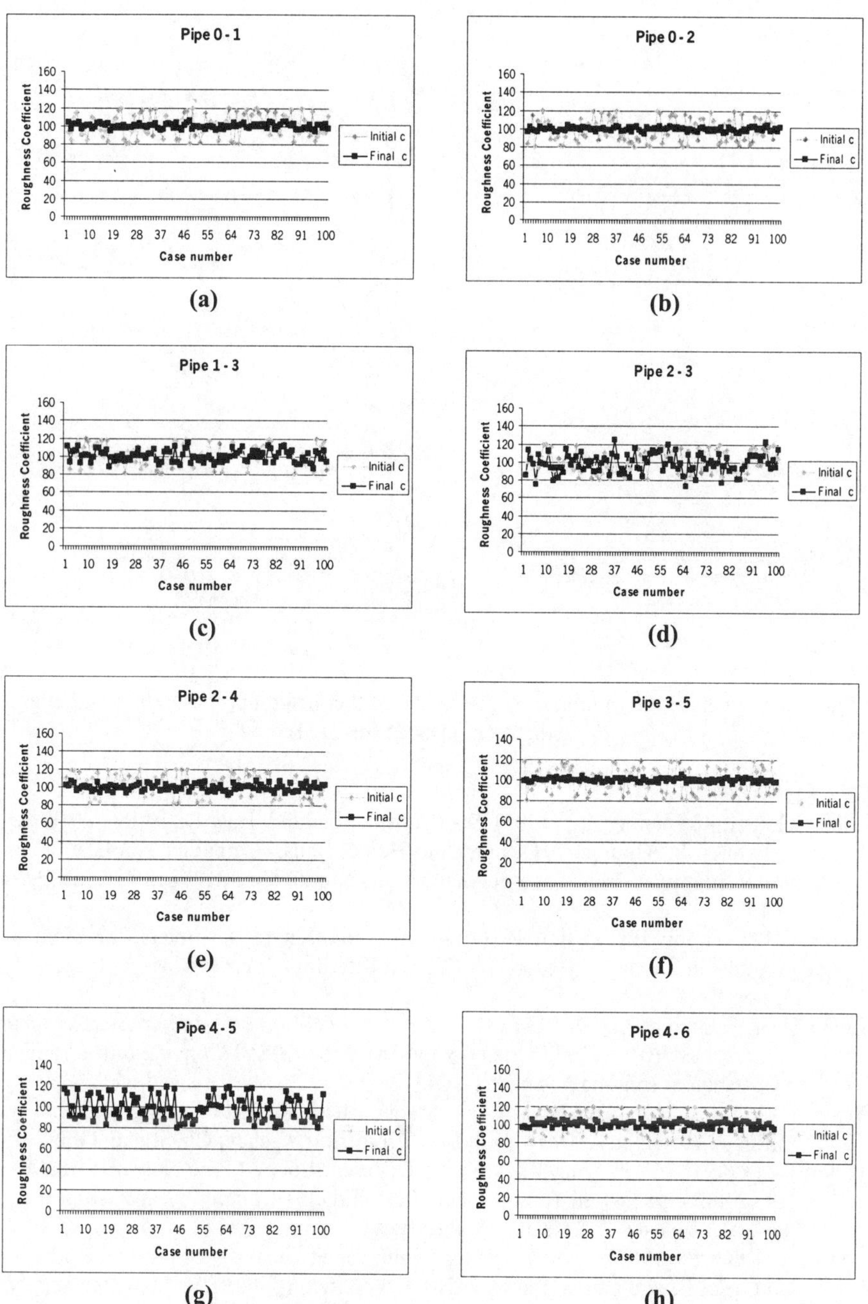

Figure 2. Values for the roughness coefficient c of the different pipes before and after the estimation phase.

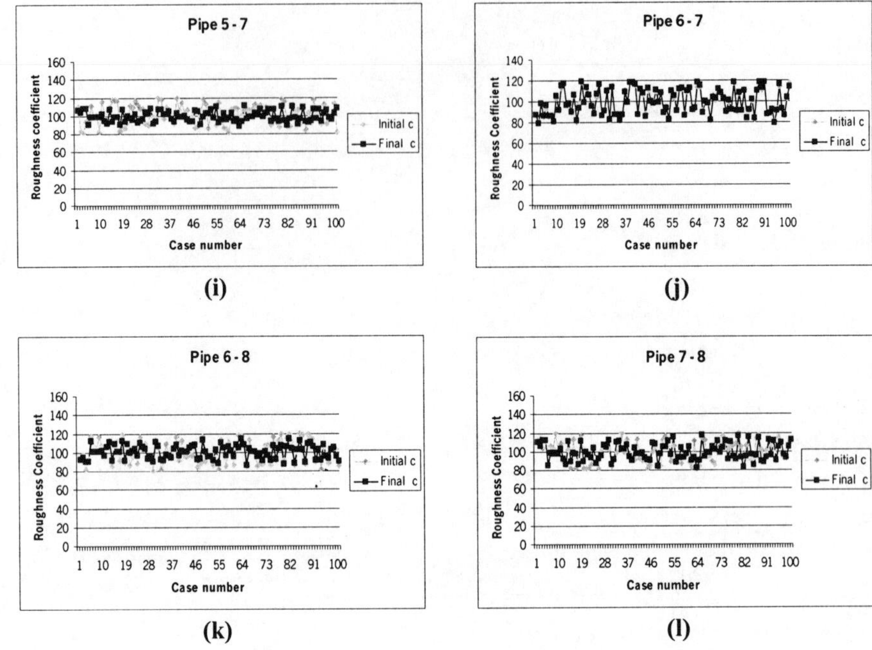

Figure 2. Values for the roughness coefficient c of the different pipes before and after the estimation phase (continued).

REFERENCES

Cohen G. and Carpentier P. (1988) State estimation and leak Detection in Water Networks. In Coulbeck B. and Chun-Hou O. (eds.) Computer Applications in Water Supply, Vol. 1 - System Analysis and Simulation, RSP Ltd, UK, pp. 317-332.

Ferraresi M., Todini E., Vignoli R. (1996) - A solution to the inverse problem in groundwater hydrology based on Kalman filtering - Journal of Hydrology, 175: 567-581.

Greco M., Del Giudice G., Di Cristo C. (1998) – La calibrazione delle portate erogate nelle reti di distribuzione idrica (in Italian). Proc. XXVI Convegno di Idraulica e Costruzioni Idrauliche, pp. 491-504.

Rossman, L.A. (1993) – EPANET, Users Manual. Risk Reduction Engg. Laboratory, Office of Research & Devt., U.S. Env. Protection Agency, Cincinnati, Ohio.

Todini E., (1999) – A Unifying View on the Different Looped Pipe Network Analysis Algorithms. In Powell R. and Hindi K.S: (Eds.) – Computing and Control for the water Industry – Research Studies Press Ltd, pp. 63-80.

Todini E., Pilati S. (1988) - A gradient algorithm for the analysis of pipe networks. In Coulbeck B. and Chun-Hou O. (eds.). Computer Applications in Water Supply, Vol. 1 - System Analysis and Simulation, , RSP Ltd, UK, pp.1-20.

Tucciarelli T. and Termini D. (1998) – Optimal valves regulation for calibration of pipe networks models. In Babovic V. and Larsen L. C. (Eds.), Hydroinformatics '98, Balkema, Rotterdam, pp1029-1036.

PART V

TRANSIENT ANALYSIS AND DESIGN

Pressure Surge Implications for the Emergency Pipeline from the Dolgarrog Hydroelectric Installation to the Bryn Cowlyd Water Supply Scheme

A.P. Boldy

ABSTRACT

Whilst this paper is best described as a Case Study it discusses the unusual features of a pressure surge investigation of the complex hydraulic system comprising the low head pipeline of the Dolgarrog Hydroelectric Power Station, the emergency pipeline from the low head pipeline to the Bryn Cowlyd water treatment works, the treatments works itself and the delivery pipeline to the clean water storage tanks.

1 INTRODUCTION

The new Bryn Cowlyd water treatment works is designed as a pressurised system. The raw water is normally abstracted from a high level impounding reservoir in the Snowdonia National Park and transported to the treatment works by a series of pipelines and break pressure tanks. In order to provide an emergency raw water supply to the treatment works if these pipelines are out of service, a pipeline has been constructed connecting the inlet to the treatment works to the low head pipeline from the Llyn Coetdy reservoir supplying water to part of the National Power Dolgarrog Hydroelectric Power Station.

A schematic drawing of the complete system is shown in Fig. 1 and the profiles, together with the junction node numbers used in the simulations, of the National Power pipeline, the emergency pipeline, the pipeline through vessel 1 in the treatment works and the pipeline to the clean water tanks are shown in Figs. SY1.Prof1, SY1.Prof2, SY1.Prof3 and SY1.Prof4, respectively.

The aims of the pressure surge investigation were:

(i) to satisfy National Power that the effects of any pressure surge scenario resulting from the operation of the emergency pipeline does not generate any detrimental pressure surge effects in the National Power pipeline with, or without, the Dolgarrog turbines in operation.

(ii) to quantify any pressure surge protection required at the treatment works in order to protect the treatment works from any detrimental effects of pressure surges generated by the tripping of the Dolgarrog turbines.

2 BASIS OF THE COMPUTER SIMULATIONS

The computer simulations were performed using the program **PTRAN** developed by Dr. Adrian P. Boldy. The program is based on the well known *method of characteristics* [Refs. 1,2,3] which converts the quasi-linear, hyperbolic, partial differential equations of motion and continuity describing the unsteady flow of a fluid in any internal flow system, into a set of ordinary differential equations which are then expressed in a finite difference form in order to produce a solution algorithm. The variant of the method used is called the *method of specified time intervals* which computes the values of piezometric pressure head and velocity at a constant time interval for each node in the system.

3 DATA

The hydraulic data for the emergency pipeline, the treatment works and the discharge pipeline to the clean water tanks was supplied by Hyder Consulting Limited. The emergency pipeline was designed to deliver 30 Mld ($0.347m^3 / s$) to the treatment works.

The Dolgarrog Hydroelectic Power Station contains four Francis turbines. The identical turbines (units 2 and 3), supplied by Litostroj, are connected to the high head pipeline from the Llyn Cowlyd Reservoir. Turbine unit 4, supplied by Kvaerner Boving, is connected to the low head pipeline from the Llyn Coedty Reservoir and turbine unit 5, supplied by Gilbert Gilkes & Gordon, is required to operate from either pipeline, either alone or in parallel with the other unit(s). In order to simulate the transient response of the complete system generated by a turbine trip scenario it is essential to incorporate the performance characteristics of the appropriate turbines.

4 SIMULATIONS

4.1 Design Phase

The results of an extensive pressure surge investigation undertaken during design of the treatment works and the emergency pipeline identified:

1. the necessity to connect a $2.5m^3$ capacity air vessel, with a steady state air volume of $1.5m^3$, to the emergency pipeline upstream of the control valve at the treatment works end of the pipeline in order to prevent the generation of very high frequency pressure head oscillations in the treatment works during the transient flow generated by the simultaneous trip of the Dolgarrog turbines,

2. an appropriate opening rate and two-stage closure relationship for the control valve. Based on these results and subsequent discussions between Hyder Consultants Limited and the valve supplier, it was decided to

incorporate the following two-stage the opening relationship for the control valve:

$0-15^o$ in 120 seconds followed by $15-90^o$ in 60 seconds and with the closure of the valve following the reverse relationship.

4.2 As Constructed System

During the detailed design and construction of the emergency pipeline significant alterations were made to the diameter of various sections of the pipeline and the method of operation of the pipeline valves was formulated by Hyder Consulting Limited in conjunction with National Power. In the light of these developments, Hyder Consulting Limited engaged Dr. Adrian Boldy to undertake a pressure surge investigation related to the operation of the ***As Constructed*** emergency pipeline. The main aspects of this investigation are summarised in the following sections.

4.2.1 Simultaneous trip of the Dolgarrog turbine units 4 and 5

Table 1 summarises a series of simulations of the simultaneous trip of the Dolgarrog turbine units 4 and 5 with the emergency pipeline in operation and discharging either zero or $0.347m^3/s$.

Table 1. Simultaneous trip of Dolgarrog turbines units 4 and 5

Run Number	Steady state discharge in ERWP (m^3/s)	Minimum air volume in air vessel (m^3)	Minimum Static Pressure Head in ERWP (m)	
			NP End	TW End
SY1-1005	0.0	No air vessel	304.4	348.9
SY1-1010	0.0	1.16	304.4	313.1
SY1-1462	0.347	1.34	300.0	221.1

4.2.1.1 Zero discharge

The air vessel located at the treatment works end of the emergency pipeline is excluded from run number SY1-1005 in order to demonstrate the pressure surge generated in the closed pipeline with no pressure surge protection.

Fig. SY1-1005.SP/27-68 shows that a maximum transient static pressure head of approximately $304.4m$ is generated at the NP end of the emergency pipeline by the simultaneous trip of the turbines. The transient static pressure head rise is propagated along the emergency pipeline and, in the absence of the air vessel, is reflected at the closed control valve with a $+1$ reflection coefficient, thereby generating a maximum transient static pressure head of approximately $348.9m$ at the valve.

Run number SY1-1010 incorporates the air vessel which acts as a damper to significantly reduce the reflected amplitude of the incident pressure surge, as shown in Fig. SY1-1010.SP/27-68, resulting in a maximum transient static pressure head of approximately $313.1m$ at the closed control valve.

4.2.1.2 Discharge of $0.347m^3/s$ *in the emergency pipeline*

Run number SY1-1462 is based on the steady state conditions corresponding to the Dolgarrog turbine units 4 and 5 generating together a discharge of $0.347m^3/s$ through the emergency pipeline, into the treatment works with eight vessels in operation and up to the clean water tanks.

Fig. SY1-1462.SP/27-68 shows that a maximum transient static pressure head of approximately 300.0m is generated at the NP end of the emergency pipeline. Due to the head losses in the emergency pipeline the steady state static pressure head at the air vessel is approximately 192.0m. Since the amplitude of pressure head rise generated at the NP end of the emergency pipeline is attenuated by frictional effects as it propagates along the emergency pipeline and then damped by the air vessel, the resultant maximum transient static pressure head at the air vessel is approximately 221.1m, as shown in Fig. SY1-1462.SP/27-65. Although the pressure surge is transmitted through the treatment works, as shown in Fig. SY1-1462.SP/194, the air vessel prevents the formation of any high frequency pressure fluctuations. [Fig. SY1-450.SP/194, from the design phase investigation referred to in section 4.2, shows the pressure fluctuations generated in the treatment works without an air vessel connected to the emergency pipeline.]

4.2.2 Opening of control valve with Dolgarrog turbines 4 and 5 generating

Run number SY1-1110 simulates the two-stage opening of the control valve in order to generate a discharge of $0.347m^3/s$ through the emergency pipeline with the turbines 4 and 5 generating.

Fig. SY1-1110.SP/27-65 shows that there are no pressure surge problems generated at the NP end of the emergency pipeline. There is a small gradual decrease in static pressure head at this point as the discharge through the emergency pipeline is established. Fig. SY1-1110.CV/O-D shows the opening relationship for the control valve and that the discharge settles to the steady state value of approximately $0.347m^3/s$ after the valve is fully open. Fig. SY1-1110.AVS/AV-D shows that the air volume in the air vessel increases to approximately 1.75 m^3.

4.2.3 Closure of control valve with $0.347m^3/s$ to treatment works

Table 2 summarises the results of the simulation of the closure of the control valve from a steady state discharge of $0.347m^3/s$ with or without the turbines 4 and 5 in operation.

Table 2. Closure of valve from steady state discharge of $0.347m^3/s$

Run Number	Turbines 4 and 5 generating	Maximum air volume in air Vessel (m^3)	Maximum Static Pressure Head in ERWP (m)	
			NP End	TW End
SY1-1866	Yes	1.28	228.0	234.0
SY1-1766	No	1.75	258.0	265.0

4.2.3.1 Dolgarrog turbines 4 and 5 generating

For run number SY1-1866 the steady state discharge of $0.347m^3/s$ through the emergency pipeline with the turbines in operation is achieved by setting the control valve fully open and controlling the discharge to each of the eight operational vessels by the appropriate setting of the valve in the discharge pipeline from each vessel.

Fig. SY1-1866.CV/O-D shows the two-stage closure relationship for the valve from the fully open position together with the discharge through the valve. Fig. SY1-1866.SP/27-65 shows that the closure of the valve does not generate any pressure surge problems at the NP end of the emergency pipeline with the static pressure increasing slightly as the valve closes. At the air vessel there is a rapid increase in the static pressure head towards the end of the first stage of the valve closure due to the corresponding rapid rate of deceleration of the water through the valve, as shown in Fig. SY1-1866.CV/O-D.

4.2.3.2 Dolgarrog turbines not in operation

For run number SY1-1766 the steady state discharge of $0.347m^3/s$ through the emergency pipeline with the turbines not in operation is achieved by setting the control valve to approximately 28 degrees and controlling the discharge to each of the eight vessels in operation by the appropriate setting of the control valve in the discharge pipeline from each vessel.

Fig. SY1-1766.CV/O-D shows the two-stage closure relationship for the valve from the steady state opening together with the discharge through the valve. Fig. SY1-1766.SP/27-65 shows that the closure of the valve does not generate any pressure surge problems at the NP end of the emergency pipeline with the static pressure increasing slightly as the valve closes. At the air vessel there is a rapid increase in the static pressure head towards the end of the first stage of the valve closure due to the corresponding rapid rate of deceleration of the water through the valve, as shown in Fig. SY1-1766.CV2071/O-D.

5 CONCLUSIONS

The results of the pressure surge investigation demonstrated:

(i) that the operation of the ***As Constructed*** emergency pipeline will not generate any pressure surge problems at the National Power end of the pipeline with, or without, the Dolgarrog turbines in operation or during the transient flow generated by a simultaneous load rejection.

(ii) that the installation of a $2.5m^3$ capacity air vessel, with a steady state air volume of $1.5m^3$, upstream of the control valve at the treatment works end of the emergency pipeline prevents the formation of high frequency pressure surge fluctuations in the treatment works pipework.

(iii) the maximum static pressure head at the treatment works end of the emergency pipeline is generated during the transient flow generated by the simultaneous load rejection of the Dolgarrog turbines when the isolating valve at the National Power end of the pipeline is open and there is zero discharge in the pipeline.

ACKNOWLEDGEMENTS

Dr. Adrian P. Boldy is grateful to Dwr Cymru - Welsh Water , Hyder Consulting Limited and National Power for there co-operation and permission to publish this paper .

REFERENCES

J.A. Swaffield and A.P. Boldy, ***Pressure Surge in Pipe and Duct Systems***, Avebury Technical, 1993, ISBN 0 291 39796 4.

M. H. Chaudhry, ***Applied Hydraulic Transients***, Van Nostrand Reinhold, 1986, ISBN 0-442-21514-2.

E.B. Wylie and V.L. Streeter, ***Fluid Transients in Systems***, Prentice Hall, 1993, ISBN 0-13-322173-3.

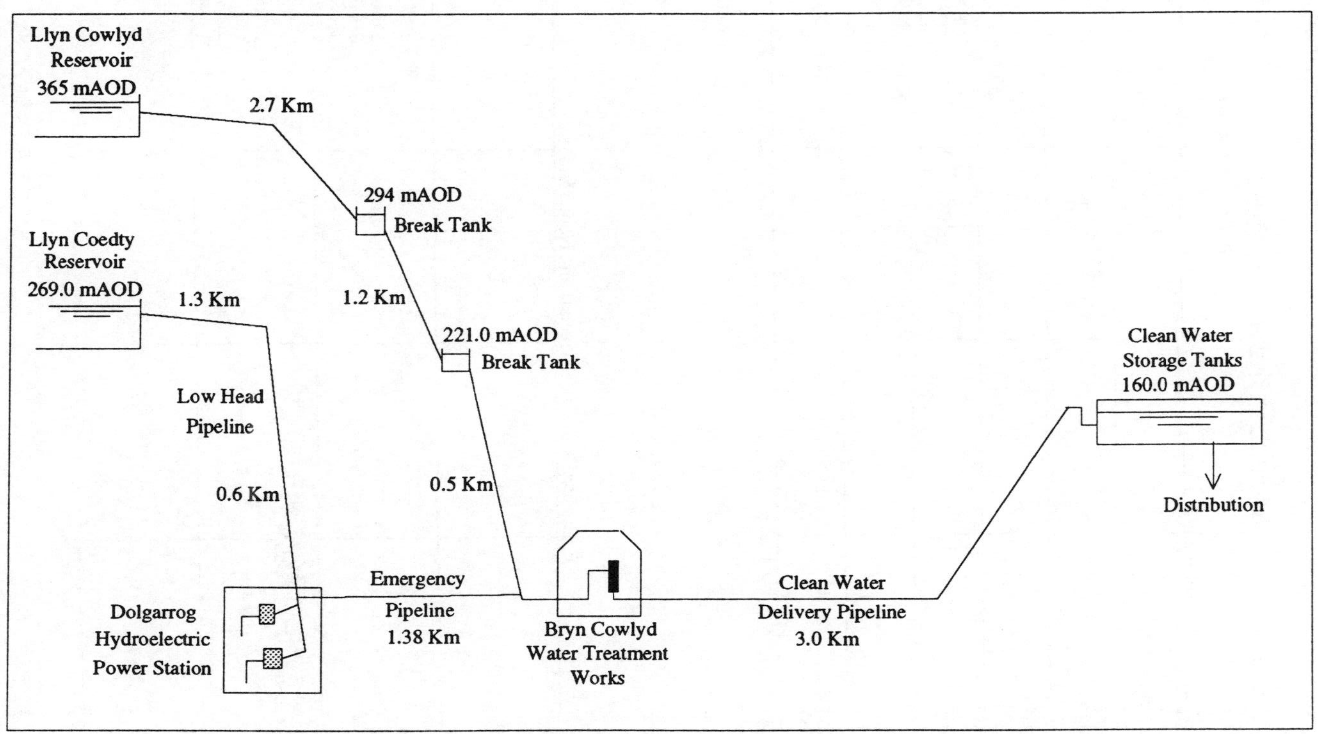

Fig. 1 Schematic Layout of Dolgarrog Low Head Hydroelectric System and Bryn Cowlyd water Supply System

System 1: National Power - Emergency RWP - Treatment Works - Clean Water Tanks

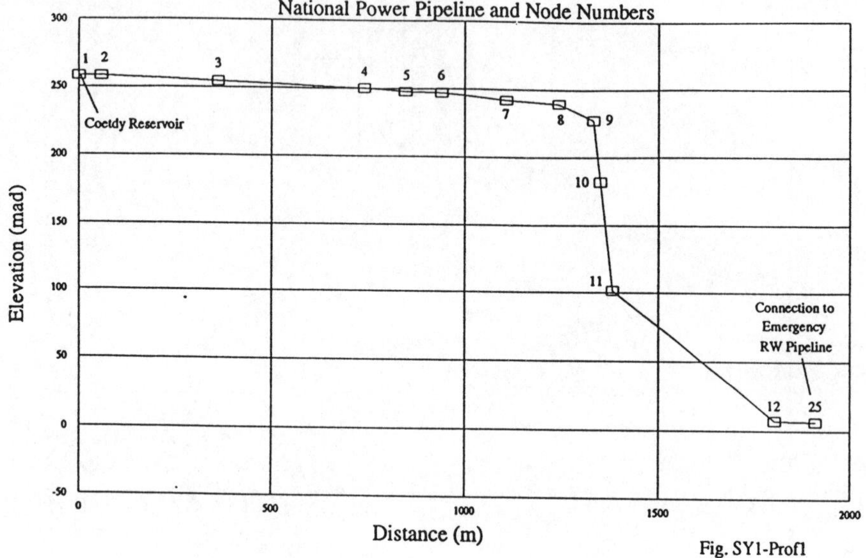

Fig. SY1-Prof1

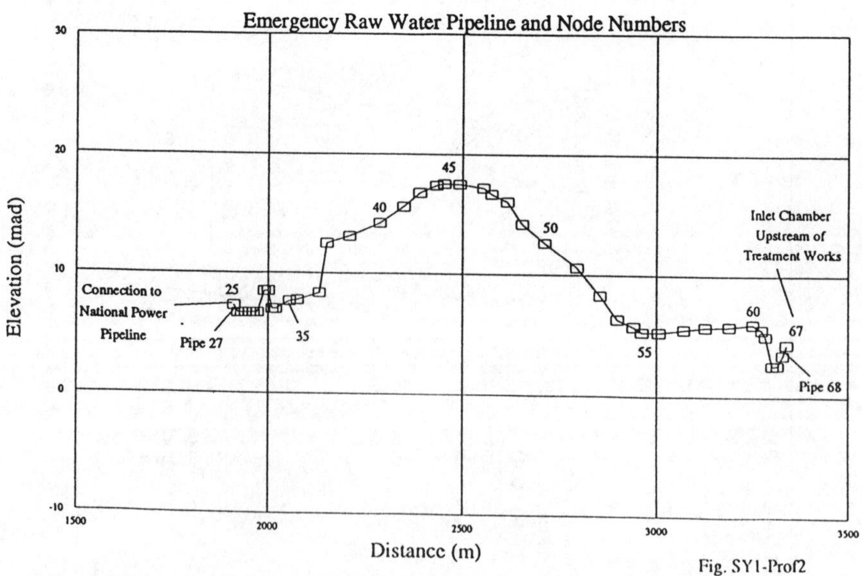

Fig. SY1-Prof2

System 1: National Power - Emergency RWP - Treatment Works - Clean Water Tanks

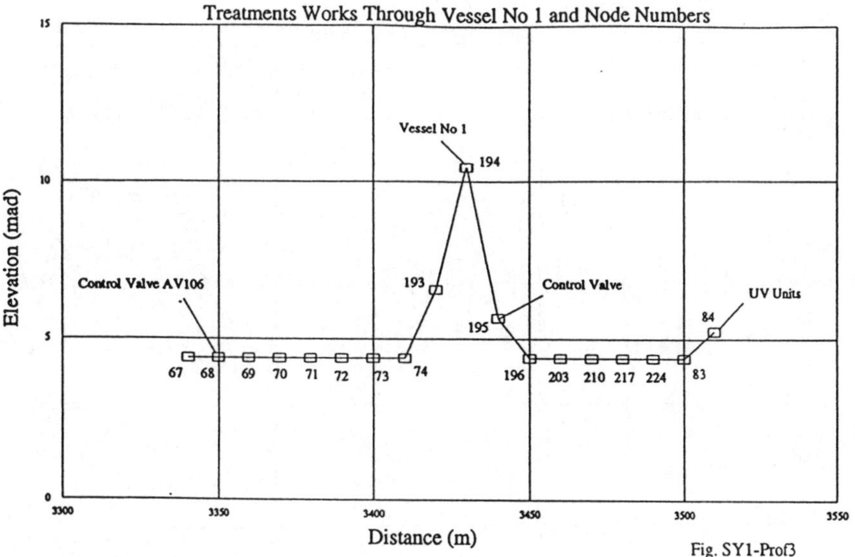

Fig. SY1-Prof3

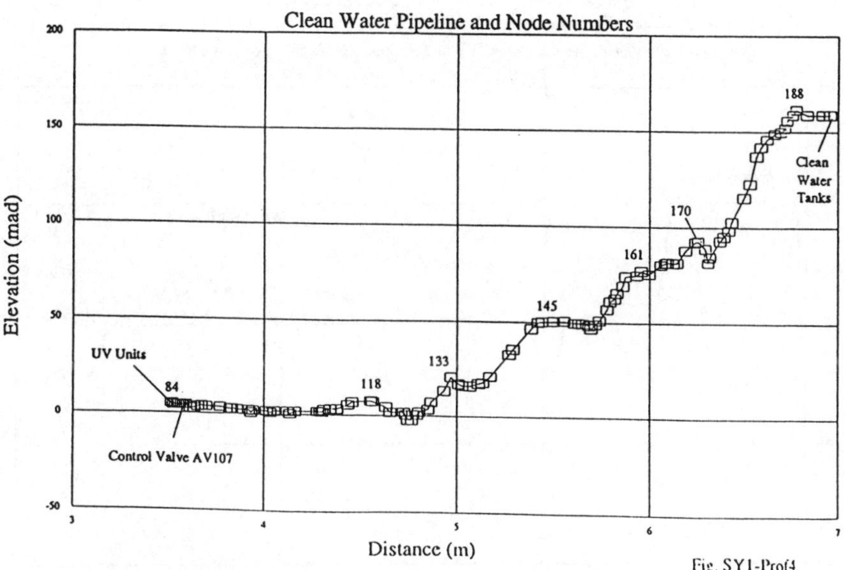

Fig. SY1-Prof4

System 1: National Power - Emergency RWP

Steady State Conditions: Turbines 4 and 5 Generating. Zero Discharge in EWRP. NP Valve open.

ERWP As Constructed. Simultaneous Trip of Turbines. No Air Vessel at TW.

Static Pressure Head in ERWP

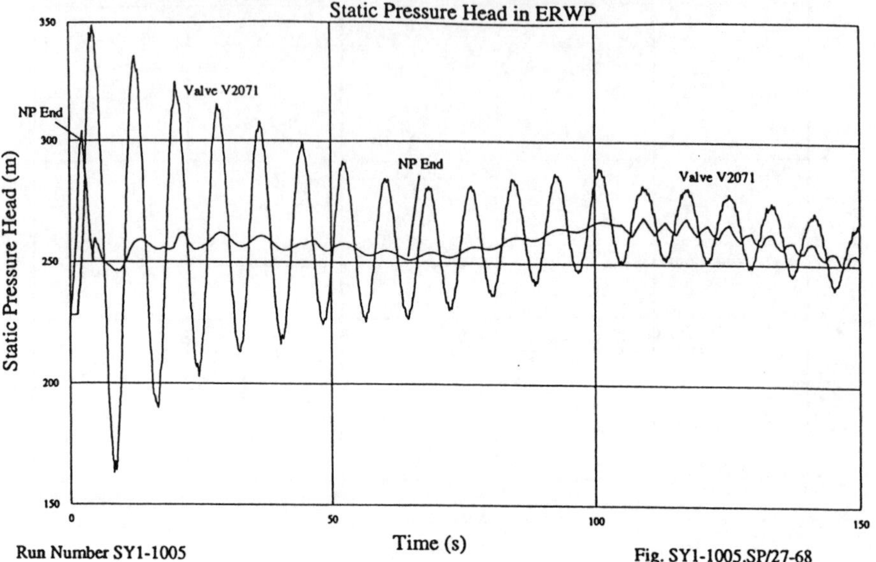

Run Number SY1-1005 Fig. SY1-1005.SP/27-68

System 1: National Power - Emergency RWP

Steady State Conditions: Turbines 4 and 5 Generating. Zero Discharge in EWRP. NP Valve open.

ERWP As Constructed. Simultaneous Trip of Turbines. Air Vessel (Vo=1.5cu.m.) at TW

Static Pressure Head in ERWP

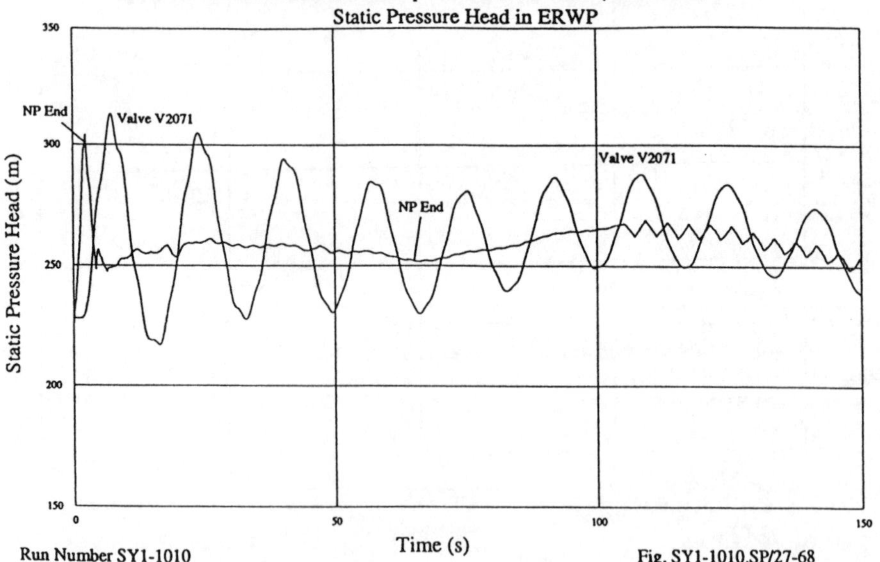

Run Number SY1-1010 Fig. SY1-1010.SP/27-68

System 1: National Power - Emergency RWP - Treatment Works - Clean Water Tanks

Steady State Conditions: Turbines 4 and 5 Generating. ERWP Discharge 347 l/s. ERWP As Constructed

Simultaneous Trip of Turbines. Air Vessel (Vo=1.5 cu.m.) at TW

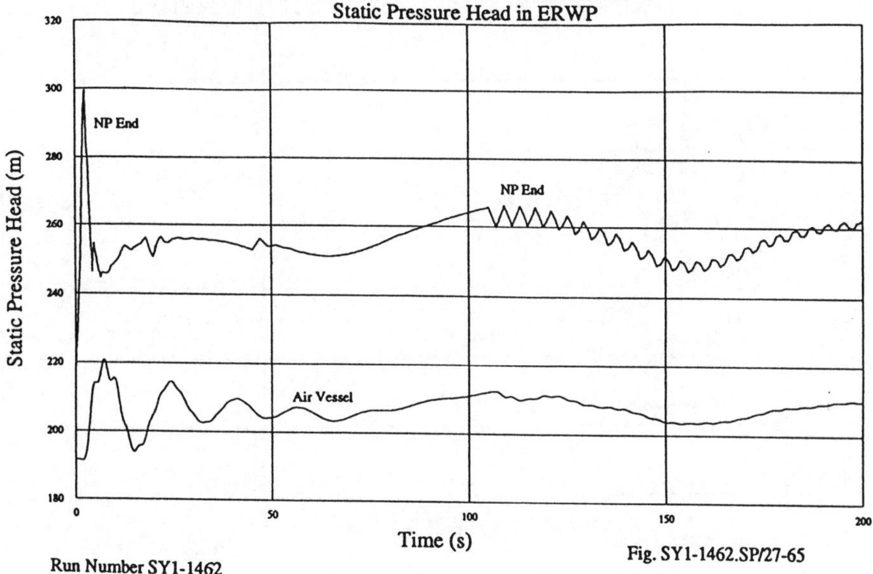

Fig. SY1-1462.SP/27-65

Run Number SY1-1462

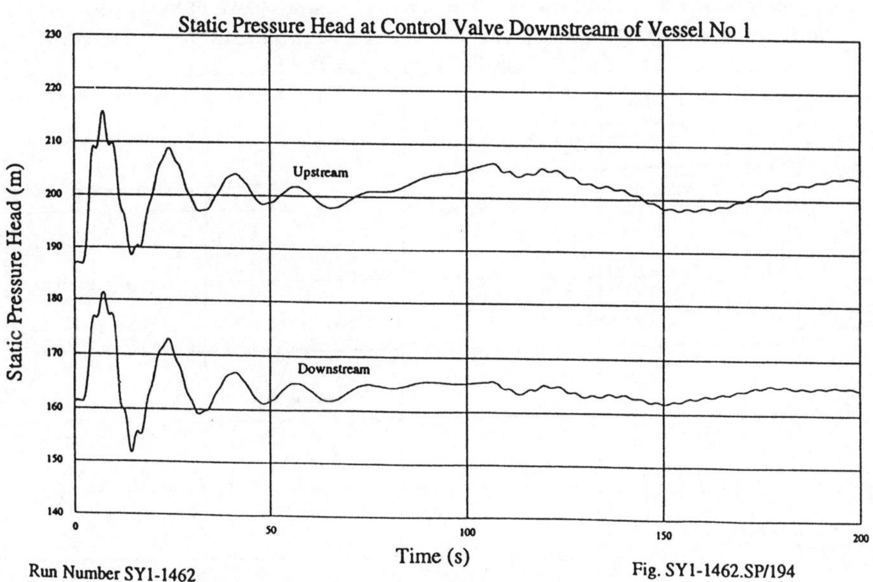

Run Number SY1-1462

Fig. SY1-1462.SP/194

System 1: National Power - Emergency RWP - Treatment Works - Clean Water Tanks
Steady State Conditions: Turbines 4 and 5 Generating. ERWP Discharge 0.347 cumecs.
ERWP: 350/300/350/300/400:0/5/250/1110/1354/1404. Simultaneous Trip of Turbines 4 and 5

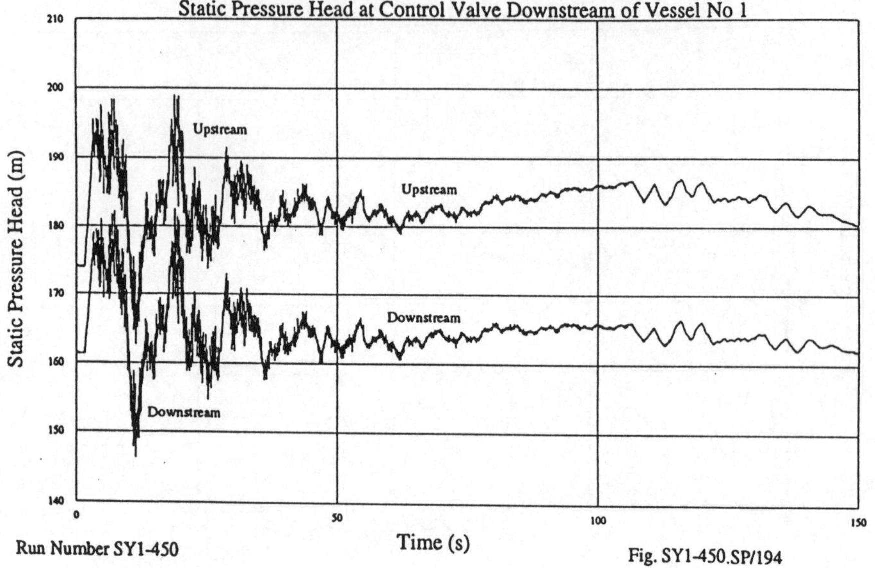

Run Number SY1-450

Fig. SY1-450.SP/194

System 1: National Power - Emergency RWP - Treatment Works - Clean Water Tanks
Steady State Conditions: Turbines 4 and 5 Generating. Zero Discharge in ERWP. NP Valve Open.
ERWP As Constructed. V2071 Opens (0/15/90:0.0/120.0/180.0) to 347 l/s. Air Vessel (Vo=1.5 cu.m.) at TW.

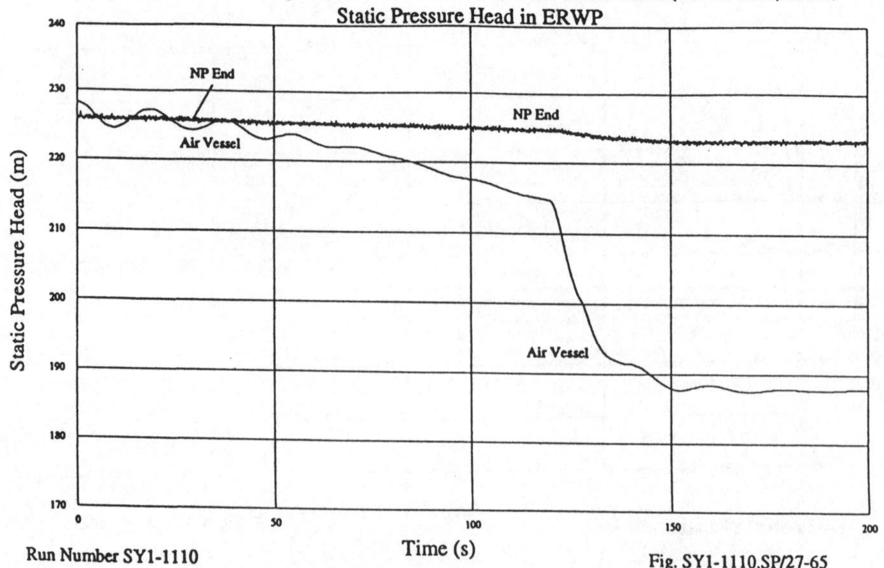

Run Number SY1-1110

Fig. SY1-1110.SP/27-65

System 1: National Power - Emergency RWP - Treatment Works - Clean Water Tanks
Steady State Conditions: Turbines 4 and 5 Generating. Zero Discharge in ERWP. NP Valve Open.
ERWP As Constructed. V2071 Opens (0/15/90:0.0/120.0/180.0) to 347 l/s. Air Vessel (Vo=1.5 cu.m.) at TW.

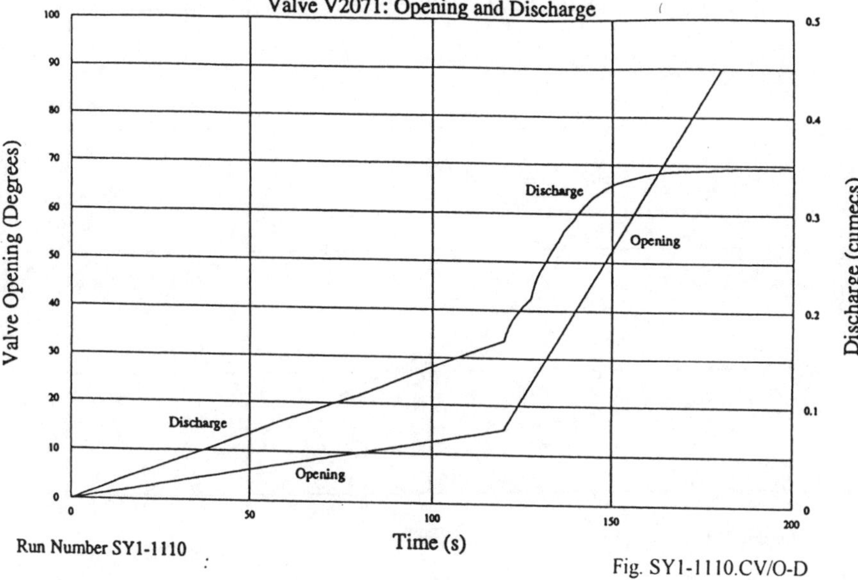

Run Number SY1-1110

Fig. SY1-1110.CV/O-D

System 1: National Power - Emergency RWP - Treatment Works - Clean Water Tanks
Steady State Conditions: Turbines 4 and 5 Generating. Zero Discharge in ERWP. NP Valve Open.
ERWP As Constructed. V2071 Opens (0/15/90:0.0/120.0/180.0) to 347 l/s. Air Vessel (Vo=1.5 cu.m.) at TW.

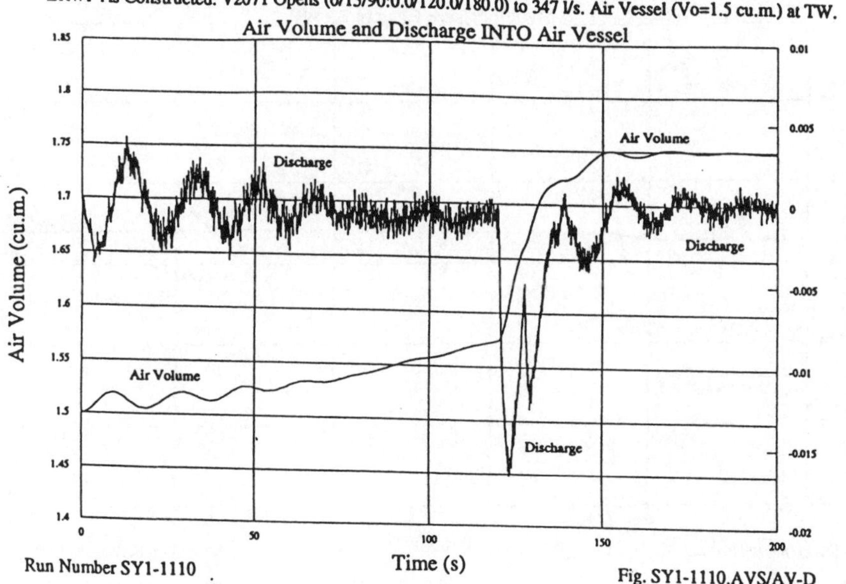

Run Number SY1-1110

Fig. SY1-1110.AVS/AV-D

System 1: National Power - Emergency RWP - Treatment Works - Clean Water Tanks
Steady State Conditions: Turbines 4 and 5 Generating. ERWP Discharge 347 l/s. ERWP As Constructed.
V2071 Closes (90/15/0:0.0/60.0/180.0). Air Vessel (Vo=1.5 cu.m.) at TW

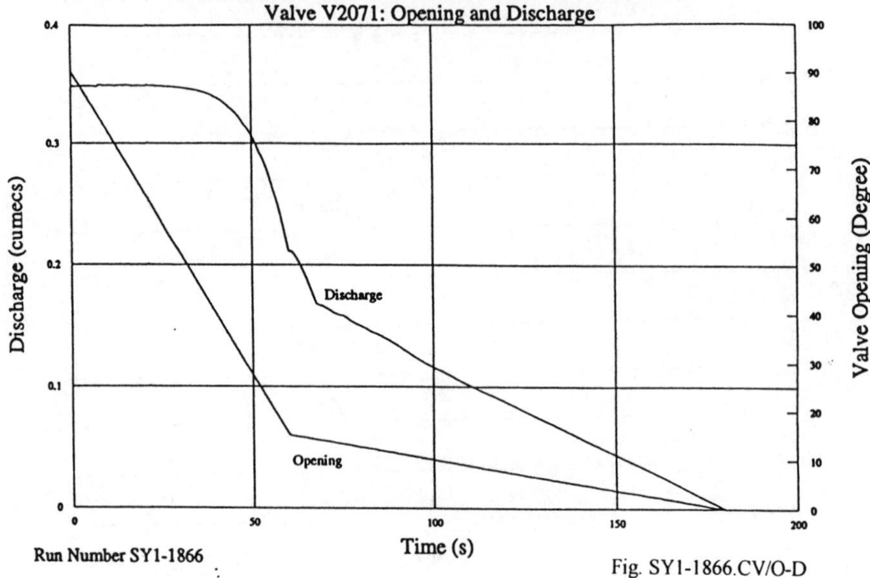

Run Number SY1-1866

Fig. SY1-1866.CV/O-D

System 1 : National Power - Emergency RWP - Treatment Works - Clean Water Tanks
Steady State Conditions: Turbines 4 and 5 Generating. ERWP Discharge 347 l/s. ERWP As Constructed
V2071 Closes (90/15/0:0.0/60.0/180.0). Air Vessel (Vo=1.5 cu.m.) at TW.

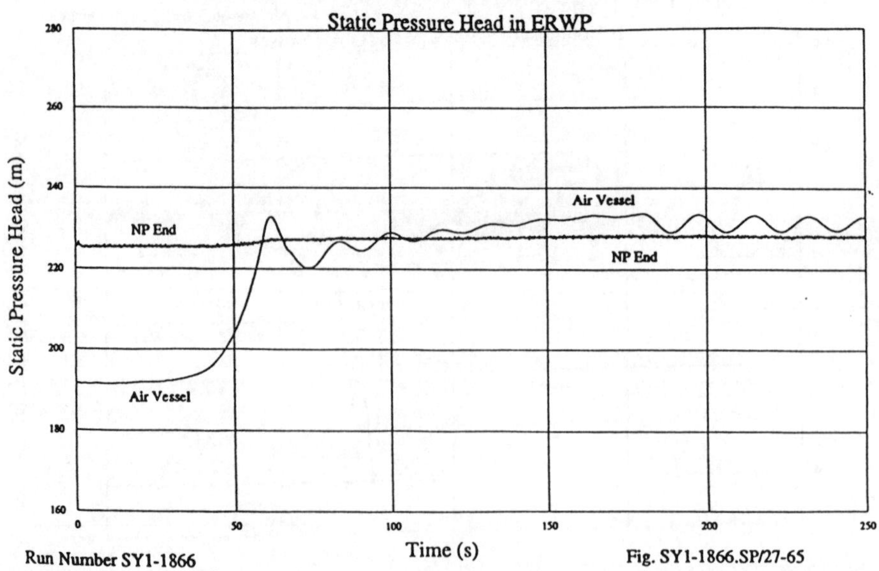

Run Number SY1-1866

Fig. SY1-1866.SP/27-65

System 1: National Power - Emergency RWP - Treatment Works - Clean Water Tanks

Steady State Conditions: 347 l/s through the System. ERWP As Constructed.

V2071 Closes at rate (90/15/0:0.0/60.0/180.0). Air Vessel (Vo=1.5 cu.m.) at TW.

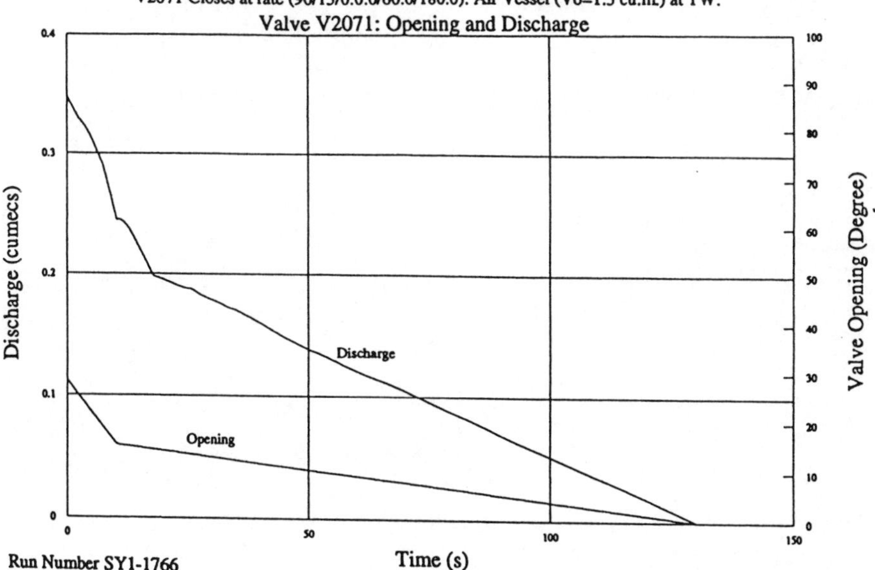

Run Number SY1-1766

Fig. SY1-1766.CV/O-D

System 1: National Power - Emergency RWP - Treatment Works - Clean Water Tanks

Steady State Conditions: 347 l/s through the System. ERWP As Constructed.

V2071 Closes at rate (90/15/0:0.0/60.0/180.0). Air Vessel (Vo=1.5 cu.m.) at TW.

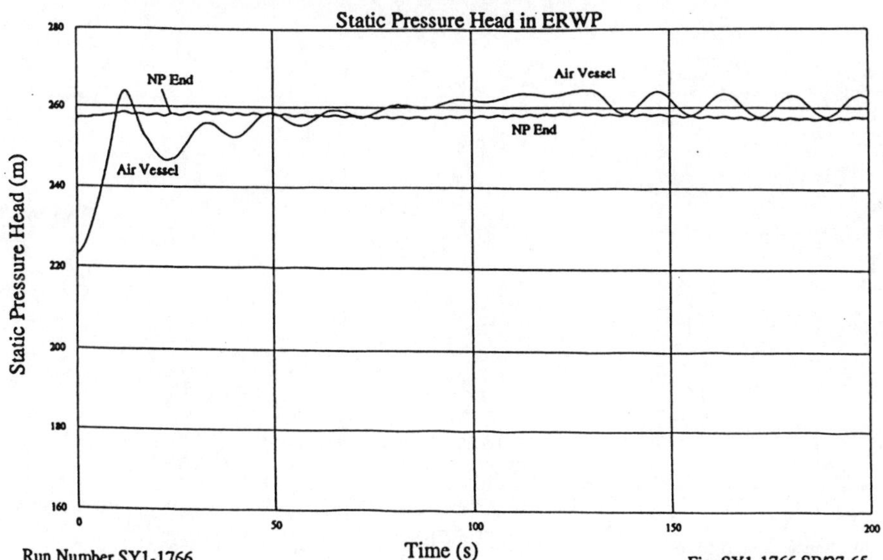

Run Number SY1-1766

Fig. SY1-1766.SP/27-65

The Interaction Between Surge and Control for Large Pipelines - Practical Implications and Use of Modelling Techniques

A.P.E. Green

ABSTRACT

Within the water industry there is an increasing need to limit the use of tanks and reservoirs to provide 'slack' in a system. One of the ways of achieving this is through better automation of valves and pumps. It has long been recognised, however, that valve closure and pump switching can generate significant surge pressures. Although calculation methods and software are available for analysing surge and rapid transients, the standard tools available are usually only suited for simulation of extreme events such as emergency closure. Similarly control system simulations typically use only a simplified representation of the hydraulic system, almost certainly excluding the effects of surge pressures.

For longer pipelines and tunnels, transient pressures can be important, affecting sensor readings and hence control actions. To avoid over pressurising mains and tunnels, valve movements may be slowed; the volume of water to be contained can be significant and special overflows and storage needed.

In this paper the practical implications of the interactions between transient pressures and control systems are discussed based the findings of recent projects. As part of the studies for uprating of the systems, control and transient simulations were carried out for a number of major water conveyors and pipelines for a number of UK cities including stored water supplies to London and Birmingham.

1 INTRODUCTION

In many situations, level and flow are controlled on the basis of a measurement of water level in an open tank, the conduit into the tank is relatively short and the dynamics of flow in the conduit are so rapid that there is no practical impact on the controller. For a longer conduit and particularly where close pressure and flow control is desired, the surge characteristics of the conduit may have a significant impact on the degree of control achieved. In the water industry this has particular impact on the requirement for balancing or break pressure tanks which may be sizeable structures requiring valuable site space and have water quality implications.

2 EXAMPLES OF SURGE AND CONTROL INTERACTIONS

2.1 The terminal control valve on a gravity supply

One of the most common situations is siting a control valve on the end of a pipeline:

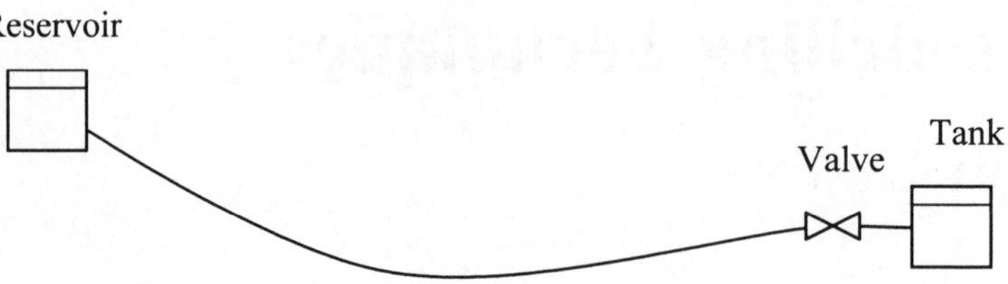

On valve movement, pressures will be generated in the pipeline in accordance with Joukowski's equation:

$$\Delta H = \frac{a}{g} \bullet \Delta V$$

Which relates the change in pressure head (H) directly to the change in velocity (V) and the speed of waves in the pipe (a). The relevant change in velocity is that occurring in one pipeline period which is defined as the time for a wave to travel from the control point to the free end of the pipe and back i.e. $T = 2L \div a$. The wavespeed or celerity in a typical water pipe or tunnel is around 1000m/s and thus a change in velocity of around 1m/s can give rise to an increase in pressure of 100m head. This pressure head is in addition to the existing static or steady head and may either take the total pipeline head outside the acceptable range or, in the case of plastic pipes, outside an acceptable excursion range.

In a short pipeline the response time may be very fast, for example a 100m pipe has a response time of only 0.2 seconds, a very short period for changes in velocity to occur. On the other hand in a 10km conduit the response time is around 20 seconds, a period during which much more significant changes may occur particularly near to the final closure of a valve.

Figure 1 Flow through pipeline, closure of butterfly valve from 50% opening pipeline length 15km

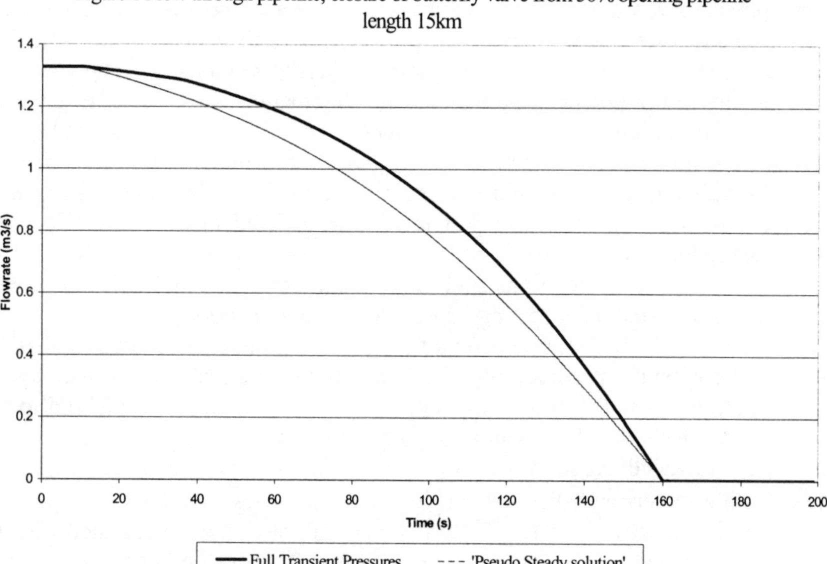

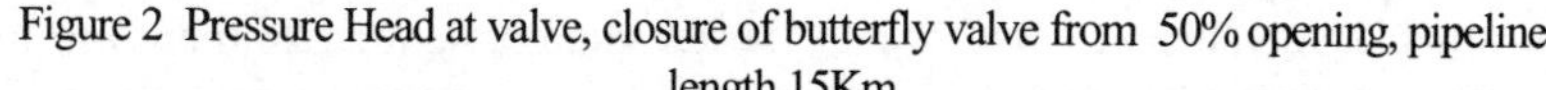

Figure 2 Pressure Head at valve, closure of butterfly valve from 50% opening, pipeline length 15Km

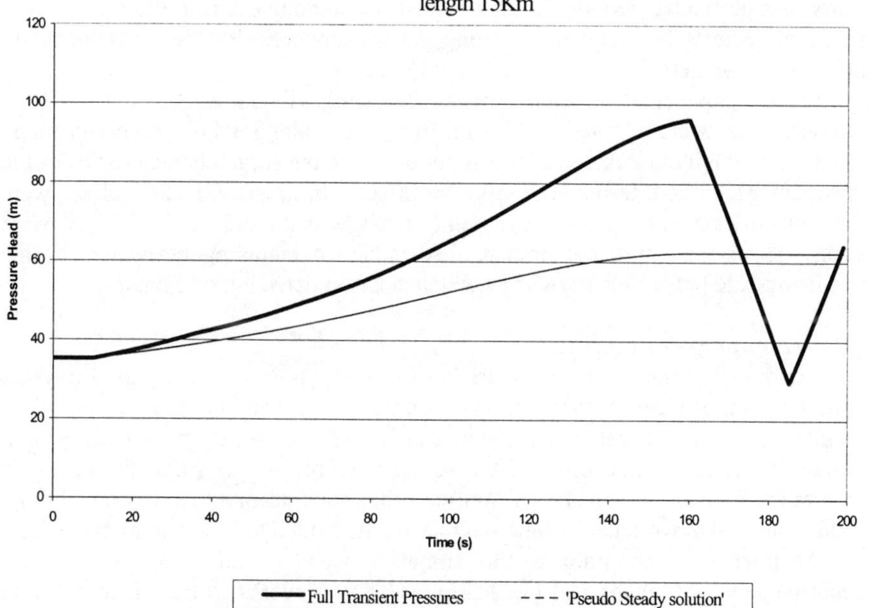

Full Transient Pressures --- 'Pseudo Steady solution'

In the simulation of valve closure the difference that surge effects have on flow and driving head across a valve are shown in Figure 1 and Figure 2. For longer pipelines the deviation between actual flow and the flow predicted by a 'pseudo steady state' model such as used in network analysis increases. When the valve is moved from one position to another where it remains for long time, the effect on flow may not be significant as the fluctuations in surge pressures are short lived in relation to intermittent changes over a period of hours. When the valve is controlled to give fine level control in a tank of limited volume, the effect can be significant, however, and may contribute to valve 'hunting'. The valve characteristics play an important part in achieving a satisfactory control regime and simple sluice valves have little control over a wide part of the valve movement. Other considerations such as potential cavitation and noise generated may be significant factors affecting valve selection and positioning.

The two conflicting requirements of surge protection and the need to shut off valves at the maximum rate are often satisfied by using a two stage valve closure that can help to make the time of movement of a valve more closely linked to the rate of change in flow. The recent introduction of interruption timers for actuators has largely replaced complex gearing, particularly for two stage valve closure rates. With an interrupter the actuator is moved in pulses as small as 300μs with periods of inactivity between. Different rates of movement may therefore be readily programmed in at different valve positions. The movement of the actuator in this way is obviously not the same as a smooth movement and model simulations were carried out on the inlet valves to Ashford Common WTW including surge effects to define the maximum pulse times acceptable. An unsuitable actuator will overheat or suffer excessive wear if there are too many starts in a particular period and thus, whilst for smoothness it is desirable to use small movements and periods of waiting, for actuator wear it is better to use larger pulses of movement.

For the Southern Conveyor pipeline in Cyprus (Evans & Sage)[1] four break pressure tanks were adopted for control of pressure along a 110km gravity main. The area of the intermediate tanks was set based on the surge characteristics of the upstream main and terminal discharge valve. In this case the valves were hydraulically operated globe valves and tanks with an area up to 500m^2 were used. The system was comprehensively modelled including movement of the paddle operated over velocity valves which close on detection of a burst.

2.2 Pumping to Reservoirs

The control of pumping to reservoirs has been the subject of many studies from both the point of view of surge and from a pumping cost and control viewpoint. A number of approaches can be taken and a variety of constraints applied. In general it may be assumed that the worst case scenario for which surge protection is required will be of a much shorter timescale than for control and pump scheduling. Modelling of the two facets of pumping have thus tended to be separate exercises.

An interesting exception to this simplistic view of real pumping situations occurred at the connection of the Ashford Common works to the Thames Water

Ring Main (TWRM) and the high lift pumping at the same site to Cricklewood and Fortis Green Reservoirs. The old high lift pumps have a high capacity and are fixed speed units. The surge protection system for the pumps is a series of fast acting 'Rotovalves' which allow dissipation of a high pressure return wave following a power failure through the suction main into the contact tank. When built, the Ashford works had two chlorine contact tanks supplying the high lift pumps. For the new works much better control of chlorination contact was required and the two tanks were to be replaced with a single new tank with its main connection to the TWRM via carefully controlled eccentric plug valves. During construction one of the old contact tanks was retained to enable works output to be maintained throughout. The high lift pumps were connected to the new contact tank and limitations put on the rate of change of total flow setpoint in accordance with the chlorination strategy and upstream limitations on rate of change of flow.

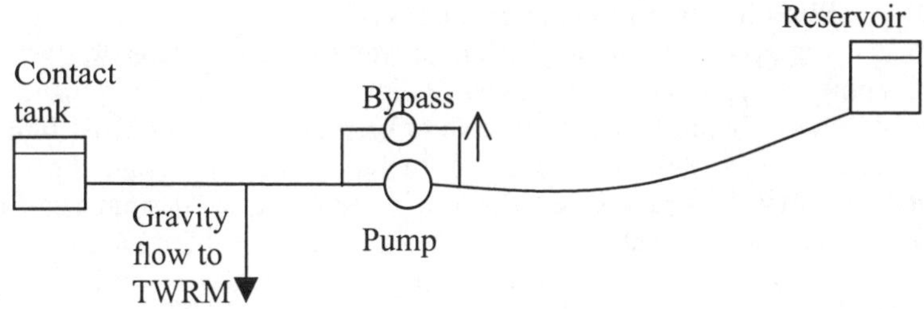

The flexibility of use of high lift pumping was overiding, however, and the original intention to abandon the old contact tank was reversed. Simulations showed that severe shocks (in terms of flow change) could be generated by start up, shut down or, worse still, power failure of one of the large pump units and the need to keep the old contact tank was confirmed. The short term changes in flow on pump start and stop thus had much wider control implications. The retention of the tank also had other advantages to the works such as maintaining some output when equipment malfunctioned and allowing significant additional short term increase in output beyond the design capacity.

2.3 Pumping into a gravity supply

There are a number of large conduits in the UK, largely developed in Victorian times, that bring water from the mountainous catchments into the urban centres without the need for pumping. One of these, the Elan Aqueduct to Birmingham consists of a series of open channel aqueduct and tunnel sections punctuated by piped syphon crossings of the valleys. In total the length is over 100km; flow up to 390Mld is released at the upstream end and conveyed to the reservoir at Frankley in Birmingham without further flow control. Along the length there are paddle operated isolating sluices to guard against bursts and a number of overflows from the open sections to prevent surcharging. Actuated sluice valves on the two 60" steel mains of the main syphons ensure that for a range of flows head is maintained at the syphon inlets so that supercritical flows into the syphon pipes do not erroneously trip the paddle operated sluices.

The longest siphon reach is the Severn siphon (32km), which has a connection to the Trimpley treatment works on the River Severn. The Trimpley pump station which was built in the 1960s has a nominal capacity of 180 Mld but had never been utilised beyond 60 Mld. Constraints on the system included the capacity and pressure rating of the air vessel at Trimpley and the size of the feeder tank at Trimpley top, which is installed to avoid air inflow and expulsion at an intermediate peak on the profile.

Models were used to examine the optimum arrangement to convey flow from Trimpley whilst maintaining high flows from Elan. This included surge and capacity simulations, again surge and operating control issues both affecting the solution adopted.

2.4 Pumping over intermediate peaks

Water engineers traditionally adopt a variety of solutions to the problem of pumping to a point below the highest point on the profile. For a variety of reasons it is desirable but not always possible to keep the pipe pressurised throughout. For potable supplies this may be due to a requirement to avoid the possibility of contamination. For raw water or waste water it may be desirable to minimise the flow of air into and out of the main as this is a frequent cause of surge pressures and even two phase slug flow as pockets of air are moved en mass. The typical situation may be approached in a number of ways:

A) Reservoir or break pressure tank at profile peak

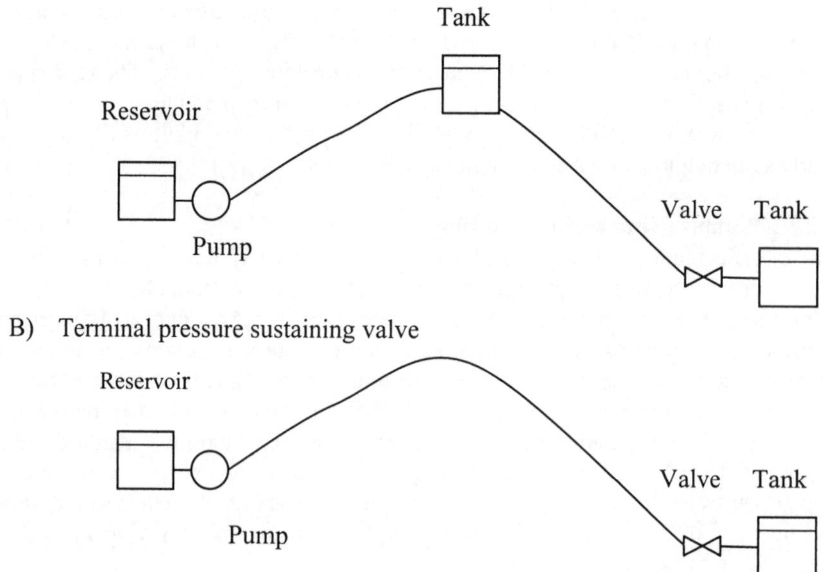

B) Terminal pressure sustaining valve

C) Allow downstream length to fill and empty or run part full

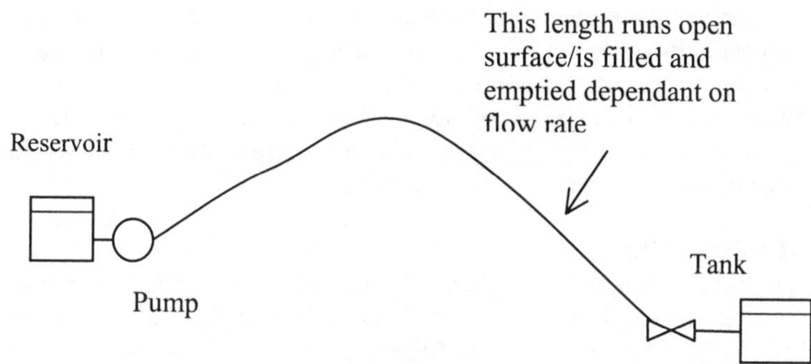

There are difficulties and options for each of these strategies which will need to be assessed and costs and risks taken into account. For example Halcrow are currently looking at a pipeline that will operate in the third type of mode. The scheme will use fixed speed pumps and minimum velocities at the inlet to the downstream tank will be achieved by ensuring that a minimum time of pumping is achieved. This can take 30-45 minutes to achieve and, without an enlarged pump sump, in normal pump cycling velocities may be only half the steady pumping condition. Another option considered was to include a terminal discharge valve and steady-state simulations suggested this to be an acceptable option. However, including the effects of surge this arrangement was shown not to be viable.

3 MODELLING TECHNIQUES

The preceding examples highlight some of the advantages in taking a wider 'holistic' type approach to pipeline analysis including effects of surge on control systems and strategies. The advantages of linking hydraulic and control models were also described in a previous paper (Evans et al)[3] particularly for water treatment works.

Development of software to enable a comprehensive model to be built that allows long and short term simulations is obviously a key factor in facilitating such an analysis. Halcrow have developed their HYDRAN program into such a tool. The original development of the modelling package is described in Pearce & Evans[2] which describes the use of a sparse matrix approach to the solution of the characteristic equations for pipelines as well as the St Venant equations for open channels. Over the years this approach has proved very adaptable and the later inclusion of control actions and signal processing has resulted in a very comprehensive package used in numerous analyses particularly for long conveyors. Recent studies have also included links and input to a water resources optimisation model looking at abstractions, storage and use of water for London from the River Thames.

4 CONCLUSIONS

For analysis of pipelines the effects of surge can be important, by changing the expected flows, for example, flow through a valve, this can have a significant impact on the desirable control system. A comprehensive modelling package that allows the incorporation of surge and control effects has been found to be particularly useful for the design of large pipelines and aqueducts, in the planning stage of upgrading and for operational studies.

REFERENCES

[1] E.P. Evans and P.V. Sage. Surge Analysis of a large gravity pipeline. 4th International Conference on Pressure Surges BHRA September 1983.

[2] R. Le B. Pearce and E.P. Evans. Surge Protection for Jubail Industrial City. Proc Instn Civ. Engrs, Part 1 February 1984.

[3] E.P. Evans, A.P.E. Green & T.M. Hoggart. Control Simulations for Water Treatment Works. In Computing and Control for the Water Industry, eds R. Powell & K.S. Hindi Research Studies Press 1999.

Water Hammer in Pipe Network: Two Case Studies

B. W. Karney *and* B. Brunone

ABSTRACT

Two case studies are presented to illustrate the behaviour of water hammer conditions in pipe networks as well as the challenges of representing this behaviour numerically. The first example demonstrates the importance of the network topology to the transient response: if branch connections are neglected, non-conservative predictions of peak pressures may result. In particular, a pressure wave created within a distribution system may propagate over various paths to produce a surprising superposition of waves at a remote location. In the second example, a dramatic pressure rise is caused by the essentially instantaneous closure of a control valve. The high-pressure wave decays more rapidly than is predicted by conventional quasi-steady friction models, and thus highlights the importance of unsteady friction. This example also provides indirect evidence of the high shear stress under transient flows through the persistent reports of red water associated with the valve closure event.

1 INTRODUCTION

Water hammer is known to have the potential for causing dramatic and sometimes serious pressure fluctuations. The hydraulic conditions that create such transient pressure waves are diverse, ranging from pump failures to simple valve closure. Two case studies are explored here to highlight the cause, control and significance of water hammer phenomena in a complex water distribution network.

The first example demonstrates the importance of a detailed representation of the pipeline network when considering water hammer conditions in transmission mains. In particular, the transients caused by power failure in the example are over predicted in the transmission system when the associated distribution system is neglected (as one would expect). However, more significantly, the distribution system response can at times be dramatically more severe than that of the transmission system.

The second case study relates to a field investigation of the pressures generated by sudden valve closure in a distribution system in Italy. The almost instantaneous valve closure in a steel pipeline system produces a high-pressure pulse, which propagates and decays within the pipe network system. What is of interest in the current study, though, is that the rate of decay of the pressure pulse is much faster than can be predicted by steady-state frictional head loss relations. In fact, recent speculations regarding the importance of unsteady friction effects are confirmed by this study in two ways, one quantitative and one qualitative. The quantitative evidence is presented by comparing the rate of decay of the transient relative to a standard water hammer model, while the qualitative evidence is suggested by water quality data (e.g., many red water and related complaints following closure). The paper contends that such effects must be considered if the dynamics and performance of water distribution systems are to be properly modelled and understood.

2 CASE STUDY 1: ROLE OF NETWORK TOPOLOGY IN THE SUPERPOSITION OF WATER HAMMER WAVES

An extensive study has recently been completed of the transient response of North Feeder main and portions of the associated Spy Hill water distribution network in the City of Calgary in Alberta, Canada. The general goal of this analysis was to determine the transient response of the system to various pump failure scenarios and combinations of network demand. More specifically, the intent was to determine whether the existing protection system has sufficient capacity to avoid excessive stress in the pipeline system following power failure and to recommend the best long-term strategy for surge protection in this line.

The entire study involved hundreds of different computer runs representing many different system configurations and protection strategies. The analysis worked with network and demand models that considered both the feedermain system alone (with demand points along its length to represent network connections) and significantly more refined models which represent not only the North Feedermain but also a considerable number of the key pipes in the distribution system.

Concern for transient conditions in this line is associated with two primary developments. The first is that the parts of the original surge protection systems installed for this line, in particular the slow closing air valves, are no longer functional. In addition, water demand originating in the Spy Hill distribution system is expected to progressively increase over the next twenty years, creating a greater hydraulic load and the need for larger pumps. Thus, future operating conditions may make the transient response more severe than it is currently.

2.1 System Description

The North Feedermain is a (primarily) 1350 mm steel pipe which runs between the Pump Station at the Bearspaw Water Treatment Plant and the Spy Hill West Reservoir. In general, the profile gradually climbs towards the reservoir along its 6.4 km length except for one relatively abrupt rise approximately one third of the distance along its length from the treatment plant. The line currently operates at approximately 100 m of head (about 150 psi).

As numerous scenarios of pumping and network demand were investigated, it became clear that this line had considerable potential for severe transients following power failure. In particular, the high wavespeeds, relatively long length of the line, the considerable static head and its unusual profile produce a system that is susceptible to water column separation at the knee in the line immediately downstream of the abrupt rise. In addition, a long length of the upper reaches of the line is subject to negative pressures immediately following power failure.

2.2 Methodology

All power failure simulations related to this study were carried out with the software program TransAM. TransAM is a general-purpose simulation model for calculating hydraulic conditions in pipeline systems. The program simulation uses the method of characteristics which is based on an established time increment and calculations are continued for a specified time interval (Wylie and Streeter, 1993; Chaudhry, 1987). Initial conditions are defined by a steady state description typically obtained from the transient model itself. In addition, the standard solution by the method of characteristics has been improved to allow flexible friction term linearization (Karney and McInnis,1992).

2.3 Protection Approaches and System Response

Various protection approaches were investigated for this line. One more unusual protection strategy that was given early attention was to introduce a check valve into the North Feeder-main approximately two thirds of the distance from the Pump Station. This valve would allow flow in the usual direction to pass through the line, but would prevent the reverse flow that causes the rapid column collapse following power failure. For this protection strategy to be successful, two specific conditions must be met: (i) sufficient network capacity must be present to supply short term flows into the demand locations during an extended power failure and (ii) insufficient network capacity must be present to prevent by-passing the check valve through the network connections, thus still allowing a severe collapse of the air cavities near the high point. Clearly, these two requirements are at odds and it is necessary to clarify the exact nature and capacity of the distribution system piping in the Spy Hill zone.

Thus a detailed layout of the piping and network connections in the Spy Hill zone was processed to construct several new network models at various levels of skeletal representation. Eventually, after conducting a variety of computer tests, an 89-pipe model was adopted as being more representative of this system than the original North Feedermain model that included only 17 primary pipe segments.

The more refined and detailed study showed two interesting results:

- The presence of the numerous but small pipe connections considerably changes the transient response of the North Feedermain system. In essence, the smaller pipe tap-offs provide escape routes for transient energy, thus partially mitigating the transient response of the primary feedermain. This first observation is expected and is well reported in the literature.
- However, more surprising is the fact that the smaller pipes in the distribution system allow a kind of "coordination" of the transient response. In effect, at least one pipe, and in many cases several pipes, are predicted in all computer simulations to experience large and destructive pressure magnitudes following column rejoinder.

The explanation for the unexpectedly large pressure rises appears to be in wave travel times: with the multiple paths that exist in a looped distribution system, some pipes invariably experience an almost equal travel time from the point of origin of the high pressure wave caused by the collapse of the air cavities at the air-vacuum valves. The convergence of waves in pipes having similar timing of wave propagation allows a "coming together" of waves from a common source, thus creating a very large total pressure through this superposition process.

The overall conclusion is that the column separation is best avoided in this system, and thus an air chamber provides the most complete and comprehensive protection. No combination of relief valves was able to relieve sufficient pressure in the system as a whole, and the check valve alternative that appeared to be so attractive in the simple model is effectively nullified by the density of the network connections.

2.4 Transient Response

Figure 1 shows the response of the simplified system to the failure of two pumps which are originally discharging approximately 1.2 m^3/s of water. Even though this system is protected by relief valves at both the pump station and at an intermediate node, the transient response is unacceptable. The maximum pressure is approaching or exceeding the 250 psi limit along the first half of the pipeline and the minimum pressures are too low on the other half. However, the pipeline model is in this case quite crude, since only the North Feedermain itself has been included in the pipe model.

If further details of the network pipeline are included in the analysis, the impression of the pipeline's response changes quite significantly. Figure 2 shows the same conditions in the network but with the 89-pipe model. Quite clearly, the transient response of the North Feedermain now looks quite acceptable, with the high pressure problems essentially eliminated and the low pressure problems considerably reduced. In fact, if the pressure relief valves in this refined network model are eliminated, the profile plot that summarizes the response of the North Feedermain system becomes essentially identical to that shown in Fig. 2.

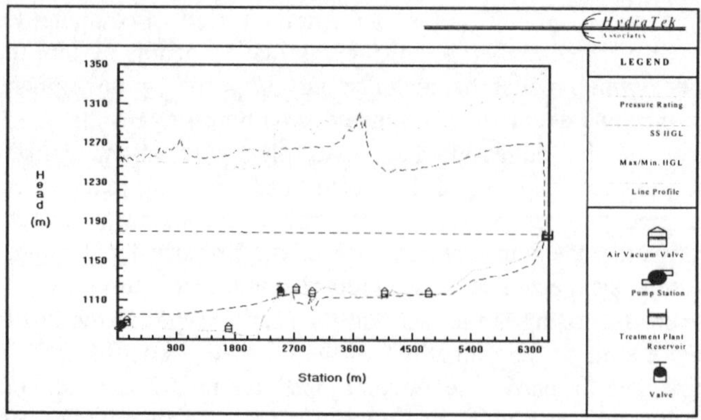

Figure 1 – Response to pump failure with crude pipeline model (1.2 m^3/s). Two 8-inch relief valves at the pump station and one 6-inch valve at node.

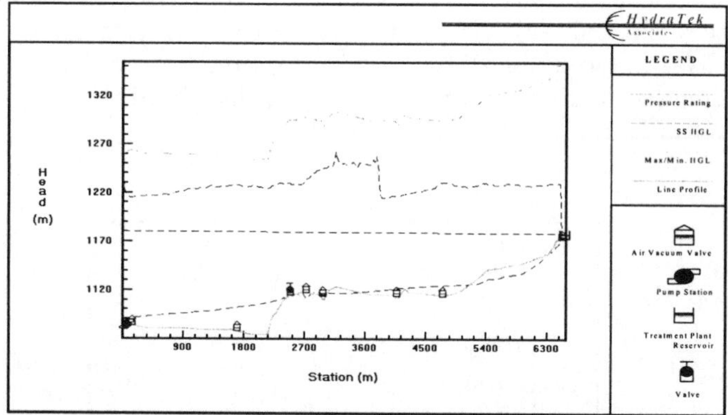

Figure 2 – Response to pump failure with refined pipeline model (1.2 m^3/s). Two 10 – inch relief valves at the pump station and one 8-inch valve at node 7.

One thing to note in Figures 1 and 2 is that although the high pressures appear to pose the most significant threat to the North Feedermain system, this is probably not the case. Rather, if the down surge following power failure is controlled, the high pressure wave will never form in the first place; since the cavities will not be created, they cannot collapse. This is the essence of the long-term transient protection system proposed for the North Feedermain system.

The events described above are typical whether there are two, three or more pumps operating at the pumping station. However, as the number of operating pumps increases, so does the severity of the resulting transient. If only two pumps fail, the transient is relatively mild and the North Feedermain requires little additional protection. If three or more pumps fail, however, the negative and positive pressures are severe enough to warrant installation of an air chamber.

2.5 Protection with Air Chamber

A transient investigation was carried out to assess whether an air chamber alone, or a combination of an air chamber with the previously determined relief valves, would provide more complete protection not only for the North Feedermain but also for the associated distribution system piping. Various combinations of relief valves and air chamber sizes have been investigated. The key of protecting the distribution piping is for the air chamber to be large enough to prevent the formation of air cavities along the line. (In addition, the air chamber should be large enough to prevent draining during a surge condition.)

The comment made earlier about the additional protection of the distribution piping that is furnished by an air chamber system is clarified in Figure 3. All three plots show the response at a node in the distribution system near to a location where transient waves converge. This node is just to the north of the feedermain and at the west end of the line. The pressure response associated with the case of two pumps failing is shown in (a) and the equivalent case of three pumps failing in (b). Protection in both these plots is by relief valves alone, and is clearly marginal for protecting this area of the distribution system, particularly in (b). However, plot (c) shows that a 50 cubic metre air chamber effectively eliminates the abrupt pressure rise associated with the relief valve strategy. The favorable behavior displayed in (c) is typical of all air chamber systems that are large enough to prevent column separation at the air-vacuum valves along the line.

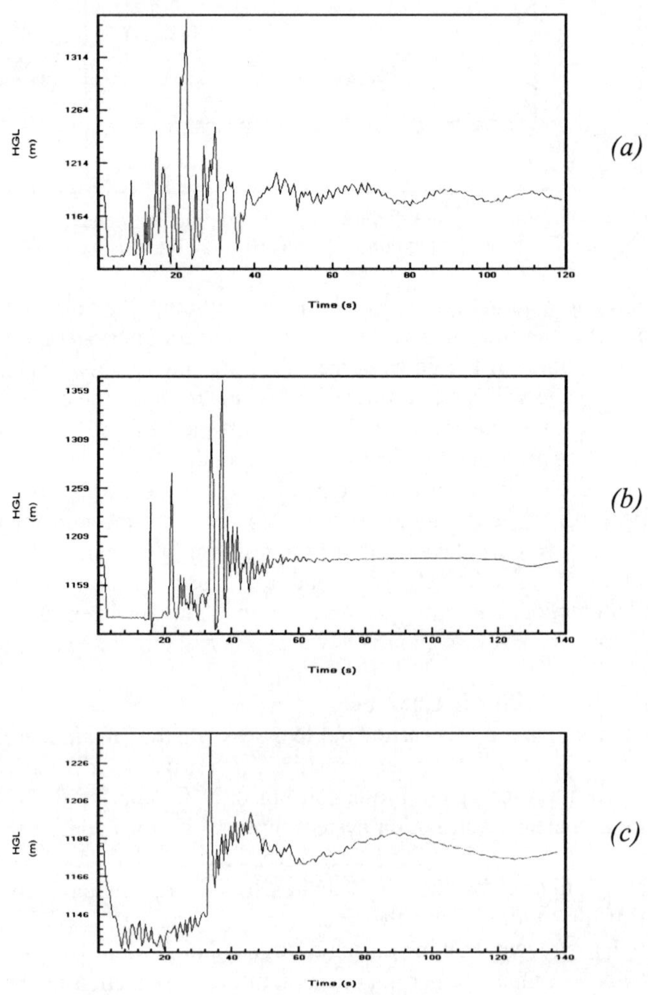

Figure 3 – The response at a node within the distribution system. (a) and (b) have protection from relief valves only, (a) with 2 pumps failing and (b) with 3 pumps failing; (c) shows the equivalent response with a 50 m^3/s air chamber.

The actual water distribution system will likely be able to damp transient pressures more quickly than the numerical model. In fact, the numerical model generally errs on the side of conservatism in that it usually predicts larger maximum and smaller minimum pressures than what would be observed in the field. Since this conservatism could be quite small, it cannot, and is not, relied upon. The rapid decay of water hammer pressures is considered in the next case study.

2.6 Summary Comments

Initially, it was expected that the details of the network connections would have a negligible impact on the transient response, only serving to slightly moderate the transient conditions in the main line. With this expected role, it was assumed it was reasonably conservative to ignore the transient response of the distribution system piping. Interestingly, this expected performance has turned out to be both accurate in one sense and misleading in another. As expected, the response of the North Feedermain itself does appear to be over predicted by a simple series representation. However, by contrast, the transients in the distribution piping have turned out to be quite considerable, with this portion of the system experiencing a kind of "magnification effect" compared to the North Feedermain itself.

3 CASE STUDY 2: TRANSIENTS GENERATED BY SUDDEN VALVE CLOSURE

3.1 System Description

The pipe network of Recanati (Italy), managed by *Consorzio Intercomunale* A.S.T., is supplied by Vallememoria well-field by means of a pumping station. In Figure 4 the complete system is shown (main pipes, all of steel, are indicated with heavy lines) while a detailed description of it is given in Brunone and Morelli (1999). In order to reduce pumping costs, one of the reservoirs of Recanati, the high level reservoir SAB, supplies the town only when, due to an increase in water demand, the pressure in the network drops below a threshold condition. To meet this constraint, an automatic control valve (ACV), of the Clayton type, has been installed on the *Le Grazie* pipe (nominal diameter DN = 200 mm, internal diameter D = 0.187 mm and length L = 630 m, with no branch) from SAB just upstream of the confluence node (Figure 5) with the *S. Pietro* pipe, from Eko, and *S. Martino* pipe, to Piottante reservoir. In Figure 6, an operation scheme of the ACV is given. By means of the ACV, both the downstream pressure, $p_d = p_C$, and the valve discharge, Q_v, can be fixed. Specifically, ACV opens (closes) when p_d is smaller (greater) than a fixed value p^*; alternatively, it closes when Q_v is greater than a fixed value Q^* (Brunone et al. 1995a). Both the threshold values p^* and Q^* along with the time rate of closing and opening of the ACV, v, can be chosen by adjusting some devices of the ACV pilot-circuit (in heavy lines in Figure 6). In the test presented here, the hydraulic control system is used to close the ACV almost instantly.

The confluence node (Figure 5) is equipped with strain-gauge pressure transducers (response time, T_R = 50 ms) for the measurement of both upstream

($p_u = p_1$) and downstream ($p_d = p_2$) pressures at the ACV. In addition, a magnetic discharge meter (T_R = 2 s) provides the flow Q_3 downstream of section 2, and a transducer (Figure 6) furnishes the position of the ACV plug (T_R = 2 s). Output signals from instruments are read directly into a PC.

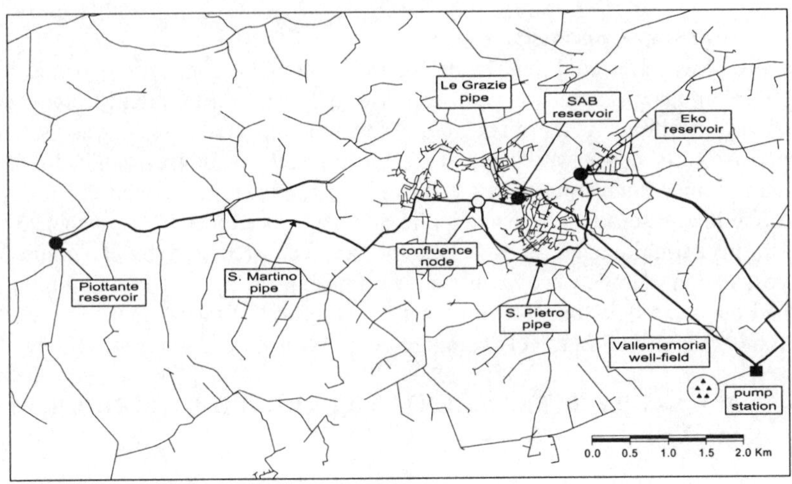

Figure 4 – Layout of the pipe system

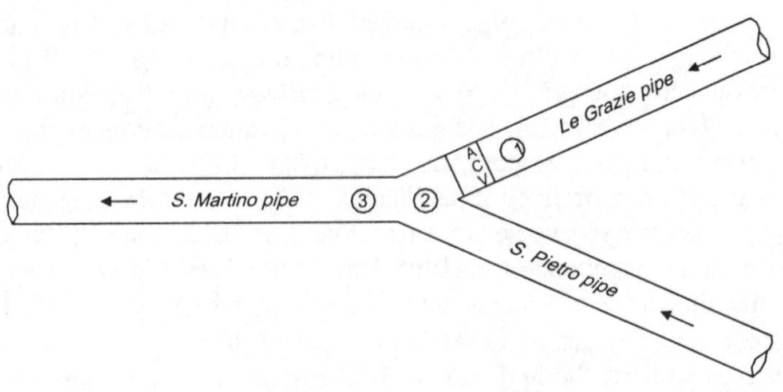

Figure 5 – Confluence indicating location of automatic control valve

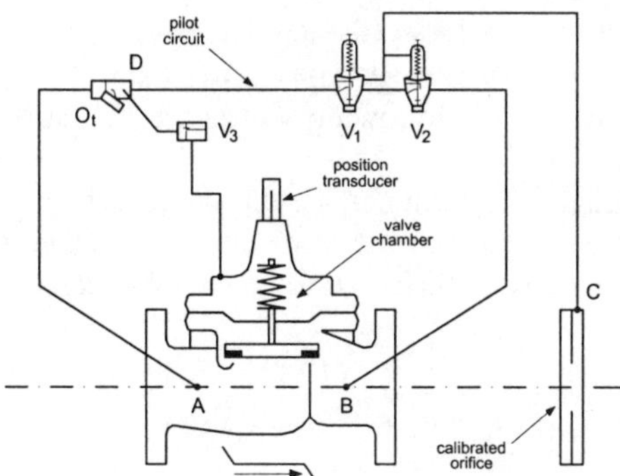

Figure 6 – Automatic control valve schematic

3.2 Transient Considerations

In order to evaluate transient effects of the opening and closing of the ACV in the *Le Grazie* pipe, field tests were carried out and a numerical model simulating all phases of the unsteady-state processes was proposed (Brunone and Morelli, 1999).

During tests on the operative pipe system, changes in the water demand which determine the operation of the ACV were simulated by the action of blow-off valves installed in the network. As threshold values, $p^* = 40$ m and $Q^* = 20$ l/s were used, whereas different values of v were considered.

In numerical modelling, based on the method of the characteristics (Wylie and Streeter 1993), attention was mainly focused on parameter estimation. In such a context, an unsteady friction model was used for the evaluation of the Darcy-Weisbach friction factor, f (Brunone et al. 1995b):

$$f = f_s + \frac{kD}{V^2}\left(\frac{\partial V}{\partial t} - a\frac{\partial V}{\partial s}\right) \tag{1}$$

where: f_s = friction factor from the Moody diagram, V = mean flow velocity, a = wave speed, s = axial co-ordinate, t = time, while the "decay" coefficient k is given by:

$$\frac{y_n}{y_{n-1}} = \left(\frac{1}{1+k}\right)^2 \tag{2}$$

in which y_n and y_{n-1} are the maximum piezometric heads, with respect to the steady-state value, in any two consecutive periods taken sometime after the end of the closing. As a consequence, three parameters had to be estimated: the characteristic roughness size ε (Colebrook-White equation), a and k on the basis of pressure and discharge data in steady-state conditions, the periodicity of pressure waves and the damping of pressure peaks after the completion of the closing, respectively. As a result, the following values were obtained: $\varepsilon = 0.37$ mm, $a =$ 1100 m/s and $k = 0.016$.

In order to properly account for the discharge behaviour of the ACV during transient conditions, the relevant flow-rate curve was determined using an *inverse problem* procedure by means of the well-known valve equation:

$$Q_v = C_d(\tau)A(\tau)\sqrt{2g(h_u - h_d)} \tag{3}$$

where C_d and A are the coefficient of discharge and the area of the ACV opening, respectively, and the function $C_d(\tau)A(\tau)$ is the valve flow-rate curve with τ = valve opening, (i.e., a dimensionless number describing valve position; valve closed if $\tau = 0$; valve fully open with $\tau = 1$). The function $C_d(\tau)A(\tau)$ may be determined by solving water-hammer equations by assuming as boundary condition at $s = L$ the time-history of $h_u = h_1$ and $h_d = h_2$, as measured during an unsteady-state test.

The current case study is closely related to Brunone and Morelli (1999), but extends the previous discussion by considering a more severe test case. Specifically, it considers a complex transient including both the opening and the subsequent closing of the ACV along with the intermediate automatic adjustment of the valve opening, which ends with a very fast closure. In Figure 7, the time history of h_1, h_2, Q_3 and τ for the above mentioned test is shown as well as the constant level of the *SAB* reservoir, $h_{SAB} = 91.60$ m. The reference datum for the piezometric heads is horizontal and passes through the confluence node ($z_2 = 198.5$ m, with z being the elevation above the mean sea level); t indicates time since the data acquisition began.

The complete closure of ACV (*fc* = final closure), has a duration T of 1.48 s, corresponding to about 1.3 Θ, with $\Theta = 2L/a$ being the characteristic time of the pipe. The corresponding time rate of closing of the ACV was $v_{fc} = (C_dA)_b / T = 0.26\ 10^{-3}$ m^2/s, with $(C_dA)_b = 0.384\ 10^{-3}$ m^2 being the value of C_dA at $t = t_b$ (= instant at which the final ACV closing begins), a larger valve of v than used previously in Brunone and Morelli (1999). At $t = 20.5$ s, the maximum value of $h_{1,max} = 109.8$ m, which gives rise to an over pressure with respect to the static head h_{SAB}, $\Delta h_{1,max} = 18.2$ m, is attained; such a value represents a much larger rise than considered previously in (Brunone and Morelli, 1999).

3.3 Water Quality Implications

As a further effect of the described transient, the quality of water supplied after the end of the tests was very poor ("red water") for a rather long period of time.

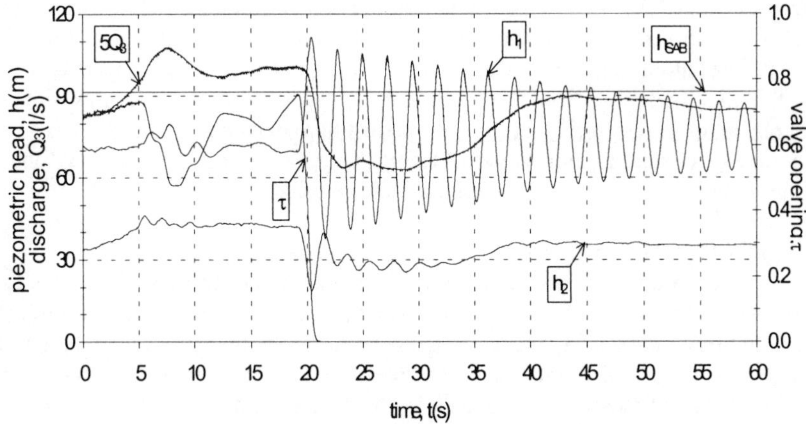

Figure 7 - Transient response due to control valve action

Specifically, tests began at about 11 a.m. and ended at about 12 a.m. At 6.00 p.m., i.e., about 6 hours after the conclusion of the test, *red water* complaints from consumers were recorded by the water supply company. Most of complaints referred to the *Le Grazie* quarter, which is crossed by the homonymous pipe but is supplied by conduits connected to the *S. Martino* pipe downstream the confluence node. Even though the duration of the water quality event may depend on the dynamics of water use in the town, the persistence of red water in the pipe system must be surely ascribed to high shear stresses associated with transients. Such a phenomenon is even more remarkable when one considers the very small values, with respect to steady-state conditions, of the discharge after the end of the manoeuvre. In other words, completely different shear stress mechanisms must take place during unsteady-state processes to take the rusty crust off the pipe wall notwithstanding the very small values of the mean flow velocity (Brunone et al. 1999).

3.4 General Comments

For a given value of V, the differences in energy dissipation between steady and unsteady flow are confirmed by the characteristic decay of the peak pressure. This decay which takes place after a complete closure is not accurately simulated by the so-called quasi-steady approach for friction forces under transient conditions (i.e., with $k = 0, f = f_s$).

As an example of the inadequacy of "traditional" friction models, Figure 8(a) compares experimental values of h_1 numerical results. Clearly, no significant reduction of pressure peaks occurs in the numerical simulation. By contrast, Figure 8(b) shows that an unsteady-state friction model (1), with $k = 0.016$, results in a much more accurate simulation of the h_1 time-history.

Both numerical simulations of Figure 9 are based on the use, as flow-rate curve of the ACV, of the same function $C_d(\tau)A(\tau)$ obtained within the described numerical procedure. In Figure 8, such flow-rate curve of the ACV is shown (heavy line) along with those (thin lines) relating to the other field tests carried out on the pipe system (Brunone and Morelli, 1999). Figure 8 points out some discrepancies between the curve from the test here considered and those from tests reported in (Brunone and Morelli, 1999). These discrepancies may be due to a different behavior of the ACV with high rates of closure.

4 FINAL COMMENTS

Water distribution systems are complex but integrated entities whose behaviour depends on the details of the pipeline system and the hydraulic state within it. Both case studies discussed in this paper indicate the modelling process must also have integrity – numerical models need to take seriously the topology of the network, the behaviour of the various devices in the system and must account for the full range of hydraulic conditions the system is subject to. In particular, transient states which mark the transition between more steady conditions introduce important mechanisms both from water quality and quantity point of view.

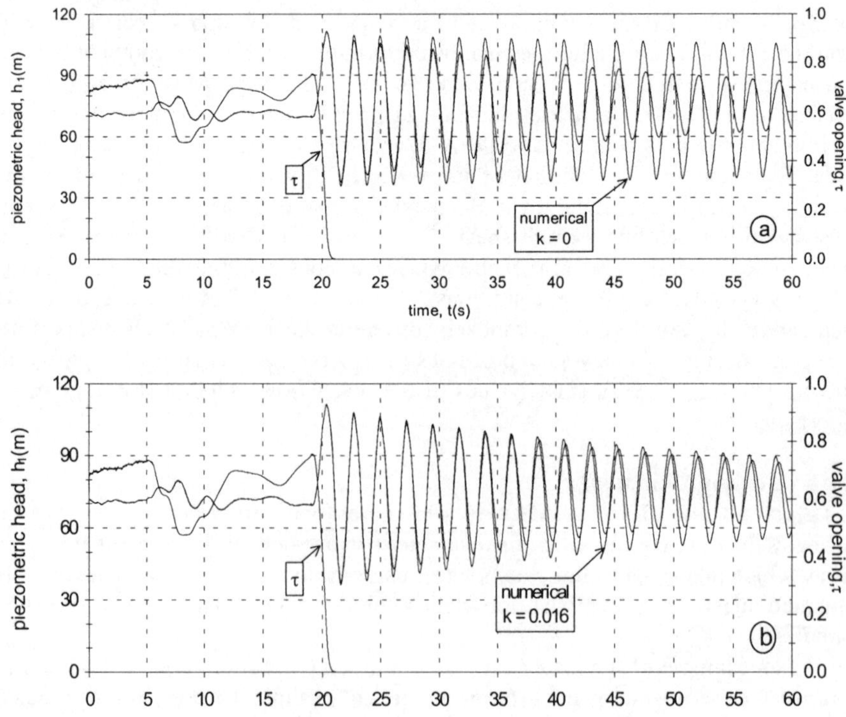

Figure 8 - Numerical simulations of valve closure

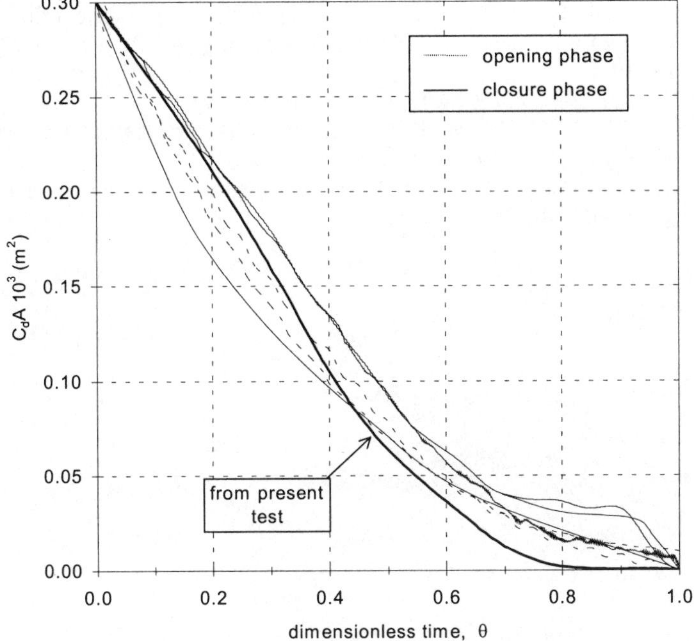

Figure 9– Automatic control valve operation (effective gate opening)

ACKNOWLEDGEMENTS

This research has been partially funded by the project "*Transient Pressures in Pressurized Conduits for Municipal Water and Sewage Water Transport*" of the European Union (contract no. SMT4-CT97-2188) and the Italian Ministry of University and Scientific Research, project "Whirling, turbulent and chaotic processes. Waterworks and environmental applications".

REFERENCES

Brunone, B., Barigelli, M., and Morelli, L. (1995a). "Effects of configuration on steady-state behavior of a pipe system". *Proc.*, *Symp.* on *Drinking Water Systems*, Bologna, Italy, 236-247 (in Italian).

Brunone, B., Golia, U. .M., and Greco, M. (1995b). "Effects of two-dimensionality on pipe transients modeling". *J. Hydr. Engrg.*, ASCE, 121 (12), 906-912

Brunone, B., Karney, B.W., and Ferrante, M. (1999). "Velocity profiles, unsteady friction losses and transient modelling". 26th Annual Water Resources Planning & Management Conference, ASCE, Tempe, Arizona (in press).

Brunone, B., and Morelli, L. (1999). "Automatic control valve induced transients in an operative pipe system". *J. Hydr. Engrg.*, ASCE, 125 (5), 534 - 542.

Chaudhry, M.H. (1987). *Applied Hydraulic Transients.* Van Nostrand Reinhold, N.Y.

Chaudhry, M.H., and Yevjevich, V. (1981). *Closed-conduit flow.* Water Resources Publications, Littleton, Colorado.

Karney, B.W., and McInnis, D. (1992). Efficient calculation of transient flow in simple pipe networks. *Journal of Hydraulic Engineering*, ASCE, Volume 118, No. 7, 1014--1030.

Thorley, A. R. D., 1991, *Fluid Transients in Pipeline Systems.*

Tullis, J.P. (1989). *Hydraulics of pipelines, pumps, valves, cavitation, transients.* John Wiley & Sons, Inc., New York, N.Y.

Wylie, E.B., and Streeter, V.L: (1993). "Fluid transients in systems", Prentice Hall, Englewwod Cliffs, N.J.

On Simplifying the Modelling of Transients in Drinking Water Distributions Systems

I. Pothof

ABSTRACT

Water hammer analyses are performed frequently if a water distribution network is extended. The criterion for the size of provisions is that under-pressures should not occur due to pump trip. The water hammer analysis for a large network becomes a tedious job, even if data about the network are available in GIS format. The network has to be simplified to overcome unacceptable computation times.

A method is presented to simplify distribution systems in such a way that the transient pressures are calculated as if the complete system would have been calculated. The method is based on identifying sub-networks and representing these sub-networks with non-return taps and extensions thereof.

The accuracy of several sub-network representations is estimated with a simulation study. Conclusions from this study are as follows. The non-return tap behaves correctly with respect to the average pressure at the sub-network junction. If the sub-network volume is small compared to the main pipeline volume, then the pressure error is less than 5 m in a test system. As the volume ratio between sub-network and main pipelines increases the influence of the storage volume of the sub-network becomes significant. In these situations the tap should be extended with a single pipe with changing diameter such that the pipe velocity equals the velocity in a selected sub-network pipeline. Large sub-networks are represented efficiently by these combinations of a single pipeline and a tap. The pressure error remains less than ± 3 m in the elaborated test systems.

1 INTRODUCTION

WL | DELFT HYDRAULICS has consulted many drinking water distribution companies (especially in The Netherlands) on the application of safety provisions such as air vessels and fly wheels in their distribution systems. The water hammer simulation program WANDA is used to calculate the transient behaviour of a pump trip at maximum capacity.

1.1 History

The most important part of a water distribution system project is the modelling of the system. It is not always feasible to model every single pipe in the network for a transient calculation. Consequently the network is simplified in a reasonable manner to a main pipeline system and a set of sub-networks. Reasonable means in this respect that the main pipelines are identified, based on pipe diameters, flows and the network layout. Now the distribution system consists of a main pipeline system and a certain amount of tap location flows, representing the flows into the sub-networks.

A TAP is a component in WANDA which has been used for many years. A TAP behaves like a normal partially opened tap, allowing the specified flow. In a dynamic situation the tap flow decreases as the upstream pressure decreases. Once the upstream pressure equals the downstream pressure the flow becomes 0. When the upstream pressure decreases further two phenomena may take place: a return flow may enter the system or a check valve closes and the flow remains 0 until the upstream pressure again rises above the downstream pressure. For this reason, two types of TAP exist: a return TAP and a non-return TAP. It is assumed that the non-return TAP models the behaviour in water distribution systems most realistically; one may think of industrial water users having check valves installed next to their flow meter or household washing-machine taps, which are equipped with built-in check valves. Another reason to use non-return TAPs in water hammer simulations is the fact that a conservative scenario is calculated if non-return TAPs are applied. The minimum pressure in the system is lower with non-return TAPs than with return TAPs, because of the return flow capacity of the return TAPs. The TAP is specified with the following parameters:

- initial flow
- downstream pressure

The initial flow is normally set to the flow of the corresponding sub-network in the maximum capacity scenario. The downstream pressure is normally set to the elevation head of the sub-network. In the steady state preceding the pump trip the upstream TAP pressure has a certain value higher than the downstream pressure. From the pressure difference across the TAP and the initial flow a quadratic resistance coefficient is calculated: $C = \Delta H / Q^2$. This coefficient remains constant during the transient calculation.

1.2 Objectives

The use of non-return TAPs becomes less realistic as the size of the sub-network increases. It is obvious that a large sub-network has a certain return flow capacity into the main pipeline system. The return flow of a large sub-network affects the minimum pressure in the main pipelines due to pump trip in a positive manner. If a return flow from the sub-network develops, the required size of an air vessel or fly-wheel will decrease.

The following questions arise from the observations above:

- What is the relation between the minimum pressure at main pipeline junctions in the complete system and at TAP junctions in the simplified system in which sub-networks have been substituted with TAPs?
- How should the sub-network be modelled or the TAP be improved to produce more accurate, less conservative minimum pressures?

- Does a relation exist between the improved TAP minimum pressure and the minimum pressure in the sub-network?

Answers to these questions are discussed in this paper. A set of 3 simple test networks has been used to investigate the behaviour of the TAP and several improvements.

The three test systems and their simplifications are discussed in section 2. Section 3 discusses the standard TAP behaviour and its limitations. Section 4 describes several approaches leading to an improved TAP model. Simulation results to compare the different approaches are discussed in section 5. Finally the conclusions and recommendations are described in section 6.

2 APPROACH

Three test systems are used in this paper. The test systems are comparable with respect to the pumping station, the maximum capacity and the diameter of the main pipelines. They differ in the layout of the sub-networks. Two extreme sub-network layouts have been modelled. Test system 1 contains equally distributed water demands in the sub-network. Test system 2 contains water demands only at the end of the sub-network branches, all demands are located at the same distance from the junction. The third test system contains two sub-networks: an equally distributed and a concentrated sub-network. Test system 3 has been used to check interactions between sub-networks.

The pumping station is capable of supplying 1000 m^3/h at 44.5 metres. The specific speed of the pumping station equals 51 $m^{3/4}/s^{1/2}\cdot min$. The polar moment of inertia equals 3 kgm^2. There are no elevation differences present in the network. A check valve with $\xi = 5$ to account for the losses in the pumping station is present downstream of the pump. The delivery head in the sub-networks downstream of the taps is 4 m above the pipe level. The pumping station delivers 1000 m^3/h into the test systems. The layouts of the test system sub-networks are described in the sections below.

The hydraulic schemes consist of two networks: the so-called complete test system and the simplified system. The complete test systems contain a full sub-network model with branches and taps at the end of the branches. In the simplified system the complete sub-network model is substituted with a single tap or improved taps, representing the complete sub-network. Both systems are equipped with virtual pressure sensors to monitor the pressure at the sub-network junction and at the tap in the simplified system. A WANDA control component is used to subtract both pressures in order to monitor the pressure error between the complete and the simplified system directly: $e(t) = H_{real}(t) - H_{simple}(t)$. A positive error means that the pressure in the simplified system is lower than the pressure in the complete system. A positive error is considered better than a negative error, because in that case a conservative minimum pressure estimate is calculated in the simplified system. If the conservative minimum pressure remains above the criterion value (atmospheric pressure usually), then it is guaranteed that the real pressure remains above the criterion value as well.

The behaviour of the standard tap and several improvements as a substitute for a sub-network is investigated in a full pump trip scenario in the three test systems. The system behaviour of the tap is presented by means of the time series of the pressure at the TAP junction in the main pipeline and the flow from the main pipeline or the flow into the sub-network. The time function of the monitored pressure error is observed in detail and described by means of the following three quantities:

- band width of the error while the primary pressure wave travels through the system,
- the amplitude of secondary pressure waves (due to reflections) and
- the remaining systematic error.

The time frame related to the primary pressure wave is defined as the travel time required to reach the end of the most remote branch of the system. Furthermore the minimum pressures in the complete and simplified systems are tabulated. The behaviour of the standard tap and the improved modelling approaches is compared on these quantities. The three test systems are described in more detail in the paragraphs below.

2.1 Test system 1: equally distributed sub-network

The main pipeline consists of a single transportation pipe (D = 600 mm) of 10 km in length. The sub-network consists of a main pipe of 10 km in length as well. Its diameter is 600 mm in the first 3 km of the sub-network pipeline and 400 mm in the other 7 km. A bifurcation is present every kilometer in the sub-network pipeline. The length of the bifurcating pipeline is also 1 km (D = 200 mm). Every tap in the sub-network demands 100 m^3/h. The lay-out and complete specifications are described in appendix Figures.

2.2 Test system 2: tree-structure sub-network

The main pipeline consists of a single transportation pipe (D = 600 mm) of 5 km length. The sub-network consists of a tree structure of bifurcations with three levels. Consequently the tree structure contains 8 end points and taps. Every bifurcating pipeline is 1 km in length. The diameter of the first 2 levels equals 400mm. The diameter of the 3rd level is 200 mm. Every tap in the sub-network demands 125 m^3/h. The layout and complete specifications are described in appendix Figures.

2.3 Test system 3: two sub-networks and infinite pipe

The main pipeline consists of a single transportation pipe (D = 500mm) of 2 km length. The first sub-network bifurcates after 2 km. The main pipeline continues with a diameter of 400 mm. After another kilometer the second sub-network bifurcates. Then the main pipeline continues with an infinite pipe. The first sub-network is 900 m in length with diversions and TAPs every 300 m. The first sub-network contains 5 TAPs, each demanding 20 m^3/h. The second sub-network contains a single TAP after 900 m, demanding 100 m^3/h. The diameters of the pipes in both sub-networks are 200 mm. The layout and complete specifications are described in appendix Figures.

2.3.1 Standard TAP behaviour

The TAP response is presented by means of the time series of the pressure at the TAP junction in the main pipeline and the flow from the main pipeline or the flow into the sub-network. The TAP behaviour is investigated on a complete pump trip (power failure). The pumps trip after 0.5 seconds.

2.4 Test system 1

The pressure wave speed needs approximately 28 seconds (c = 370 m/s) to reach the sub-network junction; see also Figures 2 - 4. In the complete system the under-pressure wave enters the sub-network. The first 3 pipes in the sub-network have been modelled with the same parameters as the main pipeline. For this reason the under-pressure wave continues into the sub-network without any reflection. Thus the flow from the main system must still decrease. After 28 seconds the pressure head in the junction of the complete sub-network remains constant at 15 m for 15 seconds, because small over-pressure reflections compensate the further decrease of the pressure. The consequence of this temporarily almost constant pressure at the junction is the constant flow into the sub-network (only line-packing effects are responsible for the slower rate of decrease). The pipe diameter decreases from 600 mm to 400 mm after 3 km in the sub-network (double travel time equals 16 seconds). After 43 seconds the primary under-pressure wave has reflected on this smaller diameter and returned to the junction to draw the junction pressure further downward. Further on the diameter in the sub-network remains at least 400 mm and the pressure head at the junction remains slightly above 4 m (delivery head) until all non-return taps in the complete sub-network have closed. Only then the pressure head further decreases to 0 m.

The simplified model shows a slightly different behaviour. The pressure and flow at the junction are in accordance with the complete system in the first 25 seconds, because the under-pressure wave has not reached the junction yet. The under-pressure arriving at the sub-network junction reflects at the resistance in the TAP. The resistance allows a fraction of the original discharge, because the downstream pressure is still more than 4 m. This means that an additional under-pressure returns from the TAP into the main system, while the complete system is only subject to line packing effects. The pressure at the junction slowly decreases further between 28 and 80 seconds because the tap is still delivering a decreasing amount of water. After 80 seconds the pressure wave has travelled back and forth through the main pipeline and returns at the tap junction as a smaller under-pressure wave, causing the pressure to drop below the delivery head of the tap and the tap to close. The tap remains closed during the rest of the simulation. The remaining small over-pressure wave is now travelling between two closed boundaries (check valve and tap) causing small over-pressure shocks at the tap. These over-pressure shocks are compensated by line packing effects.

The pressure error (Figure 4) is exactly 0 during the first 28 seconds. After 28 seconds the flow decrease in the complete system is larger than the flow decrease in the simplified system, because the complete system has a larger hydraulic storage volume at the junction because of the storage volume in the adjacent sub-network. The simplified system only incorporates the storage volume of the main pipeline and therefore shows a smaller decrease in discharge. This observation means that the simplified pressure decreases further than the real pressure, resulting in a positive pressure error. The maximum pressure error is +7.3 m. Once the pressure wave in the complete sub-network has reflected on the smaller diameter, the pressure error drops down to -1 m. Due to secondary reflections the pressure error fluctuates between -2 m and +2 m. Once the minimum pressure is attained at the junction the pressure error is small in this test case. It may be concluded that the standard tap describes the basic time-averaged behaviour of the sub-network correctly, although the dynamics related to the storage volume of the sub-network are ignored. Nevertheless this deficiency does not lead to systematic errors in the pressure at the junction. The imperfections of the tap will be discussed in detail in section 2.7.

2.5 Test system 2

Test systems 2 and 3 will be discussed in less detail, because the basic phenomena are comparable with test system 1. Special attention is paid to the network characteristics of these test systems. Test system 2 is the network containing the tree-structured sub-network. The main pipeline is 5 km instead of 10 km, leading to a larger under-pressure wave arriving at the junction than in test system 1 (see Figure 9), because the under-pressure wave is decreased in the shorter main pipeline. As opposed to test system 1 the primary pressure wave due to pump trip now leads to a lower initial pressure at the junction in the complete system compared with the simplified system. The initial under-pressure arrives at the junction after 14 s. There is one reason for the sign of this initial pressure error. The two pipes of the sub-network in the complete system have a diameter equal to 400 mm, while the main pipeline has a diameter of 600 mm. Hence the cross section area decreases at the junction and the discharge is decreased less than when using the standard non-return tap. One kilometer into the sub-network the 400 mm pipe splits into two 400 mm pipes. This area increase leads to a reflecting over-pressure wave arriving at the junction after 20 seconds. The returning over-pressure wave is overruled by the pressure wave reflecting at the end of the tree structured sub-network, resulting in a sudden pressure decrease at the junction after 26 seconds. The pressure decreases suddenly because all demands are located at the same distance of 3 km from the sub-network junction. Meanwhile the temporary over-pressure due to the cross section area increase in the sub-network travels back and forth through the main pipeline and arrives at the junction again after 45 s. This temporary over-pressure damps quickly when entering the sub-network.

The standard tap in this system behaves comparably to the tap in test system 1. The reflections mentioned above are generated in the sub-network and are neglected when using the tap. The tap closes after 25 seconds and then shows a behaviour with a period of 26 s (corresponding with the double travel time in the main pipeline).

The maximum pressure error equals +6.5 m due to the temporary over-pressure generated in the sub-network. When the minimum pressure is reached at the junction in the complete system the pressure error attains its minimum of -4.2, meaning that the tap overestimates the minimum pressure. Consequently the non-return tap does not necessarily model the complete system in a conservative way with respect to minimum pressures, as illustrated in this case. Nevertheless it can be concluded that the tap models the global time-averaged behaviour correctly without systematic errors, but lacks the reflections generated in the sub-network. In this case these reflections prove to result in the critical pressures.

2.6 Test system 3

In test system 3 the distance from the pumping station to the sub-networks is only 2 km. As opposed to the other test systems, test system 3 contains 2 sub-networks demanding only 100 m^3/h each and an infinite pipe demanding the remaining 800 m^3/h. To prevent interaction with cavitation phenomena, cavitation has not been included in this simulation and, as a consequence, pressures may drop below vapour pressure (app. -10 m). This case is best compared with test system 1. Both taps in the simplified system close due to the primary under-pressure, while the sub-networks return a certain flow into the main pipeline; -100 m^3/h from the first and -10 m^3/h from the second sub-network. As a consequence the pressure error at the junction increases (see Figures 14 to 16). The error is made at the first tap in the simplified system and transferred to the second tap. The maximum pressure error is +3.8 m after 10 respectively 12 s. A small over-pressure reflects on the infinite pipe (diameter change from 400 mm to 500 mm). This over-pressure is absorbed by the sub-networks in the complete system between 15 and 20 s, while the over-pressure passes by the closed taps without being disturbed. For this reason the pressure error drops to -4.8 m (first sub-network) and to -4 m (second sub-network) after 18 s.

Again the final average pressure error is 0, but the minimum pressure at the first junction is a consequence of the storage volume and a reflection in the first sub-network. This minimum pressure is not calculated by the simplified system.

The imperfections of the standard tap are discussed in the next section.

2.7 Imperfections of the standard tap

It has been reasoned above that the standard tap has three basic imperfections leading to inaccurate minimum pressure calculations:

- reflections, generated in the sub-network, are not revealed by the tap,
- the storage volume, capable of supplying a certain return flow to the main system, is not present in the non-return tap.
- the minimum pressures in the simplified system using non-return taps are not necessarily lower than the minimum pressures in the complete network.

The first and second aspect become more important as the relative volume of the sub-network increases. The effect of a return flow is negligible if the sub-network is small compared to the main pipeline system. On the other hand if the sub-network is very large, the return flow will dominate the system response. The size of a sub-network relative to the main pipeline system can be expressed in terms of a relative volume. The relative volume of a sub-network in a distribution system will be the total sub-network volume divided by a fraction of the total main pipeline volume. Define:

$V_{m,i}$ = Volume of main pipe i, [m^3]
$V_{m,j}$ = Volume of main pipeline network dedicated to sub-network j, [m^3]
$V_{s,j}$ = Total volume of sub-network j, [m^3]
$q_{i,j}$ = Steady flow through main pipe i, to sub-network j, [m^3/h]
q_i = Total steady flow through main pipe i, [m^3/h]
R_j = Relative volume of sub-network j in the complete system [-]

$$V_{m,j} = \sum_i \frac{q_{i,j}}{q_i} \cdot V_{m,i}$$

$$R_j = \frac{V_{s,j}}{V_{m,j}}$$

Using these definitions the following relative volumes are calculated for the test systems:

Test system	relative volume sub-network 1	relative volume sub-network 2
1	0.72	-
2	0.71	-
3	0.16	0.05

The larger the relative volume the more important the modelling of the sub-network becomes. From the table above it is expected that an improved tap, taking into account a more accurate storage volume, will affect test systems 1 and 2 and have less influence on test system 3, because of the values of the relative volumes. It is also assumed that sub-network 1 in test system 3 is more dominant than sub-network 2 in this test system. This assumption is confirmed by the results of the standard tap, because the error that occurred in the first sub-network is transferred to the second sub-network.

The following error ranges have been found using the standard tap to represent the sub-networks. To check the importance of the storage volumes test system 1 has been simulated with several main pipe lengths ranging from 1 km to 20 km (original value is 10 km). The error width is the difference between the maximum and pressure minimum error.

Test system	relative volume largest sub-network	Error range [m]	Error width [m]
1, 1 km	7.2	[-2; 20]	22
1, 5 km	1.4	[-4; 11]	15
1, 7.5 km	0.96	[-3; 9]	12
1, 10 km	0.72	[-2; 7.3]	9.3
1, 20 km	0.36	[-6; 2.8]	8.8
2	0.71	[-4.2; 6.5]	10.7
3	0.16	[-4.8; 3.8]	8.6

As expected the error width decreases as the relative volume of the sub-network decreases. The error width does not tend to 0, because the reflections of the primary pressure wave, generated in the sub-network, can not be calculated using the standard tap. It is concluded from the table above that an improved tap model should always take reflections into account. The storage volume should be taken into account if the relative volume of the sub-network is more than 0.5. Two extended tap models will be discussed in chapter 4.

3 EXTENDED SUB-NETWORK REPRESENTATIONS

The reflection behaviour of a sub-network depends on the travel times in the sub-network. In order to model these reflections a pipe should be included in the tap model. The extensions discussed in this chapter are limited to extensions which consist of a single pipeline and one tap, as the objective is a significant simplification of the system.

The first extended representation is based on an appropriate choice of the pipe length of a sub-network, which has the same pipe specifications as the first pipe in the sub-network. The second extended representation is slightly more sophisticated and allows diameter variations in the sub-network. Both approaches are discussed in the sections below.

3.1 Centre-point of gravity pipe length

A fraction of the water flows to the end of the sub-network branches, another fraction is demanded in the first pipe of the sub-network. The effect of all demands and distances from the junction can be accounted for by choosing a centre-point of all demands. Define the centre-point of gravity (*cpg*) length of sub-network *j* as follows:

$$l_{cpg,j} = \frac{\sum_{i=1}^{N_j} l_{j,i} \cdot q_{j,i}}{\sum_{i=1}^{N_j} q_{j,i}}$$

where

$l_{j,i}$ = Distance from the sub-network junction to tap location *i* [m]

$q_{j,i}$ = Flow to tap *i* in sub-network *j* [m^3/h]

N_j = Number of taps in sub-network *j* [-]

Represent the sub-network by a pipe with *cpg*-length and specifications equal to the first pipe in the sub-network. At the end of this pipe, the tap allowing the sub-network flow is modelled. In this way the volume and a weighted reflection point of the sub-network are characterised. This approach is simulated in the three test systems in cases testX_2, where 'X' refers to the test systems.

3.2 Constant velocity pipe diameter

Water hammer pressures and reflections are strongly determined by fluid velocities (Joukowsky). In order to model an additional improvement to the *cpg*-length the velocities along the *cpg*-pipeline should equal the velocities in the sub-network. The most important path in the sub-network is the path along which the largest amount of water flows. If several paths (or sub-paths) contain the same flow, one of these paths may be selected arbitrarily. It is not necessary to search for a path which is longer than the centre-point of gravity length. The length of this path may be shorter than the centre-point of gravity length. Substitute pipes should be modelled along this sub-network path such that two conditions are fulfilled. First the velocity in the substitute pipes at a distance x from the junction should equal the velocity in the sub-network path at the same distance x from the junction. It should be taken into account that the velocity in the complete sub-network decreases as sub-network taps withdraw water from the sub-network, while all water flows through the substitute pipes to the single substitute tap at *cpg*-distance at most. Secondly the wall thickness in the substitute pipes should be adjusted such that the wave speed velocity in the substitute pipe equals the corresponding wave speed velocity in the sub-network path. At the end of these substitute pipes the tap is allocated with a downstream head equal to the delivery head of the sub-network and a demand flow equal to the sub-network demand flow. In many situations the water distribution in the sub-network is not known exactly. If the water distribution is not known, velocities in the substitute pipe can not be calculated. In this case the local water distribution should be estimated based on available statistical data and knowledge of the local situation. Then the velocities can be calculated in the substitute pipeline. This approach is simulated in the three test systems in cases testX_3.

4 TESTING THE EXTENDED TAP MODELS

Calculation of the centre-point of gravity lengths in the three test systems results in the following lengths:

Test system	*cpg* length sub-network 1 [m]	*cpg* length sub-network 2 [m]
1	5500	-
2	3000	-
3	660	900

Calculation of the constant velocity diameters leads to the following diameters and wall thicknesses:

Test system	distance from junction [m]	diameter sub-network 1 [mm]	wall thickness sub-network 1 [mm]
1	1000	600	30
	2000	632	31
	3000	667	33
	4000	478	24
	5000	516	26
	5500	566	28
2	1000	566	28
	2000	800	40
	3000	566	28
3	300	200	8
	600	258	11
	675	447	18

The results of both extended sub-network representation models are compared with the behaviour of the complete test systems and the simplified tap test systems. The results are compared on the following quantities:

- Pressure error range between the complete and simplified models
- Pressure error amplitude during secondary reflections
- Systematic error during secondary reflections
- Location of minimum pressure in the complete test system
- Value of the minimum pressure in the complete test system
- Location of the minimum pressure in the main pipelines
- Value of the minimum pressure in the main pipelines

The time graphs of the pressure error are drawn in Figures 4 to 6 for test system 1, in Figures 10 to 12 for test system 2 and in Figures 16 to 18 for test system 3.

Test 1	Simplification error characteristics			Minimum pressure			
	range [m]	ampl. [m]	dev. [m]	overall location	[m]	Main pipeline	[m]
real	-	-	-	(AE)	-2.4	Preal1 (1045)	-1.9
test1_1	[-2.7; 7]	1.5	-0.5	-	-	Psimpl1 (0)	-0.1
test1_2	[-4.3; 3]	1.2	-1.2	-	-	Psimpl1 (149)	-1.1
test1_3	[-2.8;1.6]	1.5	-1.0	-	-	Psimpl1 (896)	-1.5

The complete system of test 1 attains its minimum pressure in the sub-network in node AE (see Figure 1). The minimum pressure in the main pipeline is -1.9 m (1.9 m below atmospheric pressure) and is found after approximately 1 km. The simplified models are not able to reproduce the minimum pressure in the system, although the deviation from the minimum pressure decreases from 1.8 m to 0.4 m. The simplification error range decreases from 9.7 m in test1_1 to 4.4 m in test1_3. The systematic error varies between -0.5 m and -1.2 m.

Test 2	Simplification error Characteristics			Minimum pressure			
	range [m]	ampl. [m]	Dev. [m]	overall location	[m]	Main pipeline	[m]
real	-	-	-	Preal1 (3485)	-3.3	Preal1 (3485)	-3.3
test2_1	[-4; 6.7]	2	0.0	-	-	Psimpl1 (303)	-2.0
test2_2	[-7: 13]	3	-2	-	-	Psimpl1 (3333)	-5.4
test2_3	[-2; 2]	1.5	0			Psimpl1 (3485)	-2.6

In this test case (tree-structure) the minimum pressure occurs in the main pipeline. Test2_2 shows a systematic deviation from the complete system. The reason is the large diameter of the long centre-point of gravity pipe in the simplified model. Test2_3 remains within a deviation of 2 m and the minimum pressure is overestimated by 0.7 m. The simplification error range decreases from 10.7 m in test2_1 to 4.0 m in test2_3.

Test 2	Simplification error Characteristics			Minimum pressure			
	Range [m]	ampl. [m]	Dev. [m]	overall location	[m]	main pipeline	[m]
Real	-	-	-	(H)	-36.2	Preal1 (2000)	-11.9
						Preal2 (231)	-12.6
						Preal3 (1000)	-16.3
test3_1	[-5; 4]	0.2	0	-	-	Psimpl1 (2000)	-11.2
	[-4; 4]	0.2	0	-	-	Psimpl2 (462)	-12.1
				-	-	Psimpl3 (1000)	-16.3
test3_2	[-4.5; 4]	0.2	0	-	-	Psimpl1 (2000)	-11.6
	[-4; 4]	0.2	0	-	-	Psimpl2 (846)	-12.1
				-	-	Psimpl3 (1000)	-16.3
test3_3	[-3.3;3.2]	0.25	0	-	-	Psimpl1 (2000)	-11.3
	[-2.3;2.8]	0.2	0	-	-	Psimpl2 (615)	-13.2
				-	-	Psimpl3 (1000)	-16.3

Test 3 is the test system with 2 sub-networks. The sub-networks are small expressed in relative volume. Nevertheless the extended sub-network representation shows better results with respect to the pressure error decreasing from 5 m to 3 m.

It is immediately concluded from the tables above that the results improve significantly after application of the extended sub-network representation consisting of a pipe with constant velocity diameter and a single tap. The minimum pressure in the main pipelines deviates less than 1 m if the extended model with constant velocity diameter is applied.

5 CONCLUSIONS AND RECOMMENDATIONS

A simulation study has been performed to estimate the accuracy of several sub-network representations, based on a single non-return tap. Conclusions from this study are as follows:

- The non-return tap behaves correctly to an accuracy of 2 m in the minimum pressure in the main pipeline, although temporary deviations in the pressure at sub-network junctions of 7 m may occur. The accuracy of the non-return tap improves with respect to the minimum pressure in the main pipeline to a level of less than 1 m if the relative volume of the sub-networks is smaller than 0.5. The minimum pressure in the complete systems is lower than in the simplified system.
- Reflections generated in the sub-network can not be accounted for in a standard non-return tap.
- As the volume ratio between sub-network and main pipelines increases the influence of the storage volume of the sub-network becomes significant as well.
- In these situations the tap should be extended with a single pipe with changing diameter such that the pipe velocity equals the velocity in a selected sub-network pipeline. Large sub-networks are represented efficiently by these combinations of a single pipeline and a tap. The pressure error remains smaller than 3.5 m in the elaborated test systems. The minimum pressure in the main pipeline of the simplified test systems deviates less than 1 m from the minimum pressure in the complete test systems.

The following recommendations are made:

- Test the behaviour of the improved TAP in highly looped systems, for many sub-networks have two connecting junctions to the main system.
- Test a large distribution network with many small sub-networks with standard taps and extended models.
- Consider the extended tap behaviour with respect to valve closure.
- Estimate the influence of relative sub-network volume on the minimum pressure accuracy in a more general context.

APPENDIX FIGURES

Test 1

The simulations in test 1 have been based on a time step of 0.4 seconds. The hydraulic scheme is depicted in the figure below.

Figure 1: Hydraulic scheme Test 1

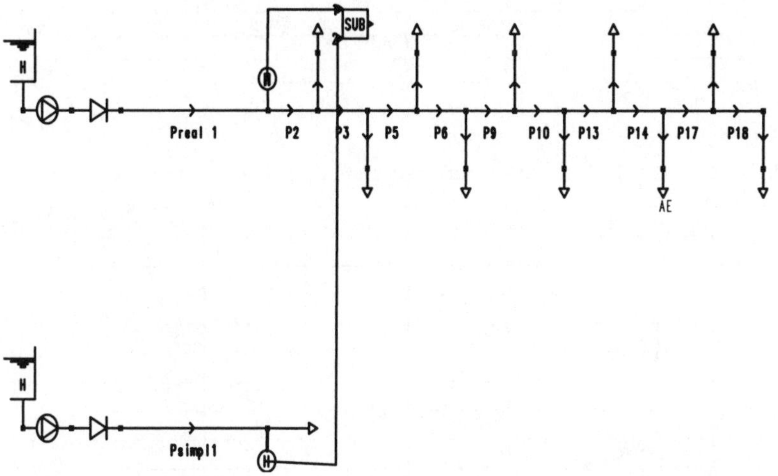

Pipe specifications

Preal 1, Psimpl1, P2, P3, P5:

L	=	10	km
D	=	600	mm
k	=	0.2	mm
e	=	30	mm
E	=	3	GPa
c	=	373	m/s

Other pipes without labels:

L	=	1	km
D	=	200	mm
k	=	0.1	mm
e	=	8	mm
E	=	3	GPa
c	=	357	m/s

P6, P9, P10, P13, P14, P17, P18:

L	=	1	km
D	=	400	mm
k	=	0.2	mm
e	=	20	mm
E	=	3	GPa
c	=	357	m/s

Figure 2: Test 1_1. Real and standard TAP flow to junction [m³/h]

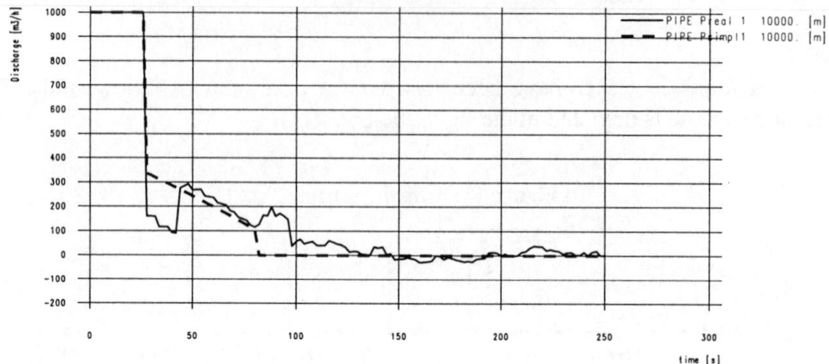

Figure 3: Test 1_1. Real and simplified pressure head at junction [m]

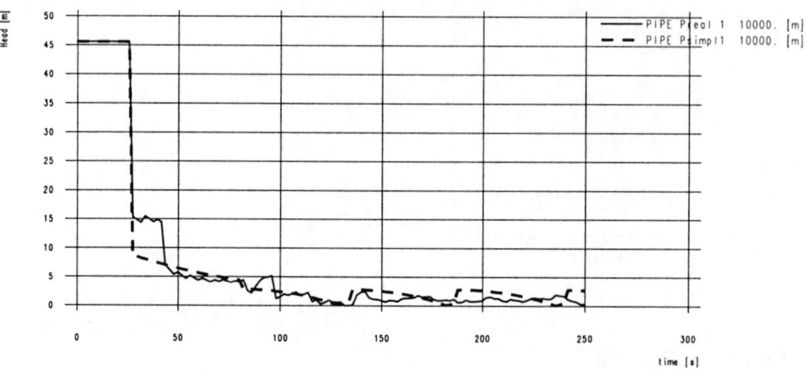

Figure 4: Test 1_1. Head error at junction using standard TAP [m]

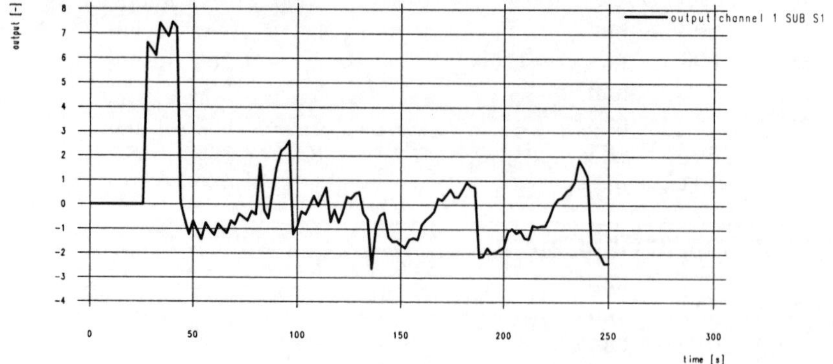

Figure 5: Test 1_2. Head error at junction using *CPG* pipe length and TAP [m]

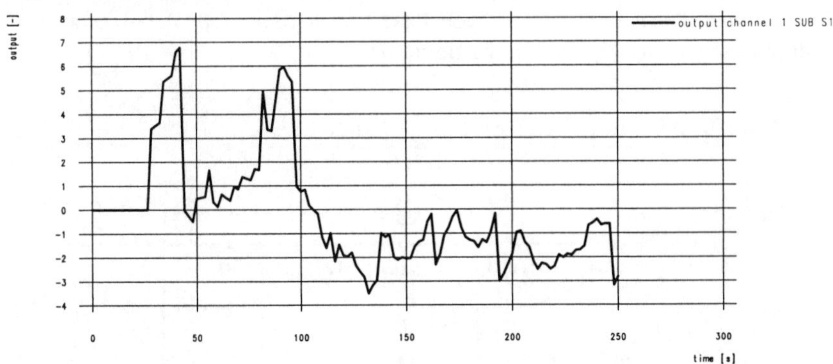

Figure 6: Test 1_3. Head error at junction using constant velocity diameter [m]

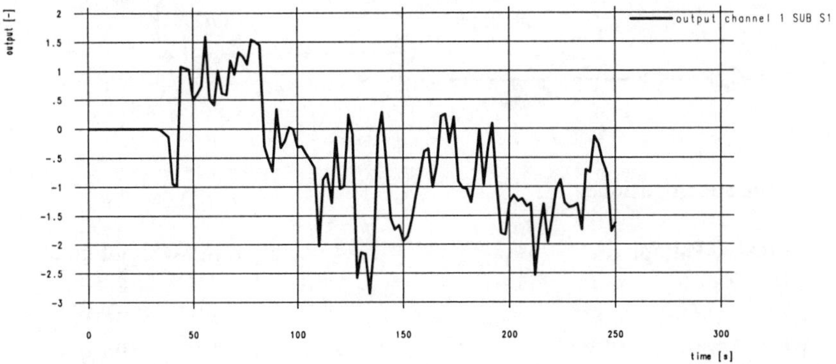

Test 2

The simulations in test 2 have been based on a time step of 0.4 seconds. The hydraulic scheme is depicted in the figure below.

Figure 7: Hydraulic scheme Test 2

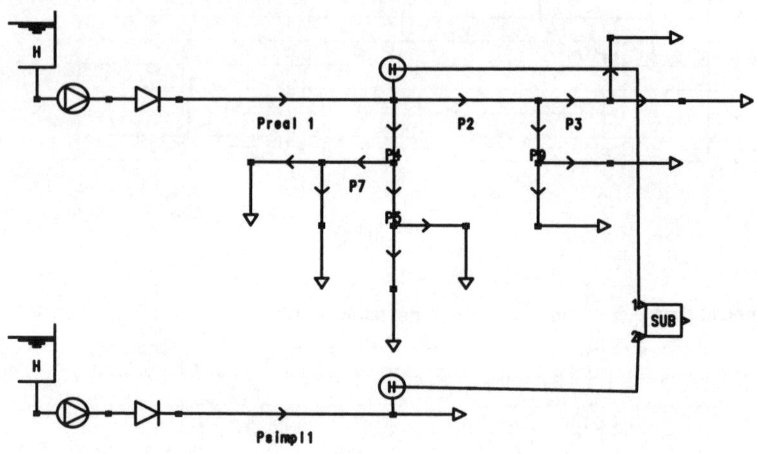

Pipe specifications

Preal 1, Psimpl:

L	=	5	km
D	=	600	mm
k	=	0.2	mm
e	=	30	mm
E	=	3	GPa
c	=	379	m/s

Other pipes without labels:

L	=	1	km
D	=	200	mm
k	=	0.1	mm
e	=	8	mm
E	=	3	GPa
c	=	357	m/s

P2, P3, P4, P5, P7, P9:

L	=	1	km
D	=	400	mm
k	=	0.2	mm
e	=	20	mm
E	=	3	GPa
c	=	357	m/s

Figure 8: Test 2_1. Real and standard TAP flow to junction [m^3/h]

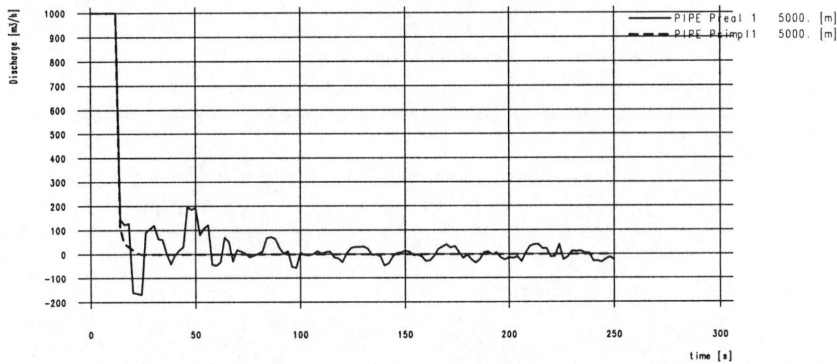

Figure 9: Test 2_1. Real and simplified pressure head at junction [m]

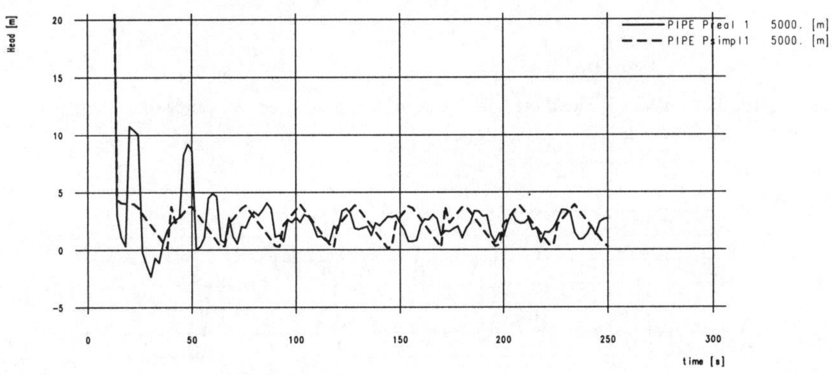

Figure 10: Test 2_1. Head error using standard TAP [m]

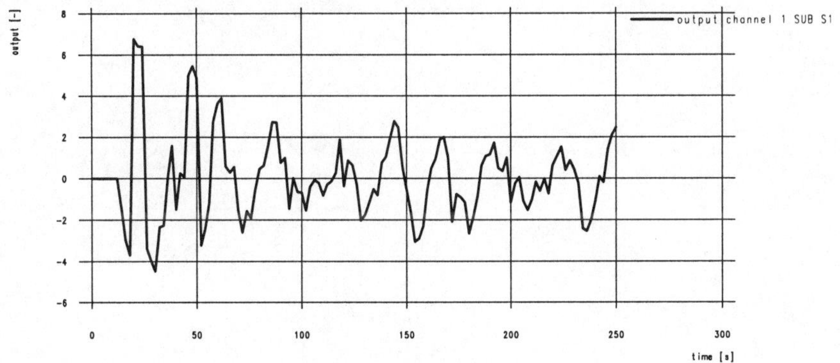

Figure 11: Test 2_2. Head error using *CPG* pipe length and TAP [m]

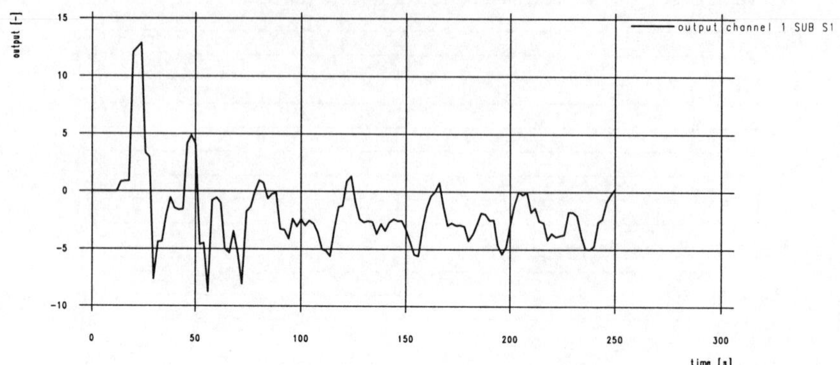

Figure 12: Test 2_3. Head error at junction using constant velocity diameter [m]

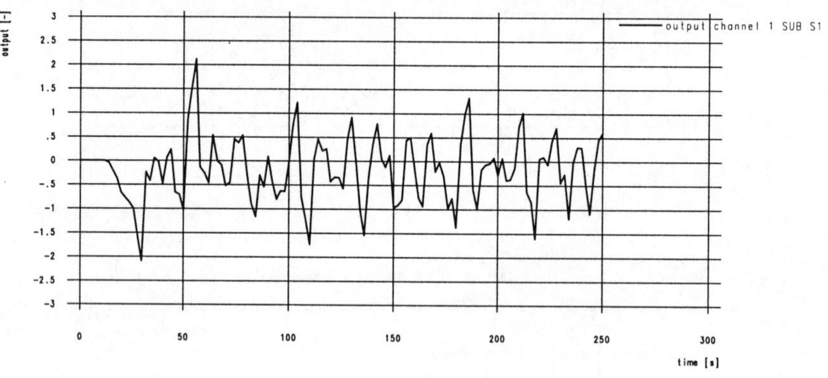

Test 3

The simulations in test 3 have been based on a time step of 0.2 seconds. The hydraulic scheme is depicted in the figure below.

Figure 13: Hydraulic scheme Test 3

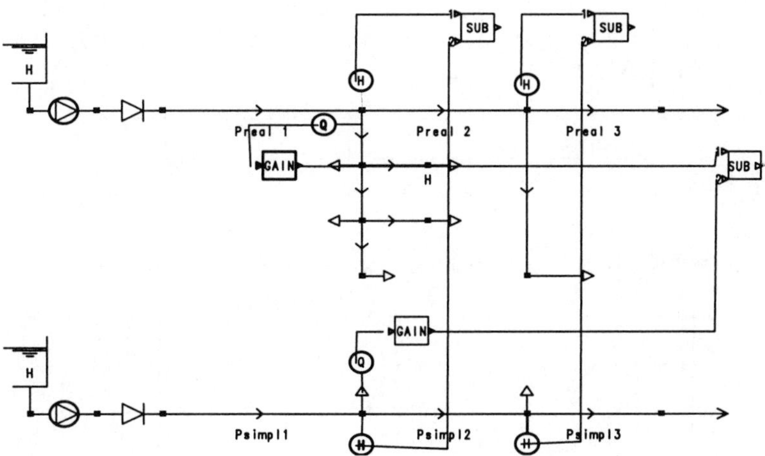

Pipe specifications

Preal 1, Psimpl1:

L	=	2	km
D	=	500	mm
k	=	0.2	mm
e	=	25	mm
E	=	3	GPa
c	=	370	m/s

Preal 2, Preal 3, Psimpl2, Psimpl3:

L	=	1	km
D	=	400	mm
k	=	0.2	mm
e	=	20	mm
E	=	3	GPa
c	=	385	m/s

Other pipes without labels:

L = 0.3 km (single pipe in sub-network 2 has length 0.9 km)

D	=	200	mm
k	=	0.1	mm
e	=	8	mm
E	=	3	GPa
c	=	375	m/s

Figure 14: Test 3_1. Real and standard TAP flow to junction [m^3/h]

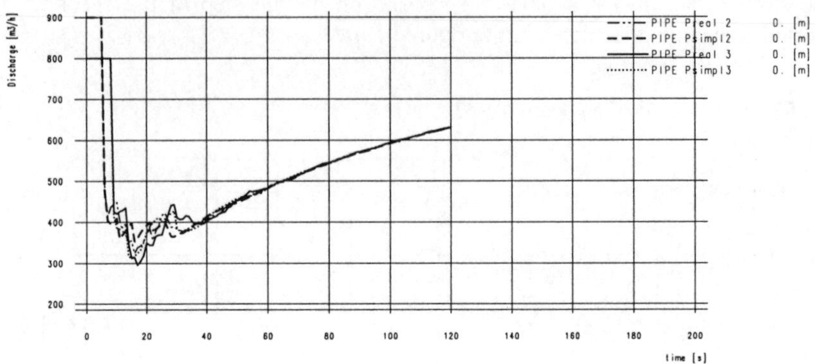

Figure 15: Test 3_1. Real and simplified pressure head at junction [m]

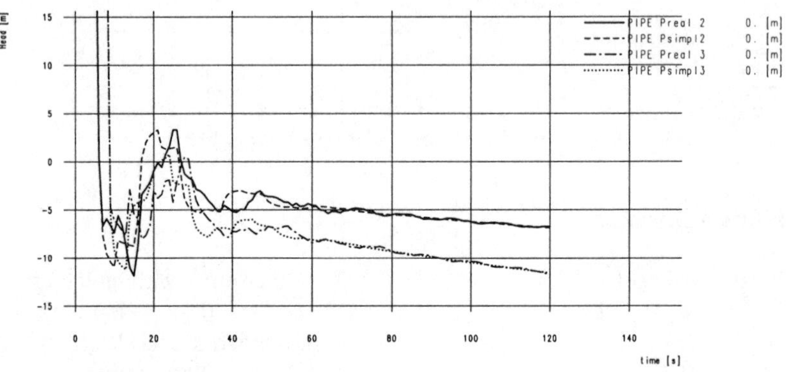

Figure 16: Test 3_1. Head error using standard TAPs [m]

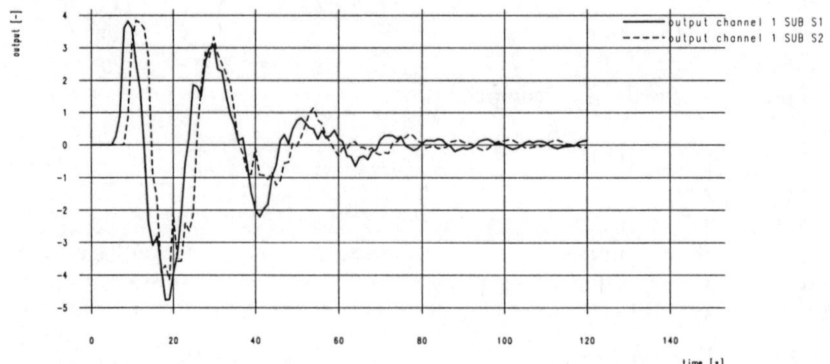

Figure 17: Test 3_2. Head error using *CPG* pipe lengths and TAPs [m]

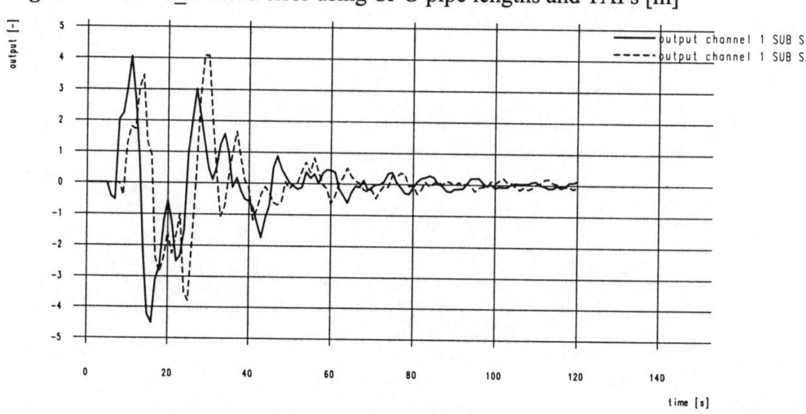

Figure 18: Test 3_3. Head error using constant velocity diameter [m]

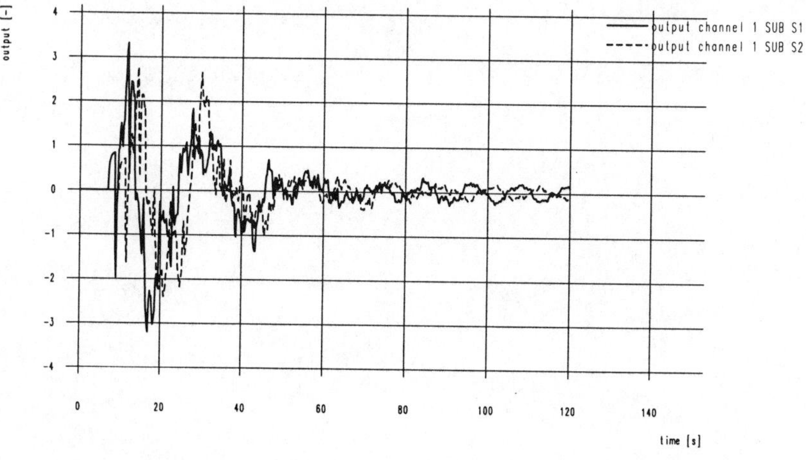

Review of Standards and Ground-rules on Transients and Leak Detection

I. Pothof

ABSTRACT

The European Commission has commissioned a group of universities and hydraulic companies to formulate groundrules for standardising all issues relating to the manufacture, design and construction of water and sewage pipelines, all with specific reference to pressure transients and leak detection. The project consists of 6 work packages:

1. survey on existing standards and guidelines with respect to transients and leak detection
2. construction of a database which contains hydrodynamic field and laboratory experiment data
3. collection and analysis of transient data
4. collection and analysis of leakage data
5. collection and analysis of fluid-structure data
6. benchmark analysis of hydraulic software

The purpose of this article is to present the current status of Work Package 1 (*Survey on existing standards and guidelines with respect to transients and leak detection*) and to attract the attention of the hydraulic community. The hydraulic community will be invited to supply the consortium with their field and laboratory data on transients and leak detection to feed the databases.

1 INTRODUCTION

There is a need for a set of standards within the European Union for the design and construction of municipal water or sewage pipelines to avoid or counteract the adverse effects of transient pressures. A group of European hydraulic engineering companies and universities have, with the support of the Standards, Measurements and Testing programme of the European Union, joined forces to establish the ground-rules from which these standards can be developed. The project is also looking into the novel use of transient pressure events for detecting leaks in pipelines.

The project, entitled *"Transient Pressures in Pressurised Conduits for Municipal Water and Sewage Transport"* is managed by the BHR Group Ltd. (BHR) from the United Kingdom in association with: WL | DELFT HYDRAULICS (DH) and TNO Building and Construction (TNO) from the Netherlands; the

Institute of Hydraulics of the University of Perugia (UoP) in Italy; the Instituto Superior Technico of the Departamento de Ingenharia Civil of the University of Lisbon (UoL) in Portugal; and Flowmaster International Ltd. (FMI) and the Industrial Statistics Research Unit (ISRU) of the University of Newcastle-upon-Tyne in the United Kingdom.

1.1 Scope

The main objective of the project, with reference SMT 4-CT97-2188, will be to provide the Commission with groundrules for standardising all issues relating to the manufacture, design and construction of water and sewage pipelines, all with specific reference to pressure transients. This will have the effect of reducing the risk associated with pressure surges in these pipelines by enabling standardisation of pipeline design practice that will include provision for transient behaviour. Such provision may take the form of a design strategy of transient avoidance or some sets of actions contingent upon the appearance of the transients.

A second objective is to assess the possibility of using transients as a means of determining the operational state of a water or sewage conduit. This may result in the ability to detect system leaks remotely.

The project is structured as Work Packages each with a project manager and project team drawn from the partners. The complete project is split into 6 Work Packages (WP).

Table 1: Overview of Work Packages

WP	Task	Task leader
WP 1	Standards and groundrules	DH
WP 2	Collate existing data and develop database	BHR
WP 3	Collect and analyse transient data	UoP
WP 4	Collect and analyse leakage data	UoL
WP 5	Collect and analyse fluid structure data	DH
WP 6	Benchmark analysis	FMI

1.2 Position of Work Packages

The primary objective of the project, formulating groundrules with respect to transients, is also the primary objective of WP 1. WP 2 through 6 result in valuable information for the formulation of the groundrules.

In WP 2 a database model is set up to store field and laboratory data: measured data as well as relevant geometric and operational characteristics. In WP 3 transient field and laboratory data are collected and analysed. In WP 4 leak detection field and laboratory data are collected and analysed. Furthermore a methodology will be developed to detect leaks using transients with disturbances due to hydraulic devices (valves etc.). In WP 5 field and laboratory fluid structure data are collected and analysed. Data from WP 3 through 5 are put into the database structure developed in WP 2. The developed database is used in WP 6 to test transient software against the field and laboratory data, resulting in a benchmark of transient software.

WP 1 relies upon the experience and results gained in WP 2 through 6 to formulate a methodology in groundrules to take into account transient pressures during the design and operation of a pipeline systems. The leak detection method based on transient analysis might result in an additional groundrule as well, if the developed method proves to be valuable.

1.3 Objectives of Work Package I

The following central questions will be answered in Work Package I:

- When should designers and operators evaluate transient pressures in water supply, waste water or other fluid transportation systems?
- How should designers and operators evaluate transient pressures in water supply, waste water or other fluid transportation systems?

The answers to these questions are given in the form of groundrules for future European standards which contain requirements on when and how transient phenomena and leak detection principles have to be taken into account during the design and operation of a pipeline. These groundrules have to be embedded in the (huge) available amount of standards and rules applied in the design stage of a pipeline project. The methodology to be developed applies to other fluid pipeline systems as well.

This paper focuses on the actual state of Work Package 1. Work Package 1 consists of 4 tasks:

1. List existing standards
2. Assess existing standards in detail
3. Develop safety design model
4. Specify groundrules

Tasks 1 and 2 have been elaborated at this moment. The results of these tasks are discussed in this paper. Section 2 discusses the search strategy to find potentially relevant standards with respect to transient flow conditions or leak detection. Section 3 describes the results of the search for potentially relevant standards. Furthermore these standard are assessed on their relevance with regard to leak detection methods and methods to take into account transient pressures during the design of pipelines. The conclusions of the review of existing standards with respect to transients and leakage detection methods are discussed in section 4.

2 APPROACH

Standards, codes and guidelines can be divided into obligatory and non-obligatory rules. In general the standards are defined by standardization organizations. The codes and guidelines are defined by trade organizations (TO).

The collection of standards, codes and guidelines can be entered from different points of view. Transient flow conditions and leakage detection can be discussed in standards, which focus on safety, test circumstances or engineering practice procedures.

2.1 Search strategy

Each partner in the project had the responsibility to investigate the standards, codes and guidelines in a number of countries. Standards, codes and guidelines are searched for in catalogues of standardization bodies and other involved organizations (trade organisations). The following search methods have been applied:

- standards catalogue search on ICS-codes (ICS = International Classification of Standards); 18 ICS codes have been used.
- standards catalogue search on keywords
- personal contacts with employees of standardization and other organizations

The following key words (or subsets) have been applied in their specific languages.

Table 2: key words

transient	leak detection	Under-pressure
surge	Leak proof	Hydraulic calculation
Pressure	leak testing	Conveyance of fluids
Water hammer	Pipeline	Sewage
Leak	Pipe	Water transport

3 EXISTING STANDARDS

The search and review results are discussed in this chapter. References to handling transients or leakage detection methods in pipeline systems have not been found in the standards, codes and guidelines of Belgium, Luxembourg, Denmark, Finland, Sweden, France, Greece, Portugal and Spain.

References to the subjects, although limited, have been found in standards or guidelines from Austria, Ireland, Italy, The Netherlands and United Kingdom.

The standards, codes and guidelines issued by ISO and CEN are discussed in separate paragraphs. Most interesting references have been found in standards and guidelines from North America and Germany, which are also discussed in separate paragraphs.

3.1 Standards with limited references

References to transients have been found in Austrian standard OENORM B 2531, part 1: Drinking water supply systems on premises; rules for design, construction and operation. OENORM B 2531, part 1 in par 5.4 states that "only devices have to be installed that do not cause an excessive pressure and that closing does not lead to a pressure increase due to water hammer of more than 2.0 kgf/cm^2 (approx. 2 bar) over the steady state network pressure". Par. 9. is about noise originating from pipes or water hammer. One of the steps to be taken is to "avoid high pressures and high velocities in pipes; the optimal velocity is 1-2 m/s and for special cases it can be 3 m/s". Furthermore the same paragraph states that "to avoid high pressures and high velocities in pipes, the valves for instant stop are not to be used".

Irish standards state that pressure due to transient events should be considered in defining the design pressure of a system.

No Italian standards have been found, but some laws and regulations contain details about or references to leak detection and transient phenomena in water system design and operations. Furthermore in Italy the Public Works Ministry gives instructions, not in force as laws or regulations but comparable to guidelines. One ministerial publication specifies that all possible ways of functioning of the considered pipe system must be examined in order to properly evaluate water hammer overpressures. If no specific transient calculation has been performed, values in Table 3 can be applied.

Table 3: Pressure in kgf/cm^2

Hydrostatic pressure up to	6	6 – 10	10 - 20	20 - 30
Water hammer overpressure up to	3	3 - 4	4 - 5	5 - 6

The emphasis in Dutch standard NEN 3650 (Requirements for steel pipeline transportation systems) is put on the requirements for steel pipeline transportation systems, where transient pressures are interpreted as a type of load onto the system; each type of load is an input for the stress calculations which are presented extensively. This standard states that the maximum incidental pressure, which is the pressure limit for transients, equals 1.15 times the design pressure. No guidelines are presented to calculate the maximum pressure as a result of emergencies or extreme operational conditions.

No single British standard discusses the subject of transient flow conditions further than stating that pressure due to transient events should be considered in defining the design pressure of a system. BS 806 states that "Specific allowance shall be made for water hammer if the results of pressure surges exceed the design pressure by more than 20%". BS 8010 states in several sections (2.5, 2.8 and 3) that surge pressure should be calculated. BS 8010 Part 3 furthermore states that the transient pressure may not exceed the maximum allowable operating pressure by more than 10%. BS 8010 Part 3 also mentions ten leak detection methods.

3.2 ISO standards

The investigation of ISO standards on ISC-codes and keywords has resulted in a list of about 60 standards and draft standards. These standards have been screened on abstracts, resulting in 9 standards, which have been evaluated in detail.

No single ISO standard discusses the subject of transient flow conditions in more detail than stating that transients should be prevented. Leakage detection is not discussed at all in ISO standards. Surge pressures have been globally discussed in only 1 ISO standard: ISO 4413: 1998 (Hydraulic fluid power - general rules relating to systems). It states that surge pressures and intensified pressures shall not cause hazards. It is not elaborated how inadmissible pressures are to be prevented. Furthermore standard 4413 contains recommendations for flow velocities and the ranges of pressure gauges.

3.3 CEN standards

The investigation of European standards has resulted in a list of 42 standards that have been examined in detail. This list has been obtained by means of the CEN Catalogue and CEN Technical Committees (CEN TC's) as well as via catalogues of EU countries. Surge pressures have been globally discussed in only 1 CEN standard: prEN 805: 1992 (Water supply - Requirements for external systems and components).

No single CEN standard discusses the subject of transient flow conditions or leak detection methods in detail. In many European standards definitions for the maximum allowable pressure or the maximum operating pressure can be found. Sometimes the maximum allowable pressure is referred to as maximum incidental pressure or the design pressure; see EN 639, EN 773, EN 805, EN 1456 and EN 12008 and EN 12186. In some of these definitions surge is explicitly included. Although surge pressures are mentioned in the definition sections of these standards, the design recommendations do not discuss surge in more detail than stating that surge should be taken into account during design, maintenance and operation.

PrEN 805: 1992, issued by TC 164 (Water supply) comes closer to an integrated approach in the design and operation recommendations. PrEN 805 provides definitions of surge and the maximum design pressure with a fixed or calculated allowance for surge. However, it does not elaborate how the allowance for surge should be calculated. PrEN 805 furthermore states that transient loads are to be taken into account and that consideration shall be given to surge limiting equipment, although it is not explained how. Nevertheless this standard qualifies as one of the few that discusses the subject of pressure surges in detail.

3.4 Standards in North America

Contact with the Standards Council of Canada (SCC) showed that no national codes or standards with respect to transient flow conditions or leakage detection have been developed.

Standards in the USA with respect to hydraulics are being developed by many organizations. The catalogues of these organizations have relatively good search facilities. In order to get more detailed information, the standardization organizations have been contacted by e-mail as well. The catalogue search results and the e-mail contacts resulted in a list of 10 standards which has been examined in the full text.

Table 4: Most promising North American standards

Standard	Title
ASTM E1211-97	Leak Detection and Location Using Surface-Mounted Acoustic Emission Sensors
AWWA M36 (1990)	Water Audits and Leak Detection. Manual of Water Supply
ASME B31.8-1995	Gas Transmission and Distribution Piping Systems
API 346 (1998)	Results of Range-Finding Testing of Leak Detection and Leak Location Technologies for Underground Pipelines
API 1149 (1993)	Pipeline Variable Uncertainties and Their Effects on Leak Detectability

No single American standard discusses the subject of transient flow conditions in detail. ASME standard B31.4 (1992) states that maximum pressure due to surge is 10% greater than internal design pressure; it does not elaborate how the maximum pressure is kept below this value. Leak detection methods are described in the standards in Table 4.

The following leak detection methods are discussed in these standards:

- Acoustic emission to distinguish a leaking from a non-leaking situation
- Audible leak detection by experienced personnel
- Annual water balance (water audit)
- Zone flow measurement in an isolated section; also known as mass conservation (possibly with linefill correction)
- pressure drop test
- constant pressure volume
- chemical tracer
- Accidental leak detection during normal maintenance operation
- Comparison of transient measurements with transient model results

ASME standard B31.8 furthermore discusses four methods, typically for gas transportation systems: vegetation survey, bubble leakage test, surface and sub-surface gas detection. API 1149 (1993) provides a thorough discussion on the effects of parameter, variable and measurement uncertainties on leak detectability. The zone flow, pressure drop, constant pressure volume and transient methods apply typically to single pipelines only. The other methods can be applied to networks as well.

3.5 Standards in Germany

The investigation of German standards has resulted in a list of 48 standards, mainly DIN EN standards. 15 DIN EN standards have been examined in detail. Since all of these standards are European standards, the results have been discussed in paragraph 3.3. DVGW (German Union for Gas and Water), VDI (Union of German Engineers) and VDMA (Association of German Machine and

Plant manufacturers) have been contacted about existing standards and guidelines related to transients and leak detection. VDMA has replied that the subject of transient flow conditions and leakage detection are not discussed in any DIN standard.

DVGW provides guidelines (Arbeitsblätte) for the gas and water industry in Germany. Three DVGW guidelines provide concrete and comprehensive criteria and recommendations for design and operation of water pipeline systems, related to transients or leak detection. These are:

- W303, Dynamic variations of pressure in water supply systems,
- W391, Losses of water in water distribution networks (translated from German), 1986
- W393, Methods for leak location in water supply pipelines, 1991

Guideline W303 gives an extensive and comprehensive overview of many relevant aspects of transient flow conditions for the designer and operator of pressurized water supply systems. The language of the document is German. The following subjects are described subsequently in W303:

- Physical phenomenon and calculation procedures,
- General design recommendations and pressure criteria,
- Recommendations for design and operation in gravity driven systems, in pump driven systems and combined systems. For each of these 3 types of systems the critical characteristics are described, as well as possible causes and measures,
- Ranking of pressure reducing measures,
- Requirements for surge calculations,
- Measurement of transients.

The recommendations for design and operation of W303 are briefly described in this paragraph. Three basic types of hydraulic systems are distinguished and discussed extensively. A fourth type is the network and is in fact a combination of the three basic types. The three basic system types are:

1. Gravity driven systems (delivery head < suction head)
2. Gravity driven systems equipped with pumping stations (delivery head < suction head)
3. Pumped systems (delivery head > suction head)

The systems are described in more detail in W303. The following causes of transients are mentioned for each of the systems:

1) Gravity driven systems (delivery head < suction head)
    - valve opening or closing, use of control valves beyond their original limits of use, vent blow off, entrapped air in the system, change in turbine load, including switch on/off, (sudden) large extraction from the system, pipe rupture.

2) Gravity driven systems equipped with pumping stations (delivery head < suction head)
   - Pump start, stop and trip, entrapped air in pumping station (leading to excessive loads on pump motor), valve manipulation, resonance due to unstable pump motor,
3) Pumped systems (delivery head > suction head),
   - Pump start, stop and trip, unstable pump curve, resonance due to energy storing components (such as air vessels), dynamic check valve behaviour.

The following measures are explained and recommended, including which measure is most appropriate in the situations above:

- Increase of valve closure and opening times
- Air vessels
- Fly wheels
- Surge tower
- By pass pipe
- Pressure relief valves
- Air valves

Consequently DVGW guideline W303 gives a comprehensive overview of pressure surges.

4 CONCLUSIONS

Many standardization organizations have made their catalogues available on Internet. In general the search facilities are poor. The catalogues can be searched on keywords and sometimes on ICS codes, but combined keywords and wild characters or Boolean logic in search facilities are only rarely supported. ICS codes are supported in the catalogues of ISO, Austria, Denmark, Finland, Ireland, Italy, Netherlands, Spain and Sweden, but not supported in the catalogues of, among others, USA, France, Germany and the United Kingdom.

No single standard has been found which discusses a methodology to take into account transient pressures during the design stage of a pipeline system. Only German guideline W303 (issued by DVGW) provides criteria and recommendations to account for transient pressures; it states that transient pressures should not exceed the pressure class of the system. The closest references in standards to transient pressures state that transient pressures should be prevented. In many standards definitions for the maximum allowable pressure or the maximum operating pressure can be found. Sometimes the maximum allowable pressure is referred to as the design pressure. Surge is explicitly included in a few definitions of maximum incidental pressure. In some standards a maximum incidental pressure is expressed as a percentage of the maximum operating pressure: 110% in ASME B31.4 (1992), IS 328 and BS 8010, 115% in NEN 3650, 120% in BS 806 and 125% to 150% in an Italian ministerial publication. Austrian standard OENORM B2531-1 states that the pressure increase due to water hammer shall be less than 2.0 kgf/cm^2 (approx. 2 bar).

Many standards exist addressing the subject of leak testing, instead of leakage detection. Leakage detection methods are described and evaluated in standards and related documents. Several documents describe or evaluate one or more leakage detection methods.

- An Italian law (*Public Works Ministry Decree January 8th 1997, n.99)*,
- An Italian ministerial publication (*Circolare Ministero dei Lavori Pubblici, n. 27291* about regulations concerning criteria and methodology for leaks evaluation in water and sewer systems),
- German DVGW guideline W391 (1986), losses of water in water distribution networks
- German DVGW guideline W393 (1991), methods for leak location in water supply pipelines
- British standard BS 8010 Part 3 (1993), code of practice for pipelines. Pipelines subsea: design, construction and installation
- ASTM E1211-97, leak detection and location using surface-mounted acoustic emission sensors
- AWWA M36 (1990), water audits and leak detection. manual of water supply
- ASME B31.8 (1995), gas transmission and distribution piping systems
- API 346, results of range-finding testing of leak detection and leak location technologies for underground pipelines
- API 1149 (1993), pipeline variable uncertainties and their effects on leak detectability

It has been found that almost all standards with respect to hydraulics can be classified in two categories: standards which are dedicated to single components in an hydraulic system and standards which are dedicated to an hydraulic system as a whole. The subjects of this project (transient flow conditions and leakage detection) do not fit in this classification, because both transient flow conditions and leakage detection are induced by single components (emergency pump shut-off, pipe rupture), while the related phenomena occur in the complete hydraulic system. Furthermore the effects are diminished by other single components (air vessel, surge tower, SCADA control system). Standards, dedicated to hydraulic systems as a whole, seem most suitable to treat the subjects of inadmissible transient flow conditions and leakage detection.

Consequently a more integrated approach is required by addressing the subject of transient flow conditions in the design of fluid pipeline systems. The scope of European draft standard prEN 805:1992 (Water supply - Requirements for external systems and components) seems suitable to address a methodology to take into account pressure surges during the design of a pipeline system. This standard has been issued by CEN Technical Committee 164: Water supply.

ACKNOWLEDGEMENTS
The author would like to thank all of the partners for permitting the publication of this paper and in particular Peter Baker (FMI), Marco Ferrante (UoP), Helena Ramos (UoL) and Dave Stewardson (UoN). The author also acknowledges the support of the European Union who part-funded the project through the Standards Measurement and Testing Programme (project reference SMT4-CT97 2188).

Micro Hydropower Plant Behaviour in Water Conveyance Systems

H. Ramos, A. Borga *and* D. Covas

ABSTRACT

Favourable conditions created sometimes in water conveyance and irrigation systems as characterised by excess flow energy along pipes or canals can be used to yield energy by the installation of micro-turbines.

Some particular issues concerning load variation must be considered in order to control instability occurrence that would induce the interruption of the supply system. The dynamic characterisation of micro-turbines operating under normal and abnormal conditions must be well defined to improve the effectiveness of the system.

The energy production will be strongly influenced by the daily demand. Depending upon the discharge and the available head, the choice of the most suitable turbine type will condition the dynamic effects, such as overspeed and discharge variation. In these types of facility, one can use not only impulse turbines, assuming the available head allows it, where the discharge control can be carried out by the nozzles, but also reaction or centrifugal reverse-pump turbines, which have a more complex dynamic behaviour. The analysis of turbine-flow interaction and the elastic behaviour of the hydraulic circuit, regarding pressure or free-surface fluctuations, will constrain the supply and the system design (e.g. pressure classes in pipes or height of canal walls and weirs).

1 INTRODUCTION

The installation of micro-turbines in water conveyance and irrigation systems might be an alternative solution for the dissipation of excess energy that would be, otherwise, lost in pressure reducing valves (PRV) or head loss chambers, depending upon the type of hydraulic circuit (e.g. pressurised conduits or canals). These combined solutions - hydro schemes within water supply and irrigation systems - are very important for an integrated management system, in order to address the optimisation of water uses, operating conditions, safety and economic factors.

Since rainfall can vary widely from year to year as well as daily consumption, correct water management must be, therefore, currently a high priority to assure a sustainable development in terms of water resources and uses.

2 SYSTEM TYPES

The use of micro-turbines will depend on the characteristics of the hydraulic circuit that can be composed, fundamentally, of three different types: a) a totally pressurised circuit; b) a pressurised circuit with intermediate reservoirs (e.g. treatment plants or ground reservoirs); c) a totally free-surface flow (e.g. irrigation canals). Ho is the net head that will be utilised by the turbine.

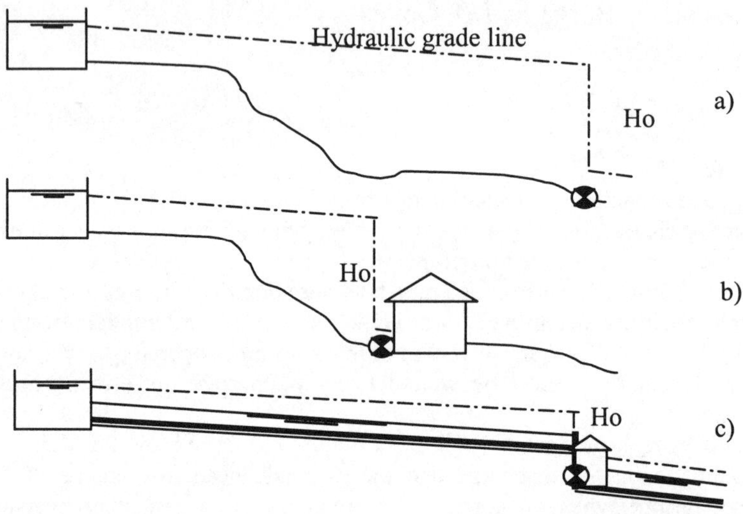

Fig. 1 Typical layouts for different micro-turbines installation

Generally speaking, pressurised systems demand reaction turbine types (e.g. Francis and Kaplan or propeller) or reverse pump-turbines. These types of turbine present the main advantages by maintaining the flow under pressure avoiding, thus, eventually dangerous environmental contamination problems for drinking water systems. In the case of variable high or intermittent heads, the impulse turbine solution (e.g. Pelton, turgo and cross-flow) is recommended and more suitable and can replace loss chambers, only requiring a free surface tailrace.

These turbines in drinking water systems are not advisable since they can contribute towards the occurrence of contamination.

In free-surface flow systems (e.g. irrigation canals) reaction turbines of open flume type for low or medium head can be used (e.g. high specific speed turbines – Francis, Kaplan and bulb).

3 ENERGY PRODUCTION

The use of micro-turbines in zones with significant differences in topographic levels, not only avoids the use of higher pressure classes of pipes, with the consequent cost reduction, but, above all, has the benefit associated with renewable energy production. This solution enables the control of pressures and presents mitigation measures for energy losses.

The development of a zone planning study allows better control of the water conveyance system through dividing the system into different pressure zones, according to their topography and demand constraints. In fact, the use of pressure reducing valves or load loss chambers have in common the dissipation of excess energy in a localised zone of the system, being one of the most efficient and easily applicable alternatives to regulate and control pressures and elastic wave propagation.

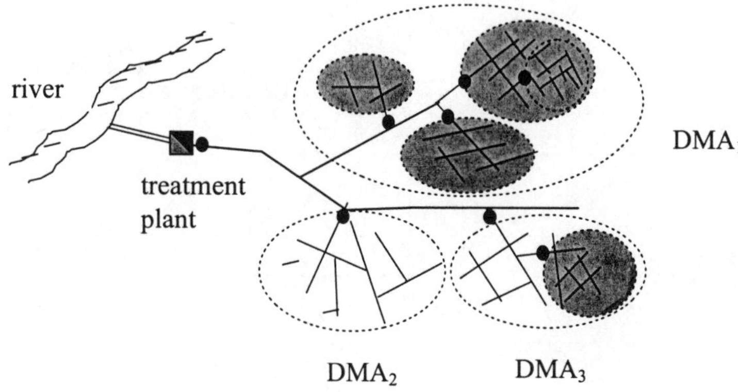

Fig. 2 Zone planning by District Meter Area (DMA) splitting

A specific study zone can be split up into several smaller areas (District Meter Area -DMA) according to the topology of the system and the quantitative discharge customers. DMAs are normally controlled by closed valves or gates and monitored by pressure transducers and flow meters (Figure 2). The importance of establishing DMAs is to allow the identification of the most favourable zone for turbining, as well as to define excess pressure during the day, and, consequently, the most suitable micro hydropower performance.

A micro-turbine, when installed in a multiple-use system, as a water supply system associated with power generation, often needs operating with variable discharges, which can vary between low values, normally with high heads (e.g. during night) and high values associated with low heads (e.g. in peak hours) (see Figure 3).

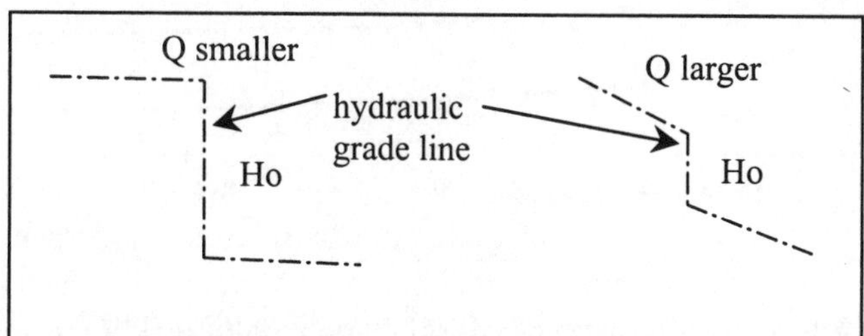

Fig. 3 Influence of consumption law in turbine operating

Therefore, important constraints will characterise these systems by, sometimes, inducing infeasible economic analysis from the point of view of energy production, since they are strongly influenced by daily consumption (Figure 4).

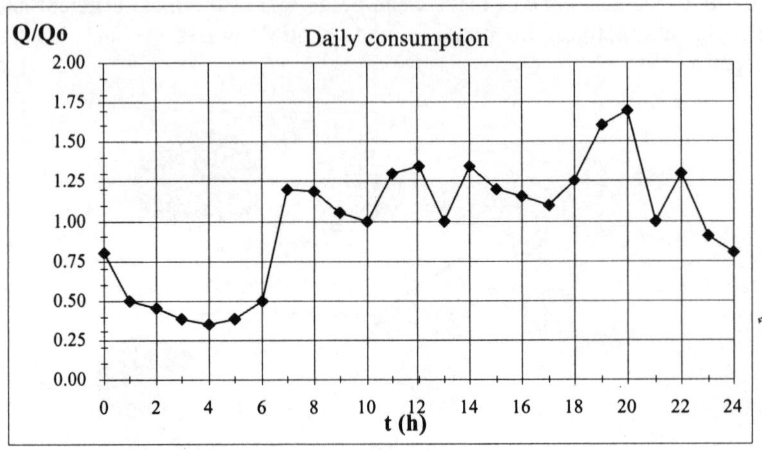

Fig. 4 Example of typical daily consumption

As the unit cost of a micro hydro plant is a non-linear function, depending on the turbine type installed, a brief estimate of average cost is presented based on manufacturers' information (Figure 5).

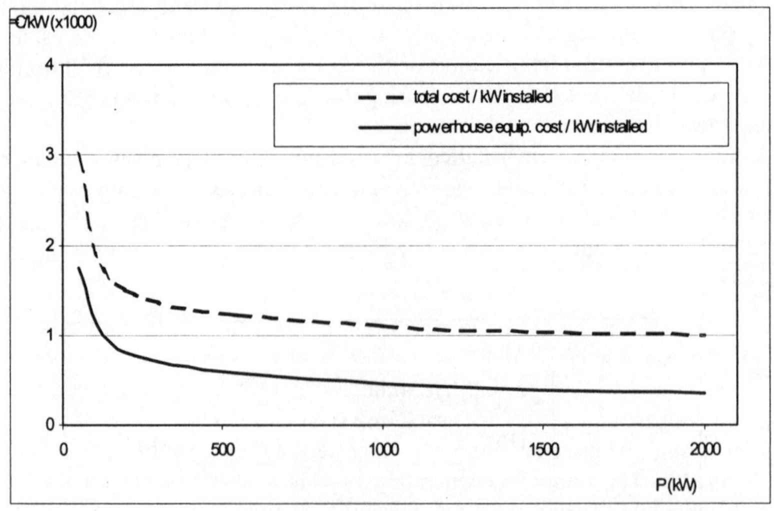

Fig. 5 Approximate unit costs (in Euro x 1000) per kW installed

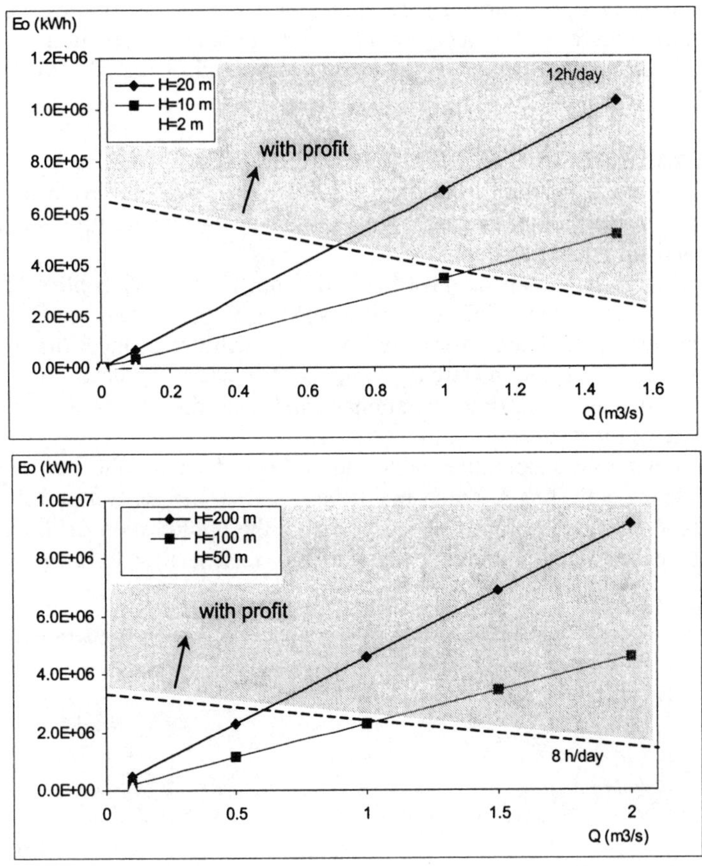

Fig. 6 Estimation of energy production for low and high heads

Sensitivity analyses were developed for low and high output power systems in order to estimate the efficiency of these systems. A system with small hydroelectric capacity (e.g. low output) low head (e.g. from 2 till 20 m) and medium range of discharge (e.g. 0.001 till 1.5 m^3/s) was considered. It is appropriate to deal with energy estimation for an average year (Eo), using a minimum period in a day of around 12 hours for generating. Afterwards, a similar analysis was developed for high head systems (e.g. varying between 50 and 200 m) with medium turbine discharge values (e.g. from 0.1 to 2 m^3/s) corresponding to a greater energy capacity system, where a smaller generating period (i.e. 8 hours) can be considered (see Figure 6). The minimum generating time is a restrictive factor that must be carefully calculated during economic feasibility analysis.

According to a profit analysis developed on the basis of some case studies, as presented in Figure 6, the increase in energy production resulting from some parameters, e.g. output power (P (kW)) and turbine discharge (Q (m^3/s)), can be assessed.

4 DYNAMIC EFFECTS OF MICRO-TURBINES

There are some constraints that can condition the best powerhouse operation imposed by space, minimum discharge requirement for energy production, safety, stability and automation systems.

A hydrotransient analysis must be carried out for a better control of dynamic behaviour of the system. The dynamic response of the system can have remote effects, namely, phenomena associated with hydraulic resonance due to different harmonic phases from different components, when induced by transitory instabilities that occur during operational conditions due to normal or abnormal manoeuvres of the discharge control systems.

Reaction turbines operating with variable speed enable the improvement of their efficiency, adjusting to the available head and discharge. With no fixed speed, it induces also any particular frequency, easing the optimisation of the number of pairs of poles leading to a smaller generator and, consequently, to lower costs.

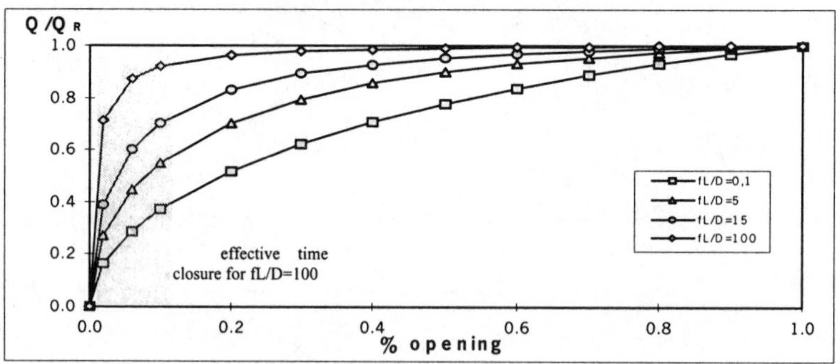

Fig. 7 Influence of hydraulic circuit on discharge variation of an impulse turbine

5 IMPULSE TURBINES

The time of nozzle closure (T_C), and the parameter fL/D of the upstream hydraulic circuit (where f is the Darcy-Weisbach resistance factor, L pipe length and D pipe inner diameter) are the main factors for transient analysis. For high values of fL/D, the discharge variation, during a closure of the nozzle, is only effective near the completely closed position which can transform a slow closure manoeuvre into a fast one (Figure 7).

6 REACTION TURBINES

Applying Euler's equation to a control volume between inlet and outlet of a reaction turbine runner, the discharge variation can be obtained as a function of rotating speed in accordance with the following equation:

$$\frac{Q}{Q_R} = \frac{A}{N/N_R} + B \cdot N/N_R \qquad (1)$$

in which A and B are parameters that depend on the type of turbine (e.g. characterised by N_s as showed in Figure 8) and Ns is the specific speed of a turbine that is defined by:

$$N_s = N_R \frac{\sqrt{P_R}}{H_R^{1.25}} \qquad (2)$$

in which N_R = rated wheel speed (r.p.m.), H_R = rated head (m), Q_R = rated discharge (m^3/s) and P_R = rated turbine power (kW).

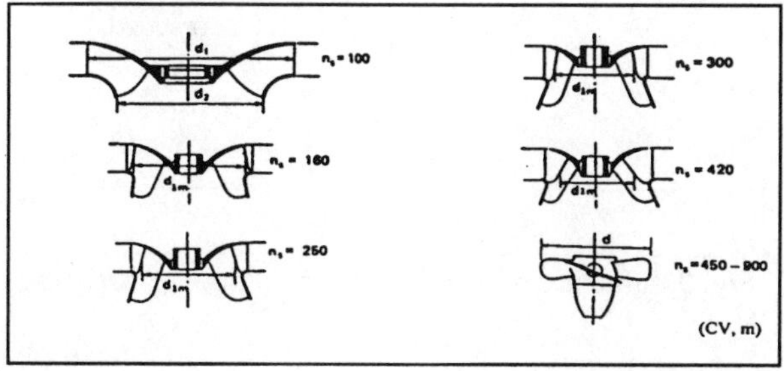

Fig. 8 Different runners of reaction turbines (N_s(CV, m)=1.17 N_s(kW, m))

Based on experimental research development for an open-flume reaction turbine and on dimensionless equation (1), the parameters A and B were estimated by correlation, yielding for a low specific speed turbine (Ns = 135 (m, kW)) A = 1.22 and B = -0.22, and for a high specific speed (Ns = 342 (m, kW)) A = 0.54 and B = 0.46 (Figure 9). These parameters are obviously related to the turbine runner type (e.g. A >1 or A<1 and B<0 or B>0) showing the real tendency of turbine discharge variation when runaway occurs.

According to the type of turbine, the maximum overpressure will depend on the effective duration of the discharge variation (Figure 10).

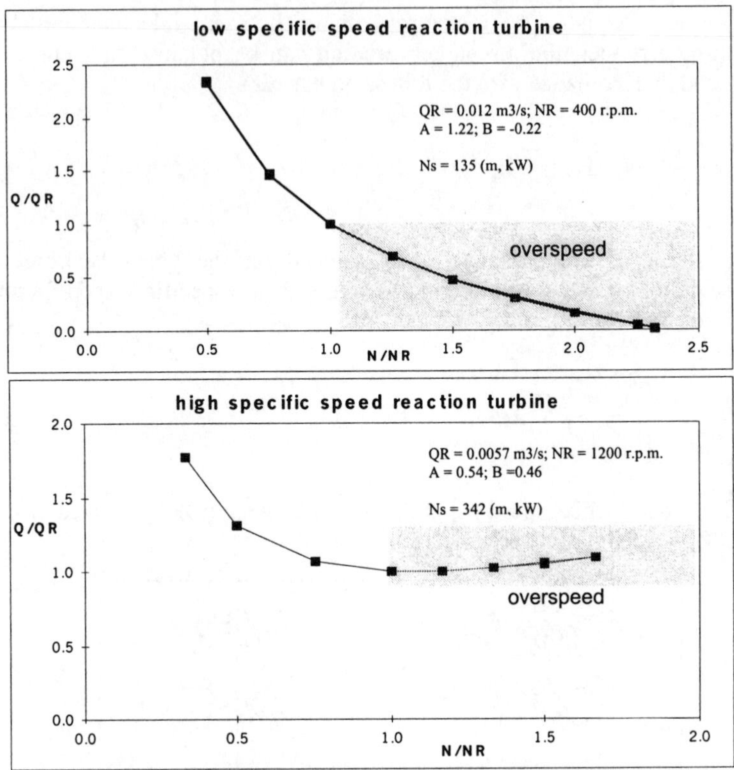

Fig. 9 Influence of speed variation on turbine discharge

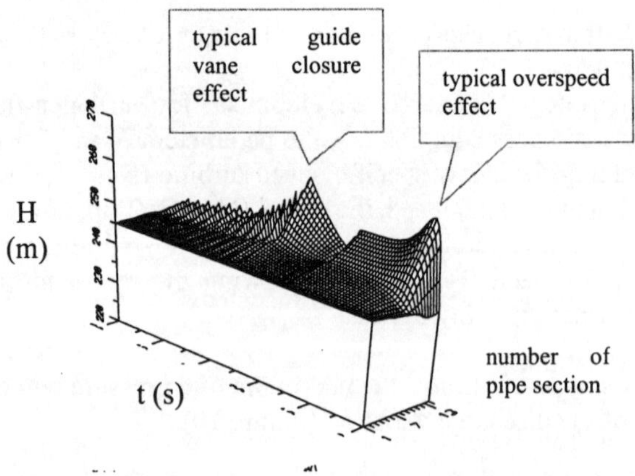

Fig. 10 Dynamic response of the system at powerhouse when equipped with low specific speed reaction turbine

In the case of reaction turbines, the dynamic behaviour of the integrated system due to interaction between the powerhouse and long conveyance systems yields relevant overpressure variations that can be induced as much by guide vane closure effects as over-speed effects (Figure 10). This latter effect must be conveniently quantified for low specific turbine speed because it can be the major conditioning factor for safety design purposes.

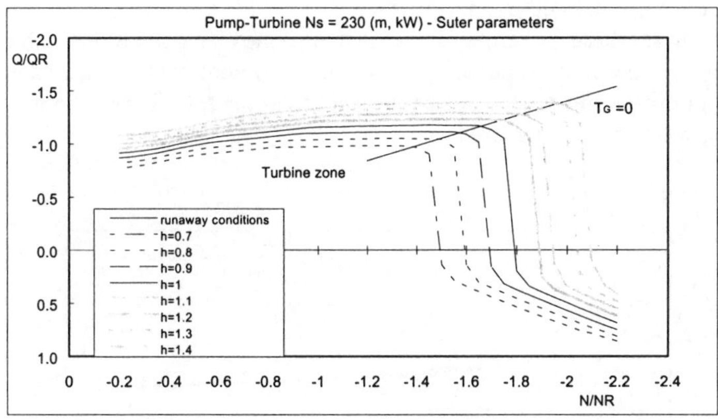

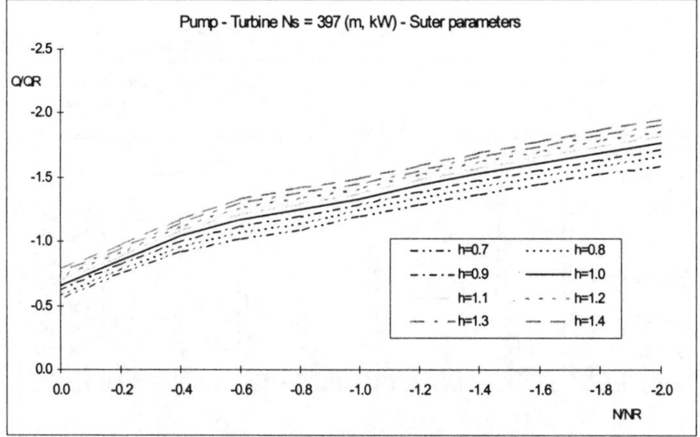

with $h=H/H_R$

Fig. 11 Turbine zone for a reverse pump-turbine based on Suter parameters

In the same way, the use of reverse pump-turbines will influence the upsurge, depending on the type of runner. Figure 11 is a result of Suter's parameters analysis that shows the discharge variation with rotational speed, when operating in the turbine zone. An abrupt reduction in turbine discharge is noticed for a low specific speed runner, the converse being true for a high specific one. This dynamic effect, known by over-speed effect under runaway condition, can condition pipeline design and system operation stability.

7 CONCLUSIONS

Special considerations must be taken into account in the installation of micro turbines in water supply systems, with regard to protection devices against water hammer effects due to runaway conditions when low specific turbine speeds are installed, at turbine start-up or shut-off.

Nevertheless, these integrated systems, with innovative solutions regarding the association of micro hydropower plants with different water uses, have a big advantage in presenting no special additional environmental impacts and avoiding, in this way, waste of water energy.

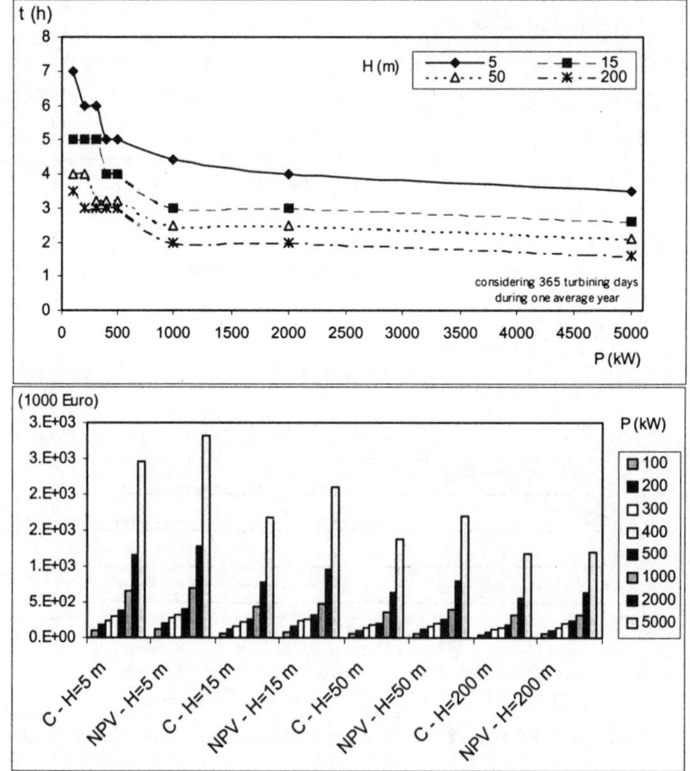

Fig. 12 Development of economic sensitivity analysis for a range of installed power and head

Drinking or water supply systems have a greater guaranteed discharge, generating duration and can have higher heads, yielding consequently greater output power. Based on economic sensitivity analyses, these systems with long circuits and high topographic differences in levels between upstream and downstream of the powerhouse, seem to be more suitable for energy production compared with irrigation systems, due to seasonal discharge variations and according to costs per kW installed.

Nevertheless, for these systems an algorithm was applied based on economic considerations in order to verify the minimum number of generating hours per day, enabling more suitable investment criteria based, fundamentally, on project end time, Cash flow (C) and Net Present Value (NPV) (see Figure 12).

REFERENCES

Ramos, H; Covas, D. - O Benefício Económico e Ambiental da Produção de Energia Renovável em Sistemas Adutores. IV Silusba, 24-26 Maio, 1999.

Ramos, H; Covas, D. – O Aproveitamento da Energia em Excesso em Sistemas de Abastecimento de Água. 8º Encontro Nacional de Saneamento Básico, Barcelos 27 a 30 de Outubro, 1998.

Covas, D.; Ramos, H – A Utilização de Válvulas Redutoras de Pressão no Controlo e Redução de Fugas em Sistemas de Distribuição de Água. 8º Encontro Nacional de Saneamento Básico, Barcelos 27 a 30 de Outubro, 1998.

Ramos, H.; Covas, D. – Pressure Reducing Valves and Micro-Turbines for Leakage Control in Water Supply Systems. Review "Manutenzioni on Industrial Plants". Bologna, February 1999

Ramos, Helena e Betâmio de Almeida, A. - Lei de Variação do caudal Turbinado para Pequenas Turbinas Instaladas em Caixa Aberta - 4º Congresso da Água/Viii SILUBESA - II-527, pag.. APRH Lisboa, 23 a 27 de Março de 1998.

Ramos, Helena e Betâmio de Almeida, A. - Efeitos Induzidos pelo Comportamento Dinâmico em Centrais do Tipo Pé de Barragem - 4º Congresso da Água/Viii SILUBESA - II-527, pag.. APRH Lisboa, 23 a 27 de Março de 1998.

Ramos, H. and Almeida, A. B. - Hydrotransients Induced by Dynamic Behaviour of Turbo-generators at Small Hydroelectric Power Plants - Pressure Surges and Fluid Transients. Editor: A. Boldy. Harrogate, England - pp 417-429 BHR Group April, 1996.

Betâmio de Almeida, A. and Ramos, Helena - Hydraulic Transients in SHP (Small Hydropower Plants) Systems. ESHA (European Small Hydropower Association). No 17, pp 34-37 - 1995/96.

The Simulation of Transients in Hydro-Automatic Systems under Flow Control

T. Y. Sheronosova *and* V. V. Tarasevich

ABSTRACT

The paper is devoted to mathematical modelling and simulation of the influences of control actions on piping systems.

The difficulties arising under modelling of transients in large complex pipe systems, caused by the influence of control devices, are discussed. The problem of decomposition of complex pipe networks is discussed. The scheme of decomposition named by the authors as "the mathematical test stand" has resulted. The approach offered allows the control unit and regulated object (the rest of the pipe system) to be considered separately as two subsystems. The exchange of signals between a regulator and the regulated object is simulated by the concept of Riemann invariants.

This approach is demonstrated on an example of the pressure control system in a return pipeline of a heat supply system. The mathematical model of the control system operation is given. This model consists of equations describing the valve movement, hydraulic characteristics of pipelines and mounted equipment, and pump performance equations.

The results of calculating the system response to disturbances of various amplitudes and durations are demonstrated.

The numerical experiments have allowed the contribution of the various factors (such as inertial properties and compressibility of fluid etc.) to be specified in the dynamics of the automatic control process. Also these experiments have allowed the influence of a predetermined mode of operation on the recovery time to be observed.

The approach offered directly enables the form of a signal on an input to be set, and the form of a signal on an output to be received, taking into account all hydraulic parameters and non-linear effects.

1 INTRODUCTION

1.1 Pipe Systems

Piping systems are an important part of many industrial facilities, and determine the operating mode of industrial and utility water usage.

1.3 The role of simulation

There is often not enough information for complete modelling in practice, since the researcher usually has only nameplate data of devices, which do not contain or contain only part of the information on dynamic characteristics of the device. Therefore the granularity of a practical model is determined by the amount of really accessible information concerning elements of the simulated system.

It is necessary to have a suitable model for operation of the regulation and control devices to estimate correctly the results of the influence of the control elements on the functioning of the system as a whole.

The most satisfactory results can be obtained by modelling the complete system, into which the systems of automatic control are included as separate subsystems. However, the modelling as a whole requires significant computational and mathematical resources. Therefore the problem of decomposition of the system is an essential one, that is, how subsystems (in particular, systems of automatic control) can be independent from each other but take into account their interaction with the general system.

The model of a control unit operation can be quite complex (such units can, in turn, represent complex systems); the pipe system can contain many such control units (often of one-type). It is possible to obtain the response function of the regulation system for possible disturbances by varying values of input parameters in the given range. Further, for the analysis of an external system, instead of a model of a regulation unit, it is possible to use the function simulating the reaction of the control unit to various external disturbances. Thus, the problem of the analysis of complex system operation is split into two more simple stages. The disturbance influences on separate units of the system (units of regulation and control) are first simulated separately, and then the derived integrated reaction characteristics for the tested units are used as simplified models in the structure of the system considered.

2 THE SIMULATION PROBLEM FOR SYSTEMS WITH AUTOMATIC REGULATORS

2.1 System decomposition

It is inconvenient to consider the response of a control system to external disturbances "from a pure view". This is because the resulting unsteady process represents the sum of the disturbance which has arisen in the system, with the reaction of the automatic control system to this disturbance and the response on the part of an external part of system to this reaction. Considering the system as a whole, it often fails to determine how far each of the system components exerts influence on the system functioning. There is a need for decomposition of the system, i.e. segregation of regulated object from regulator into a separate subsystem considered independently from the remaining system. In Figure 1 the control unit *CU* is shown as a part of system *S*; in Figure 2 the scheme of decomposition is submitted, where control unit *CU* and the other part of system *S'* (as controlled object) are displayed as separate systems.

It is obvious that the operational stability of such objects depends on the stability of the piping system function. Nowadays such systems are applied in various branches of industry. Water supply networks, gas conduits, heating systems, industrial pipeline systems, oil pipelines, hydraulic drivers, systems of fuel supply etc., are examples of piping systems. Pipelines are also the structural part of such complex industrial systems as power plants.

We consider two-parameter piping systems, i.e. for which the process parameters in the pipes are defined by two quantities: pressure p and water discharge Q. In the case of non-isothermal flow, when the liquid stream is also characterised by the temperature T, we will talk of three-parameter systems (flow parameters are p, T and Q). The case of non-isothermal two-phase flow in the pipes when the phase concentration φ is included in the parameters belongs to a four-parameter system (flow parameters are P, T, Q, φ), etc.

It is inconceivable to provide for high performance of piping systems without the use of automatic regulators and automatic control units, which are components of piping systems also. For example, various throttling regulators keeping a designated pressure or discharge in specified points of the system are such units. The operational efficiency of performance of the system as a whole depends on the functioning of systems of automatic control in many respects, including in extreme events and emergencies.

1.2 Automatic Control Systems and Optimal Control

The schemes of automatic control used nowadays (the so-called "proportional-and-floating controllers", "proportional regulators", "integral regulators" etc.) are based on the analysis of the current information (sensor indication at present) and/or the information from some previous period (for example, integral regulators or proportional-and-floating controllers). Such control will be effective in the sense that the system supports process parameters close to the given normative values under a small deviation of the process parameters from a balance point.

The problem of improvement of operational reliability and effective control in piping systems requires the transition from the concept of automatic regulation and control to the concept of optimal control. Such a concept means a search for the best control not only on the basis of the past i.e., of stored information up to the present, but also on the basis of the forecast of the future state of the controlled system. The use of mathematical models of the controlled object and the control systems is expedient for obtaining such a forecast. The essential demand from such models and algorithms for their implementation is their quick-action, since on-line control must have time to operate on the basis of the forecasts computed by these models. The rapidity of calculation is dependent not only on the speed of the computer but also on the complexity of the model implementation algorithm, which is connected to the granularity of the model. Therefore there is a need to have also simplified "high-speed" models alongside detailed models.

It is necessary to subdivide the parameters of interaction into input signals and output signals for the completion of the decomposition procedure. These parameters of interaction are shown as arrows in Figure 2.

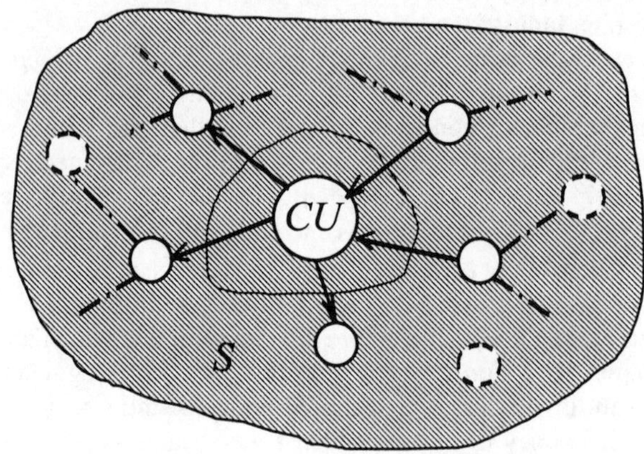

Fig. 1 The scheme of system *S* with control unit *CU*.

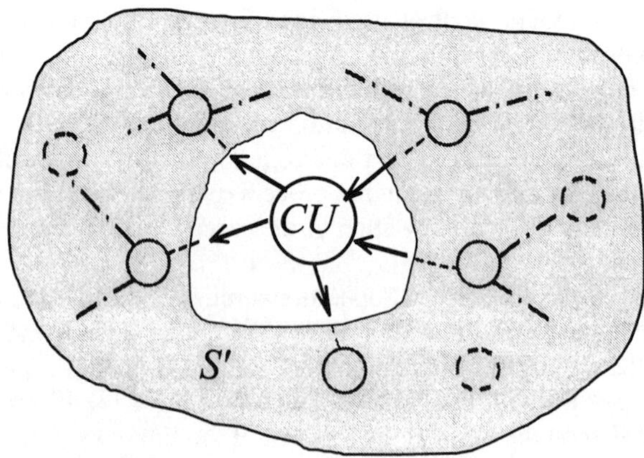

Fig. 2 The scheme of system decomposition. Here *CU* is considered as a separate system, with *S'* the residual part of *S*.

2.2 Mathematical test stand

In the case of two-parameter piping systems the "natural" physical parameters p and Q are the parameters of common hydrodynamic processes, and are the sum of the result of control system *CU* operation and the reactions on this from other parts of the system. As the process of disturbance propagation has a hyperbolic nature, the "natural" parameters of the process are a combination of the input disturbance and the reflected signal, which propagate along the appropriate input and output characteristics. In this case it is possible to describe input signals and output signals as input and output Riemann invariants [1].

It is possible to consider this control unit *CU* as if extracted from the system and placed on the test stand, where the disturbance acting from an external part of system *S'* is replaced by input Riemann invariants. Therefore such an approach was named as "the mathematical test stand" by the authors.

3 SYSTEMS OF PIPELINE PRESSURE CONTROL

In pipe networks (including water distribution systems) of large size and of adverse relief there is a need to construct pumping substations, which usually require an additional level of regulation for the hydraulic regime of the water network after the substation.

The pumps in the pumping substations are often automated. Also, in a number of cases the regulation of inlet pressure to the pumps and high-pressure protection for the return pipeline are necessary.

3.1 An example of the system of automatic control

As an example to which the technique considered was applied, the system of automatic pressure adjustment in the return pipeline of a heating system on the suction pipeline for pumps (*WP*) in a pumping sub-station is chosen. The scheme of the system is shown in Figure 3.

The objective of the control system is to keep a constant pressure p_{reg} on the suction pipeline of pumps *WP* (point *A* in Figure 3). The inclined regulating valve *AW* is an actuating mechanism of the system (see Figure 3). The control element of the system is the pressure relay *PSR*.

The principle of operation of the automatic control system consists of the following:

If the pressure p_A on a suction pipeline of the pump (point *A*) equals p_{reg}, all the control system is in balance. If the pressure falls at point *A*, the balance is broken and pressure in the pulse chamber *2* diminishes. The valve rod blocks the top nozzle under the influence of the spring and high pressure p_B is transmitted from point *A* into the chamber *1*. The rod of the actuator valve *AV* begins to move downwards, reducing the discharge through it and increasing its resistance. As a result the pressure begins to increase at point *A*.

The bottom nozzle is blocked along with the access of high pressure to control chamber *2* and chamber *1* stops the under pressure increasing at point *A*. Therefore liquid overflow occurs from chamber *1* to atmosphere, and the rod of the actuator valve moves upward under the influence of the counterbalance *M*, reducing resistance of the valve and increasing the discharge through it. Consequently the pressure falls at point *A*.

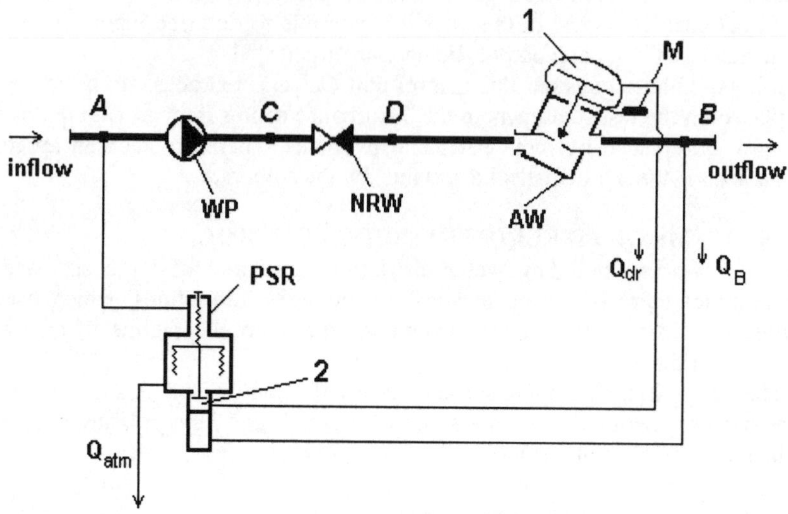

Fig. 3 The scheme of automatic pressure regulating system

3.2 The partitioning of the system

It is important to note we cannot assign, for example, a pressure disturbance or a flow disturbance as an arbitrary function of time, as p_A, p_B and Q depend on each other. However, it is possible to set as an input disturbance their combination in the form of input invariants r and s from the external part of a piping system S', as shown in Figure 4. The Riemann invariants are defined as

$$r = \left(p + \rho c \frac{Q}{\omega} \right) \Big/ 2 , \qquad s = \left(- p + \rho c \frac{Q}{\omega} \right) \Big/ 2 ,$$

where c is velocity of perturbation propagation ($c \approx 1000$ m/s approximately); ω is area of pipe cross-section; p is pressure; Q is water discharge; ρ is density of liquid.

The "natural" physical variables are expressed through r and s as follows

$$p = r - s , \qquad Q = \frac{\omega (r + s)}{\rho c} .$$

Thus we can give an external disturbance in the form of known functions $r_A = r_A(t)$, $s_B = s_B(t)$. Not decreasing the generality of a problem, we can specify the disturbances just at one point, for example, at point A, supposing the incoming signal at point B is a constant:

$$r_A = \varphi(t), \; s_B = const , \tag{1}$$

i.e. the undisturbed state is given as an output of the regulation valve.

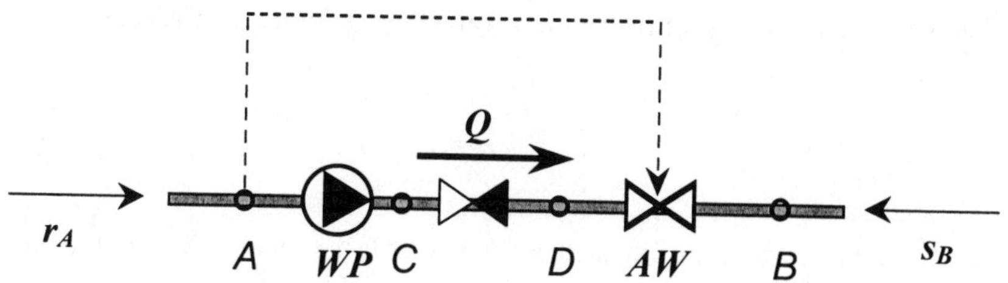

Fig. 4 Modelling scheme. ***Q*** - the valve flow; r_A, s_B are Riemann invariants.

4 MATHEMATICAL MODEL OF CONTROL SYSTEM

The mathematical model of the control system represents the combined system of the non-linear algebraic equations and ordinary differential equations. It includes the equations describing liquid flow along a path *A* - *B*, allowing for the action of the pump, variable resistance *AW* and other local resistances, the equations of dynamics of a moving plunger, the equations describing the dynamics of the liquid flow in internal pipes of control system at interaction with *PSR*.

4.1 Assumptions made

The mathematical model of the automatic control system is based on the following assumptions:

- The elastic properties of a chamber *1* of valve *AW* and the liquid contained in it, are taken into account;
- The inertial properties of a liquid in the interconnecting pipelines of the control system are taken into account;
- The flow of the liquid in the system is taken as turbulent.

4.2 Flow dynamics in control system pipes

The pressure differential between points *A*, *C*, *D* and *B* is described by the following system of equations:

$$p_A - p_C = -f(Q,\nu) + \rho\frac{l_{AC}}{\omega}\frac{dQ}{dt} + \xi_{AC}|Q|Q \tag{2}$$

$$p_C - p_D = \xi_{CD}|Q|Q + \rho\frac{l_{CD}}{\omega}\frac{dQ}{dt} \tag{3}$$

$$p_D - p_B = \xi_{av}|Q|Q + \rho\frac{l_{DB}}{\omega}\frac{dQ}{dt}. \tag{4}$$

where p_A, p_B, p_C, p_D are pressures in an corresponding points; ξ_{CD} is the resistance of a non-return valve (*NRW*); l_{AC}, l_{CD}, l_{BD} are the lengths of pipes between corresponding points; $f(Q, \nu)$ is the pressure-discharge characteristic of the pump.

The variable resistance of the actuator valve ξ_{av} is given as a function of the relative valve movement:

$$\xi_{av} = \xi^{n}_{av}\left(\frac{x^{n}_{av}}{x_{av}}\right)^{N}, \tag{5}$$

where ξ^{n}_{av} is the resistance of the actuator valve at full opening; x_{av} is the valve stroke; x^{n}_{av} is the maximal valve stroke; N is an experimental power exponent.

The liquid flow through interconnecting pipelines from chamber *1* to chamber *2*, from point *B* to chamber *2* and through a drainage branch pipe, are described by the following equations:

$$p_1 - p_2 = \xi_{dr}\left|Q_{dr}\right|Q_{dr} + \rho\frac{l_{dr}}{\omega_{dr}}\frac{dQ_{dr}}{dt} \tag{6}$$

$$p_B - p_2 = \xi_B\left|Q_B\right|Q_B + \rho\frac{l_B}{\omega_B}\frac{dQ_B}{dt} \tag{7}$$

$$p_2 - p_{atm} = \xi_{PSR}\left|Q_{atm}\right|Q_{atm} + \rho\frac{l_{atm}}{\omega_{atm}}\frac{dQ_{atm}}{dt} \tag{8}$$

where p_2 is the pressure in chamber *2* of pressure relay *PSR*; ξ_{dr}, ξ_B, ξ_{PSR} are resistances of the drainage line from chamber *1* to chamber *2*, the force pipeline from point *B* to chamber *2*, and the drain pipeline from chamber *2* to atmosphere, respectively.

Here Q_{dr}, Q_B, Q_{atm} are the discharges in the corresponding pipelines, in this connection

$$Q_{dr} + Q_B = Q_{atm}. \tag{9}$$

4.3 Pump dynamics

The characteristic of the pump $f(Q)$ can be specified as

$$f(Q,\nu) = -a_{wp}Q^2 + b_{wp}Q\nu + c_{wp}\nu^2, \tag{10}$$

where a_{wp}, b_{wp}, c_{wp} are the coefficients describing the operating characteristics of pump ([1]), ν is angular speed of pump.

An angular speed ν satisfies the equation

$$I_p\frac{d\nu}{dt} = T_{em}(\nu) - T_p(Q,\nu), \tag{11}$$

where I_p is total inertial torque of all revolving masses, $T_{em}(\nu)$ is torque of electric motor and $T_p(Q,\nu)$ is anti-torque moment characteristics of pump ([2]).

4.4 Equations of valve dynamics

The equations of a valve rod dynamics assume the form:

$$M_{eq}\frac{d^2x_{av}}{dt^2}=S_m(p_{atm}-p_1)+S_v(p_d-p_b)-F_{fr}+F_{cb};\ \left(0\le x_{av}\le x_{av}^n\right) \tag{12}$$

where M_{eq} is the equivalent mass of moving parts of the valve (pressure regulator valve, valve rod, diaphragm, counterbalance); S_m is effective area of diaphragm; S_v is area of pressure regulator valve; F_{fr} is force of friction in stuffing box seal under the movement of a valve rod; F_{cb} is force of a counterbalance.

The variation of pressure of a liquid p_1 in a chamber *1* depends on a valve stroke, elastic properties of the chamber and outflow Q_{dr} in the following way:

$$\frac{W_1}{E}\frac{dp_1}{dt}=\frac{\pi D_0^2}{4}\frac{dx_{av}}{dt}-Q_{dr}, \tag{13}$$

where E is the integrated modulus of elasticity of valve chamber *1* and the liquid contained in it; W_1 is the volume of valve chamber *1*.

4.5 Model of a control element

The model of action of the pressure relay is based on the hypothesis of linear dependence of the conduction $1/\xi_{PSR}$, $1/\xi_B$ in the chamber *2* on changing of pressure p_A.

Resistance of a drain line (from chamber *2* to atmosphere) we shall find from the formula:

$$\frac{1}{\xi_{PSR}}=\begin{cases}0, & \text{if } p_{reg}-\Delta\ge p_A;\\ \dfrac{1}{\xi_0}\dfrac{p_A-p_{reg}+\Delta}{2\Delta}, & \text{if } p_{reg}-\Delta\le p_A\le p_{reg}+\Delta;\\ \dfrac{1}{\xi_0}, & \text{if } p_{reg}+\Delta\le p_A;\end{cases} \tag{14}$$

where ξ_0 is resistance of a drain line at the full open valve *PSR*; Δ is the control accuracy.

Resistance of pressure line (from point *B* to chamber *2*) is found from formula:

$$\frac{1}{\xi_B}=\begin{cases}\dfrac{1}{\xi_B^0}, & \text{if } p_{reg}-\Delta\ge p_A;\\ \dfrac{1}{\xi_B^0}\dfrac{p_{reg}-p_A+\Delta}{2\Delta}, & \text{if } p_{reg}-\Delta\le p_A\le p_{reg}+\Delta;\\ 0, & \text{if } p_{reg}+\Delta\le p_A;\end{cases} \tag{15}$$

where ξ_B^0 is the resistance of pressure pipeline at the full open valve.

5 CALCULATION TECHNIQUE AND RESULTS

5.1 Calculation Technique

So the operational modelling of the pressure adjustment system is reduced to the solution of the system of equations (2) – (15) with given external disturbances (1). The system is in balance at the initial instant. The calculations were made by a numerical method. The temporal axis was partitioned into steps with a constant time interval. All the differential operators in the equations (2) – (4), (6) – (8), (11) – (13) were replaced by finite differences and all unknown quantities of the right-hand parts were obtained from the top (unknown) time layer. Further, if one linearises all the non-linearity in equations (2) – (15), a system of linear equations is obtained for which solution presents no difficulty. In this way, it is possible to find values of the required quantities for a new time step, and to pass to the following step.

However, in view of the presence of singularities under valve operation (full closing) and constraints (full opening of the valve) such a scheme gives very strong computing disturbances near the singularities. Therefore, the authors tried to avoid the linearisation of the quadratic terms, especially with variable resistance of valves. In this case the system is reduced to a system of quadratic equations in the unknown discharges Q, Q_{dr}, Q_B. It is easy to calculate the other process parameters after finding these quantities. Such a method of solution is steady with respect to the instants of valve closing and opening.

5.2 Results of calculations

The mentioned technique allows any disturbances to be specified on input to the control system and the response of the system to them to be observed. As an example we shall consider a disturbance in the form of a smoothed step pulse with amplitude A_{imp} and steepness of disturbance front $V_{imp} = A_{imp}/t_{imp}$, where t_{imp} is the duration of increase of a pulse (see Figure 5).

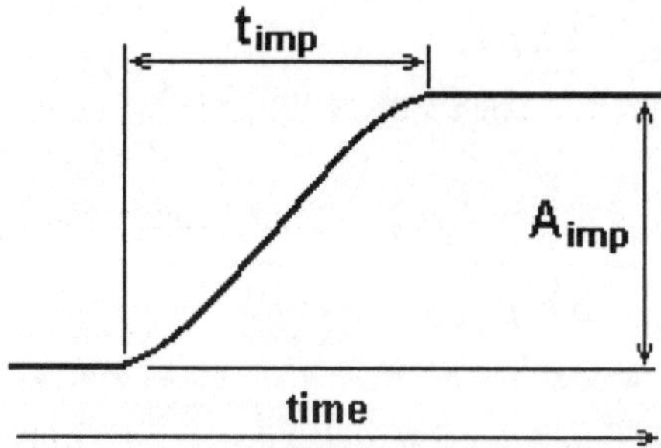

Fig. 5 The pulse shape.

In Figure 6 the results of calculations are represented as the response function of the control system $t_{resp} = t_{resp}(A_{imp}, V_{imp})$ to the disturbance. The function shows how much time is necessary for the system to restore an adjustable quantity in the given range of regulation. The amplitude of disturbance as a proportion of the undisturbed signal on input is plotted on axis y, the steepness of disturbance in $1/sec$ is plotted on axis x and the response time of the system in seconds is plotted on axis z.

From Figure 5 it can be seen that with an abrupt disturbance front the response time of the system depends practically only on amplitude. With a small steepness of disturbance $t_{resp} = 0$, i.e. the system keeps the adjustable parameter in the given limits during all transients. Only with intermediate range of values V_{imp} does the response time of the system, t_{resp}, depend both on the amplitudes and on the steepness of the disturbance.

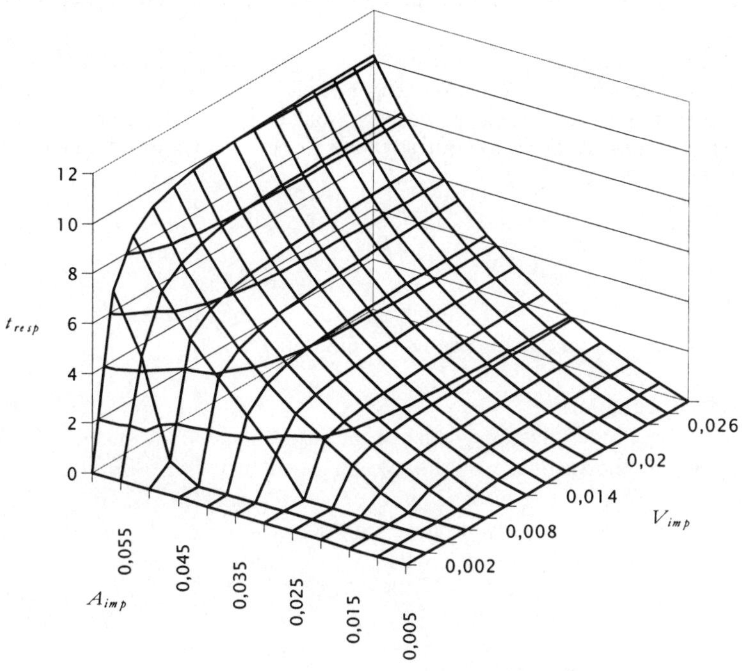

Fig. 6 The surface of the response function

6 CONCLUSIONS

The application of "direct" mathematical modelling enables disturbances to be given arbitrarily and the response of the system to be observed "directly" over the whole range of emergencies. Systems of regulation have some normal range for adjustment. Disturbances of large amplitude can be the reason for falling outside the limits of this control range. The approach considered is quite effective for calculations in such situations. It takes into account the functional constraints on the working of operating equipment and execution units of control devices.

Effective control should be based on the forecast of the future state of the controlled system. For production of such a forecast it is proposed that mathematical models be used of both the control object and the control systems. Alongside complete detailed models, simplified "fast" models for operational forecasting are also required.

The research problem of the response of pipe systems to changes in hydraulic parameters is that the implementation of a full-scale test of the considered automatic control system represents large practical complexity. In this regard, a way of mathematical modelling seems the most effective solution. It allows one to predict the response of the system to disturbances with a sufficient degree of accuracy. The simulation allows one to estimate the state of the system and to specify necessary control actions for maintaining the required modes of operation.

REFERENCES

1. B.L.Rozhdestvensky, N.N.Yanenko, The systems of quasilinear equations and their applications to gas dynamics. Moscow, Nauka, 1978. (in Russian)
2. J.A.Fox. Hydraulic Analysis of Unsteady Flow in Pipe Networks. – London, 1977.

Optimal Control of Slow Transients in Water Distribution Networks Using Linear Quadratic Regulator Theory

R. S. Souza *and* F. H. Chaudhry

ABSTRACT

This paper formulates the problem of continuous optimal control of slow transients in a water distribution network using linear quadratic regulator theory to derive an expression for the optimal opening of pressure reducing valves. As few state variables are observed in practice, an estimator of missing state variables is included in the continuous feedback system. The practical application is illustrated for the case in which an example network moves from a given equilibrium condition to another. The results of these simulations show that linear quadratic regulator technique is efficient in determining the optimal control valve settings.

1 INTRODUCTION

An efficient water distribution of urban water supplies involves safe operation of water networks and favourable pressures. Such reliability and efficiency of distribution is obtained through conveniently located pressure control valves in the network by complete or partial automation of the system. Various authors [1,6,7] have applied optimal control theory concepts to the operation of irrigation canal systems. These techniques have some desirable characteristics which can be equally useful in the design of hydraulic networks as in [12].

The theory of Linear Quadratic Regulator (LQR) is applied in this paper for the analysis of multivariable optimal control of a water distribution network. The non-linear partial differential hydrodynamic equations expressed in the form of ordinary differential equations for "slow transients" or the unsteady flows of incompressible liquid in rigid conduits by coincidence matrix method were linearized around equilibrium flow conditions and were arranged in appropriate form for the application of control theory as in [10]. LQR is applied to derive an expression for optimal opening of the pressure reduction valves. This paper represents a preliminary step towards the use of optimal control theory concepts in practice.

2 HYDRAULIC MODEL

The gradual changes in the flows in a conduit brought about by changes in demand or by the action of control valves can be reasonably modelled by the unsteady flow equations for incompressible fluid in a rigid conduit. In this formulation, the system responds instantaneously to any disturbance in pressure or velocity which is equivalent to admitting an infinite celerity of propagation of disturbances. The differential equation for unsteady flow q, in a rigid conduit with turbulent resistance obtained by integration of the equation for conservation of linear momentum along the conduit length l is given by:

$$L\frac{dq}{dt} = H - F_W q|q| \tag{1}$$

where $L = l/ga$ and $F_W = fl/(2gda^2)$. In these equations, g is the acceleration due to gravity; a, the cross sectional area of the conduit; t, the time; f, the friction factor; and d is the diameter of the conduit.

This equation, together with the continuity equation and boundary conditions, describes the dynamics of transient flow in a rigid conduit.

2.1 System of Equation for a Network

For a network consisting of different kinds of elements and specifications, the following system of state-space equations was proposed in [10] for the vectors of independent pipe discharges q_I and heads h_1:

$$f_1 = \frac{dq_I}{dt} = W_I E_Y + P_I Q_T \tag{2}$$

$$f_2 = \frac{dh_1}{dt} = -C^{-1}[A_1 q + Q_1(h_1)] \tag{3}$$

where $E_Y = P[-f(q) + A_1^t h_1 + A_3^t h_3]$ and $Q_T = L^{-1}A_2^t R^{-1} dQ_2/dt$. f(q) is the vector of head losses whose jth element is given by $F_j q_j |q_j|$ where $F_j = f_j l_j / 2gd_j a_{ij}^2 + f_{vj} / 2ga_{ij}^2$. In these equations, f_j is the friction factor, f_{vj} is the valve head loss coefficient, d_j is the diameter and a_j is cross-sectional area of the jth pipe. L is a diagonal matrix whose jth element $L_{jj} = l_j/(ga_j)$. C is the diagonal matrix whose ith diagonal element $C_{i,i}$ corresponds to the water free surface area of variable level reservoir at the ith node. $W = L^{-1}(I - S)$, $S = A_2^t R^{-1} A_2 L^{-1}$, $R = A_2 L^{-1} A_2^t$, I is the identity matrix, $PWP^{-1} = [W_I | W_D]$. $P = [P_I | P_D]$ is a linear transformation matrix, q_I is the vector of independent discharges, $q_D = -A_{2D}^{-1}(A_{2I} q_I + Q_2)$, $A_2 P^{-1} = [A_{2I} | A_{2D}]$, and finally, $A_1, A_2, A_3, h_1, h_2, h_3, Q_1, Q_2, Q_3$ are respectively the incidence matrix, head and water demand vectors for nodes (1) with unknown head and demand, nodes (2) with head unknown but demands known and nodes (3) with head known but demands unknown.

2.2 Linearization of State-Space Equations

The non-linear state-space equations (2) and (3) of the continuous system can be arranged in vector form as:

$$\frac{dX}{dt} = f(X,U,t) \tag{4}$$

where f is a non-linear vector function, U is the input vector as the control mechanism, and X is the vector of state variable expresses as:

$$U = \begin{bmatrix} f_v \end{bmatrix} \quad ; \quad X = \begin{bmatrix} q_1 \\ h_1 \end{bmatrix} \quad ; \quad f = \begin{bmatrix} f_1 \\ f_2 \end{bmatrix} \tag{5}$$

In order to analyse the transient states near the final equilibrium state, the system in (4) is linearized as :

$$\frac{dx}{dt} = Ax + Bu \tag{6}$$

using the following definitions for perturbed state and control variables:

$$X = x + X_f \quad ; \quad U = u + U_f \tag{7}$$

where x is the vector of state perturbations, u is the vector of input perturbations. The final equilibrium states X_f and U_f satisfy the following relationships:

$$f(X_f, U_f, t) = 0 \tag{8}$$

and

$$A = \left. \frac{\partial f(X,U,t)}{\partial X} \right|_{X_f,U_f} \quad ; \quad B = \left. \frac{\partial f(X,U,t)}{\partial U} \right|_{X_f,U_f} \tag{9}$$

Note that the constraints $\|x\| < \|X_f\|$ and $\|u\| < \|U_f\|$ must be met.

3 LINEAR QUADRATIC REGULATOR THEORY

The LQR specifies an integrated quadratic cost function in continuous time as:

$$J(u) = \frac{1}{2}\int_0^{\infty}\left[x^t(t)Qx(t) + u^t(t)Ru(t)\right]dt \tag{10}$$

where the weight matrix, Q, is denominated as the matrix of state costs and the weight matrix R is called the matrix of control costs. The matrices Q and R are respectively semi-positive-definite and positive-definite.

The determination of optimal control sequence by minimising (10) seeks to minimise the deviations of the state variables (discharges and heads) concomitantly with the deviations in the valve operation. It prevents specially the large fluctuations in state and control variables thus controlling the instabilities generated by the internal and external perturbations.

It can be shown [4] that the control vector u has the following simple structure:

$$u(t) = -Kx(t) \tag{11}$$

where $K = R^{-1}B^tP$ with P given by $A^tP + PA - PBR^{-1}B^tP + Q = 0$. The resulting feedback system is finally obtained as:

$$\dot{x}(t) = \left[A - BK\right]x(t) \tag{12}$$

3.1 State Estimator

In equation (11), all the state variables are supposed to be known through observation of the system. However, in practical situations, not all the variables can be measured either due to an impossibility or prohibitive costs. In order to implement a design based on the state feedback, it is necessary to estimate the unobserved state variables as $\hat{x}$ using the so-called "estimators".

The design of a LQR control can thus be divided in two phases. In the first phase, a control is designed assuming that all the state variables are available for feedback. In the second phase, one designs the state estimator which evaluates all the state variables (or only those not measured) necessary for the feedback.

The estimator is driven by the errors between the measured and predicted values of the selected state variables of the system as:

$$\dot{\hat{x}}(t) = A\hat{x}(t) + Bu(t) + L\left[y(t) - C\hat{x}(t)\right] \tag{13}$$

where the observer matrix L is determined from the dual problem for the feedback control matrix K of the regulator problem. The estimated state $\hat{x}(k)$ is used to construct the control vector u(k) as:

$$u(t) = -K\hat{x}(t) \tag{14}$$

3.2 Optimal Control System

Figure 1 shows the expanded control system in which an estimator of states is incorporated into a feedback control [3].

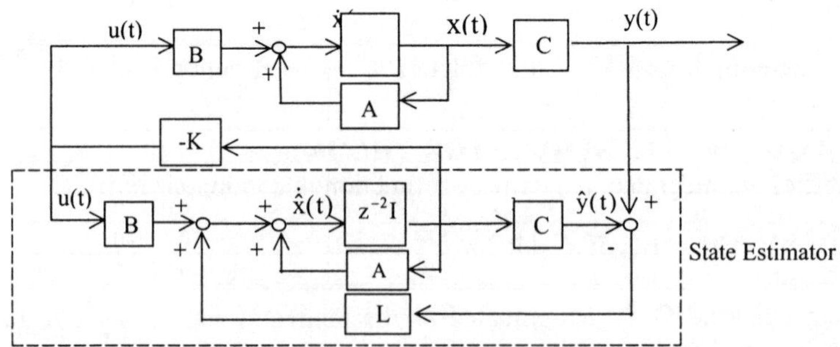

Fig. 1 Block diagram for feedback control system with the state estimator.

The complete state of the feedback control system is obtained by Eqs. (6), (13), and (14) expressed in the form:

$$\begin{bmatrix} \dot{x}(t) \\ \dot{\hat{x}}(t) \end{bmatrix} = \begin{bmatrix} A & -BK \\ LC & A - BK - LC \end{bmatrix} \begin{bmatrix} x(t) \\ \hat{x}(t) \end{bmatrix} \tag{15}$$

where $y(t) = Cx(t)$ and C is the output matrix.

4 RESULTS

The LQR is applied to a simple example network using the rigid conduit model within the framework of incidence matrix formulation. Simulations of continuous linear control are conducted for given initial states.

4.1 Example Network

The simple network used in [10] is chosen for implementing the LQR. The formulation is quite general, however, for application to large size networks.

The example network (Fig. 2) has 5 pipes, 2 variable level reservoirs with overflow weirs (nodes 1 and 2), 2 demand nodes (3 and 4), 2 constant level reservoirs (nodes 5 and 6) and 4 pressure reducing valves in pipes 1, 3, 4 and 5.

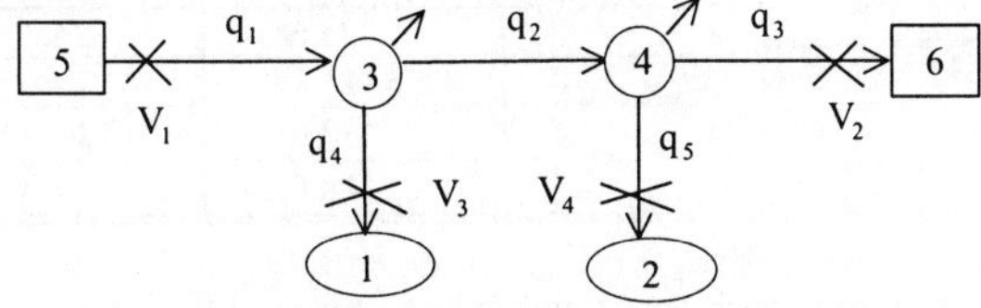

Fig. 2 Example network

The pipe and node data are presented in Tables 1, 2 and 3. Tables 4 and 5 present the initial and final states of the network calculated by algorithm in [11].

Table 1 Pipe characteristics of the example network.

Pipe	Length, l_i (m)	Diameter, d_i (m)	Friction Factor, f_i
1	4000	1.5	0.02026
2	3500	1.2	0.02520
3	2200	1.0	0.02720
4	1500	1.0	0.02840
5	1300	0.8	0.02820

Table 2 Characteristics of variable level reservoir nodes 1 and 2 (type 2).

j	Weir Crest Level h_{crj} (m)	Area $C_{i,i}$ (m^2)	Weir Discharge Coefficient, C_j^*
1	24	10	2.71
2	18	10	2.71

* $Q_j(t) = C_j(h_j - h_{crj})^{3/2}$ se $h_j \geq h_{crj}$ ($j = 1,2$) e $Q_j(t) = 0$ se $h_j < h_{crj}$.

Table 3 Characteristics of constant level reservoirs nodes 5 and 6 (type 3)

Node, j	Reservoir Heads, h_j (m)
5	30
6	18

Table 4 Initial and final steady-state nodal heads.

Node j	Head, h_j (m)	
	Initial	Final
1	24.35	24,37
2	18.35	18,37
3	25.66	26,08
4	21.85	21,86
5	30.00	30,00
6	18.00	18.00

Table 5 Initial and final steady-state discharges and valve head loss coefficients.

Pipe i	Discharge q_i (m^3/s)		Valve Head Loss Coefficient, f_i	
	Initial	**Final**	**Initial**	**Final**
1	1.71	1.80	36.75	20.00
2	1.14	1.20	-	-
3	0.57	0.60	83.42	70.07
4	0.57	0.60	06.09	15.14
5	0.57	0.60	07.50	2.35

The demands at nodes 3 and 4 are assumed to be zero.

4.2 State Matrices

Final steady-state conditions were used to calculate the state matrix A and the control matrix B obtained by evaluating Eqs. (9) on the basis of the rigid conduit model in Eqs. (2) and (3) as:

$$A = 10^{-3}\begin{bmatrix} -27.2394 & 1.9971 & 0.2153 & 0.9708 & -0.9306 \\ 4.5498 & -24.7646 & -0.1168 & -2.8747 & 0.5047 \\ 9.2921 & 1.0384 & -44.5782 & 0.5047 & -2.3029 \\ 0.0000 & 100.0000 & 0.0000 & -245.9241 & 0.0000 \\ 0.0000 & 0.0000 & 100.0000 & 0.0000 & -245.9241 \end{bmatrix} \quad B = 10^{-3}\begin{bmatrix} -0.0434 & -0.0256 & 0.0289 & -0.0677 \\ -0.1008 & 0.0139 & -0.0856 & 0.0367 \\ -0.0225 & 0.0409 & 0.0150 & -0.1674 \\ 0.0000 & 0.0000 & 0.0000 & 0.0000 \\ 0.0000 & 0.0000 & 0.0000 & 0.0000 \end{bmatrix}$$

The dimensions of these matrices depend upon the number of the state variables (5) and the control variables (4) in view of the fact that the state variables x_1, x_2, x_3, x_4 and x_5 are respectively the flow discharges in pipes 2, 4 and 5 and heads at nodes 1 and 2.

4.3 Control and Estimator Matrices

The control matrix K was obtained from the solution of Eq. (10). It requires the specification of matrices Q and R as well as A and B. The choice of elements of the matrices Q and R is dictated by the type of feedback system response desired. In the present case, the matrices were specified on the basis of the method presented in [8]. This method determines feedback such that the system has a predefined set of eigenvalues (poles). The estimator was designed for the measurements of nodal heads at reservoir nodes with unknown demands (1 and 2). The control matrix K that places all the poles of the feedback system for the example problem at -0.05 and the estimator matrix L with the poles at -0.1 in the complex plane are given below:

$$K_{-0.05} = \begin{bmatrix} -132.6201 & 966.2176 & 315.3171 & -2187.1462 & -630.0976 \\ -103.5623 & -195.4598 & -243.7585 & 401.2027 & 491.2978 \\ 57.5906 & 783.1300 & 16.6455 & -1723.5318 & -36.8428 \\ -69.8904 & -107.8439 & 987.4971 & 224.7543 & -1974.8168 \end{bmatrix} \qquad L_{-0.1} = \begin{bmatrix} 0.1341 & 0.3481 \\ 0.0724 & 0.0435 \\ 0.0340 & 0.1058 \\ -0.0568 & 0.0284 \\ 0.0284 & -0.0317 \end{bmatrix}$$

4.4 Application of Linear Quadratic Regulator

One of the main objectives of automatic control in a water distribution system is that it responds automatically in a desired manner to the arbitrary variations imposed in it. After these oscillations occur, the system will have to adjust itself and tend to a new equilibrium state minimising the objective function (10).

In this application, the initial condition was specified in terms of independent discharges in pipes 2, 4 and 5 and heads at nodes 1 and 2 as in Tables 4 and 5.

The network response in terms of the variations in the state variable x (with and without the estimator) is illustrated through discharge in pipe 4 and head at node 2 in Fig. 3. It is seen that the variations approach zero (equilibrium state) asymptotically and that the estimator, following closely the full knowledge response, is capable of reproducing reasonably well the dynamic behaviour of the network despite the lack of observed information. One observes further that the oscillations around the final state are relatively small showing that positions of the poles chosen here were appropriate.

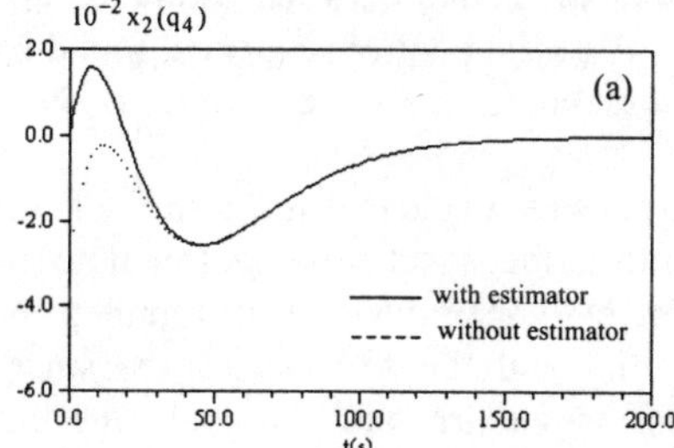

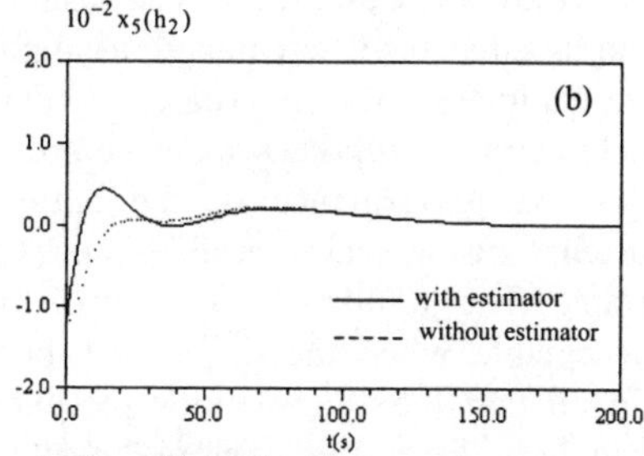

Fig. 3 Variations in state variables calculated by LQR with and without estimator: (a) discharge in pipe 4 and (b) head at node 2.

The valve control obtained by LQR for valves 2 and 3 is exemplified in Fig. 4 which shows the sequence of their openings in time with and without estimator. It is observed that the valves assume stable position when the network approaches equilibrium. Although the estimator produces the valve operation which, in the final stages, closely follows the full knowledge control, it presents greater oscillation in the beginning.

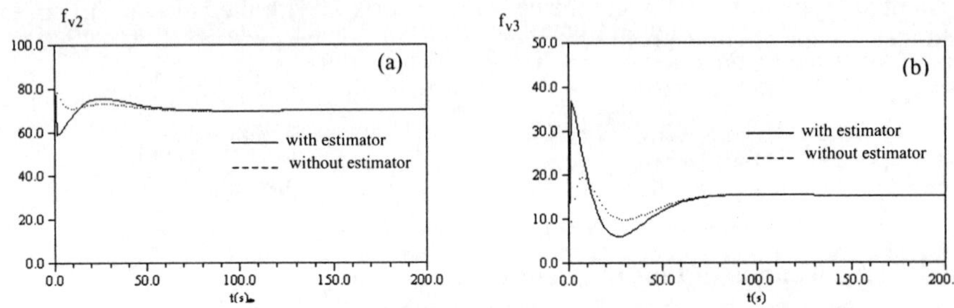

Fig. 4 Head loss coefficients calculated by LQR with and without estimator: (a) valve 2 and (b) valve 3.

5 CONCLUSIONS

This paper studied the application of the continuous optimal control theory methods in the operation of an example water distribution network through pressure reducing valves in which the system moves from a given equilibrium condition to another.

The results of the numerical simulation show that the linear quadratic control is efficient in leading the network to the desired final condition without undue oscillations. The valve head loss coefficients, constituting the sequence of control, found by the optimal theory were found to be in the acceptable range. For the case in which only few variable are measured, an estimator of unobserved state variables was designed whose results show that the errors between estimated and measured states decay to zero asymptotically. Also the control sequence, found by the estimator, for the valves or the head loss coefficients were near to those obtained from full knowledge control.

The performance of the optimal control was evaluated in terms of the oscillations in state variables and valve control, the system being found to be stable. The simulated response of the system by linear theory was found to be acceptable when the differences between initial and the final states are small. Given that the formulations presented in this paper are quite general, one can conclude from the example results that the feedback control of real water distribution systems can be accomplished using the linear regulator control theory.

REFERENCES

Balogun, O. S., Hubbard, M., De Vries, J. J. - Automatic Control of Canal Flow Using Linear Quadratic Regulator Theory. *Journal of Hydraulic Engineering*, v. 114, n. 1, p. 75-101, 1988.

Chaudhry, M. H. - *Applied Hydraulic Transients*. New York, Van Nostrand Reinhold Company, 1979. 503p.

Isermann, R. - *Digital Control Systems*. Berlin, Springer-Verlag, 1981. 334p.

Kwakernaak, H., Sivan, R. - *Linear Optimal Control Systems*. New York, John Wiley & Sons, 1972. 575p.

Onizuka, K. - System Dynamics Approach to Pipe Network Analysis. *Journal of Hydraulic Engineering* , v. 112, n. 8, p. 728-749, 1986.

Reddy, J. M. - Local Optimal Control of Irrigation Canals. *Journal of Irrigation and Drainage Engineering*, v. 116, n. 5, p. 616-631, 1990.

Reddy, J. M., Dia, A., Oussou, A. - Design of Control Algorithm for Operation of Irrigation Canals. Journal of Irrigation and Drainage Engineering, v. 118, n. 6, p. 852-867, 1992.

Saif, M. - Optimal Linear Regulator Pole-Placement by Weight Selection. *International Journal of Control*, v. 50, n. 1, p. 399-414, 1989.

Shimada, M. - Graph-Theoretical Model for Slow Transient Analysis of Pipe Networks. *Journal of Hydraulic Engineering*, v. 115,n. 9, p. 1165-1183, 1989.

Shimada, M. - State-Space Analysis and Control of Slow Transients in Pipes. *Journal of Hydraulic Engineering*, v. 118,, n. 9, p. 1287-1304, 1992.

Souza, R. S. – Computational Aspects of Steady-State Analysis of Water Distribution Networks with Hydraulic Components. Master´s Thesis - São Carlos School of Engineering, University of São Paulo, São Carlos, Brazil, 1994.

Souza, R. S. – Optimised Operational Control of Water Distribution Networks by Linear Regulator Theory. Doctoral Thesis - São Carlos School of Engineering, University of São Paulo, São Carlos, Brazil, 1998.

Wylie E. B., Streeter V. L. - *Fluid Transients*. New York, McGraw-Hill, 1978. 384p.

PART VI

MONITORING AND CONTROL

MONITORING AND CONTROL

[illegible]

URBAN WATER SUPPLY SYSTEM OPERATION PROBLEM

[illegible]

In addition to the above, technical design [illegible] treatment plant and pipe failures and economic considerations (e.g. in treating and transporting water) should be taken into consideration.

In this context, the urban water supply system operation problem is defined as the supply of water with certain physical and chemical properties, at given pressure

Fuzzy Expert System Model for the Operation of an Urban Water Supply System

P. L. Angel R., J. A. Hernández R. *and* J. J. Agudelo R.

ABSTRACT

From a technical point of view, the main objective of an urban water supply system is to supply good quality potable water to the users of the system in adequate quantities and pressures. To achieve this goal, various operational parameters should be determined, including valve and pump settings and storage tank volumes. These operating decisions are based on the present state of the system and the operation restrictions, e.g. the proximity of a peak hour demand or pipe failures. Operational decision making in an urban water supply system often relies on experienced operators and on the use of a Supervisory Control And Data Acquisition (SCADA) system that obtains values for the system's state variables.

In this paper, we introduce a fuzzy logic model for the operation of an urban water distribution system with multiple pumping stations, storage tanks and reservoirs. The model considers the operation philosophy and the contingencies of the system under actual operating conditions. The model was implemented in a computer tool used for diagnosis and prediction, consisting of a fuzzy expert system. It bases its decisions on the processing of a knowledge base derived from information provided by experts and input data from the SCADA system. The model was validated using the water supply system of the City of Medellín, Colombia.

1 THE WATER SUPPLY SYSTEM OPERATION PROBLEM

Water consumption in modern cities presents a combination of periodic and irregular behaviour. Potable water distribution thus requires continuous decision making on the manipulation of the physical components of the system, allowing control under extremely dynamic situations. Factors like weather conditions, holidays, variations in socio-economic conditions in different areas of the city and the presence of industrial areas may significantly affect the consumption and therefore, the distribution of potable water.

In addition to the above, technical design restrictions (e.g. the capacity of treatment plant and pipe failures) and economic considerations, (e.g. in treating and transporting water) should be taken into consideration.

In this context, the urban water supply system operation problem is defined as the supply of water with certain physical and chemical properties, at given pressure and volume, incorporating the variation of consumption over time. The operation should avoid excessive storage tank retention times and incorporate water resources management. The various components of the water supply system should be operated within their appropriate safe ranges.

Currently water supply system operators base their methods on their knowledge and experience of the layout and operational behaviour of the water supply system and the dynamics of water consumption in the city.

It is important that any operational model of the system should operate as an expert system incorporating the knowledge base of operator experience. In the light of the complexities of modelling the system by traditional techniques, state variable behaviour, component diversity, high frequency of required in- and outputs and the application of the operation rules, fuzzy logic was selected as the most appropriate form of modelling for the water supply system.

2 THE MODEL

2.1 Model Elements

An urban water supply system is made up of more than one natural source of water. The raw water is pumped to the treatment plants for purification. It is then gravity or pump fed to storage tanks, which function as balancing storage to regulate the load on the treatment plants.

In our model, bulk pipes are defined as pipes between water sources, treatment plants and storage tanks. Distribution pipes, on the other hand, connect the storage tanks with the distribution networks. Bulk networks are supplied with different types of valves for controlling the flow. Storage tanks may be fed directly from other tanks, and were categorised in terms of the number of tanks in series between them and the treatment plant. Parallel tanks were reduced to a single tank. Because several treatment plants may feed a given tank, it may be classified into different categories simultaneously.

2.2 Model Objectives

The main purpose of the water supply system is to provide the users with potable water. To safeguard public water supply, the stored volume in a given tank should be adequate for the needs of both the distribution and downstream tanks fed by it.

The first objective is thus to determine the appropriate tank storage level at any given moment. This volume varies with time and is determined by the application of criteria that take into account the maximum storage capacity of the tanks and the consumption from it. The goal is to ensure adequate tank volumes, rather than keeping the tanks as full as possible.

The second objective is the calculation of the optimal feeder flow rate pattern for a given tank. This calculation is based on water treatment and transportation costs, unless other restrictions exist in the system. In the gravity fed tanks it is possible to regulate the quantity and the time of filling continuously. For pumped storage tanks, however, it is only possible to determine the start and end of the filling period. The total volume supplied will be a function of the characteristics of the pumping system.

2.3 Knowledge Representation

The relationships between the components of the system are modelled by means of frames, because the objects involved in the system can be described as structured. Some of those objects are shown in Figures 1, 2, and 3. The knowledge representation method incorporates goal-oriented production rules, i.e. with backward linking. Within the production rules fuzzy terms are included to capture the uncertainty of the conditions in which the experts make decisions. However, the establishment of some objectives depends on forward linking.

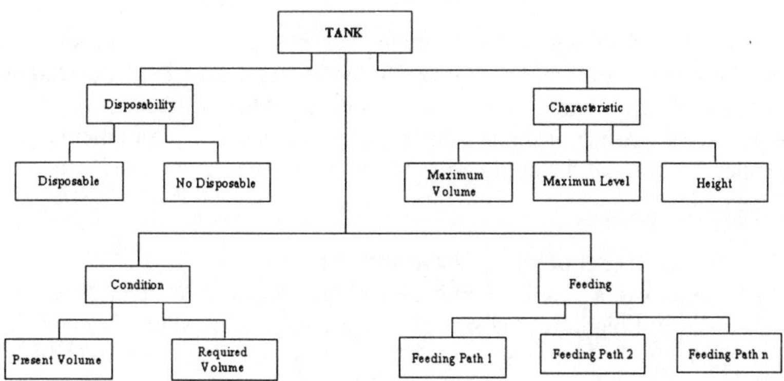

Figure 1. Structured Object Representation of Tank Characteristic

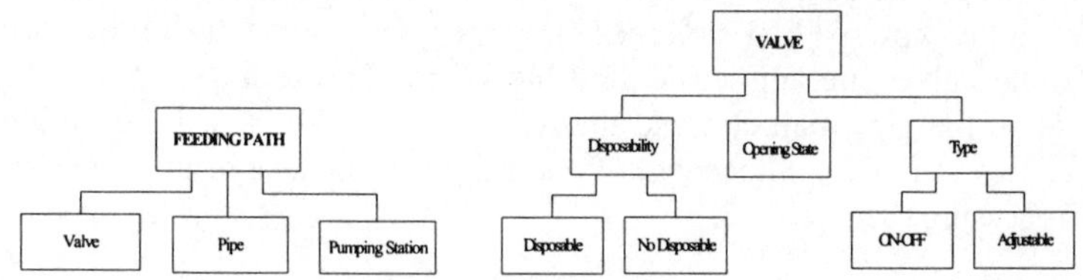

Figure 2. Representation of the Feeding Path and Valve Structured Objects

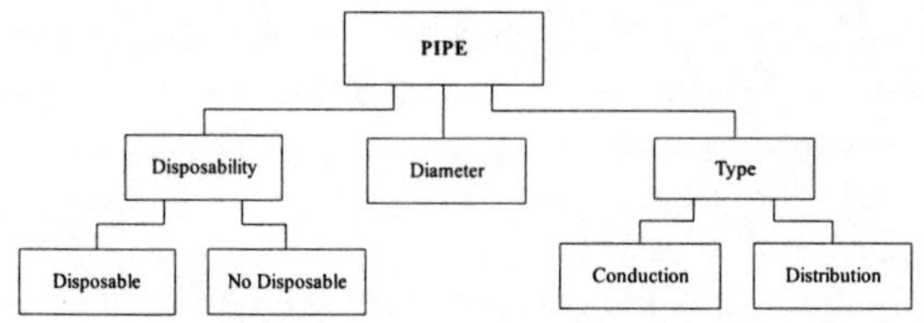

Figure 3. Structured Object Representation of Pipe

2.3.1 Model Functioning

In the first instance, the projected volumes of the lower category tanks are calculated. Backward linking is used to determine the volume of tanks in the superior category, relating the actual state and the water demand of each tank after determining the volume of all the tanks. In the case of pumped feeding, optimisation of the pump system is considered. After calculating the desired level of the tanks in the superior category, the required treatment plant load is determined.

The entire system is then recalculated, optimising the feeding pattern of tanks over time to secure the desired volume, aiming for a minimum cost and a stable working load in the treatments plants. The model is supported in a SCADA system that allows the identification of pipe pressure, the volume in the tanks, and the condition of pumping stations permitted to make predictions and watch the water distribution.

In the determination of the storage volumes in the optimisation operation, production rules are applied; these rules contain in their antecedents and consequences fuzzy terms and semantic fences. After processing the knowledge base, the mass balance of the system is checked using fuzzy arithmetic. In Figure 4, a general outline of the application of the fuzzy rules in the water supply system is illustrated.

2.3.2 Fuzzy Model of a Representative Element

Fuzzy systems use a smaller number of rules compared to other types of expert systems. The full model of the solved real-life case is, however, beyond the scope of this paper. In the limited space available, only one of the most significant rules, i.e. the operation of throttle valves, will be discussed. The definition of fuzzy variables is presented in Table 1, and their graphical representation is shown in Figure 5. A partial diagram of valve allocation is shown in Figure 6.

When the required tank volumes have been determined, the production rules of the throttle valves are applied to find the appropriate settings. The valve setting depends on the pipe diameter and the required flow rate and is determined by the de-fuzzyfication of the antecedent of the triggered rule as a result of the knowledge base processing. See Table 2.

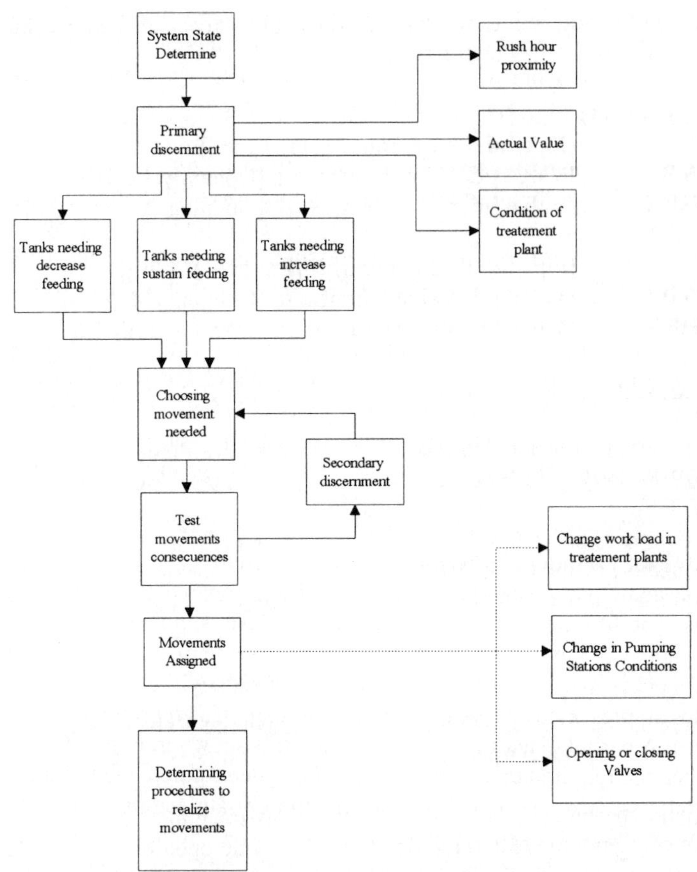

Figure 4. Conceptual Scheme for Fuzzy Operation of an Urban Water supply system.

Table 1. Definition of Linguistic Variables of the Element Adjustable Valves

Input Variables	Linguistic Variables	Rank	Universe of Speech
Flow	Low	0 – 35 %	0 – 100 %
	Middle	35 – 65 %	
	High	65 – 100 %	
Pressure	Low	0 – 50 %	0 – 100 (100 = Nominal Service Pressure + 10%)
	Middle	35 – 65 %	
	High	50 – 100 %	
Opening	Low	0 – 50 %	0 – 100 % (Opening Percentage)
	Middle	30 – 70 %	
	High	50 – 100 %	

The Following are examples of adjustable valve rules in pseudo-code and rewritten in Fuzzy Clips instructions.

Pseudo-Code

```
IF        {Pipe [id. Pipe (Tramo)] [Pressure (Low)]}
AND       {Valve [id. Valve (Number)], [Opening (Low)]}
THEN      {Valve [id. Valve (Number)], [Opening (a-few Middle)]}

IF        {Pipe [id. Pipe (Tramo)],  [Pressure (Low)]}
AND       {Valve [id. Valve (Number)], [Opening (Middle)]}
THEN      {Valve [id. Valve (Number)], [Opening  (High)]}
```

FuzzyClips Code

```
(deftemplate Presion-Tuberia                ; Pressure-Pipe
  0 100 Porcentaje-Presion                  ; Percentage-Pressure
  (
  (bajo-presion  (z 0 50 ))                 ; low-pressure
  (medio-presion (pi 35 50 ))               ; middle-pressure
  (alto-presion  ( s 50 100 ) )             ; high-pressure
  )
)

(deftemplate Apertura-Valvulas              ; Opening-Valves
  0 100 Porcentaje-Apertura                 ; Percentage-Valves
  (
  (bajo-apertura (10 1) (30 0.5) (50 0) ) ; low-opening
  (medio-apertura (30 0) (50 1) (70 0))  ; middle-opening
  (alto-apertura (50 0) (70 0.5) (90 1) ) ; high-opening
  )
)

(deftemplate Valve
  ( slot Presion   (type FUZZY-VALUE Presion-Tuberia))
  ( slot Apertura  (type FUZZY-VALUE Apertura-Valvulas))
)

(defrule Control-Calidad-Presion-Distribucion-02
  ( Valvula
    ( Presion   ?Presion & bajo-presion)
    ( Apertura  ?Apertura & medio-apertura)
  )
=>
  ( printout t " Apertura "  (maximum-defuzzify ?Apertura) crlf)
```

Table 2. Results of Adjustable Valves Rules Application

Input Variables	Evaluation	Output Variables	Value
Low Pressure	10 %	Middle Opening	25 %
Low Opening	10 %		

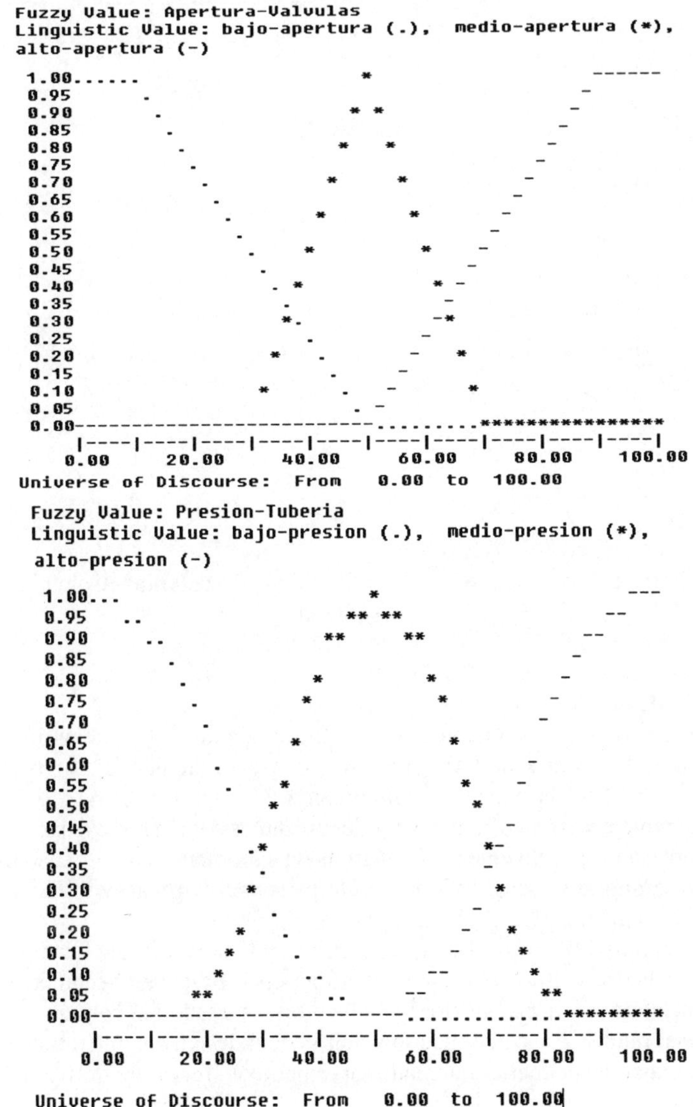

Figure 5. Linguistics Variables Graphics

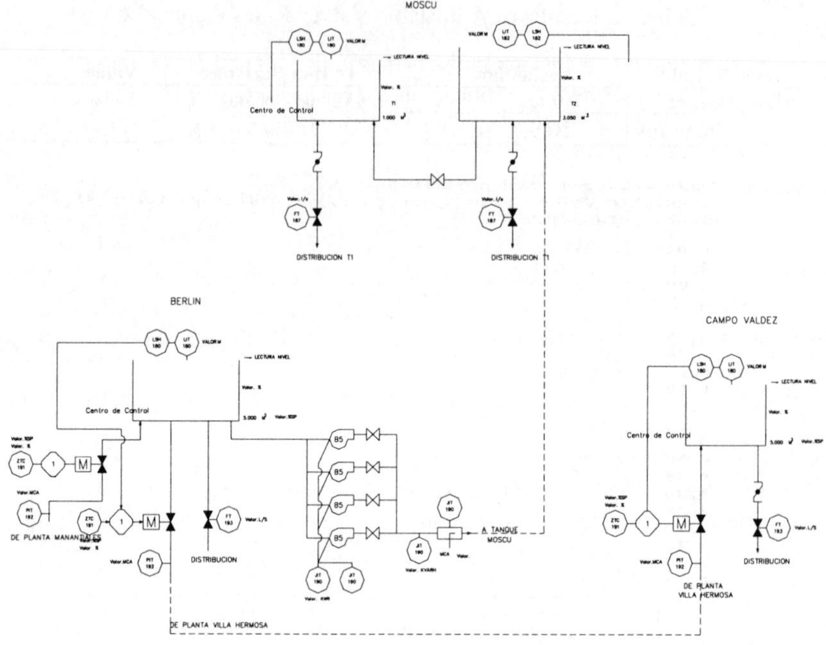

Figure 6. Water supply system of the City of Medellín – Colombia, Partial Diagram

3 CONCLUSIONS

Fuzzy logic is a very valuable tool for the modelling of systems in which the application of human expertise is required, or where the natural language prevails in decision making, as is the case with urban water supply systems.

The implementation of a fuzzy model presumes a wide knowledge of a system. For this reason it is convenient to assign its development to a multidisciplinary co-operative group, involving both corporate personnel familiar with the system and external personnel that re-evaluate the work paradigms.

The nature of water system operation facilitates the application of fuzzy modelling because it is dependent on experience of expert technicians and the communication of decisions to the operators is made in colloquial language. Traditional models can be applied to water system design and operation, but as it is a system that is neither sequential nor repetitive, fuzzy models may be more appropriate.

Additional advantages of applying the fuzzy modelling to water supply system operation are the ease with which it can represent the expert's knowledge and the presentation of instructions in common language by the technicians.

REFERENCES.

Memorias Taller Andino En Sistemas Expertos Y Robotica, Cif Centro Internacional De Física. Universidad Eafit, Medellín Colombia, 1989.

Wendy B. Rauch-Hindin: Aplicaciones De La Inteligencia Artificial En La Actividad Empresarial, La Ciencia Y La Industria, Ed. Diaz De Santos. S. A. 1989. Madrid.

Coenen Frans And Bench-Capon Trevor: Maintenance Of Knowledge-Based Systems. The Apic Series. Academic Press Harcourt Brace Jovanovich, Publishers, London.

Rich E.: Inteligencia Artificial, Programas Educativos S. A. De C. V. 1993, México D. F.

Nebendahl Dieter: Sistemas Expertos: Marcombo, Boixareau Editores, 1991 Sant Adria De Besós, España.

Sell Peter S. Sistemas Expertos Para Principiantes. Megabyte Noriega Editores. 1984 Mexico D. F.

R.A. Orchard. Fuzzyclips Version 6.04. User's Guide

Bench-Capon Trevor: Maintenance Of Knowledge-Based Systems. The Apic Series Academic Press Harcourt Brace Jovanovich, Publishers, London.

Management System for the Control of Urban Water Supply in Córdoba City (Argentina)

F. A. Delgadino, S. M. Reyna, M. Herz, J. Torres, A. Rodriguez, C. P. Depauli, G. Barrera Buteler, E. Helmbrecht, M. L. Juarez, *and* C. Oroná.

ABSTRACT

Córdoba, a 1300000 inhabitant-city, is the second biggest conglomerate in Argentina. Drinking-water supply was provided directly by the State until 1997 when a private company took on the management of the service for 30 years, with important investment commitment to get a better and larger system (target). The Provincial State in its new role, as regulator and controller of the privatization, considered it necessary to design a System of Permanent Control for the Concession, defining the human and material resources needed.

The Permanent Control System designed has an underlying model, in which information is input. Then, the relevant information is selected and validated. There are three specific models that process information: Quality, Users and Resources Monitoring. The first model refers to drinking ability aspects, pressure and continuity. The second one refers to commercial aspects and the relationship with the customer. The last one refers to the monitoring of the financial and economic situation of the Public Utility Company (ensures the continuity of the service), the investment in tangible assets (ensures the accomplishment of the future target) and the internal process screening, including training and growth (ensures a minimum of quality during the Concession development). After processing the information, the models are interrelated. They generate Actions and Recommendations for the Control Authority.

1 MANAGEMENT CONTROL SYSTEM OF THE CONCESSION

1.1 Design of the System

It is the State's responsibility to guarantee the accessibility and drinking qualities of water for the present and future population. In order to do this, it must regulate the market conditions to insure efficiency in the production and equity in the distribution.

Thus, the Concession contract fixed objectives and goals related to spatial and temporal variables, and it also specified, as a contractual obligation, minimum means to reach them. The vastly different problems that may arise requires that the

State adopts an efficient Management Control System of the Concession, that would be able to respond in this new role, not as the Public Utility Provider but as the Public Utility Controller.

First, two objectives were identified, both related directly to the interest of the State to monitor the Provider:

1. Quality (Technical Goals and Objectives related to drinking qualities and accessibility).
2. Users (Commercial Goals and Objectives related to the satisfaction of the customers, and the treatment given by the Provider, simulating conditions of a free market).

If these objectives were accomplished, it would be assumed that the contract had delivered the required results. But the contract also established that minimum means were to be used, this had to be monitored not only because of legal reasons but also to evaluate special situations that might arise (e.g.: goals and objectives that were not met). Then, a third objective appeared:

3. Resources Monitoring.

1.2 System Premises

The list of model premises are: obtain a System that permits one to know the behavior of the Concession: present and future; optimize resources for control and regulation tasks; decrease the information asymmetry; search for measurable indicators; recognize that the customers are the final objective.

1.3 Time Aspects

The control system is a strategic one. It will have to adapt and perfect its own functioning according to the changes in time. The Concession will adapt its management to the new technical and administrative processes and the Control System should accompany these changes.

2 QUALITY MODEL

The model has as its main objective the ability to obtain a complete, dynamic and continuous diagnostic of the water quality, including its pressure and continuity characteristics. It must be able to recommend actions in advance of different states of the treatment and distribution system.

The Quality Model has the following components: Fixed control point network; Laboratory; Data processing unit; Mobile control.

2.1 Spatial Aspects

A network of fixed points where measurements are made permits one to determine a series of parameters established in the Province Norms. These measurements are made combining automatic sensors and standardized sampling.

- A Reference Laboratory that analyses the water samples.
- A Mobile Laboratory (made of two units).
- An Operative Unit in charge of processing and storing data and reports.

2.2 Information Fluxes: Pressure and Quality

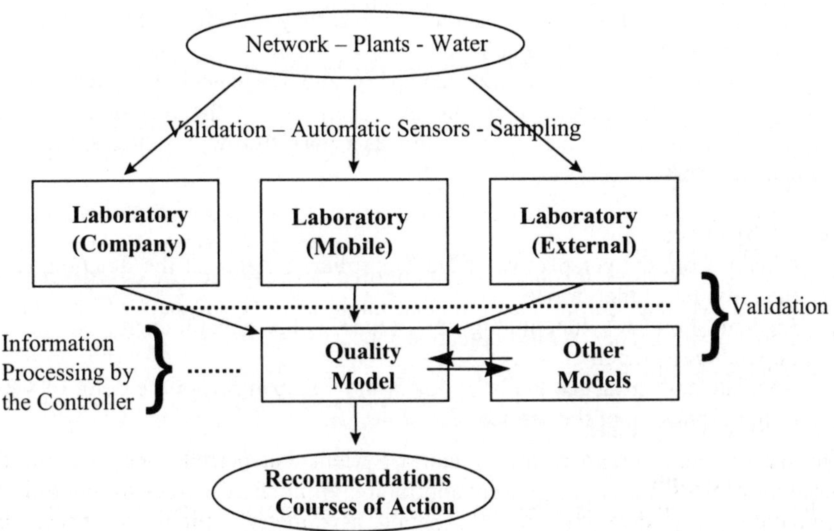

2.3 Actions and Inspection:

In the application of the Quality Model, different situations can arise that the model can consider "a priori" in some cases and thus suggest actions and give recommendations. As an example, the following are described:

a) When an important difference between the information given by the central laboratory of the Company and the external laboratories and/or the mobile laboratories in some point of the system or for some measured parameters (particularly when the possible incidence on the health of the population is direct): in this case, the Company must be notified, the data has to be checked with additional information by means of cross sampling. Preventive measures must be taken and later, an external audit is needed.

b) When bacterial contamination is detected in some point of the water network the following measures have to be taken: notification and claim to the Company; confirmation of the data supplied using additional samples and analyses by the mobile laboratory; follow-up of the zone where the contamination was detected; alert the users for the adoption of preventative actions.

c) When an insufficient pressure is detected in the network during a prolonged period (by means of the sensors): make a notification and claim to the Company; send the mobile laboratory to check quality during the pressure reestablishment.

d) When pipes break: the mobile laboratory has to be sent to make the quality control during the service reestablishment in the affected zone.

3 USERS MODEL

3.1 The problem

There exist attributes that reflect intangible aspects :

- The attributes of the product and/or service: they include the functionality of the product/service, price and quality.
- The relationship between the Company and the customers: includes the delivery of the product/service to the user, considering the response, time of delivery dimension and the opinion the user has of the product/service the Company delivers.

3.2 Objectives

a) Obtain, from the perspective of the Controller, a vision of the functioning of the provider of the service.
b) Evaluate how this functioning contributes to the general increment of the population well being.
c) Regulate and influence the behavior of the Company from the point of view of the expansion of the service.

3.3 Information

If the information that is obtained from the user in its active role (claims) is grouped, basically the claims can be considered as commercial, technical and of mixed nature.

Commercial: Mistakes in the bills, no reception of bills, reception of elapsed bills, payments non registered, bad disposition in the clients attention, problems in the measurement of the volumes consumed, repetition of the claims, errors in the information needed to produce the bill.

Technical:

Service Quality: Lack of water (affects continuity), lack of pressure (affects quantity regarding discharge available), repair requirements: house connections, sidewalks and streets, new house connections.

Product Quality: Chemical and bacteriological quality, odor, taste, color.

Mixed:

In this category fall those cases in which problems are at the same time both commercial and technical, such us claims for damaged property (houses) or delays in new connections and rehabilitation of pre-existing ones

3.4 Variables and Indicators

The goal objectives of the contract of the Company with the State are: quantity, quality, continuity and tariff. On this base, independent variables are defined related to each one of the goal objectives. Then, analysis modules are established.

- Quantity:	x5i	: claims for lack of pressure
	x6i	: claims for water leaks
	x15	: number of users with bill delinquencies
- Continuity:	x4i	: claims for lack of water
	x9i	: claims for home connection repairs

	x8i : claims for water leaks
	x15 : number of users with bill delinquencies
- Quality: (product)	x1i : claims for chemical quality
	x2i : claims for physical quality
	x3i : claims for microbiological quality
	x15 : number of users with bill delinquencies
- Quality: (technical service)	x7i : claims for damaged property reparation
	x8i : claims for sidewalk and street reparations
	x9i : claims for home connections repairs
	x10i: claims for new conn. and rehab. of preexisting ones
	x15 : number of users with bill delinquencies
- Quality: (commercial aspects)	x11i: claims for new conn. and rehab. of preexisting ones
	x13i: claims for problems in the client attention
	x15 : number of users with bill delinquencies
- Quality: (tariff)	x12i: claims for disagreements with the invoices
	x14i: claims related to arrears of payment
	x15 : number of users with bill delinquencies

i 1) = quantity
2) = answering time
3) = solution time

Data corresponding to the independent variables is input for monthly periods (annual analysis is done also). Variables are standardized and processed to apply a multiple regression analysis. The dependent variables are the Satisfaction Degree Indicators. Multiple regression analysis is done between the Satisfaction Degree Indicators (I) (obtained from the independent variables: x1 a x 15, as is shown in the example above), and the satisfaction (S) measured through polls. Coefficients that show the quality of the fit are obtained too.

In order to obtain the satisfaction values to do the regression analysis, at least two types of surveys are needed: a) Annual survey of the users satisfaction and determination of socio-economic changes in the population; b) Annual survey of the non-users satisfaction and determination of socio-economic changes in the population. The Users Model will be the result of the interaction of the statistical information obtained from: Quantitative analysis of the users database; Annual survey of the users satisfaction; Annual survey of the non-users satisfaction; Validated data from the Company users database.

Determining I, global indicator, from the satisfaction (and considering its changes in time), allows one to know the quality of the service from the user's viewpoint. This way, it is possible to make decisions with respect to the adequacy or not of the behavior of the Company following a rational approach.

Example of Satisfaction Degree Indicator

$$I4 = \frac{X_{41}}{sca\,(X_{41})} \left(\frac{med\,(X_{42})}{max\,(X_{42})} + \frac{med\,(X_{43})}{max\,(X_{43})} \right) / \; 2$$

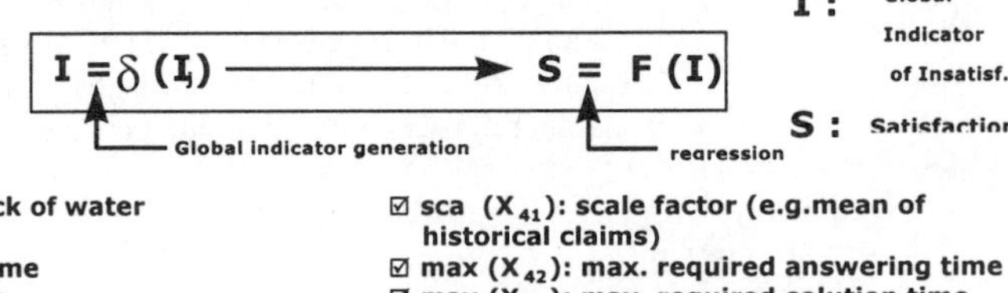

☑ X_4 : claims for lack of water
☑ X_{41}: quantity
☑ X_{42}: answering time
☑ X_{43}: solution time
☑ med (X_{42}): mean answering time
☑ med (X_{43}): mean solution time
☑ sca (X_{41}): scale factor (e.g.mean of historical claims)
☑ max (X_{42}): max. required answering time
☑ max (X_{43}): max. required solution time

4 RESOURCES MONITORING MODEL

4.1 Objective

The proposed system of the concession management control is based mainly on the Quality Model (guarantee of drinkable and available water) and the Users Model (guarantee of commercially fair treatment) which control results. But for legal reasons (Concession contract establishes minimum obligatory means) and for the evaluation of decision making scenarios, it is necessary that the Controller monitors investments, assigned resources, economic and financial health, and other aspects related to the Concessionaire Firm as a production and management unit.

4.2 Criteria

The alternatives in which obligatory construction was short of reaching the expected goals, or where goals were attained with fewer than specified obligatory works were used as a case study. It was considered that investment can be measured in physical units (works) and in financial units (assets), also.

To evaluate situations of unsatisfied goals within a city area, delays in the investment schedule, technical remarks to investment accounting or substitution proposals, three approaches can be made:

- Literal reading of the contract.
- Behavior approach, that attributes the deficiencies to causes that could be solved by the Company with appropriate investment, going further than what the contract required specifically.
- Situational approach that attributes the deficiencies to causes exogenous to the Company.

Each different approach would lead the Controller to different courses of action. For any of the options, information must be available.

4.3 Method

Considering the characteristics of the variables for this monitoring task, these sub-models are included: Financial-Economic, Physical Investment, Growth and Training and of Internal Procedures.

The **Financial-Economic Sub-model** considers indicators for: (a) Comparing the structure of costs with the international level, (b) Controlling the level of vulnerability with respect to the financial resources management policies required by the needed investments, (c) Monitoring of the tariff levels regarding financial behavior arisen from the monopoly situation.

Indicators Monitoring: comprises classical economic-financial ratios of solvency, leverage, liquidity, profitability and general tendencies, that come from audited balances of the Company. From the analysis of the tendencies, changes in the behaviour of the Company can be predicted, alerting the Controller. A tri-monthly periodicity (provisory) and annual (definitive) suffices.

Concession Risk Analysis: A method of sensibilization of the cash flow is proposed. For this purpose, the real reported actuation parameters (the number of users, mean applied tariff, collection percentage, expenditures percentage, etc.) are applied to the projections of the Company. The method will permit the determination of an Internal Rate of Return, which can be compared with the one considered by the Company in the bidding offer.

Tariff Analysis: The tariff has a strong real state data basis and an additional element due to extras above the basic consumption allowed. It is important for the Controller to know the relation of the users tariff (considering point cases and globally) at a point in time and how it has evolved (dynamics). At present the number of connections is 350000 a stratified sample of no more than 500 users is proposed for this follow-up. The sample error can be established simulating on the sample the total gross invoicing and comparing it with the known value. The objective of this is to distinguish between neutral tariff changes (for the Concession), and changes that would involve changes in the total gross invoicing (it will also help to identify cross-subsidies among users).

The **Physical Investment Sub-model** identifies indicators, monitoring sources and frequency for the follow-up of the plan of works and of the state of the infrastructure. The necessary infrastructure works needed for the improvement and expansion of the service traditionally were made with State planning and projects, while construction was carried out by a company (winner of a bidding process) with exhaustive inspection by the State. In this Concession, all the factors are the Company's responsibility, requiring goals both of quality and of enlargement of the service. It would not seem necessary to control the processes, but the contract establishes minimum required investments, identified with specific projects. Thus it is proposed to:

- Control the accomplishment of the required investments.
- Control the inventory evolution.
- Evaluate the relation: investments-goals.

Evaluate the relation: investments-vulnerability.

To avoid the need for an inspectors crew dedicated to checking the construction sites, it is a top priority to take advantage of the information generated by the company, and of the obligation (imposed by the contract) to have independent auditors. This way, the data needed for the indicators will be validated and the results will reflect the evolution of the items objective of the model. Other sources of information and validation are: Users and Quality Models, external information from other official agencies, and inspection programs.

In the **Sub-models: Growth and Training, and Internal Procedures**, the following actions of the Company are monitored: a) development of the human resources and the organization structure, b) growth and relations with the environment, c) systems and processes such as invoicing, collection, external information and communication.

5 CONCLUSION

The control of the Concession of the water service regarding goals and objectives gives independence to the Company in resources management. At the same time, this allows the Controller (the State) to concentrate on the results (technical quality of the water and quality as perceived by the user). Nevertheless, the contractual requirement of minimum means obliges the Controller to have enough information to monitor the investment processes and the evolution of the Company. Thus, it was decided to adopt a mixed model that dedicates two of its sub-models (Quality and Users) to goals and objectives and one to minimum means (Resources Monitoring). This model in every step takes advantage of the information provided by the Company through validation processes in order to optimize its resources.

For the Province of Córdoba State, this is the first and most important privatization that has taken place. It implies a new role for the State that is being perfected as the privatization process matures and encompasses more services. This first approach should help the State in its new role, and the experience collected should allow better contracts for future Public Utilities privatizations.

REFERENCES

Cicero, Nidia Karina (1996). Servicios públicos, control y protección. Ediciones Ciudad Argentina. Buenos Aires, Argentina.

Dirección de Agua y Saneamiento (1994). Normas Provinciales de Calidad y Control de Aguas para Bebida. Córdoba, Argentina.

OFWAT (Office of Water Services) (1995-1998). Various Reports and Documents. UK.

Provincia de Córdoba (1997). Contrato de Concesión del Servicio de Agua Potable de la Ciudad de Córdoba. Córdoba, Argentina.

Sarmiento García, Jorge (1996). Concesión de Servicios Públicos. Ediciones Ciudad Argentina. Buenos Aires, Argentina.

On-line Monitoring and Control of Water Distribution Networks

K. Edwards *and* R. Kirby

ABSTRACT

Anglian Water, along with other UK water companies, have implemented on-line monitoring of District Meter Areas to expedite the effectiveness of leakage control strategies. The Anglian Water system uses a combination of Low Power Radio and the companies existing telemetry system to provide the mechanism for data collection. Other companies have used a range of other communication media to achieve the same effect. Future technologies such as Automatic Meter Reading and on-line water quality monitoring are laying the foundations for the introduction of decision support techniques allowing Network Management to be revolutionised in the next 10 years.

1 INTRODUCTION

Anglian Water is now a global water company with a growing international operational base. The company supplies water and waste water services to some 10 million people world-wide. In the United Kingdom Anglian Water service the eastern area of England. The challenges facing the company are a generally flat and rural location which has a dry climate, rainfall levels being only half the national average. Also being within commuting distance of London and with improving communications the region has the fastest growing population in the UK, at around 1% per annum.

2 BACKGROUND

Water distribution networks are generally the largest asset the water utility possesses. Unlike the development source works and other capital installations, often the network has evolved over many years and is often not well designed and may be operated sub-optimally. Many systems use a variety of pipeline materials and are often in poor condition structurally. The poor condition of water distribution systems has led to high levels of leakage, intermittent supply, inadequate pressure, and water quality problems. Unlike modern supply installations however, traditionally the network has not benefited from the use of telemetry and on-line control technologies.

Tighter regulation, changing climatic condition, growing populations, water resource limitations and greater customer expectations are leading to change. Within the UK the water companies are investing in control technologies to aid better understanding and improved operational practices.

Traditional methods of leakage detection are costly and inefficient. Anglian Water, along with others in the UK, have adopted the District Metering approach to leakage detection.

3 BASIC CONCEPT

The strategy of District Metering requires the distribution network to be subdivided into small defined areas. Each of these zones or areas have flow and often pressure constantly monitored. The zone structure is generally hierarchical with large water supply areas being subdivided initially into zones of approximately 10,000 properties and these then being divided again in to district metering areas of around 2000 properties.

The data from these defined areas is monitored continuously and analysed to assess where leakage is occurring. Leaks can then be pinpointed to enable efficient repair work to be undertaken. Pressure reduction is a key contributor to lowering losses from the network. Dynamic pressure reduction using valves linked to sensors in the network is now being developed. The flow and pressure information from the field is now being used to calibrate sophisticated "all mains network models" so that the model can be used as both operational and design tools.

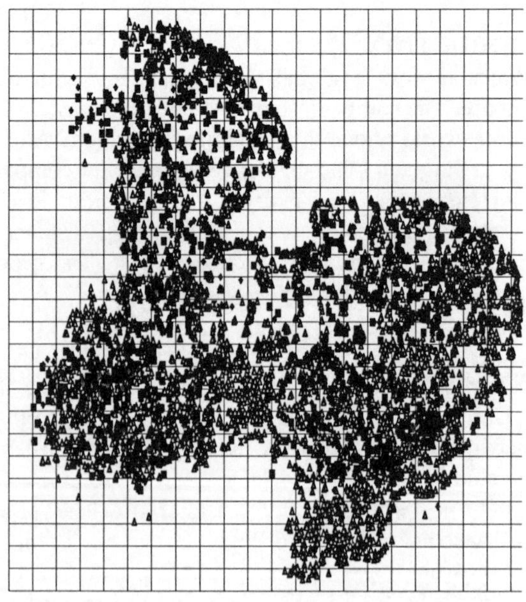

Figure 1 - Monitored Locations in AW Region

4 CURRENT TECHNOLOGIES

A number of water companies in the UK have deployed communications into the field to allow the rapid collection of flow and pressure data. Anglian Water [1] has made use of its extensive Anglian Regional Telemetry System (ARTS) to collect data from the network. ARTS is a radio based scanning telemetry system which has been deployed at virtually all the companies operational sites with over 7500 outstations and 250,000 measurement points.

Consideration was given to the cost of gathering data from up to a further 2500 meters. Fixed link telephone and cellular radio both proved too expensive and so a Low Power Radio (LPR) system was developed. This system has allowed the collection of pressure and flow readings to be taken every 15 minutes from all sites. Data is transmitted from a "remote slave unit" to the "hub" every 15 minutes. Each RSU sends its data at a slightly different time to avoid clashes with other data movements. Daily the RSUs are contacted by the hub to synchronise their clocks and recover any data lost during the day. Data is then collected from the "hub", via a telemetry outstation, and passed to the central telemetry system for archiving.

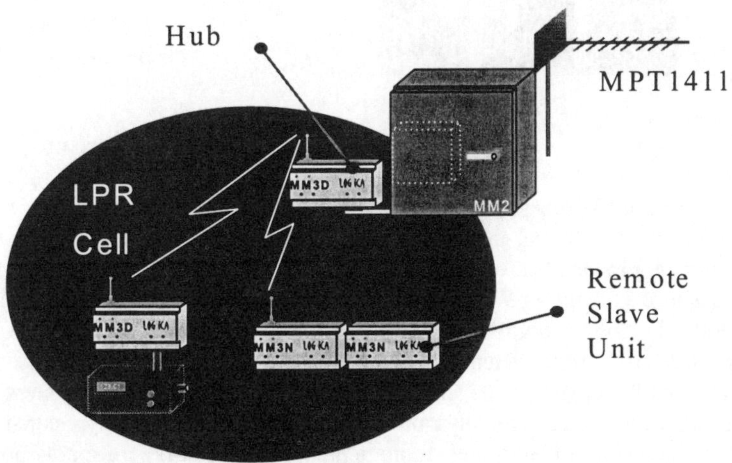

Figure 2 - Low Power Radio Cell

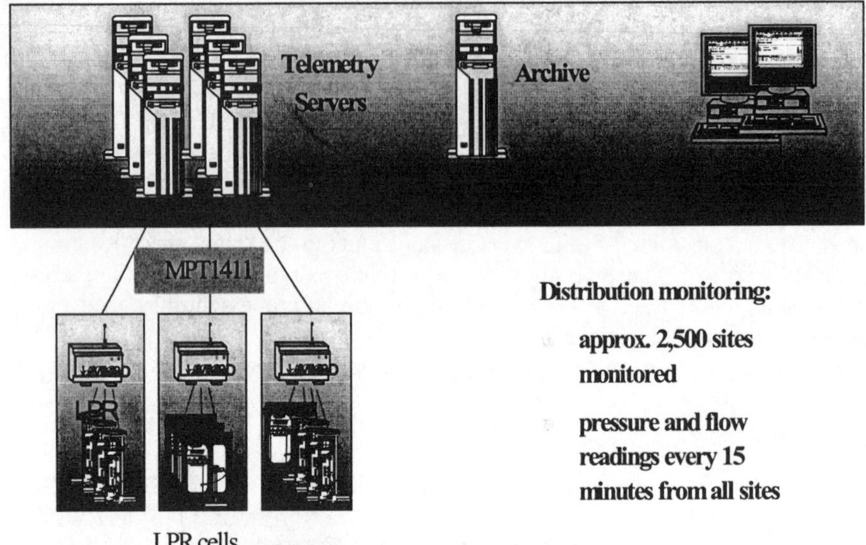

Figure 3 - Overall System Architecture

Thames Water [2], who supply the City of London, were faced with a massive problem of reducing unacceptably high leakage levels in the congested city area. They approached the problem using the same fundamental district metering structure as Anglian Water. To install a comprehensive system in London required 3000 flowlogger points (including pressure), 400 pressure reducing valves and the closure of 10,000 system boundary valves. Wherever possible both Anglian and Thames Water have used Electromagnetic flow meters, all the meters needed to be bi-directional and generally buried without a chamber to reduce construction costs. The meters and electronics equipment, where possible, are mains powered but in difficult circumstances they are battery powered.

Unlike Anglian Water, Thames have adopted the use of a cellular data network (Vodafone Paknet) for data transmission. Thames have also utilised GPS/GIS technologies to capture asset data so that the information can be graphically displayed on their corporate systems.

Other water companies in the UK have adopted the basic concepts of District Meter management of their networks. Software packages such as SOCRATES [3] have been developed to interpret flow data from district meters. The system can assist in understanding industrial use and patterns of domestic consumption, it has also uncovered unknown surge and oscillating flow effects. Improvements in the development of these software tools linked to field measurement will undoubtedly improve operational efficiency and asset management.

One of the overwhelmingly clear issues surrounding the analysis of DMA data is the huge volumes of data involved. In Anglian Water's case there will be eventually readings from some 2500 meters, each capable of recording flow in both directions. The resulting data load will be 480,000 values a day, therefore the system needs to be capable of coping with the data and assembling it in a suitable way for the user. All the data needs to be prepared for use in the middle of the next working day.

Company

Territories / Fields

Network Managers

Public Water Supply Zones

Distribution Zones **Distribution Zones**

District Meter Areas **District Meter Areas**

Figure 4 - Aggregation of data to derive information

The flows from the meters need to be resolved into the net flows for a zone. This requires mapping between the meter points and the DMA. The DMA's have been defined on the companies Geographic Information System by linking boundary valves to form polygons. The fully defined zone is held in Anglian Water's common database, a system for storing all asset and organisational reference information, called DMS. Access to the DMS archive is being developed using an Internet web server to enable end-users to view the majority of the data via a standard browser.

The following reports show the type of information available to the leakage teams

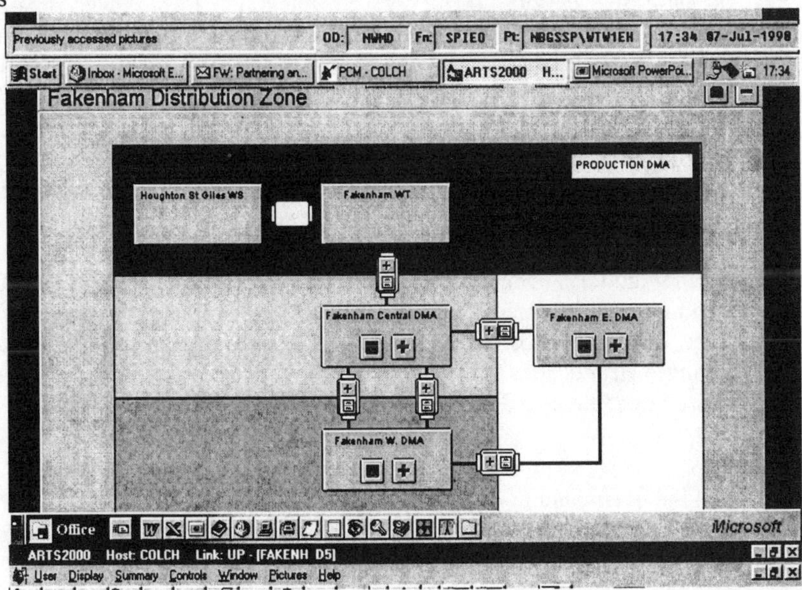

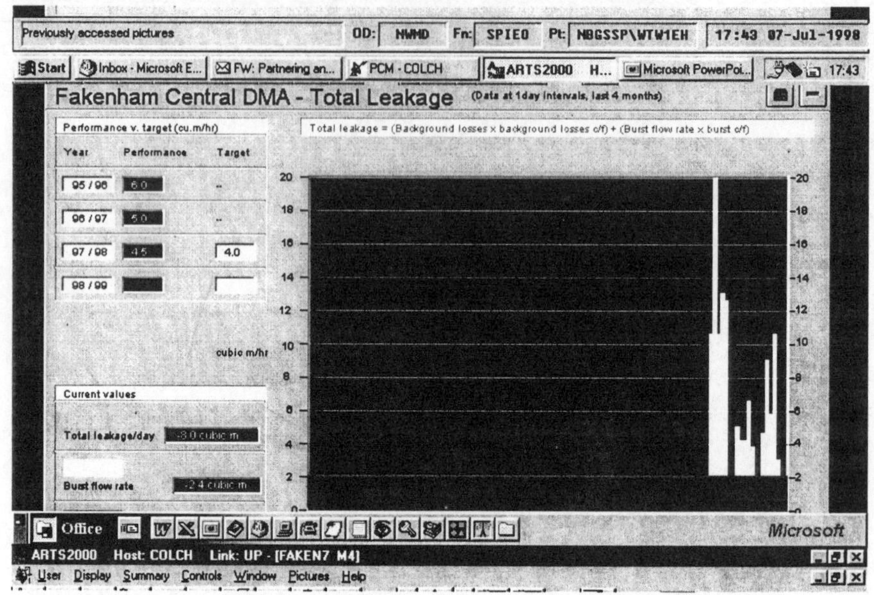

5 FUTURE TRENDS

The future introduction of Automatic Meter Reading technology will also allow greater understanding of water use and therefore losses to be assessed more accurately. Anglian Water has developed a partnership with Logica and ABB Kent meters to develop an advanced AMR system. The system under development uses a revolutionary all electronic meter with an integrated Low Power radio board transmitting at 184MHz. Meters located at customers properties will transmit usage data via a similar system as used for DMAs. The availability of such data will provide another component of the information needs of the operational manager. This system is currently undergoing trials which, when successfully completed, will allow the full scale deployment over the next few years.

Water quality monitoring within the distribution network is in its infancy and where undertaken is generally restricted to chlorine residual measurement. However, process monitoring and control equipment is improving and becoming more reliable. Putting communication technology into the field both on the network and also at customers' premises will allow much better understanding of the network. The introduction of decision support software and techniques such as "data mining", neural networks etc. will allow Network Management to be revolutionised over the next 10 years.

REFERENCES

Edwards K, Abu Judeh W and Tooke M. Online Monitoring within the Distribution Network to Aid leakage Control. *IWSA Workshop Amsterdam 1998*

Reynolds L. Case study of the Thames Experience with District Metering *Minimising Leakage Conference IQPC London 1998.*

Harris C and Bessey S. Night Line Analysis using Socrates. *Minimising Leakage Conference IQPC London 1998*

Open Loop and Closed Loop Pressure Control for Leakage Reduction

B. Ulanicki, P.L.M. Bounds, J.P. Rance *and* L. Reynolds

ABSTRACT

The UK water industry is addressing the major challenge of reducing leakage from water supply and distribution networks, driven by a mixture of economic, political and social factors. The leakage reduction problem as a whole is complex and requires co-ordinated actions in different areas of water network management, such as: direct detection and repair of existing leaks, general pipe rehabilitation programmes, and operational pressure control. It is widely accepted that a significant proportion of leakage is attributable to many small leaks (e.g. through pipe joints), where detection and repair actions are uneconomic. Operational pressure control is a cost-effective action for reducing leakage over whole sub-networks, and for reducing the risk of further leaks by smoothing pressure variations. The paper formulates and investigates on-line control strategies such as predictive control and feedback control for areas with many pressure reducing valves (PRVs) and many target points. The considered algorithms take into account explicitly a leakage model. The predictive control strategy is based on the sequential solution of an optimisation problem using an updated demand prediction and an updated control model. The feedback control strategy is based on using rules relating boundary flows and PRV set points. The two feedback strategies, centralised and de-centralised are investigated. For the de-centralised strategy the optimal pressure/flow curves calculated in the off-line mode can be implemented inside PRV valves. The results are applied to an area with three PRVs and two target points.

1 INTRODUCTION

The water companies are making major investments to reduce leakage through traditional solutions and the evaluation and application of new technology. The leakage reduction problem as a whole is complex and requires co-ordinated actions in different areas of water network management, such as:

- Direct detection and repair of existing leaks (so called "find and fix" actions).
- General pipe rehabilitation programmes.
- Operational pressure control.

Water companies undertake a mixture of these complimentary actions. General pipe rehabilitation is the most costly and long term action, but is undertaken to improve a number of different factors including leakage and water quality. Detection and repair actions are targeted at sub-networks where high levels of leakage are indicated, and tend to be most effective on larger leaks. It is widely accepted that a significant proportion of leakage is attributable to many small leaks (e.g. through pipe joints), where detection and repair actions are uneconomic. Operational pressure control is a cost-effective action for reducing leakage over whole sub-networks, and for reducing the risk of further leaks by smoothing pressure variations. Pressure control also has other important benefits in addition to the reduction of existing leakage. There is growing evidence that pressure control reduces the incidence of pipe bursts, thus reducing water loss and the costs of repairing bursts. Other benefits include reduced disruption to traffic, and reduced consequential losses (e.g. from flooding). Fewer bursts also reduce disruption to customers' supplies, which is an important water industry performance measure.

Historically, most networks have had no active pressure control and there is scope to reduce leakage through the reduction of excess pressure. Recent developments, including the restructuring of water networks into smaller sub-networks called Pressure Management Areas (PMAs), the introduction of more pressure and flow monitoring points, and the developments in valve control have increased the opportunity for sophisticated pressure control.

PMA boundaries are usually closed except for inputs via pressure reducing valves (PRVs). PRVs provide an automatically controlled reduction in pressure over a range of upstream pressures. The valve resistance is controlled by a target downstream pressure. PRV set-points may be constant, or may be scheduled. The set points must be determined with care since over-reduction in pressure may cause unacceptably low service pressures. PRV set points are determined with respect to Target Points, which are identified within the PMA. Target Points are chosen to be representative of PMA minimum service pressures. Many PMAs have only one PRV input, however, a significant number have multiple PRV inputs (which, typically, may number 2 to 4). More recently, valve suppliers have offered open-loop profile PRVs and feedback flow compensated valves. It is possible to specify a 24 hour pressure profile and load into a open-loop profile PRV. The 24-hour set-point profile can be updated more frequently based on measurements (predictive control). A flow compensated PRV changes output pressure depending on its flow. The relationship between flow and the pressure can be defined by a pressure/flow curve.

The paper formulates and investigates control strategies for both types of valves for areas with many inputs and many target points. Algorithms for calculating open-loop optimal profiles and optimal pressure/flow curves for the feedback control are formulated. The algorithms take into account a leakage model.

The basic assumptions are that the approach is valid for areas with many PRVs and many target points and is concerned with the control of steady-state pressure.

This study considers the network system's steady state over an extended period of time. It does not consider inertia of water and pressure transients in pipes. The steady state assumption is used in hydraulic extended-period simulation packages such as GINAS, EPANET, WATNET etc.

The on-line control strategies presented in the paper are using two innovative modelling tools, an optimal PRV scheduler and input-output control model that are described in the following section. These tools can be used to solve planning and on-line control problems. The objective of the planning problem is to estimate the maximum potential savings of optimal PRV scheduling. Case-studies (Ulanicki et al. 1999) indicate that the leakage can be reduced to 1/3 of its original value. This paper concentrates on the on-line control strategies of how to accomplish the potential savings in practice. This work was part of an investigation into the control of PMAs supplied by multiple PRVs for Thames Water Utilities Ltd.

2 PREDICTIVE CONTROL

The control strategy is illustrated in Figure 1. The physical system is controlled at regular intervals, e.g. every hour. At each interval optimal PRV set-points are calculated by an optimal scheduling algorithm (Ulanicki et al, 1997). The scheduling algorithm for calculating optimal PRV schedules requires a model of the network and an updated demand prediction as shown in Figure 1.

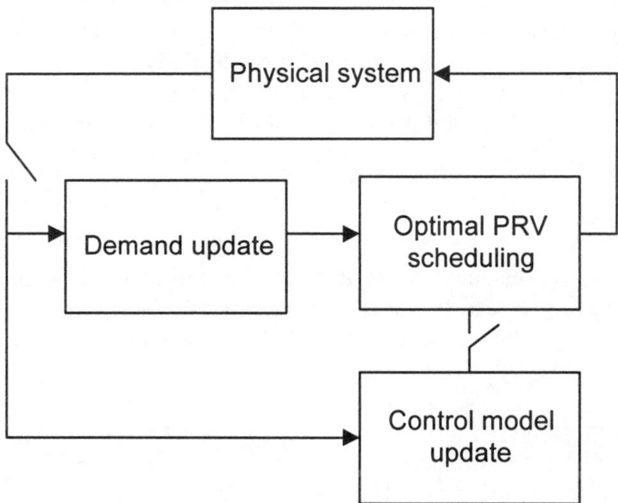

Figure 1. Predictive pressure control scheme.

Optimal PRV scheduling algorithm

The optimal PRV scheduler fulfils the following objectives:

- Minimises cost of the total water supplied into the area (including leakage flow).
- Satisfies pressure requirements at critical nodes.
- Satisfies other operational requirements. e.g. unidirectional flows.

The problem is solved using an advanced mathematical modelling language called GAMS [Brooke et al, 1992; GAMS, 1999]. GAMS automatically calls a non-linear programming solver called CONOPT (Drud, 1985) to provide the solution. The standard hydraulic equations have been expanded to include a leakage model. It was assumed that leaks are allocated to nodes and the leakage flow can be calculated from the following formula

$$q_{leak} = k \times p^{\alpha} \tag{1}$$

where k is a leakage coefficient and α is a leakage exponent. k is estimated from the night flow data (WRc-UK, 1994a) and the recommended value of α is 1.1 (WRc-UK, 1994b).

The scheduler can solve non-linear hydraulic models and is accurate over a wide range of conditions. Medium-sized networks can be solved directly from data typically found in a simulation model, while very large models can be solved after automatic simplification of the model (Ulanicki et al, 1996). Alternatively, a simple input-output model (control model) can be derived directly from measurements.

Control model identification

A control model defines relationships between input variables (PRV set points, global demand) and output variables (PRV flows and pressure at the target points). These operational measurements guarantee that the control model is observable. Parameters of the control model can be updated sequentially at each time step or occasionally (e.g. every month) or when a major change is made to the network. The model calibration task, in the proposed control structure, is formulated as a least-square problem and solved with the GAMS/CONOPT software.

Demand update

The total demands for the previous time step are known and this allows for updating prediction for the next hour. Any of the known demand prediction methods can be used provided it works sequentially and takes into account the current information. In this case a simple exponential smoothing has been employed.

Implementation of the control strategy requires a central control computer (PC) linked with PRVs and target points. Communication between PRV and the control computer is in two directions. The computer sends new set-point schedules for the next time step and reads the values of the PRV flows. Communication with target points is in one direction to read values of target pressures for the monitoring and model identification purposes. The frequency of intervention and the control horizon can vary according to the particular situation.

3 FEEDBACK CONTROL

The simplest PMA to consider is one controlled by a single PRV. In fact common practice is to control only a single variable head PRV by fixing all other boundary heads. The system has a single input flow and head set-point. The input flow must equal the demand and leakage of the system. The head set-point affects the pressure and leakage of the system. As pressure increases leakage increases according to (1). For a given level of demand, the head set-point and input flow relate to each other according to this relationship between pressure and leakage. This is called the system response or system curve. Any head-flow point on the system curve is called an operating point. The optimal operating point is the point that minimises the input flow into the network but satisfies the pressure requirements at certain target nodes. This is calculated by the optimal PRV scheduling algorithm. An optimal point is valid for a given level of demand. If optimal points are calculated for a range of demand levels then the curve that intersects these points is called the optimal curve. The optimal curve can be approximated accurately by a quadratic polynomial. The optimal curve provides the feedback rule for online feedback control. A controller measures the output of the PRV continuously and adjusts the head set-point using the feedback rule. The idea can be easily generalised for an area with many PRVs and many target points. For on-line control, a network or demand prediction model is not required because the feedback rule was derived from these offline. However, if the network model was to change significantly the feedback rule would need to be updated from a re-calibrated model.

3.1 Centralised Feedback Control

The centralised feedback control scheme is depicted in Figure 2. The scheme works continuously. All the flows into the network (q_1 and q_2) are measured and sent to a central controller. This single controller calculates the set-points (h_1 and h_2) for all the valves. These set-points are sent to the respective valves, which are actuated. It is not necessary to measure target pressures.

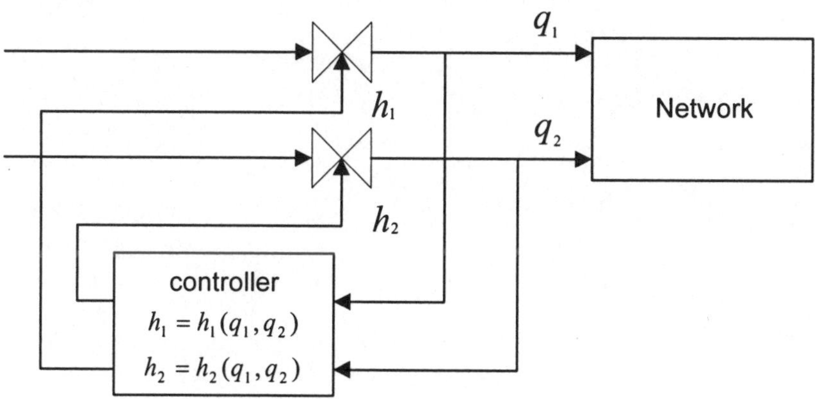

Figure 2: Centralised Feedback Pressure Control Scheme

3.2 Decentralised Feedback Control

Like centralised feedback control the process is continuous, but instead of a central controller each valve has its own local controller (Figure 3). Each flow into the network (q_1 or q_2) is measured and read by its local controller. Each controller calculates the set-point (h_1 or h_2) for its valve, which is then actuated.

Decentralised feedback control is especially convenient from a practical point of view because it is local control, inexpensive and continuously reacts to changes in demand. Because the control is local to the valve, no central computer nor communication lines are needed. Therefore this scheme provides a low cost solution.

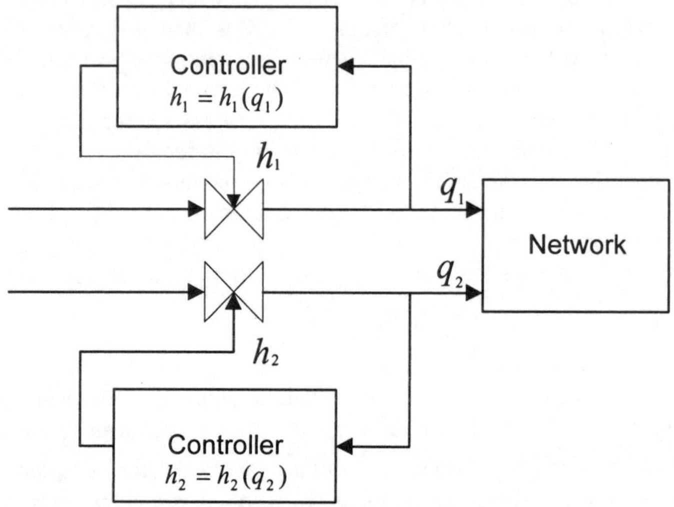

Figure 3: Decentralised Feedback Pressure Control Scheme

4 CASE STUDY

The methods for on-line control were tested off-line by using a simulation model to mimic the physical network system. All the different calculations are written in the GAMS language. This includes the simulator, which contains a leakage model.

The model used in the case study is depicted in Figure 4. Three PRV inputs (nodes 260, 255 & 250) and two target points (nodes 270 & 283) were chosen.

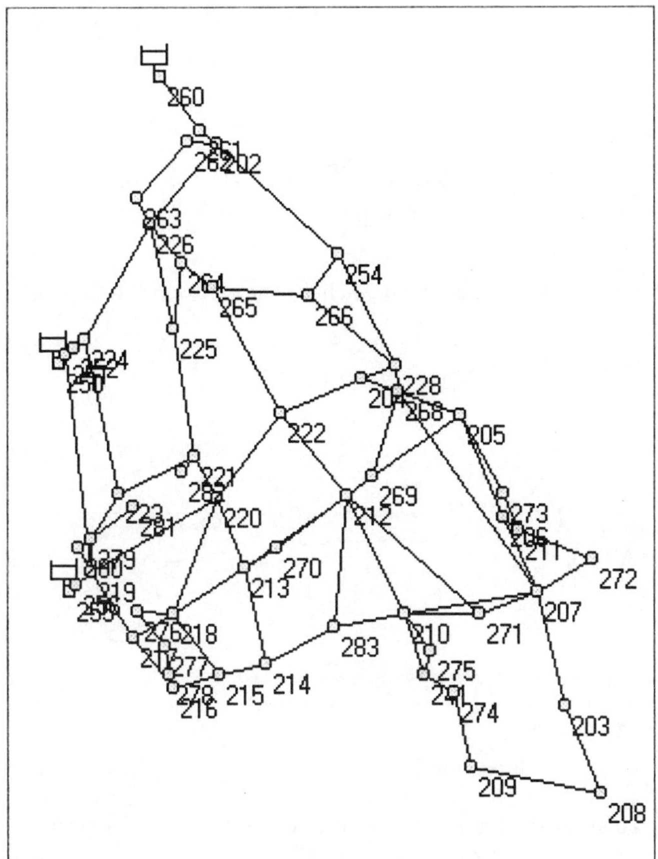

Figure 4: The case study network model

4.1 Results of Predictive Control

Target pressures are kept to almost the minimum pressure limit of 15m (Figure 5). The boundary flows follow the pattern of demand (Figure 6). The leakage and operational costs are kept low.

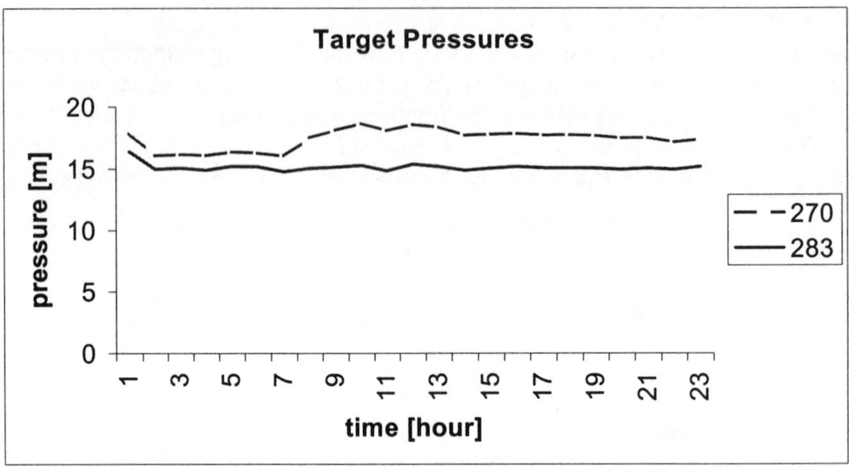

Figure 5: Target pressures during predictive control for case study model

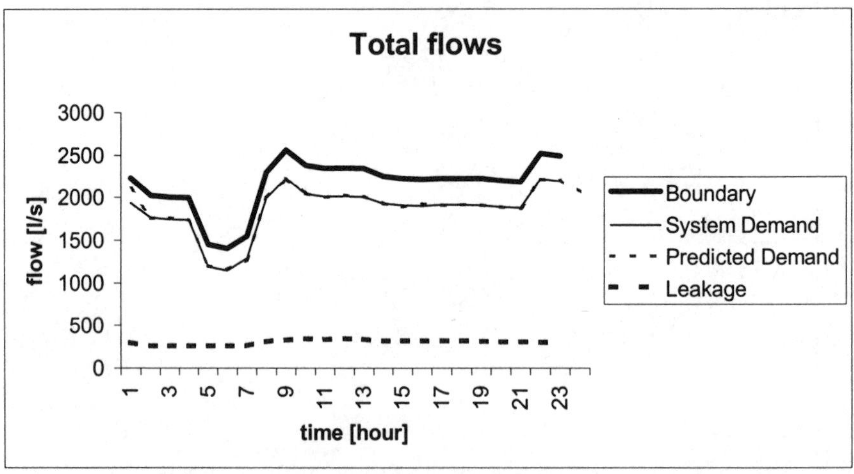

Figure 6: Total flows during predictive control for a case study model

4.2 Results of Decentralised Feedback Control

The decentralised feedback rules for calculating the valve set-points from the flow measurements are in the form of quadratic relationships. The feedback rules for two PRVs for the case-study model are shown in Figures 7 and 8. The optimal flows at the third node 250 were almost zero and so a PRV was not required. Each head-flow point plotted in the figures represents the optimal setting for the system for a given demand level. These were calculated off-line using the optimal PRV scheduler. A best fit of the plots produced a quadratic curve and its equation. Because there are two independently controlled pressure/flow curves the solution will be sub-optimal for the given range of demands.

The feedback rules were simply added to the simulation algorithm to test the method for on-line control.

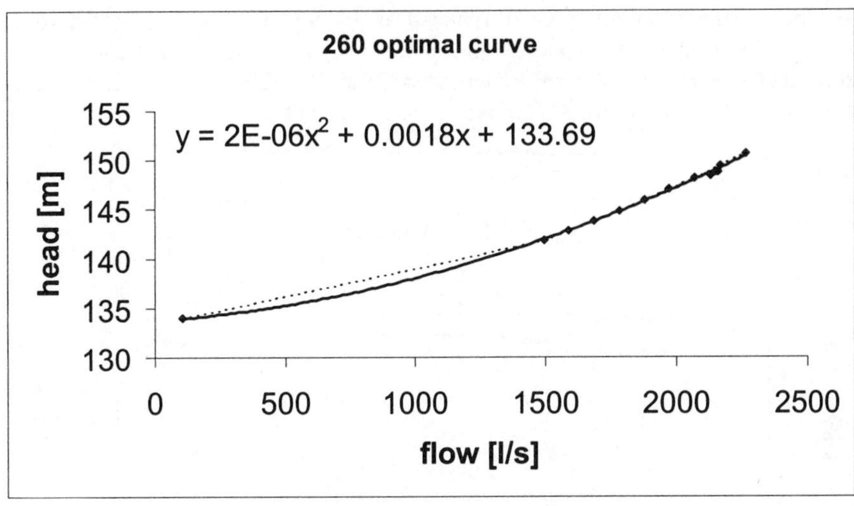

Figure 7: The feedback rule for PRV 260 for decentralised feedback control scheme for case study model

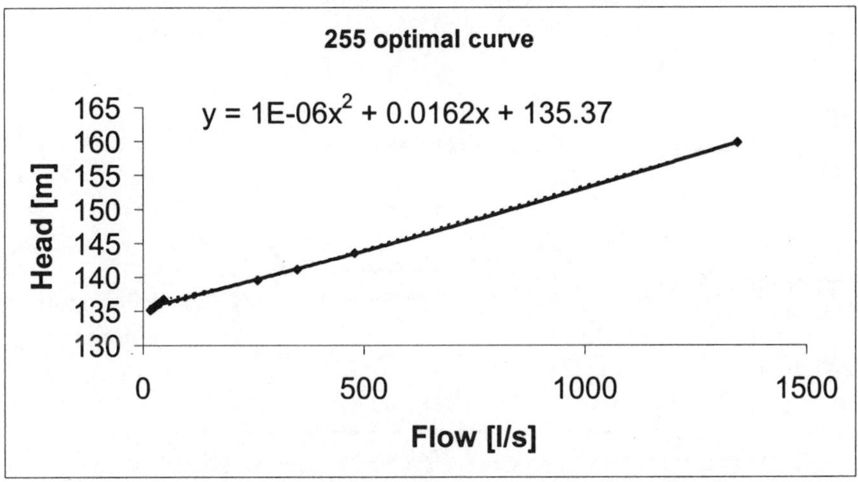

Figure 8: The feedback rule for PRV 255 for decentralised feedback control scheme for case study model

The results are comparable with predictive control. The pressures at the target nodes are kept low (Figure 9). The boundary flows are smooth and tend to follow the pattern of demand (Figure 10).

In all cases, the leakage is reduced to 1/3 of the original leakage. The total volume of water supplied into the zone is less due to the reduction in leakage and the cost of supplying the zone is reduced at all demand loads. The difference between the operational cost of the planning study, predictive control and decentralised feedback control schemes is less than 1%. Therefore, the sub-optimality of the decentralised feedback control was not significant.

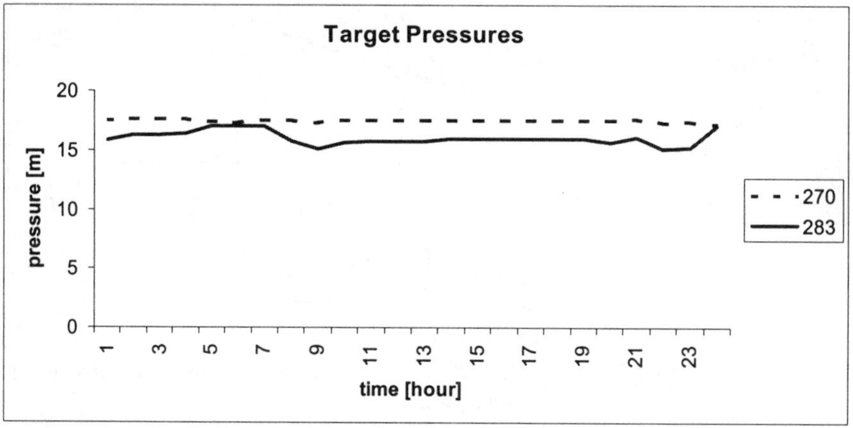

Figure 9: Target node pressures for decentralised feedback control scheme for case study model

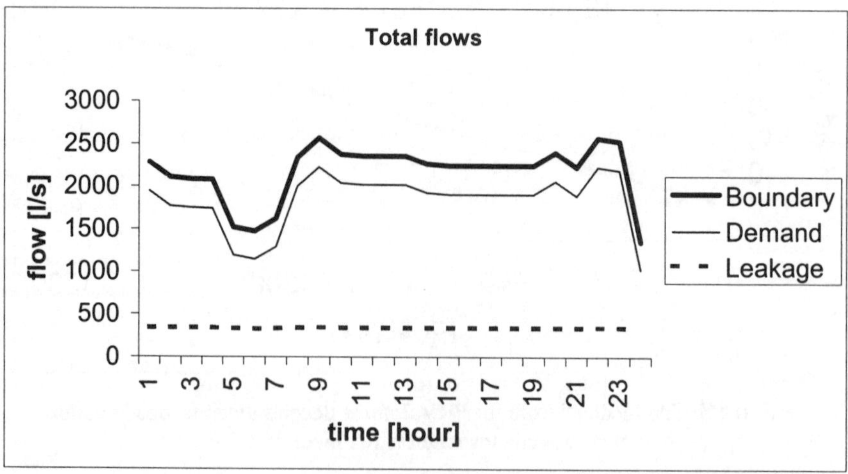

Figure 10: Total boundary, demand, and leakage flows for decentralised feedback control scheme for case study model

5 CONCLUSIONS

The paper considered the operational control of steady-state pressure in water networks. Operational pressure control is a cost-effective method for reducing leakage over whole sub-networks, and for reducing the risk of further leaks by smoothing pressure variations. The planning studies under the assumption of perfect knowledge of demand and network model (including leakage) allow for the estimation of the maximum potential savings arriving from optimal control of PRVs. These potential savings can be accomplished by using optimised on-line control strategies. The paper investigated two such strategies: predictive control and feedback control. The predictive control strategy relies on the repetitive solution of the optimal scheduling problem using the current prediction of the total demands. The scheduler can use either a detailed hydraulic model (e.g. simulation model) or an on-line control model built from current measurements. The idea of a control model is very powerful and it enables a universal control system to be created that will work for any water network. The feedback control strategy was proposed in two versions: centralised and decentralised. The feedback rules are calculated in an off-line study and then applied for on-line control. The centralised feedback provides an optimal solution with the performance comparable with predictive control. The de-centralised solution provides a sub-optimal solution that is very attractive for practical implementation. The control rules can be embedded in autonomous valves. Subsequently, the valves will be working correctly by sustaining pressure at target points for any values of demands. Only some significant changes in water network topology will require recalculating the pressure/flow curves. The simulation studies confirmed that the predictive and feedback control strategies allow for achieving leakage reduction close to that predicted by optimal planning studies, i.e. 1/3 of the original leakage level without PRVs.

REFERENCES

[1] Brooke, A., Kendrick, D., Meeraus, A. (1992) *GAMS A User Guide.* USA: Boyd & Fraser – The Scientific Press Series.

[2] GAMS Development Corporation, 1999, "*GAMS Home Page*", [On-line], USA. Available from: www.gams.com [Accessed: 22/03/99].

[3] Drud, A. 1985, CONOPT: A GRG Code for Large Sparse Dynamic Non-linear Optimisation Problems, *Mathematical programming*, 31, pp.153-191.

[4] Ulanicki, B., Coulbeck, B., and Ulanicka, K (1997). Optimisation Techniques Can Enhance Water Network Analyses, *Proceedings of the 3RD International Conference on Water Pipeline Systems*, BHR Group Conference Series, Publication No.23, MEP, London, UK

[5] Ulanicki, B., Zehnpfund, A. and Martinez, F., (1996). Simplification of Water Network Models, *Proceedings of Hydroinformatics 96 International Conference*, 9-13 September, International Association for Hydraulic Research, ETH, Zurich, pp. 493-500.

[6] Ulanicki, B., Rance, J., and Bounds, P. (1999), Control of Pressure Management Areas Supplied by Multiple PRVs, Research Report, Water Software Systems, De Montfort University, Leicester, LE1 9BH, UK.

[7] WRc-UK Water Industry. (1994a). Managing Leakage – Using Night Flow Data, Report F.

[8] WRc-UK Water Industry. (1994b). Managing Leakage – Managing Water Pressure, Report G.

D.AN.A.I.S. - An Original Real Time Expert Model Controlling the Water Supply Aqueduct of the Greater Athens Area

L.S. Vamvakeridou-Lyroudia *and* S.Politaki

ABSTRACT

In this paper the telecontrol-telecommand system of the Mornos water supply aqueduct is presented and in particular, the real time expert model DANAIS for system operation on demand. Extensive measurements under controlled flow conditions have been carried out for the estimation of the hydraulic parameters of the model, but it operates satisfactorily even if there are missing or erroneous data, due to complex signal conflict patterns and the implementation of fuzzy aggregators leading to fuzzy expert rules. Moreover linguistic variables have been added to describe desired operation conditions and hydraulic parameter autocorrective procedures have been included, so that the model can autocorrect itself in case there are changes of hydraulic parameters in time. DANAIS has been operating permanently since May 1999.

1 INTRODUCTION - MORNOS AQUEDUCT

Water supply for the greater Athens area (about 5 million equivalent inhabitants) is mainly carried out through the Mornos/Evinos aqueduct system (about 190 km long), a gravity conveyance system that carries water from Mornos/Evinos dam to Athens, managed and supervised by the Athens Water Company (EYDAP). This aqueduct consists of open channels - canals (112.5 km total length), 10 siphons (7.2 km total length) and 15 tunnels (70.8 km total length) and 1 hydropower plant. Water flow is controlled by 23 regulating constructed works spread along the aqueduct (18 normal regulating gates, called L-shaped gates and 5 large regulating gates used also for energy dissipation as stilling basins, called E-shaped gates -or EKE by the Greek acronym) and the hydropower plant, as shown in Figure 1. Depending on hydraulic conditions, it takes about 18-25 hours for the water to "travel" from the most upstream to the most downstream point of the aqueduct.

Nowadays there are only three possible water inlet points to the system, one at the upstream end (Mornos dam) and two others along the aqueduct. On the other hand there are many water outlets: to the main treatment works of Athens (Menidi) at the downstream end of the system, to another new treatment works for the greater Athens area, to the Marathon dam (used as a large reservoir close to Athens) and to various villages and small towns alongside the canal. In fact, water demands for Athens, Attica and the surrounding region have increased steadily during recent years, as water distribution networks expand to new urban areas, new treatment works are added and generally the system becomes larger and more complicated.

Since the late 70s, the system has been automatically supervised and regulated by a central master station installed at the downstream end of the aqueduct at Athens/Menidi. Demand, that is hourly fluctuating demand forecasting mainly at the Athens treatment works, together with the hydraulic conditions along the canal are used as input to the system model, that decides the desired regulating gate positions along the aqueduct and performs the necessary gate movements every 15 minutes. The initial hardware, software and hydraulic model (Dynamic Regulation type model [2,3]) were developed and installed by the Societe du Canal de Provence.

In 1994/1995 it was decided by the Athens Water Company (EYDAP) that the existing hardware/software, using out-of-date technology for the hardware (PDP11/34), was aging badly. Spare parts could not be easily found any more, while the software could not meet satisfactorily the operating conditions of the system (e.g. the hydropower plant was not simulated at all, because it did not exist when the model was initially developed, gate movements at the most upstream regulation work had to be performed manually) and was not flexible and parametric, so as to be easily modified. So it was decided that both hardware and software should be replaced.

The new automatic supervising/regulating system consists of 20 PLCs (Programmable Logic Controllers), a SCADA (Supervisory Control And Data Acquisition System) and a real time expert model called D.AN.A.I.S. (Dynamic ANalysis Aqueduct Intelligent System) for hydraulic simulation, hydropower plant operation and gate movements. D.AN.A.I.S. was assigned as a research project at the National Technical University of Athens.

2 TELECONTROL AND TELECOMMANDS

The Mornos aqueduct starts at the Mornos dam and ends at the treatment plant of Menidi near Athens. Essentially it is a one-dimensional system, with many point inlets and outlets. Flow at the Mornos aqueduct is mainly free surface, but there are also tunnels and inverted siphons, where pressure flow exists. Flow regulation and control is carried out by gates telecommanded or manually operated.

The main objective of the aqueduct operation is to provide adequate water for the large treatment plant in Athens-Menidi, the treatment plant in Mandra and to supply the Marathon lake, used as a large reservoir near Athens, operating under demand.

The aqueduct is supervised by a SCADA system. The state of the system (water levels, rate of flow at various points, opening position of gates) is also an input to the system through a large number of measurements along the aqueduct. These measurements along the canal in the form of analog and digital signals are transmitted by 20 local PLCs (PLC CS31-ABB) to four peripheral stations (PCs) and to the SCADA Central Control Station (ALPHA STATION 200-DEC, VMS operating system) in Athens/Menidi every 30 inches, where records are kept and graphics display takes place. The Transmission Network is in the form of multidrop configuration through dedicated cable.

Every 15 feet a real time mathematical simulation algorithm (DANAIS) is activated by the SCADA main operational module. Input data (desired rate of flow, recent signals, historical data bank, "intelligent" data automatically built by the model) are used for signal processing and validation, hydraulic computations and new gate position estimation (Figure 2). Orders are passed along as output to the SCADA, and through the SCADA to the PLCs, for the new gate openings of all telecommanded gates. The local PLCs are FB (Feed backwards) controllers.

The system may be classified as MIMO (multiple inputs, multiple outputs) controller [3]. In all there are:

- 1000 Digital Inputs (ON/OFF information)
- 200 Analog Inputs (Rate of flow measurements, water level, gate position)
- 100 Digital Outputs (Commands to gates)

Apart from the SCADA system, separate records of all analog measurements and orders are being collected and stored at the Central Station by a PC every 5 feet. These are used both as back-up storage and for graphics display of signal time series, because the SCADA does not store old signals.

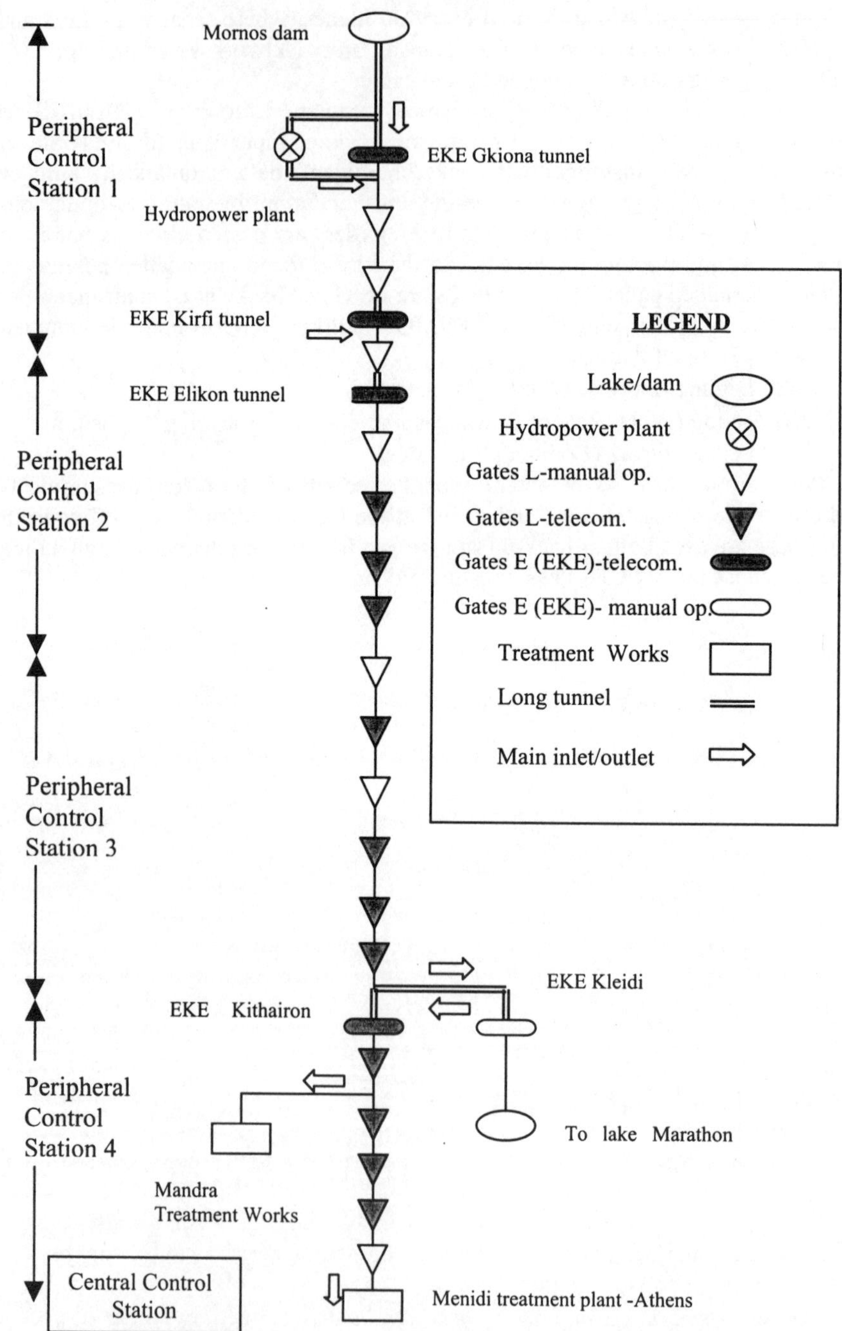

Fig.1: Schematic layout of the Mornos aqueduct (not to scale)

3 SCHEMATIC LAYOUT OF THE SYSTEM

In the old mathematical model, developed by the Societe du Canal de Provence, the system was divided in 15 parts or subsystems. Upstream and downstream limits for each "subsystem" were defined by two consecutive telecommanded gates existing at the time the model was calibrated. So each subsystem consisted of the part of the aqueduct between two predetermined gates, without uniform geometry and hydraulic characteristics. Pressure flow and free surface flow could coexist in each "subsystem", together with intermittent manually operated gates and predefined inlet and outlet points. There was no uniform parametric mode of reference for reaches and relations between them, while level-meters, flow-meters and gate position signals were assigned to each reach by their reference code number, not even by name. Moreover hydraulic computations were not carried out explicitly by the use of mathematical formulas, but by interpolations in various pre-defined tables of numerical results, that could not be modified. So the mathematical model was not flexible and parametric.

In recent years various events altered the system operation:

(a) A hydropower plant was constructed, operating through a by-pass at the most upstream tunnel of the system, so that the first gate could be by-passed or operated in combination with the hydropower plant.

(b) New main outlets (e.g. to the treatment plant at Mandra) are operating at points that were not defined by the previous model.

(c) Gates that were previously manually operated and therefore ignored by the model are being turned into telecommanded ones.

(d) A new parallel pressure pipe is under construction downstream the EKE Kithairon gate.

(e) The company wanted to add new flow meters and level meters, and take them into account for the system operation.

(f) No electronic operational data were kept for further use as data bank.

There was no way to simulate these changes using the old model. So a totally new schematic layout and coding for the system was needed. While designing DANAIS, the new model, flexibility and parametric configuration of all system items were of prime concern. The model should not only simulate present operating conditions, but also include possible future alterations easily, through simple data file modifications.

Because the aqueduct is essentially one-dimensional, the position of any computational element, point, gate, analog and digital signal, inlet and outlet along the aqueduct may fully be defined by one parameter only: its distance from the most upstream point (node) of the system (in kilometres). So any addition/removal or position change of any flow or level meter, gate, inlet, outlet etc. is simply defined by just one numerical parameter.

The Mornos aqueduct is simulated as a one-dimensional system consisting of a series of computational elements called reaches, connected by nodes. Any reach is defined by the following characteristics:

- *An upstream node and a downstream node* (defined by their respective distance from the most upstream point of the system).

- *Uniform way of flow within the reach.* Three types of flow are supported and simulated: free surface, pressure flow and chute (free surface flow with significant longitudinal slope of comparatively small length).
- *Uniform geometry within the reach.* Free surface reaches are considered to be prismatic trapezoidal channels, their cross-section described by the bottom width B, the side slope z, the maximum allowed water depth hmax and the bottom slope J. Pressure flow reaches are assumed to be single or double circular pipes or tunnels, described by cross diameter D.
- *Uniform hydraulic characteristics within the reach.* For free surface reaches a uniform n Manning's coefficient is assumed, while pressure flow reaches are considered to have uniform equivalent roughness coefficient ks.
- *No lateral inflows and/or outflows are allowed within the reach.* In case the user defines an inlet or outlet at any mid-point, it is automatically assigned to the nearest node.
- *No regulating gates are allowed within the reach.* Regulating gates are only simulated at nodes.
- *Level meters and flow meters are allowed at any point within the reach.*

According to the above specifications, the Mornos aqueduct is now divided into 61 reaches, that form the computational elements of the model. Assignment of any signal or gate to the proper reach is performed automatically by the model.

4 DANAIS-THE REAL TIME COMPUTATIONAL EXPERT MODEL

4.1 General description of structure and modules

Signal processing, hydraulic computations and calculation of new gate positions are carried out by the real time computational expert model, called DANAIS, that is activated every 15 feet. DANAIS consists of four modules, as shown in Figure 2.

The first module, called START, reads, sorts and arranges data. There are three kinds of data: The first group consists of stationary data, that is data that need not be changed every 15 feet, such as geometrical data for the aqueduct and the regulation works, code names for signals and general operational data. The second group consists of data that generally change every 15', such as recent analog/digital signals and rate of flow demands at the outlets. The third group consists of the model's own data bank, that is data that are built, stored and updated automatically by the program each time it is activated. These data include

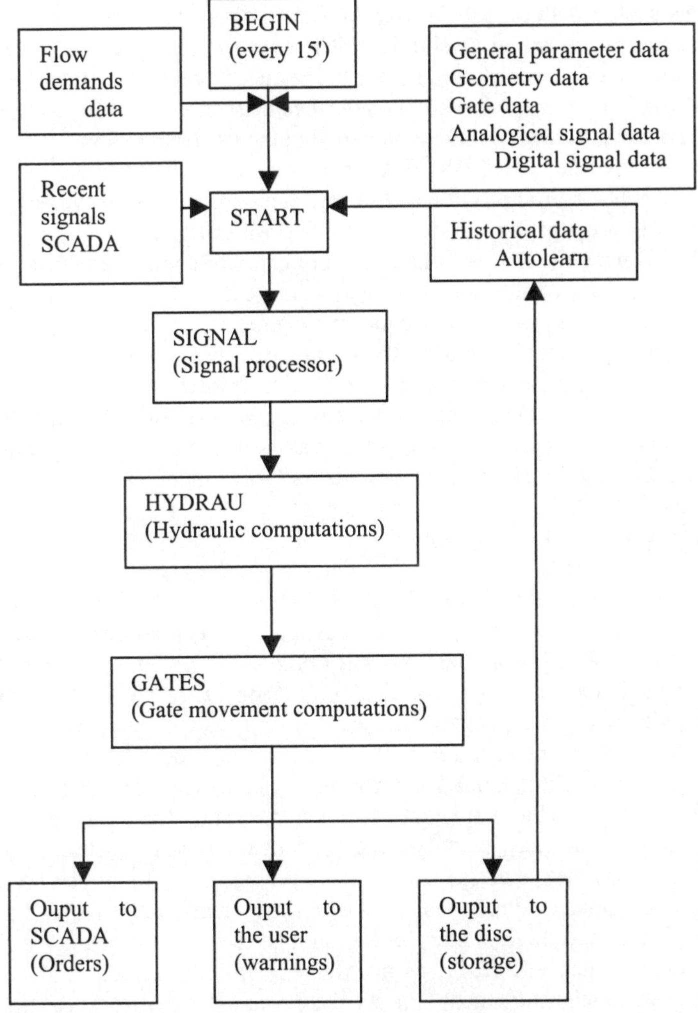

Fig.2: DANAIS Flowchart

(a) "recent history", that is detailed description of what happened within the last 8 times the model run, and (b) a real time autocorrective data bank, that allows the model to learn by its own mistakes and autocorrect itself while operating in real time.

The second module, called SIGNAL, is used for signal processing and evaluation. All recent analog and digital signals are examined, checked and cross-checked, so as to decide whether they are valid or not. The model tries to complete missing signals (if possible), while signal conflicts are detected, through predetermined complex "conflict patterns" data, that is combinations of signals that have been proven to be conflicting, during calibration. If and when the program recognizes such a pattern, proper warnings are shown onscreen to alert the user. Conflict patterns are stored as simple ASCII data files, for the user to be able to add new conflicting patterns at any time, if needed. Actually, after the first 2-3 weeks of trials, no new patterns have ever been detected or added to the conflict pattern data bank, but this feature is included in the model anyway.

The third module, called HYDRAU performs hydraulic computations, so as to determine the desired (target) rate of flow at each regulation gate every 15 feet. An inverse wave propagation model, using the Muskingum-Cunge equations, is applied [4]. During the first months of the project, while the program code for this module was being written and during the first trials, there were many lengthy discussions involving almost everyone even remotely connected to the project, from both the National Technical University of Athens and the Athens Water Company EYDAP, as to whether this kind of hydraulic model was adequate, or should another, more complicated one (e.g. an inverse Saint-Venant model) be applied to improve accuracy. In the end it was proven both by measurements and by practice, that the Muskingum-Cunge inverse wave model was reliable, fast and accurate enough. DANAIS's real problems lay in missing signals, erroneous data and gate movements, not in hydraulics.

The fourth module, called GATES determines the necessary gate movement orders to be sent to the local PLCs. This module is the most complex and complicated one. Theoretically, once the desired (target) flow through each gate is determined by hydraulic computations, the relative gate position may be computed by simple mathematical functions, that have been calibrated by measurements. Then the desired gate position is sent to the PLCs and the gate moves automatically to obey the order. Actually, in the whole procedure many things may go wrong. The data (signals) used for gate movement computations may be erroneous or missing, the mathematical function applied for the gate may not be accurate, or the gate may not obey. Moreover, additional constraints make the problem more complex, e.g.: (a) Gates should not be forced to move very much or very often to opposite directions, so as to avoid malfunctions. (b) In case of emergency (e.g. too high water level), the gates should perform emergency movements (c) Water level had to be maintained within predefined limits at certain points (e.g. at tunnel entrances) -unless there was an emergency (d) In case of complex gates with more than one opening, hierarchy had to be established, as to which gate moves first, including operation of the hydropower plant, so as to maximize energy production. To overcome all these problems, patterns for hierarchical movements had to be designed, fuzzy linguistic terms added and autocorrective data banks applied. In the following chapters, some of these features will be discussed in detail.

4.2 Operating in autocorrective mode

Although hydraulic parameters for the channel and the gates have been calibrated by measurements and trial and error techniques, it has been noticed that some parameters (e.g. friction coefficient, shape parameters of gates and weirs) either change over time, or calibration efforts did not lead to satisfactory results. In this way systematic or random deviations may occur, leading to serious operational problems. The autocorrective procedure included in the model estimates deviations from operational targets and "corrects" selected parameters accordingly. Thus the expert model learns while operating in real time mode and modifications are included in the data bank of the model.

Originally this mode of operation was designed in order to make corrections easier during the trial operation of the model for calibration, but it has proven to be such a useful feature that it was decided to be included in the model permanently.

The real time autocorrective data bank is a file called AUTOLEARN, that allows the model to learn by its own mistakes. DANAIS takes into consideration this file every time it operates in "autolearn" mode, as specified by an ON/OFF switch by the user.

This file contains inaccurate estimations, mistakes and mischiefs that occurred with the model in previous regulation cycles. In reality, there are two groups of unwanted operational "mistakes": (a) inaccurate rate of flow at the treatment works (demands are not met in time or in value by the model) and (b) too high or too low water levels in the aqueduct, especially upstream from gates or tunnels. So, the user may additionally select the kind of "mistakes" to be corrected (rate of flow and/or water-levels) for each regulatory work and gate separately.

The algorithm scans AUTOLEARN data each time it operates and modifies orders for gate movements accordingly, so as to correct its previous mistakes. Of course the model may "overreact", e.g. in order to correct a trend for too high water level upstream a gate, it may order such a gate position that water level becomes too low after some hours. In that case a new "mistake" occurs, needing fresh corrective orders. The model automatically keeps track of corrections (type, magnitude and the time it happened) for each gate. On the whole, operation is stabilized after 1-2 days in continuous autocorrective mode under similar or nearly similar hydraulic conditions. Of course, in case conditions are significantly different (e.g. target flows for the treatment plants are much lower or higher), the autocorrective procedure should start all over again.

Automatic corrections of hydraulic parameters may only be effectuated if the flow in the neighbourhood of the gate is steady-state. So DANAIS has to estimate initially whether steady-state criteria apply to the channel and then perform corrections. Steady-state conditions, at least partly along the aqueduct, are not so hard to come by, because often the treatment works operate for many hours without modifying the desired (target) rate of flow. In this case, if no other "crisis" has occurred upstream (e.g. gate malfunction, too high or too low water levels, that lead to emergency gate movements) steady state conditions are established in many parts of the aqueduct, which are automatically asserted by the model through its own historical data file. Water-level trends, rate of flow deficiency or excess may then be noticed by the model, resulting in automatically applied corrective measures and changes to the hydraulic parameters involved. Of course when target flows are modified by the user, gates move, wave propagation starts along the aqueduct and the flow becomes unsteady. Autocorrective mode is then automatically disabled by the model.

Automatic parameter corrections yielded very satisfactory results, especially for L-shaped gates, that could not be calibrated by any other means.

4.3 Gates

Flow regulation is achieved by gate movements at 23 regulatory works along the aqueduct and 1 hydropower plant. There are two types of regulatory works and gates:

Large regulatory works with gates, called E-gates (or EKE by the Greek acronym), that are placed at the downstream end (exit) of long tunnels operating under pressure. These works are also equipped with stilling basins, to be used for energy dissipation, as shown in Figure 3. All these works also include a pressure surge tower, for water hammer effects. In all there are 4 telecommanded E-gate works and 1 manually operated . Four of these works have double gates, operating in parallel. The 5th is the biggest of all, with 3 parallel gates (2 large and 1 small), constructed at the exit of the uppermost tunnel of the aqueduct (EKE Gkiona tunnel).

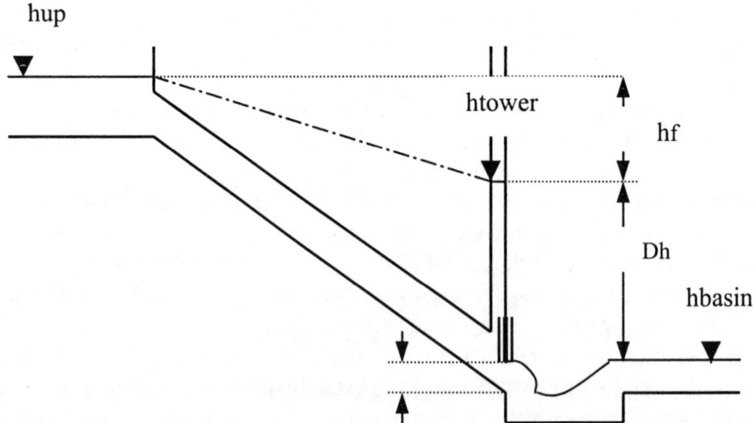

Fig. 3: Large regulatory works with stilling basins - E-gates

The rate of flow Q through the E-gate, the gate opening position PV (expressed as opening fraction $0<PV<1$) and the water depth difference $Dh = h_{tower} - h_{basin}$, are related as follows:

$$PV = a \cdot Q^b / Dh^c \tag{1}$$

Parameters a,b and c are estimated and calibrated by measurements, under steady-state conditions. Extensive measurements were performed for each work and gate separately, because these works are the main flow control devices for the aqueduct. Unfortunately, operational conditions are not always ideal. In case surge tower level meters are out of order, friction losses hf in the tunnel cannot be measured and have to be estimated approximately. Besides, as there is no flow meter near the works (except in the case of the EKE Gkiona tunnel), the rate of flow estimation cannot be accurately estimated during calibration. Moreover level meters at the tower and upstream, the tunnel are not always accurate. On the other hand, because of the size of the gates, and the importance of these works in the overall successful operation of the aqueduct, erroneous estimation of gate openings was dangerous. So fuzzy logic was introduced to the gate movement estimation. Level meter readings, friction losses and rate of flow are treated as fuzzy numbers, combined by the use of an overall fuzzy aggregator (harmonic mean was best) and especially designed de-fuzzification techniques.

Ordinary sluice gates, single or double, equipped with two long side weirs (bec de canard). Because of the shape of their weirs, in Greek they are called *Λ-gates*, or more conveniently *L-gates*. There are 18 L-gates along the aqueduct. Nowadays 11 L-gates are telecommanded, whereas 7 are manually operated. However all gates are simulated in the same way by the model. The only difference is that telecommands are sent only to automatic gates. To switch from one mode of operation to the other, the user must only change 1 parameter in the gate data file.

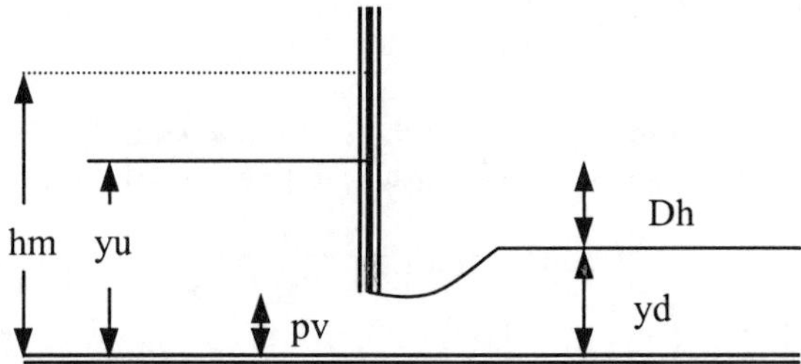

Fig. 4. Regulating operational position for L-shaped gates

There are three possible states of operation for this kind of gate. (1) The gate is partially open (Figure 4), water depth upstream yu is less than the weir crest level hm and the rate of flow passing through the gate is regulated by the gate position pv and the water depth difference upstream and downstream of the gate Dh=yu-yd. (2) Water depth upstream exceeds the weir crest level (yu>hm), so that the flow passes over the weir. The gate may be partially open or shut. In this state of operation the rate of flow cannot be controlled by the gate. Water depth is at its highest, so that the volume of water contained upstream is maximum and the flow wave propagation is considerably delayed by the gate (3) The gate is open, water passes below the gate edge (free flow) yu<pv, water level and volume contained upstream is minimum, while wave propagation is faster than the previous states of operation, but the rate of flow cannot be regulated by the gate.

Sometimes it is useful to "tend" to operate the aqueduct gates in either of the 3 above mentioned states, to simulate "medium", "full" or "low water" state of operation. So a general operational parameter is included in the model, defined by the user, that leads to different states of operation for L-gates, resulting in "normal", "slower"or"quicker" wave propagation along the canal respectively.

Because the Mornos aqueduct is not an irrigation canal, water level need not be kept stable upstream, any L-gate. The optimal or "best" position for any automatic L-gate is the first state of operation, that is to be partially open, operating without overflow or free flow, so that the rate of flow passing through the gate may be modified by adequate gate movements.

In this case, the rate of flow Q through the gate, the gate opening position PV (usually expressed as opening fraction $0<PV<1$) and the water depth difference Dh=yu-yd, are related as follows:

$$PV = a \cdot Q^{b} / Dh^{c} \tag{2}$$

where a is the shape parameter and b,c are exponential parameters. Usually for L-shaped gates $b \approx 1.00$ and $c \approx 0.50$. Theoretically, parameters a, b and c for any gate may be defined by measurements. Unfortunately, in the case of the Mornos L-gates this was not possible to be carried out within acceptable accuracy, despite repetitive trials. So a different way of approach, with the use of the autocorrective data bank was tried out:

Step 1: An initial "guess" for the shape parameter was introduced through the program data. The autocorrect mode was enabled.

Step 2: Steady-state (or quasi steady -state conditions) were established around the gate, automatically detected by the program.

Step 3: In case of a trend for too high or too low water levels upstream of the gate a correction for the shape parameter was defined as follows:

$$a_{new} = a_{old}\left(\frac{PV + STEP}{PV}\right) \tag{3}$$

where anew, aold are the new and old value for the shape parameter respectively, PV is the current gate position and STEP the corrective step. In case of a too high water level trend, STEP=0.02. In case of a too low water level trend, STEP= -0.02.

Similar corrective factors are estimated in case of inaccurate rate of flow estimations through equations (1) for E-gates and (2) for L-gates, but rate of flow corrections cannot occur in the same time as water level corrections.

In practice rate of flow corrections were significant only for E-gates, especially the E-gate closer to Athens (EKE Kithairon tunnel). Water level trend corrections were applied to all L-gates, and E-gates (water level upstream of the tunnel), except the uppermost E-gate (EKE Gkiona tunnel), for which autocorrective mode is always disabled.

The approach applied in DANAIS for autocorrection of gate shape parameters differs from the classical PI or modified PI controllers [1] used for irrigation canals, because no standard or target water level is assumed upstream of the gate.

5 CONCLUSIONS

In this paper the telecontrol-telecommand system of the Mornos water supply aqueduct is presented and in particular, the real time expert model DANAIS for system operation on demand. Extensive measurements under controlled flow conditions have been carried out for the estimation of the hydraulic parameters of the model, but it operates satisfactorily even if there are missing or erroneous data, due to complex signal conflict patterns and the implementation of fuzzy aggregators leading to fuzzy expert rules. Moreover linguistic variables have been added to describe desired operation conditions and hydraulic parameter autocorrective procedures have been included, so that that the model can autocorrect itself, in case there are changes of hydraulic parameters in time. As far as hydraulic conditions are concerned DANAIS uses an inverse wave model algorithm, based on the Muskingum Cunge equations.

ACKNOWLEDGMENTS

The telecommand/telecontrol infrastructure and software (SCADA and PLC's) have been developed at the Athens Water Company (EYDAP), supervised by the second author, while DANAIS was assigned as a research project to the National Technical University of Athens (NTUA), supervised and developed by the first author. Students of NTUA, whose graduate thesis has also been based on DANAIS, have worked long hours on the model, especially A. Makri, G. Kleftoyiannis and H. Giovanopoulos. On the part of EYDAP various people have been involved, but special thanks are due to N. Papigkiotis, A.Velios who developed the separate signal records data base and -last but not least- C. Kyriazis, general director and N. Dikis, director of water supply respectively, for their continuous support.

REFERENCES

Burt C.M., Mills R.S., Khalsa R.D. and Ruiz C.V, (1998), *Improved Proportional-Integral (PI) logic for Canal Automation*, Jour. Irrig. and Drain. Eng., ASCE, Vol. 124, No 1, pp. 53-57.

Lefebre J.,(1977), *La regulation dynamique, sa mise en oeuvre au canal de Provence,* La Houille Blanche, No 2/3, pp. 265-270.

Malaterre P-O, Rogers D.C. and Schuurmans J. (1998), *Classification of Canal Control Algorithms*, Jour. Irrig. and Drain. Eng., ASCE, Vol. 124, No 1, pp. 3-10.

Rodellar J., Gomez M. and Martin Vide J.P., (1989), *Stable Predictive Control of Open-Channel Flow*, Jour. Irrig. and Drain. Eng., ASCE, Vol. 115, No 4, pp. 701-713.

PART VII

RESERVOIR MODELLING AND CONTROL

Use of Genetic Algorithms within a Reinforcement Learning Model for Multiple Reservoir Operation

F.J.-C. Bouchart *and* E. Hampartzoumian

ABSTRACT

A Genetic Algorithm (GA) approach is proposed to identify appropriate inflow sequences for use in the training of control systems based on Reinforcement Learning (RL) theory. RL models provide a mechanism by which the operational rules of multiple reservoir systems can be defined such that the likelihood of discrete failure events can be reduced. Unfortunately, these models require the use of historic and synthetically generated inflow sequences, and operational simulations of the system. Since no techniques are available to identify the inflow sequence that will optimise the training of the RL model, the operational performance of the RL model remains sub-optimal. Preliminary investigations were undertaken to establish the viability of using GA approaches to identify near-optimal inflow sequences, which could subsequently be used to train the RL model. The results from these preliminary investigations suggest that improvements in the operational rules for multiple reservoir systems may be achieved through the use of GA.

1 INTRODUCTORY COMMENTS

The operation of any multiple reservoir system requires a sequence of release decisions to be made, where the resulting performance of the system is derived from the cumulative effect of these decisions rather than any one single decision in isolation. Hence, while it is possible to establish the set of optimal releases from the system at a specific point in time such that, for example, the expected net benefits are maximised, a failure of the system at a later date cannot be directly attributed to that specific control event. The inability of the reservoirs to meet water demands is due to the cumulative deterioration of the system over time, and not the result of any one release.

Bouchart and Chkam (1998) developed a Reinforcement Learning (RL) model for the operation of multiple reservoirs where the control system adapts to discrete failure events. System failures are fed back to the control system and the operational rules modified, without the need for an explicit relationship between the performance of the system and the individual release decisions. The neuron-like adaptive elements in the RL model identify appropriate control strategies through an on-line (or simulated on-line) unsupervised learning process. Given the current state (or condition) of the physical system, the RL controller initiates a control decision. The response of the physical system to this control decision is then evaluated and a measure of performance obtained. In the case of the multiple reservoir system, performance is defined by the occurrence of a failure within the system. Failure events then trigger modifications in the behaviour of the control element to decrease the likelihood of future failures.

The on-line learning scheme, which forms the basis of the RL model approach, employs real-time information on the state of the system. Improvements in performance of the control element may be achieved, however, through the use of historic and synthetically generated inflows, and operational simulations of the supply system, to provide initial training prior to placing the controller on-line. The aim is to expose the controller to a broad range of hydrologic conditions, thereby establishing appropriate responses to potential future events and improving the performance of the supply system. However, the computational impossibility of exposing the control element to every possible sequence of events requires the development of an inflow sequence that optimises training. Over-emphasis on drought conditions may promote overly conservative operational policies, while placing too much emphasis on average or wet years may increase the risks to supply. Unfortunately, no technique exists to define the experiences required to optimise the training of the control element and ensure that this controller will be able to performance satisfactorily over the full range of potential hydrologic conditions.

In this paper, a Genetic Algorithm (GA) is proposed to assist in the development of an appropriate, if not optimal, inflow sequence for training of the RL control element. The remainder of the paper presents the GA formulation used within the context of the RL framework before demonstrating the application of this approach using the Burncrooks reservoir complex in Scotland.

2 GENETIC ALGORITHM FORMULATION

The task of identifying appropriate inflow sequences for training of the Reinforcement Learning (RL) multiple reservoir control system is ideally suited to a GA approach (Goldberg, 1989). The performance of the controller is dependent on the inflow sequence used during training. Hence it is possible to establish the suitability of a specific inflow sequence on the basis of the system performance achieved. Furthermore, there are no analytical search techniques available that can identify the optimal inflow sequence to be used, if such an optimal sequence even exists. The technique used to select suitable hydrologic sequences must be capable of efficient search of the inflow sequence space, without reliance on an explicit representation of the system being optimised. GA falls within this class of solvers, thereby yielding near-optimal inflow sequences.

The GA approach is based on the manipulation of coded strings, each representing a solution to the problem being solved. In the current study, a string represents a specific hydrological sequence consisting of 52 inflow values, as shown in Figure 1. The 52 weekly inflow values represent a time horizon of one year. While it is recognised that longer sequences will eventually be required, these short sequences were deemed appropriate for the investigation of the suitability of the GA approach.

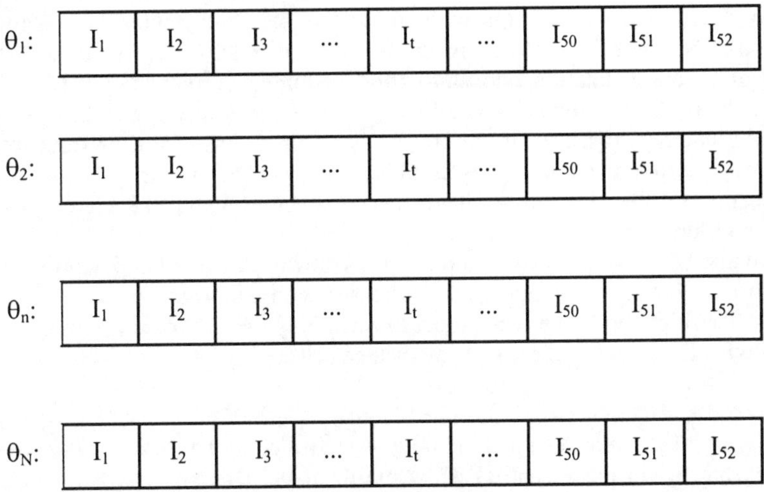

Fig. 1 Representation of inflow sequences as coded strings.

The initial population of coded strings is established from historical inflow records and from synthetically generated inflow sequences, and consists of N inflow sets, $\Omega = \{\theta_1, \theta_2, \theta_3, \ldots, \theta_N\}$. Each inflow set is then used in turn to train the RL model of Bouchart and Chkam (1998). The RL control elements are trained to eliminate the following failure events:

(a) Failure to meet water demand at the water treatment plant;
(b) Reservoir level below a specified minimum level;
(c) Reservoir spilling while excess storage capacity remains in the system;
(d) Water transfer to a reservoir that is full.

The performances of the N controllers are then evaluated using a new sequence of inflows to the system. For the purpose of this study, the performance of the system is defined as the number of system failures (as defined above) experienced. Due to the greater importance placed on a supply failure (Failure (a)), failures to meet water demands are counted as 2 failures for the purposes of system performance. This indicator of performance represents the negative of the fitness function in the GA formulation.

The fitness function values for the N inflow sets are then used to establish the next generation of inflow sets. The selection of current inflow sets to be included in the next generation is based on a roulette wheel approach. Inflow sets exhibiting higher fitness are given a greater probability of selection. N-1 members of the new inflow population are selected in this manner. The other N-1 members of the new population of inflow sets are generated by cross-over and mutation mechanisms.

The cross-over of two randomly selected inflow sets results in a new inflow sequence, which can subsequently be included in the next generation and investigated for its suitability. In this study, a single cross-over point is chosen randomly, and the new string built from the resulting segments. Random mutations are then introduced using a mutation rate of $p_m = 0.03$. One modification to the mutation algorithm introduced in the current study is that rather than randomly generating a new inflow value, an inflow for that time step is selected randomly from one of the other inflow sets. This use of existing inflow values ensures that such an inflow is likely. However, it is recognised that this modification to the mutation mechanism may severely hinder the search process for good inflow sequences.

The number of inflow sets contained in the newly generated population is N-2. This implies that the population size decreases with every generation of the GA process. Termination of the search occurs when only two inflow sets remain in the population. The better of these two sets is then selected.

3 CASE STUDY

The Genetic Algorithm (GA) methodology was applied to the Burncrooks reservoir complex located in the Kilpatrick Hills, north of the City of Glasgow, Scotland (Hampartzoumian, 1998). This application of GA relied on the Reinforcement Learning (RL) model of this reservoir system developed by Bouchart and Chkam (1998). The aim was to determine if the GA approach would yield improvements in the performance of the RL model.

The Burncrooks reservoir complex, established in 1995, is composed of five reservoirs connected by a combination of gravity and pumped mains. The system consists of a pumped main transferring water from the Greenside reservoir to the joint Cochno-Jaw reservoirs. The Cochno and Jaw reservoirs are two reservoirs separated by an embankment, and are operated as a single entity with flows from Cochno to Jaw occurring through gravity. Water from Jaw reservoir is then pumped to the Kilmannan reservoir before being transferred via a pumped main to the Burncrooks reservoir. A gravity mains conveys water from the Burncrooks reservoir to the Burncrooks Water Treatment Works. The sole operational objective of the system is to supply water to these works. More detailed descriptions of the Burncrooks reservoir system are provided in Chkam (1997) and Hampartzoumian (1998).

Hydrologic data for the reservoirs include:

(a) Two years of rainfall data at the Burncrooks reservoir and water treatment works;
(b) One year of water transfer and reservoir level records for all sites in the Burncrooks complex;
(c) Ten years of daily records of meteorological data for the Glasgow airport.

Synthetic inflow generation models were developed for the reservoir system. However, due to the poor quality of the data available, these inflow models could only be used to provide indicative inflow sequences. Therefore, details of the inflow models are not provided in this paper, and the results of the study should be taken only as indicative of the capabilities of the approach. It should be noted that the inflow sequences used within the GA model and the inflow sequence used to test the performance of the resulting control systems were generated using the same inflow models. Hence model errors are not incorporated in the comparative results being presented.

4 RESULTS

Ten populations of 20 starting inflow sets were used to establish the preliminary viability of the Genetic Algorithm (GA). If the GA approach is to identify better inflow sets, then the number of failures experienced by the supply system should decrease with successive generations. Figure 2 demonstrates that even with such small populations (20 inflow sets) and a small number of GA iterations (9 generations), substantial reductions in the number of failures can be obtained. The number of failures in Figure 2 corresponds to the modified definition for failures, where a supply failure is counted as two failures. It should be noted that the high number of failures reflects the fact that the inflow sets used for training consist of only 52 time steps, corresponding to a single year. Improvements would be achieved by using longer inflow sequences.

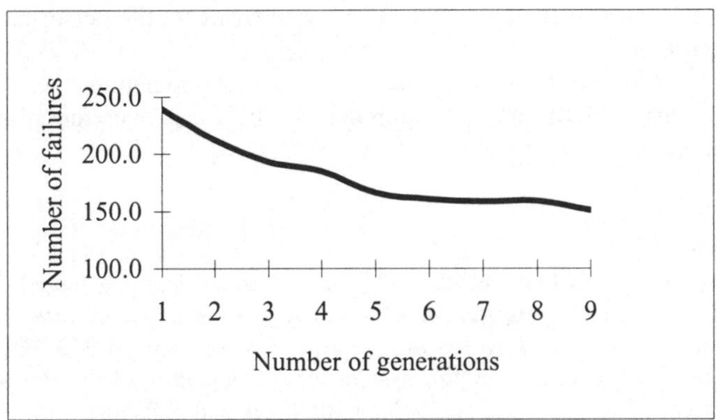

Fig. 2 Improvements in performance with each generation.

Figure 3 depicts the changes in the distribution of the system performance with each generation of the GA. There is a clear shift towards reduced failure rates and a reduction in the deviation in system performance. These results suggest that the GA converges towards similarly near-optimal solution. It should be noted that their distance from optimality, however, cannot be estimated since it is not possible to establish the optimal inflow sequence for the training of the RL model.

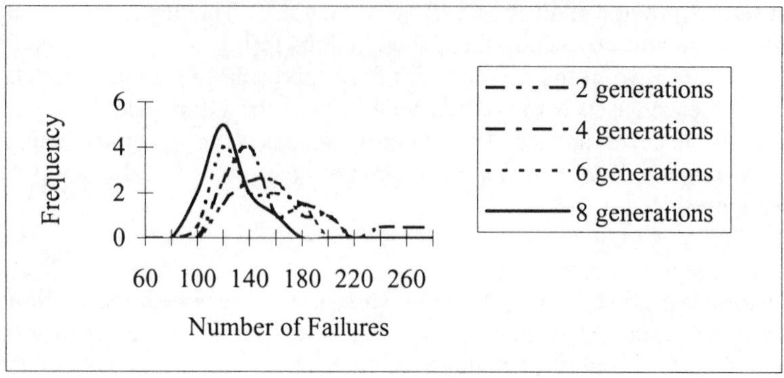

Fig. 3 Distribution of performance metric with each successive generation.

5 CONCLUSION

Preliminary results of a Genetic Algorithm (GA) approach to identify appropriate inflow sequences for the training of a Reinforcement Learning (RL) model have been presented. These results suggest that the GA approach can lead to improvements in the performance in the RL control system. Further research is now required to investigate:

(a) The influence of the length of inflow sequences on the performance of the RL model;
(b) The sensitivity of the GA approach to the initial population size;
(c) The sensitivity of the GA approach to the cross-over and mutation parameters.

REFERENCES

Bouchart, F.J.-C., and H. Chkam (1998). "A reinforcement learning model for the operation of conjunctive use schemes". *Hydrosoft'98: the Seventh International Conference of Hydraulic Engineering Software*, Como, Italy, pp. 319-329.

Chkam, H. (1997). "A decision support system for the operation of the Burncrooks reservoir complex". *MSc Thesis*, Dept. of Civil and Offshore Engineering, Heriot-Watt University.

Goldberg, D.E. (1989). Genetic Algorithms in Search, Optimisation and Machine Learning. Addison-Wesley Publishing Co., Reading, Mass.

Hampartzoumian, E. (1998). "The use of genetic algorithms as a decision aid for the management of flux transfers: The Burncrooks reservoir complex". *Research Project*, Dept. of Civil and Offshore Engineering, Heriot-Watt University.

Control of Water Levels in Polder Areas Using Neural Networks and Fuzzy Adaptive Systems

A.H. Lobbrecht *and* D.P. Solomatine

ABSTRACT

Modern water management in the Netherlands is characterised by considering water systems in their entirety together with all influencing factors and other related systems. Problems of management are posed more and more as multi-criterial problems with the strong accent on optimisation. In order to address such problems, a DSS Aquarius was built that is capable of generating several control strategies aimed at optimal control at local (regional) and centralised (inter-regional) levels. One of the problems encountered was the quite high computation time needed to generate an optimal control strategy, and its sensitivity to some of the parameters and input variables. Artificial neural network (ANN) and fuzzy adaptive systems (FAS) appeared to be efficient alternatives to using optimal control algorithms in real-time tasks. The obtained results show that ANN and FAS are able to replicate the behaviour of the Aquarius control component at one-two time steps (1 hour) ahead with the accuracy in the range 90-97%. This gives the possibility to replace the slow computational components by the fast-running trained intelligent controllers and thus to simplify the use of Aquarius in the real-time control tasks.

1 INTRODUCTION

Modern water management is characterised by considering water systems in their entirety together with all influencing factors and other related systems. Problems of management are often posed as multi-criterial problems with the strong accent on optimisation.

In the Netherlands the problem of water level management in low-land areas is quite acute and was historically determined by the "battle" against sea and river floods. Regional water control is the responsibility of approximately 65 water boards, and the problem of making optimal decisions in real-time control is taking nowadays a high priority.

Water authorities responsible for managing water systems have to cope with various problems ranging from flood prevention to recreation, and to take into account the interests of different groups of users. This makes optimal control of water resources systems a complex problem. This kind of control problem is being investigated extensively, especially for optimal operation of single or multiple reservoir systems (Fontane et al (1997), Panigrahi and Mujumdar (1997), Russel and Campbell (1996), Solomatine and Avila Torres (1996)). Optimal control of a regional water system, considering all urban and rural water subsystems is elaborated in Lobbrecht (1997). The implemented Decision Support System (DSS) *Aquarius* is used as an aiding tool for operational management of regional water systems by several water boards the Netherlands.

Deterministic models of water-related processes incorporated in Aquarius system will hence be called in short the *Aquarius model*. Aquarius uses two modes of control – centralized control and local control. Normally, centralized control is dynamic. The most important control action is pumping.

Within the Aquarius model the simulation and mathematical optimization problems are solved simultaneously in order to derive the optimal control actions – with the associated high running time and requirements to the computational power depending on the model complexity.

The objective of this study is to investigate the possibility of using AI techniques, namely artificial neural networks (ANN) and fuzzy adaptive systems (FAS), to replicate the behaviour of a deterministic model controlling the polder water levels. The mentioned techniques provide opportunities to extend the classical controls and deal with the complex systems in an integrated manner (Harris, 1994). The idea is to replace the control actions determined in Aquarius with centralized control option by building adaptive models with local information of a single regulating structure.

2 WATER SYSTEMS AND THEIR CONTROL

Operation and maintenance of a regional water resources system concern the optimal resources allocation for various interest groups in the system at the same time. In order to have an accurate overall picture of a water system state, which can be used for optimal operation and maintenance, it is necessary to take the conflicting criteria into account.

Water systems can be separated into interacting *subsystems* such as urban drainage, urban or rural groundwater, urban or rural surface water subsystems.

In the optimization problem for determining the control strategy, the requirements of each interest group can be expressed through "damage" functions, normally expressed as the dimensionless product of a *cost coefficient* and an optimization variable, representing the weighted harm to one or more interest groups. The constraints define the limits of physical water variables, which can be controlled directly or indirectly – such as surface water level, groundwater level, sewer filling, water quality level etc. The optimization problem can be formulated as follows (Lobbrecht, 1997):

$$
\begin{aligned}
&\text{Minimize: } Z(\overline{x},\overline{u}), \\
&\text{subject to: } g_i(\overline{x},\overline{u}) \le 0, && i \in 1,\dots,l; \\
&\qquad x_{l,j} \le x_j \le x_{u,j}, && j \in 1,\dots,m; \qquad (1)\\
&\qquad u_{i,k} \le u_k \le u_{u,k}, && k \in 1,\dots,n;
\end{aligned}
$$

where $Z(\overline{x},\overline{u})$ = objective function, $g_i(\overline{x},\overline{u})$ = constraints, $\overline{x}$ = vector of state variables, $\overline{u}$ =vector of control variables, $x_{l,j}, x_{u,j}$ = lower and upper limits of state variables *x*, $u_{i,k}, u_{u,k}$ - lower and upper limits of state variable *u*.

Two levels of control can be distinguished in the water system, in each of which the control actions can be manual and automatic:

- local control, that involves a single regulating structure in a water system and is executed on the basis of data gathered in the vicinity of that single structure and the standards that have been set for each subsystem.
- centralized dynamic control, in which the actions are based on time varying requirements of interests in the water system, the water system load and the dynamic processes in the water system. The dynamic control mode *with* or *without* prediction (of hydrological load) is distinguished. In dynamic control with prediction, an optimization problem is solved and the system state is predicted for specific control horizon on the basis of forecasted hydrological data.

In order to implement the mentioned optimisation, the Aquarius modelling and optimization system was developed. It is used for management of regional water resources systems in the Netherlands by building models of combined urban and rural water systems that describe water quantity and quality processes. Urban water subsystems in the Aquarius model represent fast runoff characteristics, while the rural subsystems are used for slow and delayed runoff processes. The subsystem interactions can be expressed in objects such as storage elements, regulating structures, free flow elements and also in the system description itself by formulating the boundary conditions. Different continuity equations describing hydrological processes are used for simulation of each subsystem state (Lobbrecht, 1997).

Aquarius uses a combination of simulation and mathematical optimization to determine the operational control actions and control strategy. To solve a non-linear optimization problem (eqn. 1), the Successive Linear Programming method using Taylor approximations of non-linear relationships is implemented.

In the case of a complex water resources system, the size of optimization problem, and subsequently the computational power required increases. For the time series calculation, a matrix in which the rows represent the constraints and the columns represent the physical variables defines the optimization problem. The large number of elements to be considered causes an increase in the number of non-zero elements in the constraints matrix. The total time needed for one time step simulation for a problem with 95000 variables and up to 92400 constraints is 1550s (a control horizon of 12 steps and simulation step of 2 hours is used). The results are obtained on a standard PC. In many real-time control situations these high running times are undesirable. It will be shown below how the intelligent controllers based on the ANN and FAS can be trained off-line to reproduce the Aquarius model results.

2.1 Intelligent control

Artificial intelligence (AI) techniques are widely used in solving various problems of water management and modelling (Bardossy and Duckstein 1995; Solomatine and Torres 1996; Fontane et al. 1997). AI techniques are also being used nowadays as an alternative approach for conventional controllers – see Miller et al (1990), Omatu et al (1995), Omidvar and Elliot (1997). Such controllers are called *intelligent controllers* since they employ AI techniques to produce the control actions; being combined with the conventional controllers they enable better handling of complex real-life problems. In this study the artificial neural networks (ANN) and fuzzy adaptive systems (FAS) are used for controlling the water level of polders.

3 AI TECHNIQUES USED IN CONTROL

3.1 Neural networks

Artificial neural network (ANN) (Tsoukalas and Uhrig, 1997) is an information processing system that roughly emulates the behaviour of a human brain by replicating the operations and connectivity of biological neurons. In *supervised learning (training)*, a series of connecting *weights* are adjusted in order to fit the series of inputs to another series of known outputs. Once the training is performed, *recall* (running ANN to produce an output) is very fast, in a fraction of a second; this makes ANN applicable for model-based control. In this study only supervised learning algorithms, namely feed-forward networks are considered.

The feed-forward neural networks consist of three or more layers of computing units, also called nodes: input layer, output layer and one or more hidden layers. The input vector x passed to the mapping is directly passed to the node activation output of input layer without any computation. After hidden layers provide additional computations, the output layer generates the mapping output vector z. Two types of feed-forward networks are used in this study.

3.2 Fuzzy rule-based systems

The fuzzy logic approach has been successfully applied to a wide range of control problems, beginning with the introduction by Zadeh in 1965 of fuzzy-set theory and its applications. In the fuzzy logic approach the Boolean logic is extended to handle the concept of *partial true* which implies that true takes a value between completely true and completely false. The notion of fuzzy *sets* has to be introduced, which is the collection of the objects that might belong to the set to a degree, taking all values between 0 (full non-belonging) and 1 (full belonging), instead of taking crisp value. The indication of intensity of belonging is expressed as a *membership* function, assigning to each element a number from the unit interval (Tsoukalas and Uhrig , 1997).

Fuzzy rules consist of arguments coupled by logical operators and verbally formulated, such as IF the condition is fulfilled THEN the consequence has to be true. The logical expressions are usually formulated by logical operators AND, OR, NOT and XOR. The truth value corresponding to the fulfilment rule conditions for a given premise is called the *degree of fulfilment* (DOF).

The basic structure of a fuzzy rule-based system involves four principal components: *fuzzification interface*, where the values of inputs are measured, fuzzified and the input range is mapped into the suitable universe of discourse, *knowledge-base*, which involves a numeric 'database' section and a fuzzy (linguistic) rule-base section, *fuzzy inference mechanism or engine*, which constitutes the core of the fuzzy logic control and involves the decision making logic (fuzzy reasoning such as product, max-min composition etc) and *defuzzification interface*, which maps the range of output variables into corresponding universe of discourse and defuzzifies the results of the fuzzy inference mechanism.

In a complex system, which is usually the case, the fuzzy rule-base system construction is limited (manipulation and verbalization by expert). Therefore the possibility of inducing and learning the rules from data has been investigated and implemented successfully and those systems are called fuzzy adaptive systems (FAS). Such a method, the weighted counting algorithm, is used for this study.

For given relevant variables the fuzzy rule based system has to deliver a response close to the observed results. In other words, on the basis of user defined input membership functions and input-output sets, FAS can determine the output membership functions and defuzzified outputs. If all the variables and the responses are continuous then the rules can be constructed by defining the fuzzy set that supports the fuzzy numbers $A_{i,k}$ and identifies the corresponding responses. $A_{i,k}$ is a fuzzy number $(\alpha_{i,k}^{-}, \alpha_{i,k}^{1}, \alpha_{i,k}^{+})$ where $\alpha_{i,k}^{1}$ is the mean of all possible $a_k(s)$ values which fulfil at least partially the ith rule :

$$\alpha_{i,k}^{1} = \frac{1}{N} \sum_{s \in R_i} a_k(s) \tag{2}$$

where N is the number of elements in R_i. R_i is the set of all those premise value vectors that fulfil at least partially the ith rule and it forms the subset of a training set. The method chooses the rule responses, which fulfil the certain fuzzy rule at least with a specified threshold value of DOF. The value of DOF has to be selected so that a sufficient number of elements of the training set is considered for each rule.

3.3 Model-based control

Model-based control involves a model of the controlled system or process. A specific example of indirect intelligent control of a closed loop control scheme was termed by Saerens (1995) as "model reference adaptive control". This scheme is used for the control problem case considered. The scheme involves three main components, namely a reference model, a trainable intelligent controller and the process or system under control. In this scheme the deterministic models are used as a reference and conduct the learning procedure for intelligent controllers. The intelligent controllers are the AI techniques appropriate for adaptive learning.

The simulation of a reference model is used for offline adaptive learning of an intelligent controller. At each time step the process or system state target value $y(t)_d$ passes through the intelligent controller and gets the control signal $u(t)$. When the process or system results the output $y(t)$, the measured value is passed to the intelligent controller and compared with the target value. As a response from the intelligent controller, the control signal for the next time step should be obtained.

The model-based scheme for controlling the polder water level uses the Aquarius DSS model as a reference model for the intelligent controller (Figure 1). The desired value is a target water level in the water system, a pumping rate of the drainage station is a control action and the system output is the water level in the polder area. ANN and FAS models will hence be referred to as *adaptive models*.

Note that adaptivity here is a property of training and does not characterise the use of the trained system. In fact, the trained ANN or FAS will not automatically adapt well to the changing properties of the controlled system. If the controlled system is changed, they must be re-trained on the basis of the new data.

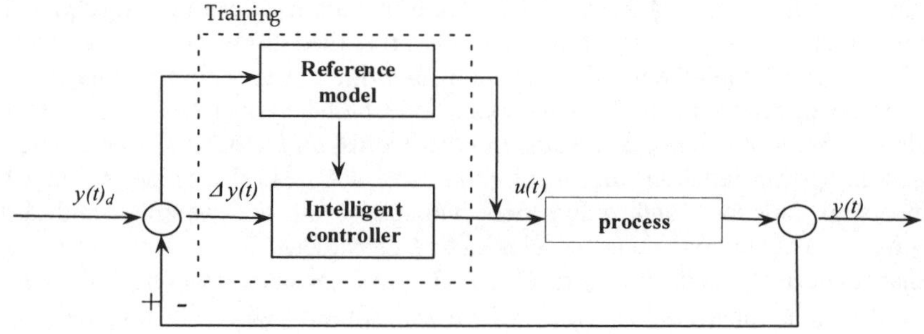

Figure 1. Model-based control

4 CASE STUDY

The study has been based on the water resources system model built on one of the areas located in the western part of the Netherlands. The water resources in this area are managed by one water board. The model represents a complex water resources system of approximately 40 000 ha including a total of 60 polders and correspondingly there are a large number of independent water level areas, with 11 pumping stations, 2 weirs and 7 inlets that serve as regulating structures. The water quantity and quality requirements of main interest groups in the modelled area, such as flood prevention, ecology, glasshouse horticulture, pasture agriculture and navigation, may be in conflict. For example, navigation requires the maximum depth of water level, while for flood prevention the minimum water level is preferable.

The controlled area has a storage basin (SB) which is the main surface water subsystem in the area. Three types of areas discharge to the SB: low-lying polders, high-lying polders and storage basin land. The SB is drained to the North Sea through the river. If the water level in the SB exceeds a certain limit, some polders have to stop pumping into the SB. Such a policy is applied in situations, when the probability of dike overflow is very high.

5 TRAINING ANN AND FAS

Adaptive learning of ANN and FAS was based on the data sets generated by the Aquarius model in both local and centralized dynamic control modes. Water levels that represent the state of a certain polder area can be controlled through a control variable - pumping rate of drainage station. For model simulation, 30 years of real hydrological data was available. The adaptive models built for each pumping station reproduce the pumping rate at time step *t*.

Before training, the correlation analysis for input/output relationships of the pumping station type of control point has been carried out. The physical variables most correlated to the output variable should enable better model performance. Therefore, the coefficients of linear correlation between output variable and possible input variable sets were identified. The moving average values for water level and precipitation variables and pumping rate at the previous time steps tend to have higher correlation to the output variable. The variables with high correlation to the output are included in the input pattern.

Adaptive models were simulated by neural network generator NNN (*www.ihe.nl/hi/sol/nnn.htm*) and fuzzy adaptive system AFUZ (*www.ihe.nl/hi/sol/fuzzy1.htm*) developed at IHE. For adaptive models, simulation 1-5 years data with time step of one hour and a tolerance target of 5% are used. Mainly, 50% of data is used for training and another 50% is used for verification. In the case of dynamic control with prediction, a control horizon of 24 hours is applied. For ANN models, from 5000 to 30 000 iterations were needed to accomplish the training. In the case of FAS models, a triangular shape of membership function, a threshold value for DOF of 0.05%, product inference and centroid defuzzification methods were applied.

Several performance indices were used: mean square error (MSE), the percentage of the output data within the tolerance target of 5%; the model accuracy (percentage of examples for which the pumping rates are accurately determined), and the total flow rate difference for the whole range of data set.

The best model and data structure for adaptive models was investigated. It was done by including the extreme hydrological events into the training data set, which should have enabled the highest variation in the training data set.

6 RESULTS

The adaptive model performances obtained have proved the applicability of AI techniques for this sort of control problem. The extreme events in the training data set including the moving average and moving sum values tend to give the best performances of adaptive models. Also training by random generated data and verification by natural data enabled reasonably good accuracy in the models. The model structures providing the best results for different control modes are presented below in Table 1 (results for control with prediction were not ready at the moment of publication). The corresponding verification plots are shown in Figures 2 and 3 for local and dynamic control respectively.

a. local control mode
   Input variables:

1. Water level at *t*
2. Water level at *t-1*
3. Pump status at *t-1*

Output variable: pump status at *t*
Number of hidden nodes (NNN) 3
Number of rules (AFUZ) 63

b. Dynamic control without prediction
Input variables:
1. Moving average precipitation, *t-1 to t-3*
2. Moving average water level in SB, *t-1 to t-3*
3. Moving average water level in SB, *t-4* to *t-10*
4. Moving average pump status, *t-1 to t-4*

Output variable: pump status, *t*
Number of hidden nodes (NNN) 4
Number of rules (AFUZ) 189

Table 1. Summary of the best results obtained

	a		b	
Tool used	NNN	AFUZ	NNN	AFUZ
MSE	0.096	0.196	0.08	0.24
Accuracy net (%)	81.16	79.34	75.25	62.37
Accuracy overall (%)	92.9	92.21	95.92	82.63
Flow through pump (difference in %)	0.044	0.53	4.62	8.97

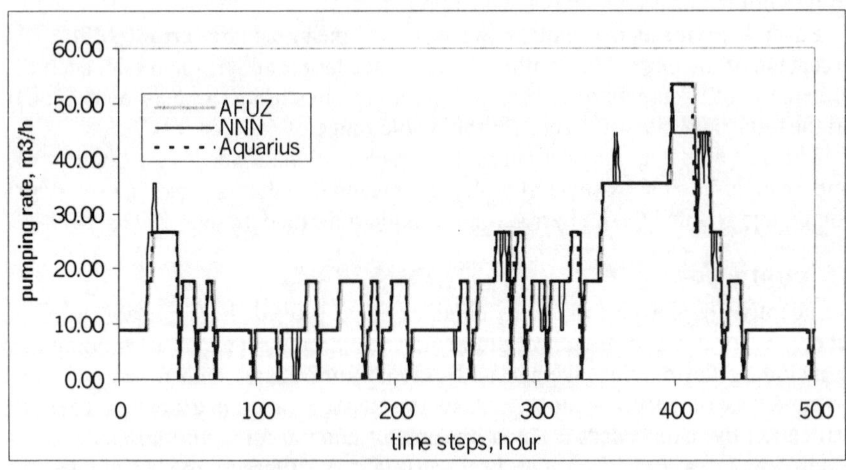

Figure 2. Performance of intelligent controllers for local control strategy

Additional analysis was performed in order to assess the performance of adaptive models in the case of extreme events. It was concluded that the adaptive models are able to learn very specific patterns of control action.

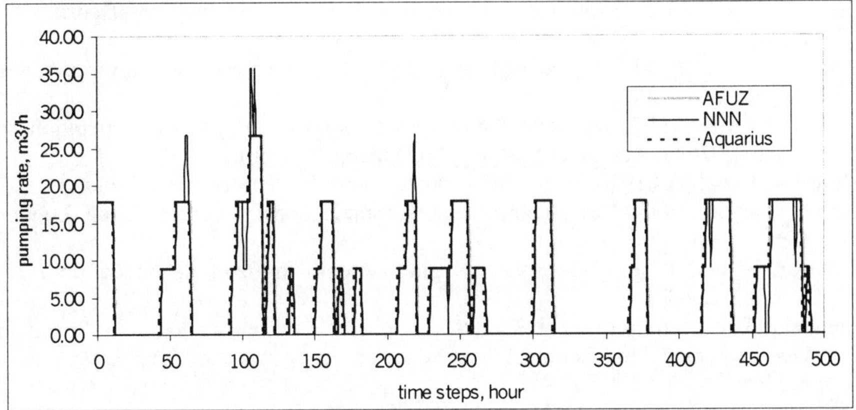

Figure 3. Performance of intelligent controllers for centralized dynamic control

7 CONCLUSION

Model-based intelligent control schemes were successfully applied for reproducing optimal control actions for polder water level control. ANN and FAS models can reproduce the corresponding control actions with high accuracy. Nevertheless, the adaptive models' performances are highly dependent on the choice of training data set so that two different data sets may result in a different level of model performance. Inclusion of variables resulting from extreme events and also the use of moving average and moving sum values in the training pattern improves the models' performance. The advantages of intelligent controllers are in their robustness and ability to reproduce the centralised behaviour of control actions by using easily measurable local information. However, it should be noted that the present control problem is model specific, consequently ANN and FAS exhibit properties of adaptation during training, but once trained on the basis of one water system model, they cannot represent the hydrologic behaviour in other model areas, and have to be re-trained.

REFERENCES

Bardossy, A and Duckstein, L., 1995. Fuzzy rule-based modelling with applications in Geophysical, biological and engineering systems, CRC press Inc

Bazartseren, B, 1999, Use of neural networks and fuzzy adaptive systems for controlling polder water level in the Netherlands, MSc thesis, IHE, Delft, The Netherlands

Dibike, Y., Solomatine, D.P., Abbott, M.B., 1999. On the encapsulation of numerical-hydraulic models in artificial neural network. *Journal of Hydraulic Research*, No. 2. .pp. 147-161.

Fontane, D.G et al., 1997. Planning Reservoir Operations with Imprecise Objectives, Journal of Water Resources Planning and Management, May/June, pp 154-162

Harris, C.J. 1994. Advances in Intelligent control, Taylor & Francis Ltd, London

Jana, A et al. 1996. Real-time Neuro-Fuzzy control of a nonlinear dynamic system, Proceedings of the 1994 Biennial Conference of the North American Fuzzy Information Processing Society, pp 210-214

Krijgsman, A.J. 1993. Artificial intelligence in Real-time control, Proefschrift. TU Delft

Lin, C.T. 1994. Neural fuzzy control systems structure and parameter learning, World Scientific Co. Ltd

Lobbrecht, A.H. 1997. Dynamic water system control: Design and Operation of Regional Water Resources Systems, Ph.D thesis, TUDelft

Miller, W.T et al (ed) 1990. Neural networks for control, The MIT Press, London

Omatu, S et al. 1995. Neuro-control and its applications, Springer-Verlag London Limited, London

Omidvar, O and Elliot, D.L. 1997. Neural systems for control, Academic Press Limited, London

Panigrahi, D.P & Mujumdar, P.P. 1997. Application of Fuzzy logic to Reservoir Operation modelling, National Conference on Fuzzy sets and Their Applications, Madras

Raman, H & Chandramouli, Y. 1996. Deriving the general operating policy for Reservoir using neural network, Journal of Water Resources Planning and Management, Sept/Oct pp 342-347

Russel, S.O. and Campbell, P.F. 1996. Reservoir Operation Rules with Fuzzy Programming, Journal of Water Resources Planning and Management, May/June pp 165-170

Sarens, M et al. 1995. Neurocontrol based on the Backpropagation algorithm, Intelligent control systems: Theory and applications, Gupta, M.M & Sinba, H.K (ed), IEEE press

Shen Y., D.P. Solomatine, H. van den Boogaard, 1998. Improving performance of chlorophyl concentration time series simulation with artificial neural networks. *Annual Journal of Hydraulic Engineering, JSCE*, vol. 42, February, pp. 751-756.

Solomatine, D.P and Avila Torres, L.A, 1996, Neural Network Approximation of a hydrodynamic Model in Optimizing reservoir operation, proceedings of Hydroinformatics '96 conference, pp 201-206.

Tzafestas, S.G and Tzafestas, C.S. 1997. Fuzzy and neural control: Basic principles and architecture, Methods and applications of Intelligent control: Microprocessor-based and intelligent systems engineering, Kluwer Academics, Tzafestas, S.C ed.

Tsoukalas L.H. and Uhrig R.E. 1997. Fuzzy and Neural Approaches in Engineering. John Wiley and Sons, N.Y., 587 p.

Computing Spatial Variation of Algae in Water Supply Reservoirs

N. R. B. Olsen, R. D. Hedger, S. E. Heslop *and* D. G. George

ABSTRACT

High concentrations of algae can be a problem in water supply reservoirs. An example is the Eglwys Nynydd reservoir in Wales, where water is abstracted for industrial purposes by a steel factory. When the algae concentration exceeds 100 mg chl./qubic meter, the steel production is affected. Although the average algae concentration in the reservoir very seldom reaches this value, wind-induced currents and other processes can cause algae accumulation near the water intake, increasing the concentration to unwanted levels. The spatial variation of algae concentration in a water supply reservoir can be calculated using a 3D CFD model, where wind- induced circulation is modelled solving the Navier-Stokes equations with a turbulence model. Equations for algal processes are also solved, including vertical movement of algae due to changes in buoyancy, algae reproduction and wind-induced turbulence. This paper presents results where the model has been tested against field measurements for three resevoir/lakes in the UK: Eglwys Nynydd reservoir, Loch Leven in Scotland, and Esthwaite Water in the Lake District.

1. INTRODUCTION

The spatial variation of algal concentration in water supply reservoirs is influenced by a number of different processes. Wind-induced circulation moves algae in horizontal directions, according to the induced velocity pattern. Velocities close to the surface are often in the wind direction, while deeper layers have velocities in the opposite direction due to return currents. The flow pattern is further complicated by thermal stratification and bottom topography. The wind also induces turbulent mixing, influencing the vertical algal concentration profiles. Additionally, the algae may move up or down in the water column by changing its buoyancy. Vertical movement depend on the algal species and may be dependent on irradiance. There is also reproduction of algae according to nutrients, temperature and light. Computing the spatial variation of algae in a reservoir, it is necessary to take the most important of these processes into account.

In the recent years, the speed and cost of computing power has changed dramatically. It is now feasible to calculate processes in water reservoirs in three dimensions using numerical models. This science, computational fluid dynamics (CFD), has produced a number of computer programs, which can be used for the purpose. CFD modelling also enables the calculation of spatial and temporal variation in algal concentration, as a function of wind-induced currents, inflow from streams and rivers, turbulence, nutrients, light etc.

Early studies of numerical modelling of lakes were carried out by Simons (1976), simulating wind-induces circulation in Lake Ontario. A three-dimensional finite difference method was used, and a constant eddy-viscosity model was used to model the turbulence. The eddy-viscosity was varied until the computational result agreed with the measurements. Falconer et. al. (1991) also used similar approach to calculate wind-induced currents in Esthwaite Water, UK. Good agreement was found with field measurements of water velocities, after calibration of a constant eddy-viscosity model.

Later studies have also included more advanced turbulence models. Gbah et. al. (1998) used a Reynolds stress turbulence model to calculate the wind-induced water currents and temperature in a thermal bar for an idealised two-dimensional width-averaged case. The advantage of the sophisticated turbulence model was that the eddy-viscosity was calculated as a part of the flow field, and did not need to be specified by the user. The model could thereby be applied to new cases without calibration. Olsen et. al. (1999) also used this approach by applying the k-ε turbulence model when solving the Navier-Stokes equations on a 3D grid of a hydropower reservoir. Guting and Hutter (1998) used the k-ε turbulence model to calculate the vertical eddy-viscosity when modelling 3D circulation in a homogenous lake. A number of advantages were highlighted compared with more primitive turbulence models, for example the clear correspondence between the water velocity and the turbulence.

The present paper describes an approach for modelling spatial variation of algae in water supply reservoirs. The paper includes a detailed description of the model and some case studies that demonstrate the applications of the model to two natural lakes and a water supply reservoir.

2. THE NUMERICAL MODEL

The numerical model calculates the water flow by solving the Reynolds averaged transient Navier-Stokes Equations. The Coriolis force is added to the equations as an extra source term. An implicit method is used to model the transient term. The convective term is solved using a first-order upwind method. The SIMPLE method is used to calculate the pressure (Patankar, 1980).

Two approaches can be used for modelling the wind-induced turbulence: a zero equation model or the standard k-ε turbulence model. The numerical models are further described by Rodi (1980), Patankar (1980) and Olsen (1997).

The wind shear stress, τ, on the lake surface is calculated as a function of the measured wind speed, U_{10}, using the following formula:

$$\tau = c_{10}\rho_{air}U^2 \tag{1}$$

The friction coefficient, c_{10}, is given by Bengtsson (1973) as $1.1\mathrm{x}10^{-3}$, and ρ_{air} is the air density.

The convection-diffusion equation for the algal concentration is solved together with the Navier-Stokes equations. Two source terms are used, describing two physical phenomena. The most important process is the vertical movement of the algae, often influenced by changes in buoyancy as a function of light. The amount of irradiance, I, decreases with the factor f, as a function of the water depth and the algal concentration. The following formula is used:

$$f = \sum e^{-k_e c_a \Delta z} \tag{2}$$

The algal concentration is denoted c_a, and z is the vertical size of a cell. The specific light transmission coefficient, k_e, is given by Bindloss (1976):

$$k_c = ac_a + b \tag{3}$$

where a and b are constants in a regression formula based on measurements in a lake.

The algal density is calculated by the following equation given by Kromkamp and Walsby (1990):

$$r_1 = r_0 + \Delta t\left(k_1 \frac{I}{I+K} - k_2 I_{24} - k_3\right) \tag{4}$$

r is the algal density, t is the time and k_1, k_2, k_3 and K are constants. I_{24} is the average irradiance over the last 24 hours. The coefficients of Kromkamp and Walsby (1990) are used. The fall/rise velocity, w, of the algae is calculated from

Stoke's equation.

If there are sufficient nutrients in the reservoir, the algal growth is light-limited. The algal growth can then be based on light extinction and a time series of the irradiance (Reynolds, 1984).

3. EXAMPLE 1: EGLWYS NYNYDD WATER RESERVOIR

Eglwys Nynydd reservoir is located in Wales, and the water is abstracted for industrial purposes by a steel factory. When the phytoplankton concentration exceeds 100 mg.chl./m^3, the steel production is affected. Although the average algal concentration in the reservoir very seldom reaches this value, wind-induced currents and other processes can cause phytoplankton accumulation near the water intake, increasing the concentration to unwanted levels. Spatial variation in algal concentration occur when the cells float to the surface and accumulate downwind (George and Edwards, 1976).

The reservoir covers an area of 1.01 km^2, and has a mean depth of 3.5 m and a maximum depth of 6 m. The main inflow stream contains high concentrations of nutrients and the summer phytoplankton is typically dominated by the blue-green algae, *Microcystis aeruginosa* Kutz. emend. Elenkin.

Fig. 1 shows the surface distribution of phytoplanktonic chlorophyll recorded by George and Edwards (1976) on one particular day during the summer of 1970. The samples were collected in the late morning when the sky was overcast and a steady wind had been blowing from the south for at least ten hours. Samples of water collected from different depths showed that most of the *Microcystis* were concentrated near the surface, and there was a very dense accumulation of algae along the downwind shore.

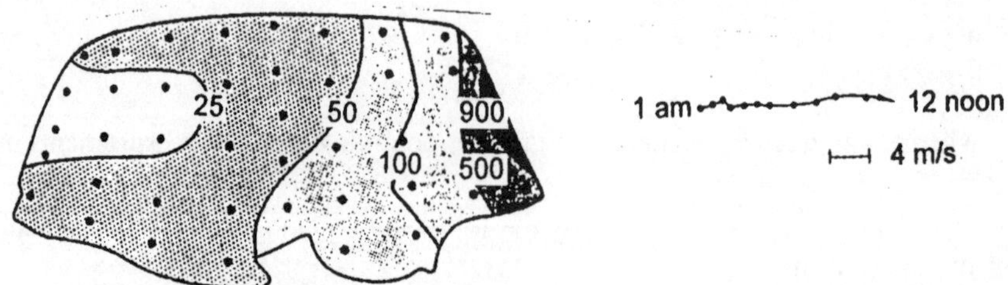

Fig. 1 Surface distribution of phytoplanktonic chlorophyll (mg.chl./m^3) in Eglwys Nynydd water reservoir, and the wind vector diagram previous to the measurements.

During the night, the algae accumulated close to the water surface. The water circulation caused the surface layer to move with the wind. There was therefore a higher concentration in the downwind direction close to the water surface, and a corresponding lower concentration in the upwind direction.

The previously described numerical method was used to model the phytoplankton concentration near the water surface. Fig. 2 shows a projection of the grid seen from above. There were up to 17 cells in the vertical direction, depending on the water depth. The vertical size of the grid cells close to the water surface was 10 cm.

The measured wind time series in Fig. 1 was extended by extrapolation and used as input for the numerical model. Initial concentration of algae was set to 50 mg.chl./m^3, uniformly distributed in the vertical and horizontal directions. A time series of irradiance was constructed, based on the time of sunrise and sunset, and formulae for light intensity. It was assumed that there had been cloud cover on the day of the measurements. The maximum irradiance was set to 700 μmol photons/m^2/s. The simulations were started on the evening before the measurements, and lasted for 18 hours. The standard k-ε turbulence model was used, and it was assumed that the algae would flocculate into groups with an effective diameter of 0.25 mm (Howard, 1993).

The calculated phytoplankton concentration close to the water surface is given in Fig. 3. The figure can be compared with the measurements on Fig. 1. There is a reasonable agreement, as the main phytoplankton concentration gradient is reproduced, although its maximum magnitude is underpredicted. A parameter sensitivity test showed that the algal colony diameter affects the results greatly. A greater diameter will give a steeper horizontal gradient and larger maximum concentrations.

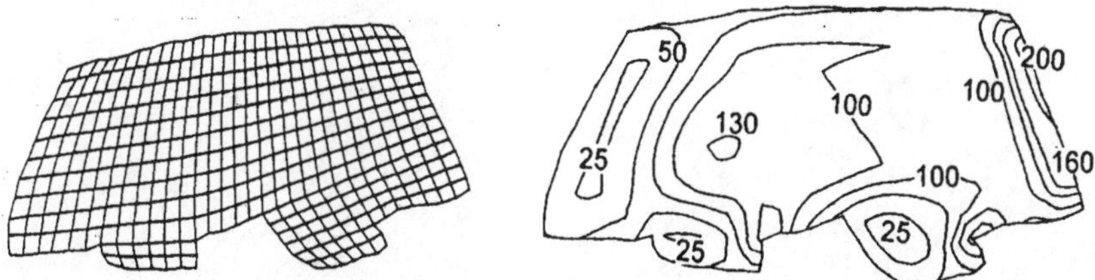

Fig. 2 Grid of Eglwys Nynydd water reservoir, seen from above

Fig. 3 Calculated surface algae concentration (mg.chl./m^3) in Eglwys Nynydd water reservoir

4. EXAMPLE 2: LOCH LEVEN

Loch Leven is a shallow nutrient-rich loch, located in NE Scotland. It is small in comparison to most other Scottish lochs, with a surface area of 13 km^2 and an average depth of 4 m. The lake receives phosphorus, nitrogen and silica, in sufficient amounts for phytoplankton blooms to occur. Many species of phytoplankton exist in the loch, with varying dominance over time.

Loch Leven is one of the most heavily investigated lochs in Scotland. In particular, many data exist from a program of intensive sampling conducted in 1985, including a remotely sensed image. The data were used to establish boundary conditions (including weather conditions, mean water quality and spatial variation in water quality) for the CFD model. In the Spring months of 1985, the phytoplankton population in this loch was dominated by diatoms. These are non-motile, negatively buoyant and rely on upwelling and turbulence to remain in the euphotic zone (the depth to which phytoplankton photosynthesis can be supported). Large-scale spatial variation, involving a consistent trend in phytoplankton concentration, was evident on 11 May 1985 from both surface sampling and the remotely sensed image.

The calculation was based on an initially uniform distribution of algae. Lake circulation was based on recorded wind and irradiance from two nearby metrological stations, for a period of two days before the sampling. A zero-equation turbulence model was used, as the k-ε model overpredicted the turbulence. The zero-equation model had previously been used to model sediment flushing of a res-ervoir (Olsen, 1999), with reasonable results. The main processes causing spatial variations in algal concentration was the fall velocity of the algae during a calm period leading to algal deposition in shallow areas. During a later increase in wind speed, turbulence caused upwelling of algae from deeper parts of the lake. Together with the wind-induced circulation, this caused higher algal concentration on the west side of the lake. The computed and measured concentrations near the surface are given in Fig. 4, showing reasonable correspondence.

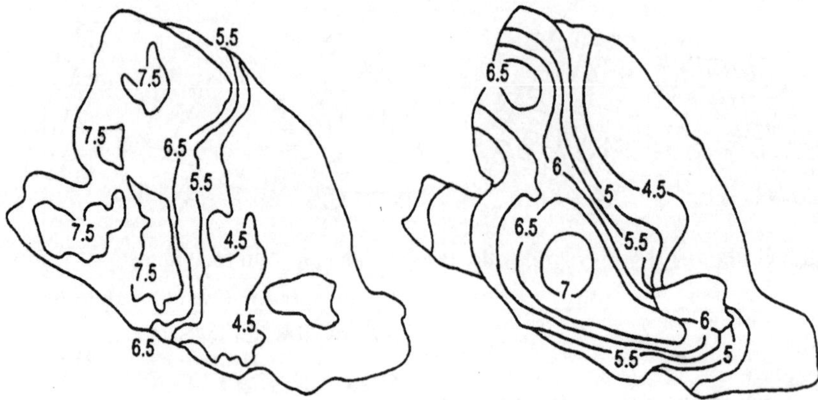

Fig. 4 Algal concentration (g/m^3) near the surface in Loch Leven, measured (left) and calculated (right).

5. EXAMPLE 3: ESTHWAITE WATER

Esthwaite Water is a small lake in the English Lake District. It has a surface area of 1 km^2, and a mean depth of 6.4 m. The maximum length is 2.5 km, and the mean width is 450 meters. The lake is eutrophic and supports dense blooms of slow-growing algae like *Microcystis, Aphanizomenon* and *Ceratium* in mid-summer. George and Heaney (1978) found that motile species, like *Ceratium*, frequently accumulated in mid-water during the day and were thus carried upwind by currents flowing above the thermocline. In contrast, buoyant cyanobacteria, like *Microcystis*, tended to accumulate near the surface and were carried towards the downwind shore in the surface drift. Two short episodes of wind-induced currents were simulated numerically, based on measured wind and irradiance. The first run modelled the situation 2, August 1973, when the dinoflagellate *Ceratium hirundinella* dominated. Surface currents were deflected clockwise from the wind direction by the Coriolis force. This caused downwelling at the downwind area of the lake, particularly in the east and south and upwelling at the upwind area of the lake, particularly in the west and north. Simulated and measured surface *Ceratium* distributions were similar (Fig. 5), although the simulation estimated an east-west trend that was not measured.

The second run modelled a situation on 9th August 1973, where *Microcystis* dominated. The positively buoyant algae accumulated over downwelling areas, resulting in a similar surface distribution to that measured (Fig. 5) though with a greater magnitude of variation.

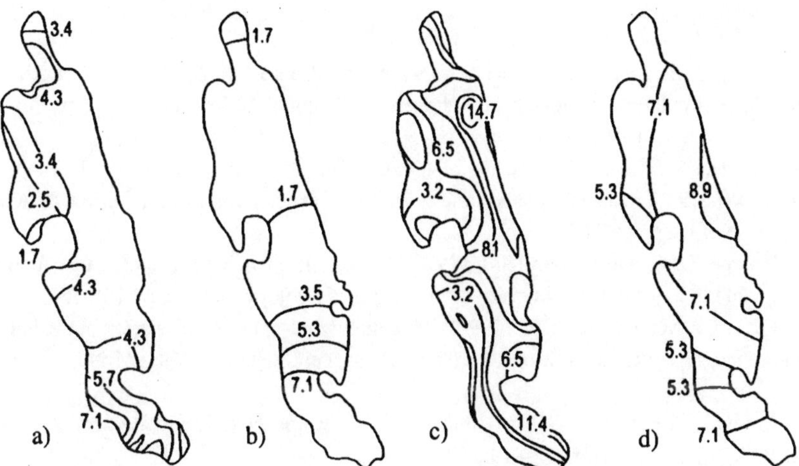

Fig. 5 Algal concentration (g/m^3) near the surface in Estwaite Water. a) *Ceratium*, measured, b) *Ceratium*, calculated, c) *Microcystis*, measured, d) *Microcystis*, calculated. North is in the upwards direction.

6. CONCLUSIONS

The three-dimensional numerical model is able to calculate the spatial variation of algal concentration in water reservoirs by modelling many of the processes occurring: wind-induced circulation and turbulence, temperature stratification and Coriolis force, algal growth and varying algal buoyancy as a function of irradiance. The model gave reasonably good correspondence with field data for three modelled lakes and reservoirs in the UK: Eglwys Nynydd water supply reservoir, Loch Leven and Esthwaite Water.

ACKNOWLEDGEMENTS

We want to thank the Norwegian Research Council for partly funding the numerical study. This project has also been supported by funding from the Natural Environment Research Council and the Institute of Freshwater Ecology.

REFERENCES

Bengtsson, (1973)"Wind shear stress on small lakes", Tekniska Høgskolen, Lund, Sweden.

Bindloss, M. (1976) "The light-climate of Loch Leven, a shallow Scottish lake, in relation to primary production of phytoplankton", Freshwater Biology, No. 6.

Falconer, R. A., George, D. G. and Hall, P. (1991) "Three-dimensional numerical modelling of wind driven circulation in a shallow homogenous lake", Journal of Hydrology, No. 124, pp. 59-79.

Gbah, M. B., Jacobs, S. J., Meadows, G. A. and Bratkovich, A. (1998) "A model of the thermal bar circulation in a lone basin", Journal of Geophysical Research – Oceans, Vol. 103, No. C6, pp. 12807-12821.

George, D. G. and Edwards, R. W. (1976) "The effect of wind on the distribution of chlorophyll-a and crustacean plankton in a shallow eutrophic reservoir", Journal of Applied Ecology, No. 13, pp 667-690.

George, D. G. and Heaney, S. I. (1978) "Factors influencing the spatial distribution of phytoplankton in a small productive lake", Journal of Ecology, Vol. 66, pp 135-155.

Guting, P. M. and Hutter, K. (1998) "Modeling wind-induced circulation in the homogenous Lake Constance using k-epsilon closure", Aquatic Sciences, Vol. 60, No. 3, pp. 266-277.

Howard, A. (1993) "SCUM-simulation of cyanobacterial underwater movement", CABIOS, Vol. 9, No. 4, pp 413-419.

Kromkamp, J. and Walsby, A. E. (1990) "A computer model of buoyancy and vertical migration in cyanobacteria", Journal of Plankton Research, pp. 161-183.

Olsen, N. R. B., Jimenez, O., Lovoll, A. and Abrahamsen, L. (1999) "3D CFD modelling of water and sediment flow in a hydropower reservoir", International Journal of Sediment Research, Vol. 14, No. 1.

Olsen, N. R. B. (1997) "Computational Fluid Dynamics in Hydraulic and Sedimentation Engineering", Department of Hydraulic and Environmental Engineering, The Norwegian University of Science and Technology. (can be downloaded from http://www.sintef.no/nhl/vass/class.html)

Olsen, N. R. B., Hedger, R. D., George, D. G. and Heslop, S. (1998) "3D CFD modelling of spatial distribution of algae in Loch Leven, Scotland", 3rd Int. Conference on Hydroscience and –engineering, Cottbus, Germany.

Olsen, N. R. B. (1999) "Two-dimensional numerical modelling of flushing processes in water reservoirs", IAHR Journal of Hydraulic Research, Vol. 1.

Patankar, S. V. (1980) "Numerical Heat Transfer and Fluid Flow", McGraw-Hill Book Company, New York.

Reynolds, C. S. (1984) "The ecology of freshwater phytoplankton", Cambridge University Press, Cambridge, UK.

Rodi, W. (1980) "Turbulence models and their application in hydraulics", IAHR State-of- the-art paper.

Simons, T. J. (1974) "Verification of numerical models of Lake Ontario: Part I, Circulation in spring and early summer", Journal of Physical Oceanography, No. 4. pp. 507-523.

Deterministic-Stochastic Modeling of Water Supply Reservoirs

S.Tomic *and* D. A. Savic

ABSTRACT

Discharge frequency analysis (DFA) is a quantitative process that assigns exceedance probabilities or recurrence intervals to river discharges of various magnitudes. The results of DFA are used in hydraulic models to predict streamflow depths and velocities. In this way, DFA produces a basic input for design and operation of water supply reservoirs. In the twentieth century, a tremendous amount of effort has been devoted to the development and improvement of extreme DFA methods. Unfortunately, the majority of this work has ignored an important fact: once a water supply reservoir is erected on a stream the discharge cannot be considered random any more. Even Bulletin 17B, the US flood frequency analysis guidelines, states "the procedures [presented in the bulletin] do not cover watersheds where floods are appreciably altered by reservoir regulation. . ." (IACWD, 1982).

Because a sample of regulated streamflows is not homogeneous, the present statistical methods are not applicable. On the other hand, deterministic methods, based on rainfall-runoff modeling, suffer from a "loss-of-variance" problem. This paper presents the use of a deterministic-stochastic methodology for the solution of the regulated DFA problem. A Monte Carlo simulation has been developed for a set of hypothetical reservoirs to illustrate this solution methodology. The model uses a reservoir routing module for the deterministic part of the solution framework. The stochastic component accounts for the reservoir initial and boundary conditions, as well as for randomness in the reservoir inflow. The results of the simulations indicate that regulated discharge distributions are extremely complex and exhibit characteristics that cannot be replicated by traditional parametric statistical distributions.

The solution methodology presented here can be used not only to describe extreme regulated discharge distributions, but also to predict the changes in regulated distribution curves that would be introduced by a change in a reservoir operating policy.

1 INTRODUCTION

A large percentage of water resources used by water supply authorities around the world comprise of complex water resource systems that combine abstraction from impounding reservoirs with pumped abstraction from groundwater, natural lakes, and rivers. The impounding reservoir, a large lake capable of gathering and storing huge volumes of water, is generally regarded as the classical form of source works. Most of them are man-made reservoirs operated according to the policies derived from the analysis of critical periods in the historic runoff record or policies derived using optimization and/or simulation which aims to minimize operating costs while maintaining acceptable levels of supply reliability. However, deterministic models that use the historical critical period or mean seasonal inflow to arrive at optimal decisions, do not encounter the hydrologic (probabilistic) uncertainty associated with flows.

Discharge frequency analysis (DFA), the determination of river discharges at different recurrence intervals, is commonly used for analyzing stochastic flow problems in hydrology. The standard procedure to determine probabilities of flows consists of fitting the observed river discharges recorded at specific probability distributions. However, this procedure only works for basins:

- that have 'long enough' streamflow records to warrant statistical analysis;
- where flows are not appreciably altered by reservoir regulation, channel improvements or land use change.

The results of DFA are used in hydraulic models to predict streamflow depths and velocities. In this way, DFA produces a basic input for design and operation of water supply reservoirs. In the twentieth century, a tremendous amount of effort has been devoted to the development and improvement of extreme DFA methods. Unfortunately, the majority of this work has ignored an important fact: once a water supply reservoir is erected on a stream the discharge cannot be considered random any more. Even Bulletin 17B, the US flood frequency analysis guidelines, states " ... the procedures [presented in the bulletin] do not cover watersheds where floods are appreciably altered by reservoir regulation..." (IACWD, 1982).

This paper introduces recent advances in DFA based on the work by Durrans (1995) and Tomic (1998). The new methodology presented here can be used not only to describe extreme regulated discharge distributions, but also to predict the changes in regulated distribution curves that would be introduced by a change in a reservoir operating policy.

2 BACKGROUND

The problem of DFA is solved using one of the three basic approaches: stochastic approaches, deterministic approaches, or combined approaches. In the following sections we introduce each approach and give its basic strengths and weaknesses.

2.1 Stochastic Approach

Early works in DFA were all stochastic in nature involving estimation of a probability distribution on the basis of the recorded discharge data. Depending on the source of the data, stochastic methods may be categorized into at-site estimation methods and regionalization methods. The stochastic methods may be further classified depending on whether they are parametric or nonparametric in nature.

The primary advantage of the stochastic approach is that it works directly with the observed data and, as such, leads to results that are consistent with the data. In this fashion a stochastic approach avoids the "loss-of-variance" problem. The main disadvantage of the stochastic approach is that it requires the input data to be random. Unfortunately, this is generally not the case for the DFA at large reservoirs. Stochastic approaches are also criticized for approaching the DFA problem in purely black-box fashion, with little or no consideration for the fundamental physics involved.

2.2 Deterministic Approach

Deterministic approaches, such as the design storm methods used commonly by design engineers, usually consist of rainfall-runoff modeling with unit hydrograph methods. The sudden progress in these approaches has been triggered primarily by the development of computers. Unfortunately, due to the complexity of the runoff generation process a runoff response behavior is beyond the abilities of models to reproduce. Thomas (1982, 1987) and Muzik (1994) demonstrate that distributions generated from output data of rainfall-runoff models display a variance that is smaller than the variance exhibited by historic data. Thomas (1982) referred to this as a 'loss-of-variance' problem. Kirby (1975) suggests that "... although this loss of variance has been attributed to various kinds of errors in and misuses of the data, theoretical analysis reveals that a substantial loss of variance is an unavoidable consequence of the [modeling] process."

The main advantage of the deterministic approach to DFA is that it tries to understand the processes that govern the transformation of rainfall into runoff. Because of this, the deterministic approach is able to account for the human intervention on this transformation process and it is extremely suitable for the DFA on the reservoirs. However, the applications of deterministic approaches are severely limited with the "loss-of-variance" problem.

2.3 Combined (Deterministic-Stochastic) Approach

Because of the strengths and weaknesses of each of the stochastic and deterministic estimation approaches, it is natural to seek some means whereby they might be integrated. In an ideal world, the best features of each of two basic approaches would manifest themselves in the combined approach. Unfortunately, this is generally not the case stochastically analyzed data that originates from continuous rainfall-runoff models calibrated to observed data still suffer from the "loss-of-variance" problem.

Despite the problems with the combined approach to DFA, there are many practical situations in which it is the only rational solution. This is particularly true in situations such as DFA on regulated streams, where recorded flow data are subject to both random and deterministic influences. The loss-of-variance problem, also encountered with the combined approach, suggests that either (1) the deterministic and statistical components are not being linked in the most effective way; or (2) that we, as scientists, are simply trying to take too large a step, and that we should be focusing on simpler problems first.

3 NEW APPROACH TO DFA

If one is to accept our argument that the cause of the 'loss-of-variance' problem in the combined approach is the profound complexity of the rainfall-runoff transformation process, then it is only natural to seek to employ runoff transformation processes that are better understood. This approach has been used in the proposed deterministic-stochastic model.

The proposed approach substitutes the rainfall-runoff process with better understood flood routing. The need to stochastically describe the entire range of possible streamflows (as opposed to just the flood peaks) in an extended period simulation leads us to consider an event-based approach for the regulated DFA. In an event-based model, there is a need to describe initial conditions in the reservoir (the water surface level in the reservoir when a flood occurs) and the boundary conditions related to the reservoir (reservoir outlet gates' positions and their discharges). This is handled through the stochastic component of the approach. The deterministic component of the approach is embodied in the simple continuity and energy principles that apply to the reservoir routing.

The derivation of the regulated DF distribution immediately downstream of a dam may be accomplished through the total probability theorem (Durrans, 1995a and 1995b). The solution framework is based on the following assumptions:

(1) regulated annual floods downstream of a dam are caused by the annual floods upstream of the dam,
(2) regulated floods are independent events,
(3) reservoir operating policy is stable.

The first of these three assumptions is the critical one. It allows us to treat the regulated annual discharge series downstream of the dam as a random discharge series. An annual flood downstream of a dam does not necessarily need to be driven by the corresponding annual flood upstream of the dam.

In the following discussion, $\boldsymbol{X} = [x_1\ x_2\ ...]$ denotes a random vector of unregulated flood characteristics at the site upstream from a dam, and $\boldsymbol{Y} = [y_1\ y_2\ ...]$ denotes a random vector of regulated flood characteristics at the site downstream from the dam. $\boldsymbol{F_X}(x) = Pr(X_1 < x_1, X_2 < x_2, ...)$ represents the joint probability distribution function of the unregulated flood characteristics. The distribution $\boldsymbol{F_Y}(y)$ is defined analogously. In practice, $\boldsymbol{F_Y}(y)$ is conditioned on the operating policy of the reservoir, but as long as the operating policy is stable, that distribution may be viewed as an unconditional one. The reservoir initial and boundary conditions are represented by a random vector $\Lambda = [\lambda_1, \lambda_2 ...]$, with corresponding probability density function $f_\Lambda(\lambda)$. Individual elements λ_i of this vector represent the initial reservoir stage, outlet gate position, and other possible reservoir initial and boundary conditions.

Input data for this solution framework are the distributions of random vectors $\boldsymbol{X}$ and Λ, and their correlation. That is, we need estimators of $\boldsymbol{F_X}(x)$ and $f_\Lambda(\lambda)$, as well as the correlation matrix between $\boldsymbol{X}$ and Λ. In the case of significant correlation between $\boldsymbol{X}$ and Λ, we need to develop an estimator for the joint distribution $\boldsymbol{X}$ and Λ, denoted as $\boldsymbol{F_{X,\Lambda}}(x, \lambda)$.

The deterministic component of the solution framework involves routing of the flood hydrographs through the regulating reservoir, to develop a conditional distribution function $F_{Y|\Lambda}(y|\lambda)$ of regulated flood characteristics, conditioned on a particular combination of reservoir initial conditions. This deterministic component of the procedure is summarized in general form as

$$F_{Y|\Lambda}(y|\lambda) = G_1[F_X(x), \lambda] \tag{1a}$$

in the case of independent X and Λ, or

$$F_{Y|\Lambda}(y|\lambda) = G_2[F_{X\Lambda}(x, \lambda)] \tag{1b}$$

for the case where $\boldsymbol{X}$ and Λ are correlated. In Equations 1a and 1b, G_1 and G_2, respectively, are functions that map the unregulated DF distribution into a conditional regulated one. This mapping from unregulated to regulated DF distribution is performed using Monte Carlo simulations.

Using the theorem of total probability, we may derive the unconditional distribution $F_Y(y)$ of regulated flood characteristics as

$$F_Y = \int F_{Y|\Lambda}(y|\lambda) f_\Lambda(\lambda) d\lambda \tag{2}$$

where the integration is performed over the complete space of the feasible reservoir conditions. Detailed definition of the applied model can be found in Tomic 1998.

4 SIMULATION MODEL

The computer model used here is based on a hypothetical reservoir having characteristics typical of these for the southeastern region of the United States. The data used for the model is based on the data availability survey reported in Durrans, Tomic, and Nix 1997 and Tomic and Burian, 1997. This model is just an illustration of the proposed framework and thus there is no need to model a complex reservoir. It is assumed that the hypothetical reservoir can be approximated by a parallelepiped with vertical sides. However, the approach presented here is capable of modeling any reservoir. The reservoir outlet structure consists of multiple controllable outlet gates, whose positions can be changed during a simulation. This situation corresponds to an attended reservoir with a different operating strategy used during extreme events than the operating policy used in the normal operation. It is also assumed that the probability distributions of initial and boundary reservoir conditions are independent of the distribution of unregulated flood characteristics. The assumptions used in the presented model are reasonable because the model intention is not to solve an actual real-word problem, but instead to illustrate the new solution framework.

Durrans (1995a) states that the simulation procedure used in this framework depends on whether the random vectors $\boldsymbol{X}$ and Λ are independent or correlated. In this research these vectors are taken to be independent, that is the reservoir operating policy does not depend on the unregulated flood. Naturally, a distinction is made between normal operation of a dam and the operation under extreme events. For the case of independent random vectors $\boldsymbol{X}$ and Λ, a step-by-step simulation procedure is given as follows (Durrans, 1995a):

(1) Develop estimators of the distribution $F_X(x)$ of the unregulated flood characteristics and $f_A(A)$ of reservoir initial and boundary conditions.
(2) Randomly sample values flood peak (Q_p) and volume (V) from the distribution of unregulated flood characteristics. Using these values construct a direct runoff hydrograph. Add base flow to the direct runoff hydrograph.
(3) Randomly sample initial and boundary conditions from $f_A(A)$.
(4) Route the inflow hydrograph through the reservoir using Equation 2 and the reservoir routing module.
(5) Repeat steps (2) through (4) N times to obtain N peak outflow values.
(6) Rank the outflow values and assign plotting positions to them. Use these values to empirically determine the regulated DF distribution, $F_Y(Y)$.
(7) Generate an empirical probability distribution of the regulated DF distribution based on the results from step (5).

Sensitivity analysis study given in Tomic (1998) suggests that the value of $N = 10{,}000$ is sufficiently large to archive the convergence of the model results. Primary model parameters are given in Table 1. The datum of the model (elevation equal to zero) is assumed to be the bottom of the "flood pool". The top of the "flood pool" is assumed to be at the bottom of the emergency spillway. The reservoir is full when the water elevation is at the top of the "flood pool". The initial water elevation is assumed to be inside of the "flood pool". The reservoir parameters are estimated and are based on the information on average unregulated discharge. For example, the reservoir orifice is designed to convey the base flow when the reservoir is full.

Table 1 Model Parameters

Symbol	**Description**
h_{max}	elevation of the top of "flood pool"
$\Delta t / T_p$	ratio of simulation time step to time to peak
N	number of model iterations
$P(h_0 = 0)$	probability of reservoir being empty at the beginning
$\overline{Q}_p$	average unregulated discharge (Q_p)
C_v	coefficient of variation of Q_p
T_{ratio}	ratio of base time step to time to peak of unregulated hydrograph
Q_{ratio}	ratio of full-reservoir discharge to average unregulated peak
V_{ratio}	ratio of average unregulated volume to reservoir volume
$\overline{h}_0$	average initial water elevation
σ_{h_0}	Variance of the initial water elevation

5 MODEL RESULTS

All the results in the following figures are plotted on Extreme Value Type 1 (Gumbel) paper. The distribution of unregulated hydrograph peaks is assumed to be EV1 and consequently, it plots as a straight line on the EV1 paper. The results are reported as relative peak flow Q_p, the ratio of peak outflow to average unregulated peak inflow,

$$Q_p = \frac{Q_{pOUT}}{\overline{Q}_{pIN}}. \tag{3}$$

In this study, a Monte Carlo (MC) simulation has been used for the integration of Equation (2) to derive a regulated DF distribution. The MC simulations have been approached in a step-by-step fashion, starting from a simple example of a detention pond and building up to a complex example of a large reservoir. Through the detention pond example, the importance of the initial reservoir condition, the elevation of the water in the reservoir at the beginning of a simulation, for the estimation of regulated DF distributions has been determined. This paper presents some of the results for the final simulation of a large reservoir. This final simulation represents a reservoir with a single outlet that changes its position during the simulation.

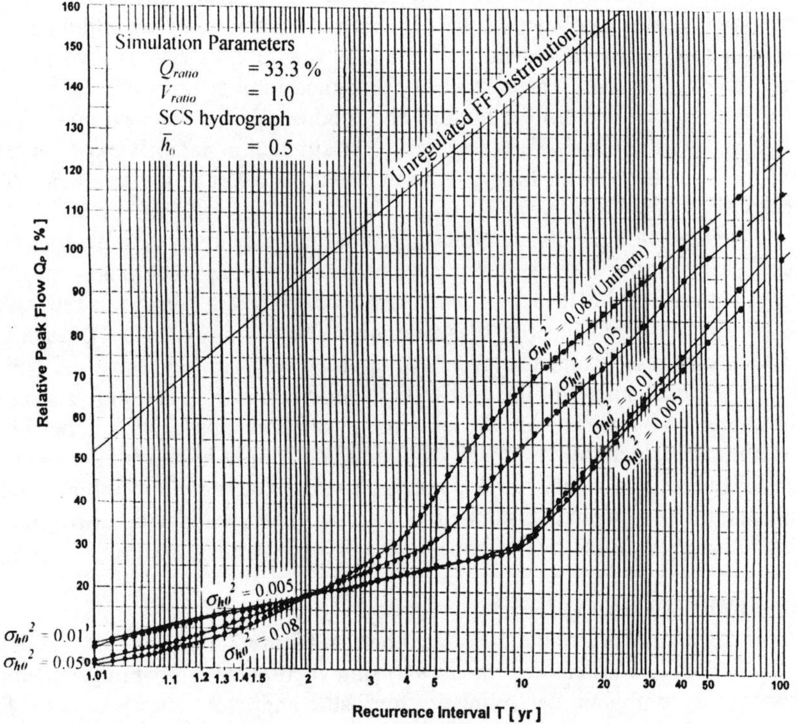

Figure 1. Regulated DF Distributions for Different ROP

Figure 1 gives the DF curves of the relative reservoir peak outflow for different initial conditions. The reservoir "flood pool" volume, given through V_{ratio}, is equal to the volume of average unregulated hydrograph (V_{ratio} = 1). Design outlet discharge (the discharge when the gate is completely open and the reservoir is full) is assumed to be 1/3 of the average unregulated peak discharge (Q_{ratio}= 33.3%). Average initial elevation of the reservoir is in the middle of the flood pool (h_o = 0.5). A detailed description of the simulation model and its results are given in Tomic, 1998. The average initial elevation of the reservoir was the same and only its variance was changed from one simulation to another. Although the simulation case described here was relatively simple, the regulated DF distributions, generated using the MC simulation, demonstrate characteristics that clearly cannot be replicated using theoretical probability distributions, as can be seen in Figure 1. Moreover, the differences between regulated DF curves become more significant as the return period of the examined event increases. This indicates that an analysis of a small set of existing data, which will describe adequately normal operating conditions (T < 10 years), might indorse large errors if we look at extreme events. The simulation approach presented here performs well in both normal operating conditions as well as in describing the behavior of the reservoir under the extreme conditions. In this way it integrates the best features of deterministic and stochastic approaches.

Figure 2 presents the resulting regulated DF distributions for 3 different reservoir operating policies (ROP). Each ROP assumes that the outlet gate will be fully opened when the elevation in the reservoir reaches critical level (h_{CR}). The critical level, h_{CR}, is represented as the dimensionless ratio of elevation in the reservoir measured from the bottom of "flood pool" and the elevation of the dam crest (h_{CR}= 1.0 indicates full reservoir). The simulation parameters of the reservoir are identical to those in Figure 1, except that the average initial elevation of the reservoir is 30 % full (h_o = 0.3).

The regulated DF distributions, given in Figure 2, are even more complex than the distributions in Figure 1. The distributions in Figure 2 contain a distinct break point between two distribution parts, and a considerable "jump" in the lower part of the distribution. Clearly, these complex shapes of regulated DF distributions prevent common parametric statistical approaches from being applied due to the unconventional shapes of the probability distributions. Figure 2 emphasizes one of the major advantages of the proposed solution approach. The proposed deterministic stochastic approach can be used not only in descriptive purposes, to model the reservoir behavior for a known set of boundary and initial conditions, but also in prescriptive fashion, to determine changes in DF downstream of a regulating dam caused by a change in a reservoir operating policy.

6 CONCLUSIONS

From the simulation results, illustrated here in Figure 2, it is obvious that regulated DF distributions cannot be modeled using theoretical probability distributions. Instead, a combined deterministic-stochastic approach must be used for the solution of the regulated DFA problem. However, to be able to successfully employ the approach described in this study, one needs to determine the variation in initial and boundary conditions of the reservoir. The variation of the initial condition is easily determined from the data on reservoir elevation, which is readily available at the majority of dams and reservoirs.

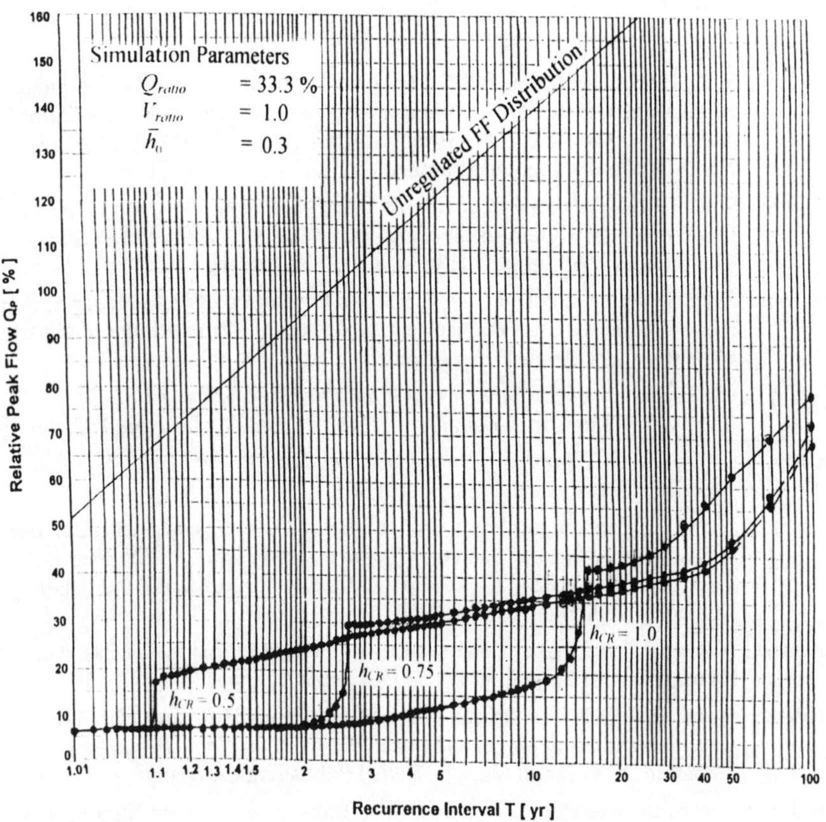

Figure 2. Regulated DF Distributions for Different ROP

The boundary condition variation can be estimated from the information on the positions of gates on a dam. However, additional data needs to be collected on the reservoirs and dams for this purpose. While most agencies collect data on individual outlet gate settings and discharges, this data is usually maintained only in handwritten, hard copy form, and thus is not readily available for use in computer modeling. Only in a few cases is this data maintained in a digital form. It appears, for the most part, that the data is stored in cardboard boxes, where accessing it would be difficult at best. The significance of the data lies mainly in the physics of the reservoir routing problem, and also in the fact that a desired outflow discharge from major dams may be obtained in any number of ways (there is a unique relationship between gate settings and the resulting discharge, but the converse is not necessarily true).

ACKNOWLEDGMENT
The research reported in this paper has been supported by the USGS, throu Alabama Water Resources Research Institute, whose support is gra acknowledged.

REFERENCES

Durrans, S. R., 1995a. Total Probability Methods for Problems in Flood Frequency Estimation. *Selected Proceedings of International Conference on Statistical and Bayesian Methods in Hydrologic Sciences*, E. Parent (ed.), Ch. 18, UNESCO Publishing, Paris, France.

Durrans, S. R., 1995b. Assessment of Human Effects Using an Integrated Approach to Flood Frequency Analysis. *Proceedings of 75th AMS Annual Meeting, Conference on Hydrology*, J13.5, Dallas, TX, January 15-20, pp. 13-18.

Durrans, S. R., S. Tomic, and S. J. Nix, 1997. An Evaluation of Data Needs to Support Flood Frequency Estimation at regulated Sites. *In:* Proceedings of XXVII IAHR Congers, August, San Francisco, CA, p. 494-99.

IACWD, 1982. *Guidelines for Determining Flood Frequency, Bulletin 17B.* Interagency-Advisory Committee on Water Data, Washington, DC.

Kirby, W., 1975. *Model Smoothing Diminishes Simulated Flood Peak Variance.* American Geophysical Union, 56(6):361.

Muzik, I., 1994. Understanding Flood Probabilities. *Stochastic and Statistical Methods in Hydrology and Environmental Engineering, Volume 1: 'Extreme Values: Floods and Droughts,'* K. W. Hipel (ed.), Kluwer Academic Publishers, Dordecht, Netherlands, pp. 199-207

Thomas, W. O., Jr., 1982. An evaluation of Flood Frequency Estimates Based on Rainfall/Runoff Modeling. *Water Resources Bulletin*, 18(2):221-30.

Thomas, W. O., Jr., 1987. Comparison of Flood-Frequency Estimates Based on Observed and Model-Generated Peak Flows. *Hydrologic Frequency Modeling*, V. P. Singh (ed.), D. Reidel Publishing Company, pp. 149-61.

Tomic, S. and S. Burian, 1997. Regulated Flood Frequency Problem: Data Availability Analysis, *Proceedings of 17th Annual AGU Hydrology Days Conference*, Fort Collins, CO, April 14-18, pp. 339-48.

Tomic, S., (1998), *Flood Frequency Analysis on Regulated Rivers*, Ph.D. Dissertation, University of Alabama, Tuscaloosa, AL, p.195.

Performance Indicators for Water Storage Tanks

J.J. van der Walt *and* J. Haarhoff

ABSTRACT

The water quality changes in a distribution system (from treatment plant to consumer) are increasingly under the spotlight. As a result, pipes and storage tanks are now not only regarded as necessary elements of a distribution system, but also as process reactors which have a significant effect on water quality changes. This paper deals with distribution storage tanks as process reactors, and methods whereby they could be benchmarked. The flow patterns and *Residence Time Distribution* of a number of large distribution storage tanks were determined by computational fluid dynamics (CFD). Performance parameters, such as T_{10}, T_{90}, the Morrill Index and the Dispersion number, were calculated from these results. Based on this analysis, new performance indicators for storage tanks are recommended.

1 INTRODUCTION

Current performance guidelines for storage tanks are only directed at the volume of available storage, usually expressed in hours at a specific flow rate. These guidelines for total volume vary significantly in their specification of the three main storage components, i.e. balancing storage, emergency storage and fire storage. In some cases additional storage elements are allowed for, e.g. bottom reserve, freeboard or a control band for automated pump control. These performance guidelines, however, only dictate the quantity of water that should be available in the storage tank at any time with little or no attention to performance guidelines pertaining to the water quality inside the storage tank. The implicit assumption had therefore been that tanks are plug flow reactors (or close approximations thereof) where the residence time of the water is approximately the theoretical retention time T.

However, water entering a tank does not flow in a plug towards the outlet. It follows the easiest route from the inlet to the outlet. This means that some water will leave the tank long before time T and other long after time T, unless it is forced to do otherwise. This is because some water might be trapped in a recirculation zone (large scale eddies), some might end up in a stagnant zone (where the water does not move), while some water will be mixed or will follow a plug towards the outlet. The implication of this imperfect plug flow phenomenon is the localised depletion of free chlorine which leads directly to associated water quality problems.

The water quality leaving the tank is therefore determined by its internal flow characteristics. This in turn is largely dependent on the way in which the water enters and leaves the tank as well as deliberate internal obstructions: geometric features which are difficult to change once the storage tank has been constructed. It is therefore important to ensure proper internal flow (hydraulic efficiency) during the design stage. As there are currently very few or no quantitative guidelines on how to ensure proper internal tank flow, the need is twofold:

- Firstly, for design guidelines to guide designers towards hydraulic efficient tanks
- Secondly, for performance indicators benchmarking the hydraulic efficiency.

This paper will demonstrate how proper geometric configurations can improve internal flow (hydraulic efficiency) and use performance indicators to benchmark the various configurations.

1.1 Water Quality as Performance Indicator

There are two primary concerns when water is stored in a tank. The water can stay in the storage tank too briefly or it can stay in the storage tank for too long. The former is often a concern in chlorine contact tanks where the water needs to be in contact with the disinfectant for a minimum time to obtain a desired C.t product. The latter is a concern in distribution storage tanks where the desired chlorine residual drops too low. Both these problems can be alleviated if the water flows as a plug from the inlet to the outlet. The objective should therefore be to manipulate the flow pattern in the tank to approximate plug flow as closely as possible. (An alternative school of thought is that all the water in the tank should be continuously and perfectly mixed to ensure that all the water in the tank is continuously exposed to the desired residual chlorine concentration [1]. Where tanks are small, it may be feasible to achieve mixing by the incoming water jet, but for large tanks, from say 10 Ml upward, this is not considered to be a practical option.)

It is not a trivial exercise to determine the concentration of a disinfectant (similar to a non-conservative tracer) at different points in a tank. Traditionally the behaviour of a process tank was inferred from the evaluation of *the Residence Time Distribution* (RTD) also known as a *Flow Through Curve* (FTC).

1.2 Residence Time Distribution

The RTD curve of a tank can be experimentally determined by injecting a tracer pulse into the inlet and measuring its concentration at the outlet. The cumulative RTD curve, which will be used throughout this paper, is normally obtained by integration of the outlet concentration over time. As the RTD reveals many interesting aspects of the hydraulic efficiency, a number of different indicators have been previously used to interpret the RTD curve as shown in Table 1.

There are some limitations in using a RTD as the only basis for evaluating tank efficiency. Firstly it is a matter of practicality. Although a RTD curve can be generated experimentally, the size of a storage tank can, in some cases, make

Table 1 - Single figure Performance Indicators for the evaluation of RTD curves.

PERFORMANCE INDICATOR	DESCRIPTION	IDEAL VALUE FOR PLUG FLOW
T	T denotes the theoretical retention time. This is the time the water takes from the inlet to the outlet, assuming perfect plug flow. The theoretical retention time is calculated as the total storage tank capacity divided by the steady state inlet flow rate. T is also used as a normalising parameter to evaluate RTD curves of different time scales.	0
T_{10}	T_{10} denotes the time at which 10% of the injected tracer has reached the outlet. A low T_{10} value is normally associated with a short-circuit between the inlet and the outlet.	T
T_{90}	T_{90} denotes the time at which 90% of the injected tracer has reached the outlet. Large stagnant zones are normally associated with high T_{90} values.	T
MI	The Morrill Index (MI), calculated as T_{90} divided by T_{10}, reflects the relative spread of the RTD curve between the T_{10} and T_{90} points. It can also be interpreted as indicating the level of mixing. The higher the value the higher the level of mixing.	1
D	The dispersion number (D) is a more complex calculation that determines the variance of the RTD curve in relation to its centroid [2]. The dispersion number is based on the shape of the *total* RTD curve while the Morrill Index is only based on two points on the RTD curve viz. T_{10} and T_{90}	0

experimental techniques prohibitively time consuming and expensive. Pilot scale tests, on the other hand, are plagued with scale effects [3]. Secondly, although a RTD provides a compact 'picture' of the temporal concentration distribution at a specific point, it does not provide the spatial concentration distribution. The spatial distribution, necessary to identify the location of stagnant zones and guide possible improvements, requires additional techniques.

1.3 Mathematical Modelling

Computational Fluid Dynamics (CFD) techniques can provide accurate information of the spatial distribution of tracer as well as the temporal distribution of tracer in the storage tank. CFD techniques are well established to simulate complex flow patterns, have been subject to extensive calibration efforts during the past decade and can accurately model steady and unsteady dispersed flow.

2 SIMULATION OF WATER MOVEMENT IN TANKS

2.1 Evaluation Procedure

The procedure that was used to investigate storage tank performance involved the following steps each of which will be explained in detail:

- Steady state simulation of the flow patterns in the storage tank.
- Unsteady propagation of a tracer and the generation of RTD.
- Calculation and analysis of various Performance Indicators.

2.2 Steady state simulation

The Reynolds averaged Navier-Stokes equations are generally considered to represent turbulent Newtonian fluid flow. The FLO++ software [4] employs the Navier-Stokes and other transport equations that can be linked by adding special user code. The FLO++ code can simulate steady and unsteady three-dimensional turbulent flow using the SIMPLE and PISO algorithms. It is based on a collocated finite volume which features structured embedded mesh refinements to model local detail. The turbulent flow was modelled using k-ε turbulence model. All calculations were performed on a 450 MHz Pentium II processor with 128 MB RAM and took about 24 hours to converge for a 150 000 cell computational grid.

The velocity profiles serve as a quick, initial check on the flow uniformity in the tank. If large stagnant zones are identified at this stage, it serves as a warning that the level of plug flow might not be ideal. The next step is to determine the effect of the velocity profiles on the propagation of tracer through the unsteady simulation.

2.3 Unsteady simulation and generation of RTD

The Navier-Stokes equations (momentum equations) can be linked to a scalar transport equation to simulate the convection-diffusion of a tracer into the storage tank. A non-reactive tracer with the same density as water was injected. This simplification means that the tracer will not influence the hydrodynamics of the storage tank and can therefore be uncoupled from the momentum equations and solved after the momentum equations are solved.

The tracer concentration at the inlet was set at 1 mg/l at the start of the unsteady simulation, while the solution for the steady state velocities was used as input. At the same time the concentration at the outlet cells was monitored. After the completion of the unsteady simulation the spatial and temporal distribution of tracer are available for the whole computational grid. An RTD curve can be generated from this information.

The scalar transport equation can further be adapted to include additional source or sink terms that can simulate for a reactive tracer, for instance the decay of chlorine. As the decay rate is dependent on the water quality, a typical first-order decay rate can be used to determine the qualitative differences between a non-reactive tracer and chlorine.

2.4 Analysis of spatial distribution and calculation of Performance Indicators

Evaluation of the RTD curve indicates the level of plug flow in the storage tank. If the RTD indicates that short-circuiting or stagnant zones are present, the spatial concentration distribution at various times can be evaluated to determine why there is short-circuiting and where the stagnant zones are. This can then assist in selecting alternatives to improve the level of plug flow.

The RTD is then reinterpreted through single figure performance indicators. Although the performance indicators attempt to compress the rich information of the RTD curve into a single number, it is much simpler to use.

3 CASE STUDIES

3.1 Existing configurations

The case studies have been selected to demonstrate the methodology used to systematically improve existing configurations and to benchmark the improvements. Five case studies will be used to demonstrate the methodology.

The tanks were all existing tanks except for Case 4, which was under consideration at the stage of the investigation. Table 2 summarises the geometric properties of each tank.

3.2 Alternative arrangements

A number of typical modifications were done in an attempt to improve the existing configurations. Some of the modifications were repeated for different size tanks to establish the scale effects (if any). Table 3 shows the configurations that will be investigated during the evaluation.

The cyclone arrangement comprised a swirl generator positioned on the side of the storage tank, pointing along the storage tank wall. The outlet was positioned at the centre.

The *manifold arrangement* comprised an inlet manifold that spanned part of the width or the circumference of the tank to uniformly direct the water towards a central outlet or distributed outlets.

The *baffle arrangement* comprised full depth internal walls that forced the water to follow a specific route through the tank.

Table 2 - Summary of existing tank geometric properties

	SIZE *Ml*	SHAPE	INLET AND OUTLET DETAIL
Case 1	25	Round	Inlet and outlet directly opposing.
Case 2	35		Side inlet with side outlet
Case 3	45		Inlet and outlet directly next to each other
Case 4	80	Rectangular	-
Case 5	120		Two inlets at one end of tank and one asymmetric outlet

Table 3 - Summary of case studies and alternative arrangements investigated for each case

Configuration	Case 1	Case 2	Case 3	Case 4	Case 5
Existing	*X*	*X*	*X*	-	*X*
Cyclone	*X*	*X*	*X*		
Manifold	*X*	X		*X*	*X*
Baffle	*X*		*X*	*X*	

3.3 Shortcomings of existing configuration

Analysis of the flow patterns of the existing configurations revealed a number of typical shortcomings viz.

- Large recirculation zones
- Stagnant zones
- Short-circuiting

Case 3 will be used to demonstrate how these imperfections can be identified by evaluating the flow patterns, the spatial and temporal tracer distribution and the RTD curve [5].

Case 3 is a typical example of a direct short-circuit between the inlet and the outlet (Figure 1). This is, in turn, the cause of a large recirculation zone and a stagnant zone. The spatial distribution of the tracer at various times (Figure 2) confirms the short-circuit, the recirculation zone along the perimeter of the tank and the stagnant zone in the centre. Even after four times the theoretical retention time, the centre of the tank shows a concentration of less than 0.6 mg/l. The short-circuiting and the stagnant areas are also indicated on the RTD curve (Figure 3a) by the low T_{10} value and high T_{90} value. The RTD does not, however, show where the stagnant zones are.

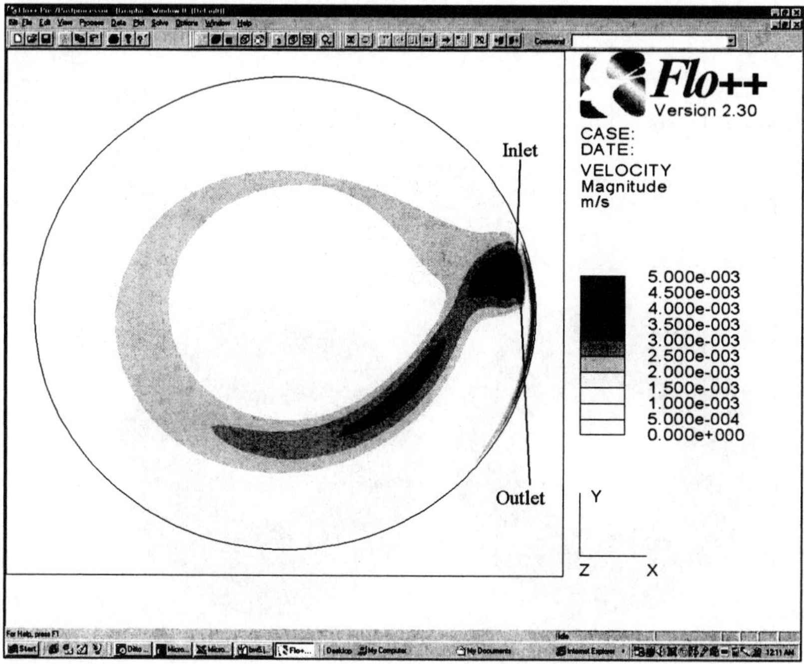

Fig. 1 Flow patterns, in plan at mid depth, of Case 3 indicating the recirculating 'tongue' (dark areas) and the stagnant zones (light areas). The black lines indicate the position of the inlet and outlet. The inlet pipe points vertically upwards. The outlet is a square box in the tank floor.

The other cases were analysed in a similar way and showed similar shortcomings. As a matter of brevity these cases will not be discussed in detail, but the performance indicators will be discussed in the next section.

3.4 Improvements through internal modifications

The analysis of the results for the existing configuration of Case 3 showed room for significant improvements that can be made through simple changes to the tank geometry. The *cyclone* option (Figure 3b) reduced short-circuiting by a large degree; T_{10} increased from 0.07 to 0.24. The dispersion number reduced from 0.98 to 0.3. Although *baffle* type arrangements are typical of chlorine contact tanks [6] it can also be used in distribution storage tanks with positive effect. The *baffle* option (Figure 3c) resulted in near plug flow conditions as reflected by the low dispersion number of 0.015. The performance indicators for all the cases investigated are summarised in Table 4.

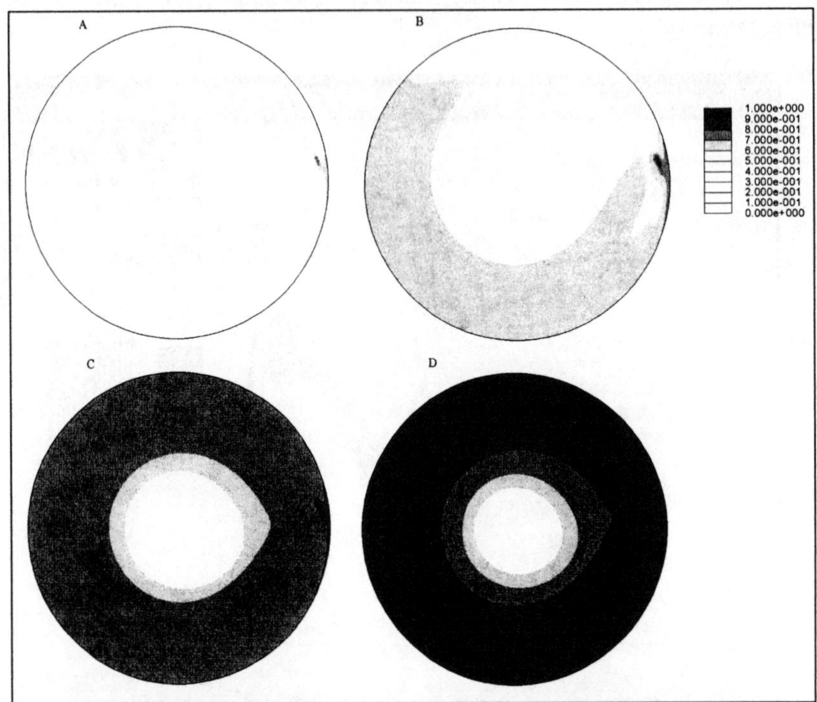

Fig. 2 Spatial distribution of tracer, in plan at mid depth, for Case 3 at normalised times a) T b) 2T c) 3T and d) 4T. The light areas indicate concentrations less than 0.6 mg/l.

4 REFLECTION

4.1 General Conclusions

From the case studies a number of general conclusions can be drawn:

- Simple changes can improve hydraulic efficiency significantly.
- The inlet and outlet hydraulics are the primary determinants of the flow pattern in a tank with unobstructed internal flow. When internal baffles are used, however, the inlet and outlet configuration becomes less important.
- Most of the hydraulic inefficiencies were caused by poorly positioned inlet and outlet pipes.

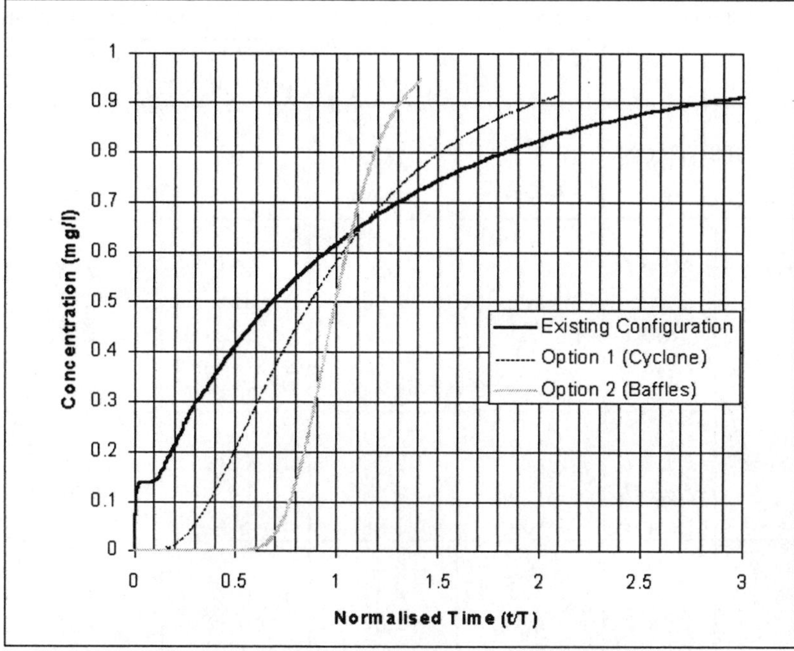

Fig. 3 RTD curves for Case 3 a) existing b) cyclone c) baffles

4.2 Guidelines for improved geometric configurations

A number of guidelines specific to the geometry can also be made in view of the tanks that were evaluated:

- To avoid short-circuiting, large recirculation zones and stagnant areas, inlets and outlets should not be in close proximity of each other and should not be directly opposite each other.
- Acceptable solutions can be achieved by mixing and plug flow. Mixing is easier to achieve in round storage tanks by a swirl action. Plug flow is easier to achieve in rectangular storage tanks by introducing inlet and outlet manifolds.
- The ultimate solution to almost all tanks requires the introduction of internal baffles [5,6].

4.3 Performance indicators

The following conclusions are made with regards to different performance indicators:

- A normalised RTD curve provides the best basis for evaluating relative improvements.
- When residual chlorine concentration in the tank is evaluated against an absolute standard, a measure of the absolute time is required and a normalised RTD curve will not suffice. An absolute RTD curve is then required.

Table 4 - Summary of Performance Indicators for existing and alternative configurations

CONFIGURATION		PERFORMANCE INDICATOR			
		T_{10} Normalised	T_{90} Normalised	*MI* -	*D* -
Existing	*Case 1*	0.07	2.37	30.9	0.98
	Case 2	0.06	2.92	61.7	1.7
	Case 3	0.01	2.8	272	2.8
	Case 4	-	-	-	-
	Case 5	0.17	3.56	20.9	0.37
Cyclone	*Case 1*	0.24	2.07	8.56	0.3
	Case 2	0.35	2.54	7.15	0.26
	Case 3	0.37	1.98	5.3	0.15
Manifold	*Case 1*	0.23	2.31	10.02	0.85
	Case 2	0.27	2.82	10.52	0.65
	Case 4	0.59	1.82	3.1	0.082
	Case 5	0.44	1.46	3.3	0.067
Baffles	*Case 1*	0.68	1.28	1.86	0.026
	Case 3	0.77	1.31	1.7	0.015
	Case 4	0.61	1.52	2.5	0.068

- The T_{10} and T_{90} values are very useful and simple parameters, but they need to be interpreted together. It was found that a low T_{10} is normally associated with a high T_{90} value. This is because short-circuiting normally leads to large stagnant areas. The converse is not always true: a high T_{90} value can be found in cases where the T_{10} value is acceptable. This can be due to an improperly positioned inlet/outlet configuration. The T_{10} and T_{90} values only look at one single point of the RTD, disregarding other parts that may bring additional insight about the internal flow pattern.
- The T_{90} value is important in quantifying the maximum time the water will stay in the storage tank.
- The Morrill Index uses two points of the RTD curve and provides in most cases a fair assessment of the degree of plug flow in the tank. The Morrill Index does, however, sometimes give misleading results. Intuitively one would expect the Morrill Index to give the same value for cases with the same amount of spread between T_{90} and T_{10}, but it does not. It is also sensitive to small T_{10} values

because of the presence of T_{10} in the denominator. The Morrill Index should therefore only be used as a quick, but not final check of the level of plug flow.

- The dispersion number provides the most balanced single figure indicator of internal tank characteristics.

5 RECOMMENDATIONS

The following recommendations are made towards constructing reservoirs with water quality in mind:

- Distribution storage tanks should not only be designed for the required storage volume, but their internal flow pattern should also be optimised during the design stage.
- Storage tanks are usually under-utilised during the first part of their design life, which means that the average residence time T is higher than the design guideline, with possible severe impact on water quality. This problem can be alleviated by initially reducing the operating level (thus the volume) in the tank.
- CFD provides a powerful, readily available tool to generate a theoretical RTD, as well as the spatial distribution of flow velocities within the tank.
- The hydraulic efficiency of the tank can be effectively specified by a few simple performance indicators such as T_{10}, T_{90}, the Morrill index and the dispersion number.
- Based on the case studies investigated, a three-tier performance rating is proposed in accordance with table 5 using the recommended performance indicators. In accordance with this performance rating **acceptable** hydraulic efficiency can be achieved without the introduction of internal baffles and **excellent** hydraulic efficiency can only be achieved with internal baffles.

Table 5 - Proposed three-tier tank classification.

	PERFORMANCE INDICATORS			
HYDRAULIC EFFICIENCY	T_{10}	T_{90}	***Morrill Index***	***Dispersion Number***
Poor	< 0.2	> 2.5	> 10	> 1
Acceptable	0.2 - 0.5	1.5 – 2.5	2.5 – 10	0.1 – 1
Excellent	> 0.5	< 1.5	< 2.5	< 0.1

REFERENCES

[1] KENNEDY M.S. et al. (1993) Assessing the Effects of Storage Tank Design on Water Quality. *Journal AWWA*, 85(7), 78-88.

[2] LEVENSPIEL O. (1989) *The Chemical Reactor Omnibook.* Chemical Engineering Department, Oregan State University. Oregan. USA.

[3] ERICSON J. E. and SOBANIK J.B. (1996) Scale Modeling: Invaluable Tool in Predicting Contact Time Improvements in Water Treatment Plant Clearwells. *Proceedings of AWWA annual conference.* Toronto, Ontario, Canada, Volume E: 473-500.

[4] LE GRANGE L. (1994) *Flo++ user manual*, Potchefstroom, South-Africa.

[5] VAN DER WALT, J.J. and HAARHOFF, J. (1999*) The Simulation of flow patterns and Mixing in the Daleside Storage tanks*, A Report Prepared for Rand Water, Johannesburg, South-Africa. (Unpublished)

[6] HANNOUN I. A., BOULOS P.F. and LIST. J.E. (1998) Using hydraulic modeling to optimize contact time. *Journal AWWA*, 90(8), 77-87.

Keyword Index